Responsive Membranes and Materials

Responsive Membranes and Materials

Editors

D. BHATTACHARYYA

*Department of Chemical and Materials Engineering,
University of Kentucky, USA*

THOMAS SCHÄFER

POLYMAT, University of the Basque Country, Spain

Co-Editors

S. R. WICKRAMASINGHE

*Ralph E. Martin Department of Chemical Engineering,
University of Arkansas, USA*

SYLVIA DAUNERT

*Department of Biochemistry and Molecular Biology,
University of Miami, USA*

A John Wiley & Sons, Ltd., Publication

This edition first published 2013
© 2013 John Wiley & Sons, Ltd

Registered office
John Wiley & Sons Ltd, The Atrium, Southern Gate, Chichester, West Sussex, PO19 8SQ, United Kingdom

For details of our global editorial offices, for customer services and for information about how to apply for permission to reuse the copyright material in this book please see our website at www.wiley.com.

The right of the author to be identified as the author of this work has been asserted in accordance with the Copyright, Designs and Patents Act 1988.

Wiley also publishes its books in a variety of electronic formats. Some content that appears in print may not be available in electronic books.

Designations used by companies to distinguish their products are often claimed as trademarks. All brand names and product names used in this book are trade names, service marks, trademarks or registered trademarks of their respective owners. The publisher is not associated with any product or vendor mentioned in this book. This publication is designed to provide accurate and authoritative information in regard to the subject matter covered. It is sold on the understanding that the publisher is not engaged in rendering professional services. If professional advice or other expert assistance is required, the services of a competent professional should be sought.

The publisher and the author make no representations or warranties with respect to the accuracy or completeness of the contents of this work and specifically disclaim all warranties, including without limitation any implied warranties of fitness for a particular purpose. This work is sold with the understanding that the publisher is not engaged in rendering professional services. The advice and strategies contained herein may not be suitable for every situation. In view of ongoing research, equipment modifications, changes in governmental regulations, and the constant flow of information relating to the use of experimental reagents, equipment, and devices, the reader is urged to review and evaluate the information provided in the package insert or instructions for each chemical, piece of equipment, reagent, or device for, among other things, any changes in the instructions or indication of usage and for added warnings and precautions. The fact that an organization or Website is referred to in this work as a citation and/or a potential source of further information does not mean that the author or the publisher endorses the information the organization or Website may provide or recommendations it may make. Further, readers should be aware that Internet Websites listed in this work may have changed or disappeared between when this work was written and when it is read. No warranty may be created or extended by any promotional statements for this work. Neither the publisher nor the author shall be liable for any damages arising herefrom.

Library of Congress Cataloging-in-Publication Data

Responsive membranes and materials / D. Bhattacharyya . . . [et al.].
    p. cm.
  Includes bibliographical references and index.
  ISBN 978-0-470-97430-8 (cloth)
  1. Membranes (Technology)   2. Separation (Technology)   I. Bhattacharyya, D. (Dibakar), 1941–
  TP159.M4R47 2012
  660′.2842–dc23

                                                                                    2012022662

A catalogue record for this book is available from the British Library.

ISBN: 9780470974308

Set in 10/12pt Times by Aptara Inc., New Delhi, India.
Printed and bound in Singapore by Markono Print Media Pte Ltd

D. Bhattacharyya: To Gale, my wife, for her encouragement and understanding; To my graduate and undergraduate students, for making academic life very stimulating; To my grandchildren, Nathan, Madeline, Lila, and Zoe for bringing additional joy in my life.

Thomas Schäfer: With gratitude to Belén, Maria, Monika, and Rainer for constant company. In memory of Conny.

S. R. Wickramasinghe: To my parents, for their encouragement; Xianghong, for her support and optimism; and Aroshe, for reminding me of the excitement each day holds.

Sylvia Daunert: I would like to dedicate this book to my students, colleagues, and mentors who constantly inspire and challenge me to be a creative scientist. Also, many thanks to all whom supported me and contributed throughout the writing and editing processes; it has been a pleasure working with all of you!

Contents

Preface: Overview of the Book Highlighting Responsive Behaviour

D. Bhattacharyya, T. Schäfer, S. Daunert, and S. R. Wickramasinghe

The integration of knowledge from the life sciences field with synthetic membranes and materials to create stimuli-responsive behaviour is an important area of science and engineering. Martien, *et al.* (*Nature Materials*, 2010) wrote: "Responsive polymer materials can adapt to surrounding environments, regulate transport of ions and molecules, change wettability and adhesion of different species on external stimuli, or convert chemical and biochemical signals into optical, electrical, thermal and mechanical signals, and vice versa". Touchscreens, light-emitting diodes, etc. are everyday devices that rely on stimuli-responsive materials. However, the response to stimuli can also be explored at the molecular scale for controlling mass transport or creating motion. The fabrication of membranes and materials which can respond to pH, temperature, light, biochemicals, and so on is an important aspect of this book. While reading this text, a multitude of stimuli and responses are taking place in the reader's body, be it for exchanging information through the dendrites and axons of neurons, or for the homeostatic control of the trillions of cells in the hosts. Our health and function entirely relies on the fine-tuned interplay of varied kinds of stimuli-triggered responses at the molecular level, which create a cascade of concerted actions that result in our body working as a perfectly fine-tuned machine. We are far from being able to reproduce the complexity of natural stimuli responsive systems, but in recent years a growing scientific community has been concerned with creating responsive systems that allow us, at the molecular level, to control global release, separation, or actuation as a response to an external stimulus. Similar to the architecture of high-rise buildings or the evolution of living systems, the underlying idea is to use a bottom-up approach for assembling individually characterized elements, molecules, or molecular constructs, which together execute a controllable function. The beauty of such systems emerges when molecular building blocks of very diverse responses are rationally designed and subsequently assembled in order to yield a fine-tuned global system response – that emulates the beauty which Nature in its mastery accomplishes incessantly.

This book is about such constructs, with a particular focus on responsive membranes and materials. It comprises contributions which range from the synthesis of stimuli-responsive membranes and colloids to their applications at very different scales; from self-assembled systems with molecular recognition capabilities within nanopores, to the combination of

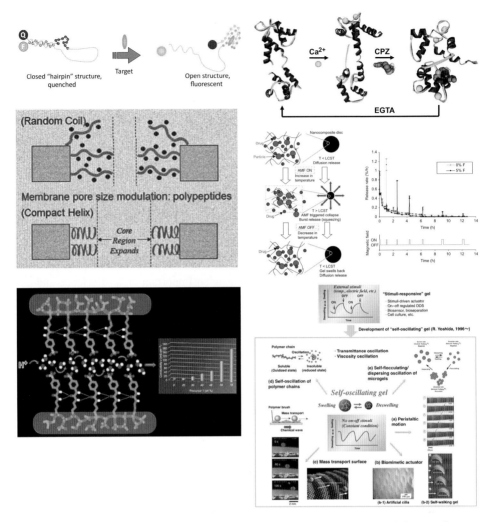

Figure 1   *Examples of some stimuli-responsive systems for tuning the permeability to selectivity of nanopores to self-organization and drug release.*

bulk materials that alter either the effective pore diameter or restrict entrance into pores. Some examples are summarized in Figure 1. Whatever their concept or final use, stimuli-responsive membranes and materials cannot be understood without bearing in mind that their response upon interaction with a stimulus results in a more favourable energetic state, translated as a decrease of the Gibbs energy. This is the basis for understanding under what conditions a system may undergo alterations, or elicit a "response", in order to release energy. The reader's attention is drawn to the role of the total chemical potential. It is the sum of the internal chemical potential – comprising parameters such as density or activity – and the external chemical potential – referring to an external force field such as an electrostatic, magnetic, luminescent, or gravitational field.

In this context, responses to stimuli are the result of driving forces that can have very different origins. As a consequence, on the one hand this provides a large degree of freedom for fine-tuning responses through the creation of subtle interplay between different kinds of forces. On the other hand, this also means that responses must not be designed or interpreted without accounting for *all* possible variable parameters, as otherwise a responsive system might naturally fail. For example, DNA-aptamers can selectively bind to target molecules, the stimulus, and thereby undergo conformational changes as a specific response. This can be explored for gating mechanisms in nanopores (Chapter 1). However, a significant increase in temperature can result in similar conformational changes as an unspecific response in which case, the DNA-aptamer loses its function. Proteins embedded in hydrogels can bind to specific ions resulting in an overall swelling of the hydrogel; however, external pressure through shear forces might strongly counteract this response (Chapter 11) and partially frustrate the responsive function. Conversely, opposing forces or interactions might also be systematically exploited for designing stimulus-responsive materials such as Janus particles (Chapter 12) or self-oscillating polymer gels (Chapter 13). These examples demonstrate that a thorough understanding of the underlying phenomena is indispensable as it provides a vast playground for creatively designing responsive membranes and materials.

Figure 2 gives an overview of the chapters of this book and their main emphasis and is intended as a quick reference. The book starts with three chapters (1 to 3) dealing with the formation of responsive hybrid materials through modification of suitable support structures. The aim is to mimic molecular transport across cell membranes, rather than relying on bulk responses. Chapter 1 explores the capacity of existing building blocks, DNA-aptamers, to undergo conformational changes upon specific binding to a target molecule. If embedded within a fine-tuned nanoporous support structure, it is shown how these receptors can trigger the release or permeation of compounds depending on the presence of a target

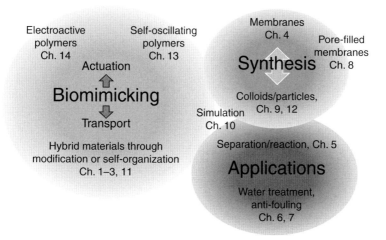

Figure 2   *Overview of the chapters of this book and a rough division by their main emphasis. Naturally, all chapters overlap in one aspect or another given that the common theme is the formation or investigation of responsive materials.*

molecule. The chapter also gives a glimpse of the analytical tools employed for verifying the dimensional changes that DNA-aptamers undergo during this process. Chapter 2 describes a methodology to create self-organized supramolecular structures in which simple building blocks are allowed to self-assemble under the influence of an external stimulus in order to achieve hybrid materials of desired selectivity, for example for selective ion-transport. Here, the concept of evolution is employed in order to upregulate the function of a membrane through adaptive adjustment in the presence of a target solute. Chapter 3 focuses on the modification of the front tip of carbon nanotubes with functional molecules to serve as "gatekeepers" and function as ideal transport channels. The proposed system benefits from the fast fluid flow through the cores of the nanotubes combined with a high density of selective receptors at their tips, mimicking molecular transport across biological membrane transporter proteins.

Modification of materials requires knowledge of the tools and methods needed to achieve the desired properties in the materials, and Chapter 4 discusses routes to surface modifications for producing responsive membranes. Different grafting methods are presented such as photo-initiated polymerization, atom transfer radical polymerization (ATRP, and reversible addition-fragmentation chain transfer polymerization RAFT), which will also be explored in subsequent chapters (5–7). Chapter 5 spans the gap from the synthesis of responsive polymers to their final application, introducing in this way the engineering component of such systems. After an overview of membrane modification techniques, it describes applications (including layer by layer assembly) concerning tunability of water flux and separation of salts (compare to Chapter 2). The chapter then goes well beyond these more obvious applications by introducing responsive (temperature and pH) membranes and hydrogels for nanoparticle synthesis and degradation of contaminants in aqueous solutions. Chapters 6 and 7 further focus on responsive membranes in water treatment, given its global importance and the fact that fouling phenomena strongly affect otherwise economic membrane separations. Linking with Chapter 4, common synthesis strategies for membrane modification as well as magnetically driven micromixers are presented in Chapter 6 and their effect on water filtration is discussed. Chapter 7 further elaborates on fouling control of the membrane surface and feed spacers, while introducing other membrane surface treatments such as ultraviolet, plasma, or surface irradiation by ion beams and chemically induced free radical polymerization.

In addition to the aforementioned emerging applications, controlled release has traditionally been the predominant field for responsive membranes and materials. While the related physico-chemical phenomena are the same, important differences can exist with regard to particular requirements for the final applications. For example, responsive membranes for water treatment must be easily available on a large scale, thus requiring membrane fabrication to be as straightforward as possible. Furthermore, the membranes should be relatively robust and maintain their level of performance despite possible variations in the composition of the (aqueous) feed solutions. In contrast, materials for controlled release mainly find their application in relatively stable physiological conditions and on an individually small scale where biocompatibility is vital. Comparing the various chapters of this book will allow the reader to become aware of one crucial aspect of responsive membranes and materials, which is often ignored at the early stage of technology developments, and that is the engineering requirements of the final application. Chapter 8 deals with pore-filled hybrid membranes capable of responding to changes in pH, and elaborates on methods to estimate the resulting pore sizes. The membranes are intended for drug release applications. The use of a magnetic field as an external stimulus is described in Chapter 9 with a focus on magnetic nanoparticles that are incorporated into a polymeric host matrix. It is shown how a magnetic field can generate responses in various ways depending on whether it is used in a static or alternating mode.

A highly important research effort is currently being made in simulating the physico-chemical behaviour of bulk materials such as polymers by molecular dynamics (MD) simulation. Although still far from substituting experimental evidence, increasingly refined MD models accompanied by enhanced computational capacities will enable experiments made *in silico*, paving the way to systematic and automated screening algorithms for the optimization of material properties and, as a consequence, diminishing considerably the need for material and time-intensive experimental trial and error approaches. Chapter 10 provides such an example for the state-of-the-art of MD simulations by describing the interaction of salt ions with a thermo-responsive polymer.

Chapter 11 extends the biomimicking concepts outlined in previous chapters to the use of hybrid bulk materials which have biological recognition moieties incorporated in their polymeric structure. The chapter describes protein-hydrogel systems which can act as sensors or valves upon interaction with targets. It draws attention to the mechanical characterization of such hydrogels which is of utmost importance for practical applications. From the previous chapters it can be seen that responsive systems can be fabricated in various forms, be it as surfaces on membranes or particles, or inside pores. The final application determines which is the most efficient or appropriate overall strategy. Chapter 12 gives an extensive overview on the fabrication of responsive polymer colloids and how their topology can be controlled depending on the final application. It also presents particular opportunities in colloidal responsive systems such as Janus and patchy particles. The use of polymer gels as self-oscillating systems is described in Chapter 13. Using the Belousov–Zhabotinsky reaction as a stimulus, its oscillation is converted into a continuous swelling–deswelling of a polymer gel and it is shown how under defined experimental conditions this can be explored by allowing the gel to "walk" autonomously. Finally, Chapter 14 introduces electroactive polymers which deform in an electrical field. The chapter gives extensive examples of such dielectric elastomers together with their characterization and a theoretical description of the underlying thermodynamic phenomena.

The book closes with Chapter 15, summarizing the developments and research needs in the predominant fields of application of responsive materials, and providing an outlook onto the vast opportunities which lie ahead for this fascinating multi-disciplinary field of materials research.

List of Contributors

Barboiu, Mihail, Adaptive Supramolecular Nanosystems Group, Institut Européen des Membranes – UMR CNRS 5635, France

Bhattacharyya, D., Department of Chemical and Materials Engineering, University of Kentucky, USA

Daunert, Sylvia, Miller School of Medicine, Department of Biochemistry and Molecular Biology, University of Miami, USA

Dickson, James, Department of Chemical Engineering, McMaster University, Canada

Dow, Elizabeth S., Directorate for Engineering, National Science Foundation, USA

Du, Hongbo, Department of Chemical Engineering, University of Arkansas, USA

Escobar, Isabel C., Department of Chemical and Environmental Engineering, The University of Toledo, USA

Gaulding, Jeffrey C., School of Chemistry and Biochemistry and Petit Institute for Bioengineering and Bioscience, Georgia Institute of Technology, USA

Gorey, Colleen, Department of Chemical and Environmental Engineering, The University of Toledo, USA

Hausman, Richard, Department of Chemical and Environmental Engineering, The University of Toledo, USA

Hawkins, Ashley M., Department of Chemical and Materials Engineering, University of Kentucky, USA

Herman, Emily S., School of Chemistry and Biochemistry and Petit Institute for Bioengineering and Bioscience, Georgia Institute of Technology, USA

Hilt, J. Zach, Department of Chemical and Materials Engineering, University of Kentucky, USA

Hinds, Bruce, Department of Chemical and Materials Engineering, University of Kentucky, USA

Hu, Kang, Research and Development, Land O'Lakes, Inc., USA

Husson, Scott M., Department of Chemical and Biomolecular Engineering and Center for Advanced Engineering Fibers and Films, Clemson University, USA

Khatwani, Santosh, Miller School of Medicine, Department of Biochemistry and Molecular Biology, University of Miami, USA

Leng, Jinsong, Centre for Composite Materials, Science Park of Harbin Institute of Technology (HIT), People's Republic of China

Lewis, Scott R., Department of Chemical and Materials Engineering, University of Kentucky, USA

Liu, Liwu, Department of Astronautical Science and Mechanics, Harbin Institute of Technology (HIT), People's Republic of China

Liu, Yanju, Department of Astronautical Science and Mechanics, Harbin Institute of Technology (HIT), People's Republic of China

Lyon, L. Andrew, School of Chemistry and Biochemistry and Petit Institute for Bioengineering and Bioscience, Georgia Institute of Technology, USA

Özalp, Veli Cengiz, POLYMAT, University of the Basque Country (UPV/EHU), Spain

Puleo, David A., Center for Biomedical Engineering, University of Kentucky, USA

Qian, Xianghong, Department of Chemical Engineering, University of Arkansas, USA

Schäfer, Thomas, POLYMAT, University of the Basque Country (UPV/EHU), Spain and IKERBASQUE, Basque Foundation for Science, Spain

Serrano-Santos, María Belén, POLYMAT, University of the Basque Country (UPV/EHU), Spain and DEEEA, Universitat Rovira i Virgili, Spain

Smuleac, Vasile, Department of Chemical and Materials Engineering, University of Kentucky, USA

Turner, Kendrick, Miller School of Medicine, Department of Biochemistry and Molecular Biology, University of Miami, USA

Wesson, Rosemarie D., Directorate for Engineering, National Science Foundation, USA

Wickramasinghe, S. R., Ralph E. Martin Department of Chemical Engineering, University of Arkansas, USA and Lehrstuhl für Technische Chemie II, Universität Duisburg-Essen, Germany

Williams, Sonya R., Directorate for Engineering, National Science Foundation, USA

Xiao, Li, Department of Chemical and Materials Engineering, University of Kentucky, USA

Yang, Qian, Lehrstuhl für Technische Chemie II, Universität Duisburg-Essen, Germany

Yoshida, Ryo, Department of Materials Engineering, School of Engineering, The University of Tokyo, Japan

Zhang, Zhen, Centre for Composite Materials, Science Park of Harbin Institute of Technology (HIT), People's Republic of China

1

Oligonucleic Acids ("Aptamers") for Designing Stimuli-Responsive Membranes

Veli Cengiz Özalp[1], María Belén Serrano-Santos[1,2] and Thomas Schäfer[1,3]

[1]*POLYMAT, University of the Basque Country (UPV/EHU), Spain*
[2]*DEEEA, Universitat Rovira i Virgili, Spain*
[3]*IKERBASQUE, Basque Foundation for Science, Spain*

1.1 Introduction

Stimulus-responsive materials can be understood as materials that change their properties upon exposure to a stimulus in various ways: they may undergo a physical bulk change, for example, as occurs in shape memory polymers upon a temperature change (Figure 1.1a); they may modify their overall (bulk) physico-chemical properties, as do, for example, ionic liquids of switchable polarity upon exposure to a gas (Figure 1.1b); or they may consist only partially of responsive segments that are incorporated into an otherwise non-responsive support structure which may change both their physical and their physico-chemical properties, such as can be observed in shrinkable polymer brushes when exposed to light (Figure 1.2a). The significant difference between the latter and the former two, which are bulk responses, is the fact that in principal only a local, selective, and specific action of a stimulus is required in order to trigger a change in the responsive part of the material. While you are reading these lines Mother Nature is continuously doing this in your body using responsive transporter proteins which are embedded in the otherwise non-responsive lipid bilayer of the cell membrane, without the need for your whole body to be exposed to light or undergo a dramatic change in temperature. In a similar manner to what Nature achieves so ingeniously in your cell membranes, so this chapter deals with

Responsive Membranes and Materials, First Edition. Edited by D. Bhattacharyya, Thomas Schäfer, S. R. Wickramasinghe and Sylvia Daunert.
© 2013 John Wiley & Sons, Ltd. Published 2013 by John Wiley & Sons, Ltd.

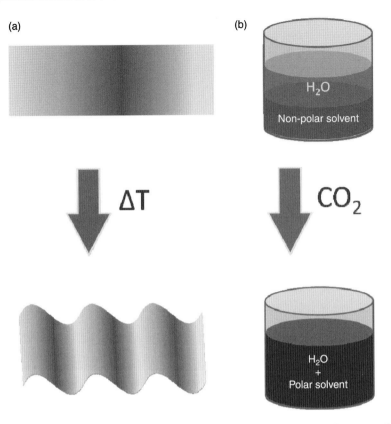

(a)   **(b)**

ΔT CO_2

Figure 1.1   *(a) Response of shape-memory polymers to a temperature change and (b) an ionic liquid–water mixture to the exposure to carbon dioxide.*

porous artificial membranes and particles which use the molecular recognition capacity of oligonuclic acids as a kind of a "gatekeeper" for triggering a local change in permeability or release of solutes.

Common strategies for designing stimuli-responsive membranes and particles that change their permeability or release rate are based on bulk stimuli such as light, pH, ionic strength, and temperature, as well as the action of an electric or magnetic field [1]. While such bulk stimuli may act locally, for example in the case of the irradiation of azo-groups of polymer brushes leading to a reversible shrinking (Figure 1.2a), they may also affect bulk solutions and materials, for example in the case of using temperature or pH as a stimulus for the reversible shrinking of hydrogels (Figure 1.2b). Acting on a bulk when indeed a local action is required means wasting energy and resources as well as limiting the degrees of freedom during the design of such systems: for example, pH and temperature as stimuli must remain within the physiological conditions when the respective responsive materials are to be used in the human body; stimulation by light may in this case not even be an option. Furthermore, if the desired response in the material is only required upon the

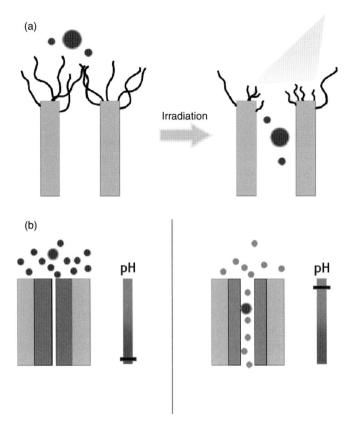

Figure 1.2   *(a) Light-responsive polymer brushes and (b) hydrogels which undergo a volume change upon a change of pH.*

appearance of a defined molecule or cell, as is the case for targeting of drugs, such bulk stimuli can in fact be highly inefficient.

Oligonucleic acids with molecular recognition capabilities, so called "aptamers", are single stranded DNA or RNA oligonucleotides that have been selected *in vitro* for binding to targets with high specificity and selectivity [2]. They occur in Nature as part of so-called "riboswitches" which are capable of binding to small molecules within the mRNA and in this way affect the gene expression [3]. The specificity of aptamers selected *in vitro* has been made use of extensively in bioanalytical devices [4] and in drug delivery systems [5]. Aptamers offer various advantages over other selective biomolecules, such as antibodies or proteins in general, for being incorporated into bioconjugated membranes:

1. they are chemically relatively stable and therefore permit straightforward modifications such as an introduction of a suitable linker chemistry;
2. they are selected *in vitro* from a pool of a large number of random sequences ($>10^{10}$) which theoretically will yield a selective sequence for virtually any kind of target;

3. the basic composition of an aptamer will always be that of an oligonucleotide, whatever the target; this facilitates its incorporation into a support structure as the basic physico-chemical interactions will remain very similar, even if the aptamer sequence is altered;
4. the *in vitro* selection procedures can be automated which allows in principal a convenient screening for any kind of target.

In the following, aptamers will first be briefly presented with regard to their structure, the way they interact with targets, how they are selected and how they can be incorporated into support structures such as membranes and mesoporous nanoparticles. This part of the chapter intends to draw attention to the interactions that aptamers undergo upon target recognition and the resulting structural changes; it also will highlight that the linker chemistry used for incorporating aptamers into structures is nothing more than common bioconjugate chemistry and therefore does not pose any greater obstacle for the design of respective functional materials than is the case for other biomolecules such as proteins. The function of aptamers needs to then be characterized in order to verify its stimuli-responsive behaviour – a subsequent section therefore deals with an overview of state-of-the-art techniques for doing this. Applications which make use of the specificity and the conformational changes of aptamers will then be succinctly presented and discussed. In this way, a context will be provided for understanding how aptamers are commonly used and with what objective, and how respective established methods and techniques can be adapted to the field of stimuli-responsive membranes and particles. Finally, stimuli-responsive membranes and particles based on aptamer gatekeepers will be described and an outlook will be given on this still very new area of research.

1.2   Aptamers – Structure, Function, Incorporation, and Selection

DNA or RNA aptamers commonly consist of less than about 100 bases, with many of those reported for bioanalytical applications not exceeding 50 bases [6]. They are therefore rather small molecules, which allows on the one hand their easy incorporation into matrices; though on the other hand, characterization of their selective interaction and related conformational change can be a challenge. Contrary to proteins, aptamers are rather simple in their composition given that their sequence can only be varied between the four bases: adenine, guanine, cytosine, and either uracil in the case of RNA or thymine in the case of DNA. The molecular recognition capability of aptamers is based on either an existing three-dimensional structure which will host only a very specific molecule with which it is capable of undergoing favourable interactions, or the folding into a well-defined three-dimensional structure upon interaction with a specific target or ligand [7]. The selective interaction of aptamers with a ligand can be manifold depending on the functionalities and the size of a ligand, comprising shape complementarity, steric hindrance, hydrogen bonding, and electrostatic interactions as well as base stacking which is particular to RNA or DNA aptamers [8]. It is evident that the highest specificities are obtained when several of these aspects are fulfilled. Figure 1.3 illustrates as an example a model of how an AMP-binding aptamer selectively interacts with two molecules of adenosine-monophosphate, AMP: the adenine stacks between bases of the aptamer fitting into a pocket already formed, and is stabilized by hydrogen bonding such that a pseudo-base pair is created. It can also be seen

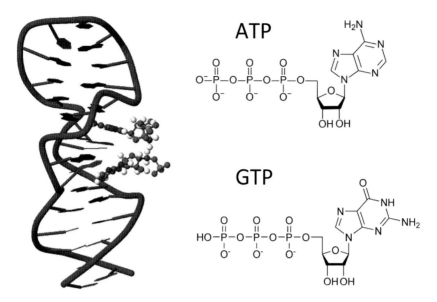

Figure 1.3 *Left: Selective interaction of an ATP-binding aptamer and two molecules of adenosine-triphosphate (ATP) as represented by Jmol; Right: Chemical structures of ATP (ligand) and GTP (no ligand) illustrating the minor differences between both molecules resulting nevertheless in a selective discrimination by the ATP-binding aptamer. See plate section for colour figure.*

that the phosphate groups of AMP stick out of the bulge region and therefore do not possess any major relevance for specific recognition, indicating that AMP, ADP, and ATP would be recognized to a similar degree. Interestingly, guanine-monophosphate is hardly recognized by this AMP-aptamer with the difference between AMP and GMP being minimal (Figure 1.3). A similar selectivity is observed for theophylline-binding aptamers which do not bind caffeine owing to steric hindrance induced by an additional methyl group only in the otherwise identical molecular structure [9]. It should be pointed out that these examples deal with relatively small molecules whose molecular weight does not exceed 500 Da, such that one can easily see how aptamers possess even higher specificities when the ligand is either a protein of a size of kDa, or a complementary strand.

Aptamers can be conveniently modified chemically, which is essential for their immobilization on surfaces such as membranes, particles, or microarrays [10]. For immobilization, common bioconjugate protocols are employed with the most frequent terminations also being commercially available, such as thiol-terminated aptamers for binding on gold substrates, biotinylated aptamers for interactions with avidin, as well as amine-terminated ones [11]. As holds for any other DNA or RNA which is immobilized on surfaces, the immobilization of aptamers is often imagined to occur such that the latter stand upright from the surface; however, it must be stressed that this is often not the case and aptamers may be found inclined, if not lying horizontally on the surface, rendering interaction with the target potentially difficult and possibly resulting in a low ratio of surface coverage:binding capacity [12].

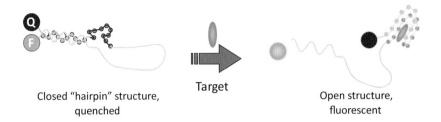

Closed "hairpin" structure,   Target   Open structure,
quenched   fluorescent

Figure 1.4 *Principle of the functioning of a molecular beacon. Stringed circles indicate the selective aptamer sequence which bind to the target, the straight line a linker; Q: quencher; F: fluorophore.*

The facile manipulation of aptamers through chemical modification has lead to a variant of this oligonucleic acid which is particularly appealing for ligand–aptamer interactions in solution, namely the so-called "molecular beacons" [13]. Molecular beacons consist of the original aptamer sequence which is extended by a linker forming a loop, followed by a complementary base sequence which allows the formation of a hairpin structure (Figure 1.4). Each end of the hairpin terminates in either a fluorophore or a quencher. In the closed state of the hairpin, this ingenious variation of an aptamer does not emit fluorescence owing to the proximity of the fluorophore to the quencher. Upon exposure to the target, however, the selective aptamer sequence binds to the ligand, in this way disrupting the stem hybridization; as the latter results in keeping flourophore and quencher apart, fluorescence is emitted that can be measured directly and in real time. The term "molecular aptamer beacon" can often be found in the literature synonymously with "aptamer hairpin structure" or "aptamer switch probe", and as is often the case with coined expressions, it should not be over-interpreted but understood within the context presented.

The function of molecular beacons obviously relies strongly on a subtle interplay between the energy required for hairpin-formation and the energy released upon interaction with the target, which warrants the indispensable reversibility. Molecular beacons must be sufficiently stable such that the majority of the population of hairpin structures remain closed in the absence of the target; on the other hand, linker stiffness and hybridization of the stem region must be weak enough to easily allow opening of the hairpin structure upon binding to the target. An interesting side-asset of beacons is the fact that the conformational change which an aptamer undergoes upon target binding is significantly enhanced, which may be exploited for gatekeeper functions in nanopores, as will be laid out farther on in this chapter.

Up to now, a limited number of selective aptamer sequences versus targets ranging from small molecules to whole cells has been known, seemingly contradicting the affirmation that a selective aptamer may be found versus virtually any kind of target if a large enough pool of sequences is provided. The reason is that automated aptamer selection procedures still require time and effort as they are conducted upon demand, the latter being due to the still limited market for these fine chemicals [14]. Existing aptamers therefore reflect research efforts in particular areas, such as the use of the thrombin-binding aptamer [15] in medicine or the cocaine-binding aptamer [16] in bioanalytics. New selective aptamer sequences are commonly identified using an *in vitro* selection procedure called SELEX (systematic

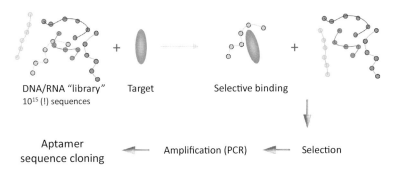

DNA/RNA "library"   Target   Selective binding
10^{15} (!) sequences

Aptamer   ⟵   Amplification (PCR)   ⟵   Selection
sequence cloning

Figure 1.5   *Principle of the "systematic evolution of ligands by exponential enrichment", SELEX, for identifying aptamers which selectively bind to a certain target.*

evolution of ligands by exponential enrichment) during which a large pool of single-stranded oligonucleic acid sequences ($>10^{10}$) are contacted under defined conditions with the target. Details on this method, continuously refined since its development in 1990, can be found elsewhere [17]. The main idea of SELEX resides in the fact that those sequences which selectively bind to the target are subsequently isolated, repeatedly contacted again with the target for refining the selection procedure, and the most adequate sequences are eventually isolated and amplified by PCR (Figure 1.5). Two aspects of SELEX are particularly noteworthy with regard to the use of aptamers as gatekeeper molecules in separation or delivery devices: first, aptamer selection should evidently take place under environmental conditions that are identical to the intended later application [18]; second, a series of sequences capable of binding to a target is commonly found, differing mainly in the degree of affinity, expressed by the dissociation constant K_D [19]. The former allows the selection of aptamers in even non-conventional environments, the latter provides a tool for fine-tuning the target-aptamer interaction with regard to reversibility and specificity, adding a further degree of freedom to the engineering of aptamer-based nanodevices.

1.3   Characterization Techniques for Aptamer-Target Interactions

The molecular recognition of targets by DNA/RNA aptamers is an equilibrium between molecular forces, resulting in conformational changes of the aptamer while generating the aptamer-target complex. The method of reference for characterizing these interactions and the resulting structural changes has therefore traditionally been NMR [20], with the need for machines of preferably higher power than 600 MHz in order to yield spectra of sufficient resolution. It is, however, not always necessary to resolve structural changes in every detail in order to verify aptamer function. In practice, information is rather required on a routine basis on the sensitivity and selectivity of the aptamer-target interaction, represented by the dissociation constant K_D, as well as the *overall* structural changes occurring during binding. For applications of aptamers it is furthermore very important to have an analytical technique at hand which allows screening of the binding kinetics and overall conformational changes under different environmental conditions in real time and possibly *in situ*, which is not trivial to achieve with NMR. Surface plasmon resonance (SPR), quartz crystal microbalance with dissipation measurement (QCM-D) and dual polarization interferometry (DPI) are techniques that allow the obtaining of these parameters for aptamers immobilized on surfaces, with SPR commonly being limited to detecting binding events rather than structural changes too [21]. Circular dichroism can be employed to detect conformational changes in solution [22] being in this sense complementary to QCM-D and DPI. Isothermal titration calorimetry allows the obtaining of important thermodynamic parameters on aptamer-target interactions as well as the associated conformational changes [23], and therefore can be seen as complementary to SPR, QCM-D, and DPI whenever the latter techniques can be operated under varying temperatures in order to derive the respective thermodynamic parameters based on the van't Hoff equation [23]. Aptamer-target binding can also be elegantly studied using target-modified cantilever tips in atomic force microscopy [24]. This approach requires great expertise in the surface modification of the cantilever tip and is therefore not straightforward, however, by moving the target toward the aptamer and subsequently retracting it while measuring the force of interaction, it can provide very useful complementary information on the acting forces which can possibly even be correlated with the viscoelasticity of the DNA molecules in solution, as is the case when they serve as gatekeepers in nanopores.

Because this chapter deals with the surface modification of particles and membranes with aptamers whose conformational change upon target binding induces a gatekeeper function, QCM-D and DPI will be discussed in the following in more detail in order to highlight how the conformational changes of such relatively small DNA molecules can be characterized with regard to their potential to serve as a gatekeeper in nanopores, particularly when the target molecules are of a size which is considered close to the limit of detection in methods such as SPR.

1.3.1   Measuring Overall Structural Changes of Aptamers Using QCM-D [25]

Quartz crystal microbalance with dissipation monitoring (QCM-D) is a surface-sensitive acoustic technique for studying a wide range of interfacial adsorption reactions. The principle of the technique resembles standard QCMs: a quartz electrode is brought into resonance oscillation at a high frequency (5–10 MHz). Any adsorption of molecules at the surface will cause damping of the frequency which is monitored and, following several assumptions [26], will be translated into a mass deposition on the sensor surface. Simple QCM measurements encounter serious limitations when the mass deposited on the sensor is not rigid, as may precisely be the case for the adsorption of biomolecules. As a consequence, the technique has been refined by repeatedly cutting off the electronic circuit for a very short time, followed by measurement of how the oscillation energy of the electrode dissipates during this time-interval. It can be intuitively understood that strong dissipation, designated as ΔD, indicates the existence of "soft" adsorption layers and weak dissipation that of "rigid" ones (Figure 1.6). Together with the standard information on the associated frequency ("mass") change denoted as ΔF, QCM-D is indeed capable of providing global but useful information on biomolecular interactions.

In the following, QCM-D will be presented with the model system being a small molecule binding aptamer that interacts with adenosine-5′-monophosphate (AMP). The ATP-binding DNA aptamer is a 27-base sequence selected by Huizenga *et al.* [27] and has been used in

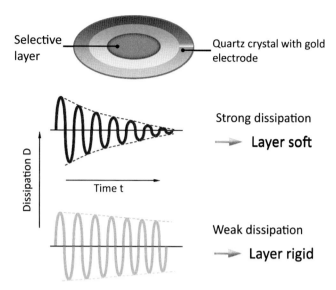

Figure 1.6   *Using the dissipation measurement of layers deposited on quartz crystal microbalances as a means to characterize density changes on the sensor surface.*

numerous nanostructures as a model system. It specifically recognizes adenine-containing small molecules such as AMP with MW = 382 Da. The tertiary structure of this aptamer is well-reported and different structure-changing functional designs [28] exist, which is an important asset when it comes to verifying novel characterization methods.

The small-molecule binding of two different forms of the same aptamer sequence was compared, namely (i) the ATP-binding aptamer sequence itself and (ii) a hairpin structure of the same aptamer sequence which was created by adding seven additional nucleotides at the 3′-end (Table 1.1). It was previously found qualitatively [29] that both forms undergo strikingly different structural rearrangements upon binding to the ligand, as illustrated in Figure 1.7. The ATP-binding aptamer (Figure 1.7a) forms a helix structure with mismatched nucleotides and changes its overall molecular conformation only slightly [30] upon binding to two molecules of AMP which intercalate into the distorted minor groove of a zipped-up internal loop segment [31]. In contrast, the hairpin structure based on the same aptamer sequence goes through an extensive molecular rearrangement (Figure 1.7b). The hairpin structure is used to disrupt the aptamer's original shape (which is essential in specific recognition of its ligand) and to obtain a double stranded duplex neck region bringing both ends of the molecule together. As a consequence, a closed structure referred to as a hairpin shape is obtained, which changes to an open structure when interacting with two molecules of AMP owing to the disruption of the neck region. Such a hairpin design has frequently been used to develop fluorescent [32] or electrochemical probes based on aptamers [33].

In QCM-D, coated quartz electrodes are commercialized with surface layers ranging from metal-oxides to polymers. However, the electrode can also be custom-modified in order to yield a high binding of aptamer to the surface [5], as was done in our case. For subsequently studying the aptamer-target interaction, the obtained aptamer film was first

Table 1.1 *The sequences of oligonucleotides used.*

ATP-binding aptamer hairpin	5'-CACCTGGGGGAGTATTGCGGAGGAAGGTT**CCAGGTG**-Bio-3'
ATP-binding aptamer	5'-CACCTGGGGGAGTATTGCGGAGGAAGGTT-Bio-3'

Bio: Biotin-C6 modification; **Bold:** nucleotides added to create hairpin.

contacted with AMP in a binding buffer at a series of concentrations (10 up to 750 µM). As can be seen in Figure 1.8a, a clear decrease in frequency occurred during subsequent AMP injections and a corresponding increase in dissipation energy was recorded (Figure 1.8a). Average values of $\Delta D/\Delta F$ from independent experiments were plotted against AMP concentrations to obtain a binding curve (Figure 1.8b), and a dissociation constant K_D of 43 µM was determined, which is in good agreement with the literature citing the affinity constants of ATP-binding DNA aptamer to be in the range of 6 to 30 µM depending on the method used in assessments [34].

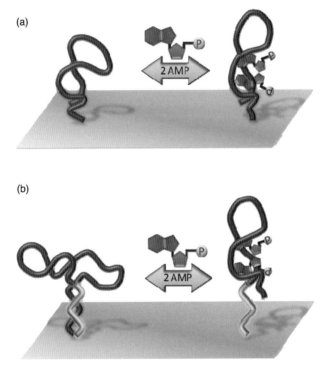

(a)

(b)

Figure 1.7 *Schematic conformational change expected upon binding to the target ATP of an (a) ATP-binding aptamer, a single 27-mer oligonucleotide (depicted as a blue line), where the overall structure is not significantly disturbed by specific intercalation of the AMP ligand and (b) the hairpin ("molecular beacon") form of the ATP-binding aptamer (blue lines) which was created by adding 7 nucleotides at the 3' end (depicted as yellow lines) forming a stem-loop structure; AMP interacts, disrupts the hairpin structure and stabilizes the same structure as in (a), whilst the additional nucleotides remain attached to the surface. See plate section for colour figure.*

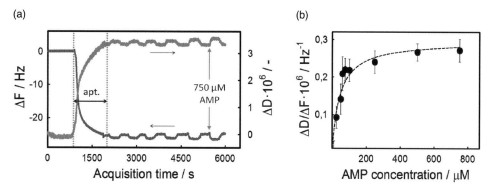

Figure 1.8 *(a) Frequency (blue lines) and dissipation (red lines) changes during binding of AMP to ATP-binding aptamer as monitored by the QCM-D, with dotted vertical lines delimiting the aptamer immobilization prior to target injection in concentrations up to 750 μM; (b) Dissipation changes normalized by the corresponding frequency change as a function of the respective AMP concentrations, and the resulting ligand binding curve (dashed line). See plate section for colour figure.*

When titrating subsequent concentrations of AMP to the immobilized aptamer *hairpin* structure instead, the results were qualitatively similar to those of the aptamer experiments, that is, with increasing AMP concentration the frequency decreased owing to binding of AMP, and the dissipation increased suggesting an increased softness of the resulting film (Figure 1.9a). However, the binding plot (Figure 1.9b) yielded a higher dissociation constant, namely 107 μM for $\Delta D/\Delta F$. This could be explained by a slight distortion of the original aptamer structure when being incorporated in the hairpin form. The excellent agreement between the dissociation constants derived from QCM-D with regard to binding of AMP to both the aptamer and the aptamer hairpin structure therefore allowed further

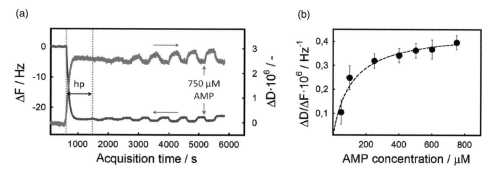

Figure 1.9 *(a) Frequency (blue lines) and dissipation (red lines) changes during binding of AMP to the ATP-binding aptamer hairpin as monitored by the QCM-D, with dotted vertical lines delimiting the hairpin immobilization prior to target injection in concentrations up to 750 μM; (b) Dissipation changes normalized by the corresponding frequency change as a function of the respective AMP concentrations, and the resulting ligand binding curve (dashed line) for the hairpin structure. See plate section for colour figure.*

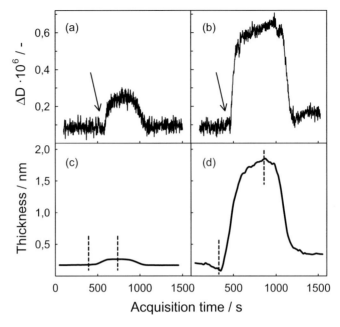

Figure 1.10 *Dissipation changes observed upon binding at 750 μM AMP in buffer solution (arrow) in the (a) aptamer and (b) aptamer hairpin structure, as well as the corresponding thickness changes observed upon binding at 750 μM AMP in buffer solution in the (c) aptamer and (d) aptamer hairpin structure; vertical lines indicate reference points from buffer solution and maximum response to the target.*

investigation into the conformational changes occurring during binding to the small target molecule in order to gain insight into the dimensional changes that these go along with.

For this purpose, the aptamer and the aptamer hairpin structure were compared at a saturating AMP concentration of 750 μM. We observed that changes in the dissipation energy were significantly different between both forms (Figures 1.10a and b), being almost twice as high in the hairpin structure. This clear discrepancy indicated that larger conformational arrangements took place in the hairpin structure upon interaction with AMP, namely in the sense of a softer film, compared to those occurring in the aptamer as was schematically depicted in Figure 1.7. With regard to the hairpin structure, AMP binding leads to the loss of the double stranded neck region while at the same time the single stranded loop region stabilizes the mismatched double helix structure of the aptamer sequence which interacts with the ligand (Figure 1.10b). "Softer" films may then be conceived as an extension of the hairpin film in the vertical direction resulting from the more open structure of the hairpin compared to that of the aptamer. As for the latter, NMR studies do not suggest any major molecular conformational change upon binding to AMP [35], but rather a relatively small rearrangement in the internal binding pocket leading to a slightly more upright position of the molecule. The latter would result in a minor change of the film density which is indeed confirmed by an only slight increase in dissipation (Figure 1.10a).

By applying a Voigt-model to the data obtained using the QCM-D by means of the data processing software provided by the manufacturer (Q-tools, Q-Sense) [25], thickness changes of both the aptamer film and the hairpin structure could be quantified, with the results being depicted in Figures 1.10c and d. As can be seen, the aptamer film increased only slightly in thickness upon AMP binding, namely about 0.1 nm. The thickness increase was dramatically more conspicuous in the case of the hairpin form of the aptamer, namely 1.6 nm. This difference could be related to the seven-nucleotide-long single-stranded linker added to the aptamer sequence in order to form the hairpin, extending the hairpin structure vertically away from the surface as it opens upon AMP binding. Recent measurements report the length of one nucleotide to be 6.3 Å33 for ssDNA in solution such that seven nucleotides would yield a length of 4.4 nm when most elongated, which is about three times what we measured in our experiments during AMP binding. However, ssDNA is flexible, its conformation dependent on the medium composition, and it has been reported that ssDNA stays partially tilted on surfaces depending on the surface density of the film and the interactions with the surface material [36]. It may therefore be speculated whether the thickness increase measured was not higher because the hairpin structure was not fully open, or whether the molecules remained partially tilted on the surface. This also proves how extremely difficult it is to carefully characterize how aptamers which are immobilized on surfaces interact and change their conformation. It can be seen that with this model system, the QCM-D technique is proved capable of yielding quantitative measures on the structural rearrangement of both aptamer sequences, even at the sub-nanometre level, and in this way provides valuable data on how aptamers would act when incorporated into nanopores. Although these insights are not based on a single-molecule detection but rather represent an average response, they are nevertheless very valuable given that in nanodevices "function" will always be related to an average performance of responsive units.

However, from the aforementioned it could also be understood that one technique alone does not yield sufficiently reliable information to be conclusive. In the following, dual polarization interferometry (DPI) will therefore be presented as a valuable, complementary technique to QCM-D.

1.3.2   Measuring Overall Structural Changes of Aptamers Using DPI

Dual-polarization interferometry (DPI) is an optical surface analytical technique which can provide multiparametric measurements of films of molecules at molecular dimensions. The principles behind the DPI technology can be found in detail elsewhere [37], but it is worth mentioning that it mainly relies on two perpendicular polarization modes of an incident laser (Figure 1.11), resulting in measurements that yield information on events occurring close to the surface (horizontal polarization) as well as those further away from the surface (vertical polarization).

In order to study conformational changes with DPI, AMP-binding aptamers were immobilized on an avidin-covered silicon-oxynitride surface via a biotin linker. The optical changes in refractive index measured during AMP binding by the aptamer at various concentrations, as described for QCM-D experiments in the previous subsection, were converted into real-time mass changes and are presented in Table 1.2. AMP injections gave detectable signals commencing from a concentration of 10 μM, and started to saturate over 100 μM of

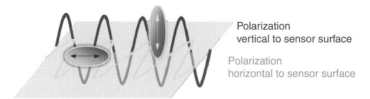

Figure 1.11   *Detection of conformational changes on a sensor surface during dual polarization interferometry making use of the different sensitivity of either polarization mode with distance from the surface.*

AMP. The association and dissociation kinetics were very fast as previously characterized by other methods [38]. GMP did not cause any increase in the mass profile at 100 μM and only a negligible increase at 250 μM (Table 1.2). Affinity analysis of the binding profile was consistent with a one to one Langmuir binding with an equilibrium constant of 46.2 ± 7.8 μM which compared well with the QCM-D data presented before. A more complete profile of AMP binding is shown in Figure 1.12. The binding of AMP to the aptamer is characterized by an increase in refractive index (RI), mass, and density and a corresponding decrease in thickness. Dissociation of AMP causes all parameters to return to their initial values. 100 μM AMP resulted in a 1.56 Å reduction in thickness of the aptamer layer. Injection of 100 μM GMP, on the other hand, did not cause any significant changes in any of the parameters. Detailed analyses of AMP-binding aptamer-AMP complex were reported by previous NMR studies, where the three-dimensional structure of the aptamer had been determined at a high resolution when the aptamer was in complex with the ligand AMP [39], showing that the aptamer possesses a stem-bulge-stem secondary structure (Figure 1.13b). Aptamers fold upon ligand binding into molecular structures in which the ligand becomes an intrinsic part of the aptamer structure [30]. The ligand-binding pockets of the aptamer provide specific hydrogen bonding sites as a result of a combination of a

Table 1.2   *Layer parameter changes during AMP binding studies. The numbers refer to changes from the start of injections.*

	Refractive index	Thickness (nm)	Mass ($ng\,mm^{-2}$)	Density ($g\,cm^{-3}$)
Binding buffer	0.0001	−0.01	0.0008	0.0016
100 μM GMP	0.0001	−0.01	0.0005	0.0012
250 μM GMP	0.0002	−0.01	0.0006	0.0020
1 μM AMP	0.0002	−0.01	0.0009	0.0011
10 μM AMP	0.0010	−0.06	0.0016	0.0058
25 μM AMP	0.0017	−0.10	0.0053	0.0094
50 μM AMP	0.0021	−0.11	0.0072	0.0119
75 μM AMP	0.0025	−0.13	0.0097	0.0139
100 μM AMP	0.0027	−0.16	0.0108	0.0148
250 μM AMP	0.0029	−0.17	0.0123	0.0156

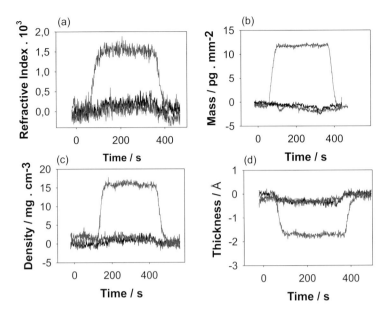

Figure 1.12 *Normalized plots of changes in layer parameters (a) RI; (b) mass; (c) density and (d) thickness during injection of AMP. Red lines: 100 μM AMP, black lines: no AMP binding buffer; blue lines: 100 μM GMP. See plate section for colour figure.*

non-Watson–Crick base stacking surface and a single base. Two molecules of AMP are recognized by hydrogen bonding between their Watson–Crick edges and the minor groove edge of guanine bases. Each AMP-G pseudo-base pair stacks between a reversed Hoogsteen G-G pair and adenine. The distinct hydrogen bonding scheme in the AMP-G pseudo-base accounts for discrimination against the three other nucleotide bases by the AMP-binding aptamer: one ligand pocket is formed when AMP forms a hydrogen bond with G21, which further stabilizes the pseudo-base pair between G5-A23 and G6-G21. The second AMP pocket similarly stabilizes the pseudo-base pair between G18-A10 and G19-G8 when the ligand undergoes hydrogen bonding with G9 (Figure 1.13b). This specific hydrogen bonding scheme, which is required for the formation of a—pseudo-base pair involving the ligand, is responsible for the selection of adenine as a ligand in the AMP aptamers. The NMR study of the aptamer–AMP complex revealed the molecular size of the complex in solution: the fit of AMP molecules in the binding pockets has been interpreted as a less than perfect shape complementarity, and binding of ligand stabilizes the final structure. This model suggests that AMP binding might result in shrinkage of the bulge region. The aforementioned shrinkage of 1.6 Å as determined by DPI suggests that the bulge opening before binding was wider and the AMP binding stabilized the bulge region of the aptamer, pulling at the opposite strands of the specific guanine and adenine nucleotides (red dashed lines in Figure 1.13b).

The lack of resolved structure in the absence of the ligand previously did not allow the determining of the quantitative dimensions of the related molecular structure changes. Using dual polarization interferometry, however, such quantitative real-time data on the shrinkage

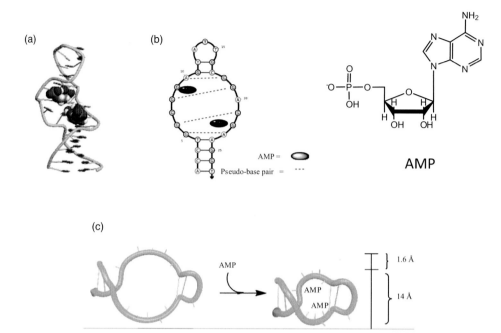

Figure 1.13 *Secondary and tertiary structure of the ATP-binding aptamer. (a) A snap-shot of the overall tertiary folding (1AW4); (b) The secondary structure alignment; (c) Schematic representation of the structural changes occurring upon AMP binding as determined in this study. The helix-depiction structures are for representative purposes. The secondary structure and helix depiction were obtained through Nupack. Drawings in b and c are not proportional to real scales, but exaggerated for representational purposes. See plate section for colour figure.*

of AMP-binding DNA aptamer immobilized on avidin surfaces are obtained directly and *in situ* providing valuable insights into the structural adaptations of aptamer molecules during the whole event of ligand binding. Again, it is pointed out that the dimensional changes quantified represent *average* data of a large population of aptamers immobilized on the chip surface, rather than single molecule detection. However, it is also stressed that it is precisely this kind of information which comes closest to predicting the performance of aptamer-based nanodevices whose performance is likewise based on an average response of all aptamer-units rather than being based on a single molecule event.

These measurements also confirm results from QCM-D discussed previously where conformational changes of aptamers were found to be far less significant than in aptamer hairpins, or so-called molecular beacons, and both methods described the conformational changes of aptamers upon ligand binding occuring in the sub-nanometre range. Hence, although aptamers are widely used in bioanalysis, this explains why for a gatekeeper function in nanopores aptamer hairpin structures appear in principle to be more adequate, as will be discussed in more detail in the following applications section.

1.4 Aptamers – Applications

1.4.1 Electromechanical Gates

Early examples of nucleic-acid functionalized stimuli-responsive polymers were electromechanical. The functionalization of porous materials with nucleic acids provided ion flux sensing which was used in various nanobiotechnological applications such as the control of the passage of ions in a channel by an external electrical signal [40]. For this purpose, a single conical-shape gold tube with 30 nm pore size was embedded within a polymeric membrane and the walls were coated with DNA oligonuceotides in order to construct the first artificial ion channel reported [41]. DNA strands with up to 45 nucleotides were used to partially block the pathway for ion transport, measured as a higher ionic resistance of the nanotube and indicating that DNA was attached at the mouth of the pores. During application of electric voltage, the negatively charged DNA attached to the nanotube pore mouth extended towards the anode, in this way serving as an electrical rectifier in the artificial ion channel. The rectification was able to be controlled by changing the length of the DNA or the diameter of the pores.

Making use of responses to pH-changes rather than electrical currents, in a similar setup the ion flux was controlled by a novel nanopore-DNA hybrid system whose gating function relied on the folding and unfolding of DNA molecules [42]. The single conical nanopores were embedded in track-etched poly(ethylene terephthalate) membranes and i-motif DNA sequences were attached inside the nanopores. The i-motif DNA sequences underwent conformational changes in response to pH changes between a four-stranded i-motif structure (pH $= 4.5$) and a random single stranded structure (pH $= 8.5$). The i-motif structure was a densely packed, rigid structure and thus caused a partial increase in the effective pore diameter. The gating performance was evaluated by measuring ionic current at different pH environments.

Protein nanopores have generally narrower pores compared to inorganic pores, and the previous prototypes of electromechanical gates were further improved by constructing protein pore–DNA hybrids [43]. α-Hemolysine (αHL) has an average pore size of 1.3 nm and single stranded DNA can block the pore of αHL. A hairpin DNA oligonucleotide was functionalized with poly(ethylene glycol (PEG)-biotin at one end and a G-quadruplex structure at the other, resulting in two dynamic sites at both ends of the nucleic acid molecule. In this way, the DNA molecule was captured in the pore after it was threaded through the pore in order to create a rotaxane structure. Perturbation of ion conductance provided a sensing signature for evaluating the functional characteristics of nucleic acids. A disadvantage of using proteins as a support structure compared with polymers is the susceptibility of the protein to change in the environmental conditions. Under physiological conditions, however, this may not be an issue. It therefore depends on the individual application which strategy is preferred.

1.4.2 Stimuli-Responsive Nucleic Acid Gates in Nanoparticles

Nanocontainers are receiving increasing interest for sensing and drug delivery applications. Nucleic acid functionalization for targeting or sensing purposes has been exploited for decades using a variety of materials such as alumina, silica, carbon nanotubes, and, more recently, graphene. However, using nucleic acids as gating molecules has been reported in only a handful of recent reports. Mesoporous silica has been the choice of material for developing stimulus-responsive nanoparticles, especially for drug delivery purposes. This originates from its unique properties such as a stable mesostucture, large surface areas and loading capacity, biocompatibility, [44] and a tunable nanometre-sized pore architecture. Another asset is the relatively straightforward nucleic acid conjugation using silanization. Considering the use of DNA-aptamers for pore blocking or as "gatekeepers", it is worth making a short consideration of the dimensions of these molecules, and what this means for the desired pore size of the support structure.

DNA is a polymer that consists of monomer units of nucleotides (e.g. deoxyribose and a nitrogen containing base) interconnected through phosphodiester bonds. Polynuceotide chains form strands of oligomers with thicknesses between 5 and 9 Å [45], the interphosphate distance of one nucleotide being between 5.9 and 7.0 Å depending on the sugar puckering when fully stretched [46]. Thus, an oligomer of ssDNA can only partially block a 2 nm size pore. A dsDNA helix has about a 2 nm width which corresponds to the pore size of a typical mesoporous silica nanoparticle. In this way, double stranded DNA can easily block the pores when they are conjugated close to the pore region even in an upright position, which explains the high suitability of mesoporous silica for aptamer-based nanodevices.

The simplest application of nucleotide capping of porous nanocarriers is through physical adsorption of single stranded DNA (ssDNA). Climent *et al.* [47] added a positive charge to the surface of mesoporous silica MCM-41 by functionalization with APTS (3-aminopropyltriethoxysilane). The negatively-charged single stranded polynucleotide interacts with the positively-charged particle surface, resulting in the nanopores closing. When contacting the system with the target, complementary polynucleotide, a double-stranded structure with the immobilized ssDNA is formed, in this way displacing the single-stranded caps and leading to opening of the pores. Although the system is not reversible and oligonucleotides as trigger molecules do not represent many applications, this gating function through a complementary oligonucleotide trigger was an example of the usefulness of polynucleotides in capping pores followed by a release of the entrapped cargo molecules. A series of DNA capping studies has recently been reported using pH or temperature as a trigger response. Figure 1.14 shows an interesting application of single stranded capping achieved by using a quadruplex forming i-motif DNA [48].

The capping of mesoporous silica pores can also be achieved by attaching the nucleic acids on the nanoparticle surface near the mouths of pores. The chemical attachment provides better control over the positioning of nucleic acids on the surface and thus leads to greater stability of the synthesized hybrid material. For example, mesoporous silica can be functionalized with double stranded oligonucleotides in order to create temperature-responsive materials [49]. Double-stranded DNA provides a capping functionality when attached to the surface of the nanoparticles, keeping guest molecules confined within the pores (Figure 1.15). In this study, the melting temperature of the nucleotide was adjusted to 47°C corresponding to the upper limit of therapeutic hyperthermia. Thus, the therapeutic cargo of the particles would be released when hyperthermia approaches this upper temperature limit. In fact, this type of design can be used to develop capping systems which can control the release at any predetermined temperature, which is the main advantage of using nucleic acid structures as thermal triggers for the pore opening in nanodevices.

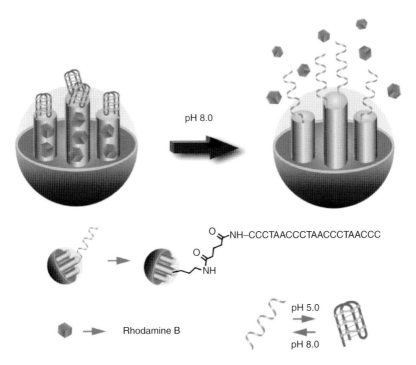

O
‖
—NH–CCCTAACCCTAACCCTAACCC

Rhodamine B

pH 5.0
⇄
pH 8.0

Figure 1.14   *Silica nanoparticles were capped with quadruplex structure DNA, and pH change released the content. (Reproduced with permission from [48] Copyright (2011) Oxford University Press).*

The melting temperature (T_m) is a quantitative measure of how much two complementary strands of DNA are separated and is defined as the temperature at which half the molecules are single-stranded and the other half are in double-stranded form. The melting process is, hence, the denaturation of strands upon the breaking of hydrogen bonds. This process is also termed an "unwinding" of the double-stranded DNA molecule and it is the basis for most basic life phenomena, from DNA replication to genetic control. T_m depends on the length and content of oligonucleotides. An increase in added nucleotide to an existing sequence increases T_m by approximately $1°C$, depending on the type of nucleotide: guanine and cytosine can establish three hydrogen bonds, while and adenine with thymine can create two hydrogen bonds during formation of a double stranded helix. A similar temperature-triggered system was developed by Schlossbauer [50]. The pores of mesoporous silica MCM-41 with a 3.8 nm pore size were capped first by avidin and then by a double stranded DNA labelled with biotin at one end. The authors reported two different length caps with 15 and 25 nucleotides which opened the pores at 45 and 65°C, respectively. As can be seen from these examples, different surface preparation methods can be employed to achieve DNA-based responsive materials; there are no mandatory recipes and the optimization of the respective protocols is mainly a function of the demands and possible restrictions of the final application.

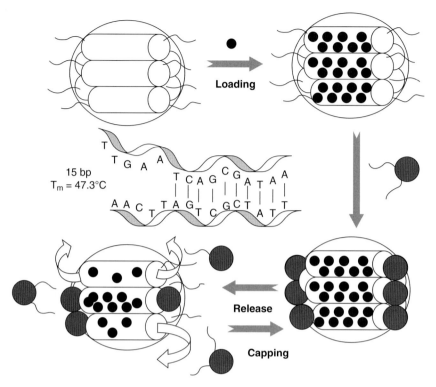

Figure 1.15 *Mesoporous silica nanoparticles loaded with guest molecules were capped with magnetic nanoparticle through DNA hybridization. The caps could be controlled through a temperature trigger. (Reprinted with permission from [49] Copyright (2011) American Chemical Society).*

1.4.3   Stimuli-Responsive Aptamer Gates in Nanoparticles

Aptamer-target interactions have been explored for novel stimulus-responsive drug delivery systems [51]. Aptamers are short nucleic acids that bind to a target with high specific affinity. Using aptamers for creating stimuli-responsive nanopores therefore extends the range of stimuli-responsive materials by providing a trigger against any preselected molecule. These biorecognition properties of aptamers have been exploited in many applications of sensors [52]. One way to make use of the molecular recognition capacity of aptamers is to cap mesoporous silica nanoparticles with gold (Au) nanoparticles which are linked to the surface of silica through DNA hybridization involving an aptamer sequence. By competitive displacement, Au particles are uncapped in the presence of a trigger molecule ATP and the guest molecules are released (Figure 1.16). Again, this capping/uncapping mechanism is irreversible.

A reversible means of creating a gating of molecules based on the specific aptamer-target interaction was proposed by Abelow *et al.* [53]. The ion transport through 20 and 65 nm glass nanopores was controlled by an aptamer sequence which underwent conformational changes in response to binding to its target cocaine. The cocaine-binding aptamer was

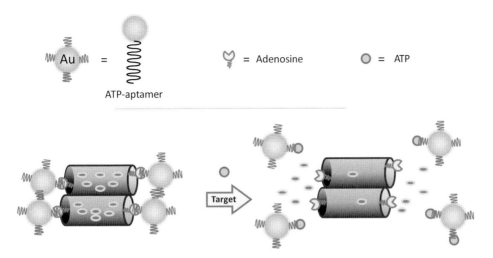

Figure 1.16 *Aptamer-target-responsive controlled drug delivery system, for details see text. (Reproduced with permission from [52] Copyright (2008) Springer Science + Business Media).*

attached in the cavities of the pore being partially unfolded in the absence of the target cocaine. In the presence of cocaine molecules, however, the aptamers formed a three-way junction structure with the target. This three-way junction occupied less space compared to the unfolded structure, in this way allowing more ions to pass the pore and creating a cocaine-induced gating function.

Rather than exploring aptamer gates in single pores, in a recent study it was shown that the molecular recognition capacity of aptamers can also be used on a larger scale and directly as a stimuli-responsive gatekeeper in mesoporous particles, even if the target is a relatively small molecule such as adenosine triphosphate (ATP) [5]. Immobilizing ATP-binding biotinylated aptamers to avidine modified mesoporous nanoparticles, it could be shown both in buffer and serum that the aptamer could reversibly trigger the release of flouresceine as a cargo molecule previously entrapped inside the pores of the particles. This study was the first to prove that aptamers can be used as "stand-alone" stimuli-responsive receptor molecules in order to trigger the opening and closing of pores based on the molecular recognition of a target.

As can be seen, nucleic acid functionalized stimuli-responsive nanomaterials present novel opportunities in drug-delivery applications. The above-mentioned examples show that the chemical properties of nucleic acids provide new tools for researchers to develop pH, temperature, and target-molecule triggered systems. However, the same properties of nucleic acids also present limitations in some applications. For example, intracellular environments are notorious for causing nucleic acid degradation. This is because of the defence mechanisms of living cells in the form of a variety of nucleases against the invasion of foreign nucleic acid materials. Blood and interstitial spaces also contain a variety of nucleases as a part of such a defence system. Thus, a real application involving nucleic acids will require increased biological stability of nucleic acids. Approaches to improve biological stability have been conceived in the form of modifying oligonucleotides or nanoparticle association structures with the aim of assigning an added protection against nucleases. However, until now these approaches have not been not been entirely successful such that stabilizing nucleic acid functionalized nanoparticles still poses a challenge for biological applications.

1.4.4   Stimuli-Responsive Aptamer-Based Gating Membranes

Approaches to creating stimulus-responsive membranes have been explored for decades for liquid separations or controlled release applications yielding materials whose permeability varies, triggered by a change of pH, temperature, or ionic strength of the adjacent liquid, or the exposure to light, an electrical or a magnetic field [1, 54]. It has remained a challenge, however, to mimic the specific and locally acting molecular recognition mechanism that Nature employs to reversibly trigger a conformational change of a membrane receptor molecule in order to bring about a variation in the permeability of a cell membrane. Bioconjugated membranes incorporating enzymes or antibodies [1] have marked an important milestone toward this goal; however, their complexity is far higher than that of, for example, oligonucleic acids ("aptamers") which in turn offer a potentially wider range of possible targets, as has been outlined above. In the following, a self-assembled stimuli-responsive membrane barrier is described which is capable of reversibly changing its permeability upon molecular recognition by an aptamer [55]. With the stimulus being a target molecule as small as adenosine -5'-triphosphate (ATP), the function of our stimulus-responsive membrane relies on the conformational change that aptamers undergo upon the specific recognition of this target molecule, rather than acting upon a bulk stimulus (Figure 1.17). The membrane is furthermore designed in a modular way using mixed materials such that it may be adapted to other target molecules without changing its principal architecture. A big advantage of this concept is the fact that aptamers can be selected against virtually any kind of target, enabling the, in principle, straightforward adaptation of the membrane design to a wide variety of sensor and separation applications.

The porous support structure of the small-molecule stimulus responsive membrane consisted of a commercially available anodized alumina membrane of a nominal pore size of 20 nm and a narrow pore-size distribution. Similarly to the modification of mesoporous nanoparticles [5], the alumina membrane surface was functionalized with

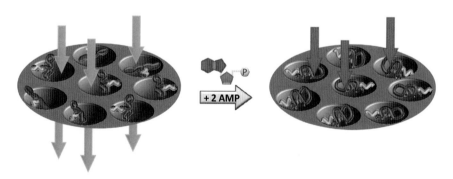

Figure 1.17   *Concept of a small molecule stimuli-responsive membrane with an ATP-binding aptamer as a gate-keeper: in the absence of the target (ATP), the pores are open; upon binding to the target, flux through the membrane pores is greatly hindered.*

avidine as a support layer through a biotin-linker, on top of which the biotinylated ATP-binding aptamer was immobilized. Owing to the strong interaction between avidine and biotin, the highest possible immobilization density of the ATP-binding aptamer as the receptor molecule of the responsive membrane was warranted. The ATP aptamer sequence used in this study (CACCTGGGGGAGTATTGCGGAGGAAGGTTC CAGGTG) was based on that reported by Huizenga and Szostak [56]. As a negative control for the selective response of the self-assembled membrane, we employed in a parallel study a mutated ATP-binding aptamer which differed only in four nucleotides from the original sequence (CACCTAGGAGAGTAATGCCGAGGAAGGTTCCAGGTG) leading, however, to a practical non-specificity toward ATP.

We tested the selective responsiveness of the membrane by measuring its permeability for a tracer molecule which was fluorescein sodium salt. For this purpose, we placed the membrane in a holder with a pump which was continuously pulling at the permeate side, and analysed the fluorescence of the filtrate on-line with a single-photon-counting spectrofluorimeter for highest detection sensitivity. At a given time, fluorescein sodium salt with a molar mass of 376.27 mg·mol^{-1} was added to the feed solution as a tracer molecule and its concentration monitored in the filtrate on-line. Figure 1.18 depicts the resulting permeation transients of the tracer molecule for the membrane modified by the avidine support layer, only, and those we registered for the final membrane with either the ATP-binding aptamer or the mutated aptamer immobilized on top of the avidine layer. With time, the filtrate concentration of the tracer molecule reached the same plateau for both the avidine modified as well as the avidine-aptamer modified alumina membranes, demonstrating that after immobilization of the aptamer the pores had remained sufficiently open to let the tracer molecules pass freely; the filtrate flux decreased, but without any retention of tracer molecules taking place. Because the stimulus, ATP, is a similar molecular size as the tracer, we could hence expect that it would also permeate freely through the membrane pores

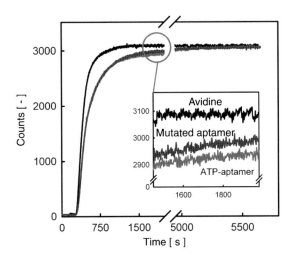

Figure 1.18 *Transient of the fluorescein permeation across the modified alumina membrane. See plate section for colour figure.*

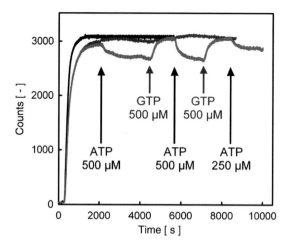

Figure 1.19 *Transient of the fluorescein permeation across the aptamer-modified alumina membrane; response to the target ATP and absence of a significant response to the GTP which is of similar structure (see also Figure 1.3). See plate section for colour figure.*

and reach any available aptamer recognition sites. We furthermore did not observe any significant difference in concentration transients between the ATP-binding aptamer and its mutation, confirming that the latter was a valid negative control for our stimuli-responsive membrane.

The selective and reversible responsiveness of the membrane incorporating the ATP-binding aptamer was tested by comparing it to the one containing the mutated aptamer upon adding a pulse of ATP to the feed solution to yield a concentration of 500 µM (the K_D of the ATP-binding aptamer with the sequence used in our study lies between 200–500 µM). As the data depicted in Figure 1.19 clearly illustrate, the membrane that was modified with the ATP-binding aptamer showed a clear decrease in the fluorescein concentration in the filtrate upon addition of the stimulus ATP to the feed solution. This indicated that the pores significantly hindered the passage of the fluorescein tracer molecule. No significant change in the fluorescence signal was observed for the membrane containing the mutated aptamer. The selective conformational change response which the ATP-binding aptamer undergoes upon interacting with the stimulus, ATP, was therefore sufficient to close the membrane pore to such a degree that partial exclusion of the tracer molecule occurred. In order to check for reversibility, we subsequently injected a pulse of 500 µM of guanosine-5'-triphosphate (GTP) to the feed solution; GTP has a similar chemical structure to ATP, but the ATP-binding aptamer we used in our study is known to practically not bind to GTP. In fact, we could observe that the fluorescein concentration in the filtrate was rising again upon addition of GTP, recovering the plateau value of the negative controls. Given that the presence of GTP in the feed solution was, therefore, insignificant for the responsiveness of the membrane, the ATP molecules previously bound to the ATP-binding receptor molecules were simply dissociated from the ATP-aptamer by the buffer solution; as a consequence, the aptamer recovered its original conformation resulting in a re-opening of the membrane pore. Repeated pulses of same amounts of ATP and then GTP gave almost identical responses

(Figure 1.19): in all cases a closing of the pores upon ATP-addition could be observed, followed by a full recovery of the permeate flux upon addition of GTP to the feed solution. When injecting a pulse of only 250 μM of ATP into the feed solution, we still observed a clear response of the membrane, but significantly attenuated, which is in agreement with the feed concentration being at the lower end of the range reported for the K_D. No significant response was observed throughout the experiment in the negative controls.

The functioning of this stimuli-responsive membrane depended strongly on a fine tuning of the average pore size and the expected conformational change the ATP-binding aptamer would undergo upon target recognition, and the results discussed reveal that this proof-of-concept still requires optimization. A pore size that is too narrow would hinder the ATP- structure from opening or, in the worst case scenario, even impede immobilization of the aptamer within the membrane pores. A pore size that is too large would inevitably result in the conformational change being insignificant for the flux through the membrane pores. We therefore deliberately chose a modular and self-assembly approach to the pore architecture, such that adjusting the resulting average pore size is relatively straightforward when other aptamer sequences are being used. In this sense, although not optimized yet, this proof-of-concept adds an important variant to conventional stimuli-responsive membranes, as the response is generated by the presence of a molecule rather than depending on a bulk stimulus such as pH, temperature or ionic strength.

1.5   Outlook

It has been shown that stimuli-responsive materials based on oligonucleic acids open up an exciting opportunity to create particles or barriers with controlled release or gating functions that mainly depend on a highly specific molecular recognition mechanism, rather than a bulk stimulus. This allows us to conceive of materials that can in principle be highly efficient and versatile. Nevertheless, it has also been highlighted that this field of research still contains challenges. For DNA-hybrid materials to function as expected, acting forces and interactions must be thoroughly understood and it is self-evident that an aptamer selected under defined conditions might not function properly once it is placed in a drastically different chemical environment. It has been shown that this fact is sometimes made use of as an efficient although totally unspecific bulk stimulus; however, when the highly specific molecular recognition capability of aptamers is to be explored, this sensitivity to bulk changes can be a limitation and should be taken into consideration. A further challenge of this research topic is the characterization of the function of the DNA-hybrid materials. Current characterization methods still yield properties of an average population of molecules and single-molecule detection and monitoring of its function still poses a challenge. One may argue that this information is irrelevant given that the function of DNA-based responsive materials represents an average response of all immobilized molecules. While this is true from an application point of view, it does not allow an in-depth understanding of the molecular recognition mechanisms as would single-molecule monitoring. The latter will link up nicely with efforts currently under way to theoretically model DNA and its interaction with targets – still a very demanding and exciting task which will allow us to truly tailor DNA-based nanodevices through knowledge rather than empirical approaches.

Acknowledgements

T. Schäfer would like to acknowledge the *European Research Council Grant 209 842-MATRIX* and M.B. Serrano-Santos the *European Commission* through the *Marie Curie Grant PIEF-GA-2009-236 628*. T.Schäfer would like to thank Prof. Pedro Miguel Echenique for hosting him in the *Donostia International Physics Centre (DIPC)*.

References

1. Stuart, M.A.C., Huck, W.T.S., Genzer, J. *et al.* (2010) Emerging applications of stimuli-responsive polymer materials. *Nature Materials*, **9**(2), 101–113.
2. Famulok, M. and Mayer, G. (2006) Chemical biology: Aptamers in nanoland. *Nature*, **439**, 666–669.
3. Breaker, R.R. (2011) Prospects for riboswitch discovery and analysis. *Mol Cell*, **43**, 897–879.
4. Iliuk, A.B., Hu, L. and Tao, W.A. (2011) Aptamer in bioanalytical applications. *Anal Chem*, **83**, 4440–4452.
5. Özalp, V.C. and Schäfer, T. (2011) Aptamer-based switchable nanovalves for stimuli-responsive drug delivery. *Chem Eur J*, **17**, 9893–9896.
6. Mayer, G. (2009) The chemical biology of aptamers. *Ang Chemie-Int Ed*, **48**, 2672–2689.
7. Vallee-Belisle, A. and Plaxco, K.W. (2010) Structure-switching biosensors: inspired by nature. *Curr Opin in Structural Biol*, **20**, 518–526.
8. Hermann, T. and Patel, D.J. (2000) Biochemistry - adaptive recognition by nucleic acid aptamers. *Science*, **287**, 820–825.
9. Zimmermann, G.R., Shields, T.P., Jenison, R.D. *et al.* (1998) A semiconserved residue inhibits complex formation by stabilizing interactions in the free state of a theophylline-binding RNA. *Biochemistry*, **37**, 9186–9192.
10. Kong, R.M., Zhang, X.B., Chen, Z. and Tan, W. (2011) Aptamer-assembled nanomaterials for biosensing and biomedical applications. *Small*, **17**, 2428–2436.
11. Singh, Y., Murat, P. and Defrancq, E. (2010) Recent developments in oligonucleotide conjugation. *Chem Soc Rev*, **39**, 2054–2070.
12. Berney, H. and Oliver, K. (2005) Dual polarization interferometry size and density characterisation of DNA immobilisation and hybridisation. *Biosens Bioelectron*, **21**, 618–626.
13. Liu, J., Cao, Z. and Lu, Y. (2009) Functional nucleic acid sensors. *Chem Rev*, **109**, 1948–1998.
14. Keeney, T.R., Bock, C., Gold, L. *et al.* (2009) Automation of the SomaLogic proteomics assay: A platform for biomarker discovery. *Jala*, **14**, 360–366.
15. Bock, L.C., Griffin, L.C., Latham, J.A. *et al.* (1992) Selection of single-stranded-DNA molecules that bind and inhibit human thrombin. *Nature*, **355**, 564–566.
16. Stojanovic, M.N., de Prada, P. and Landry, D.W. (2001) Aptamer-based folding fluorescent sensor for cocaine. *J Am Chem Soc*, **123**, 4928.

17. Mayer, G. (2009) Nucleic acid and peptide aptamers: Methods and protocols. In: *Methods in Molecular Biology*, G. Mayer (ed.) Springer, 19–32.
18. Wei, F., Bai, B. and Ho, C.M. (2011) Rapidly optimizing an aptamer based BoNT sensor by feedback system control (FSC) scheme. *Biosens Bioelectron.* doi: 10.1016/j.bios.2011.09.014
19. Drabovich, A.P., Berezovski, M.V., Musheev, M.U. and Krylov, S.N. (2009) Selection of smart small-molecule ligands: The proof of principle. *Anal Chem*, **81**, 490–494.
20. Flinders, J. and Dieckmann, T. (2006) NMR spectroscopy of ribonucleic acids. *Prog in Nuclear Magnetic Res Spect*, **48**, 137–159.
21. Sassolas, A., Blum, L.J. and Leca-Bouvier, B.D. (2011) Optical detection systems using immobilized aptamers. *Biosens Bioelectron*, **26**, 3725–3736.
22. Jing, M. and Bowser, M.T. (2011) Methods for measuring aptamer-protein equilibria: A review. *Anal Chim Acta*, **686**, 9–18.
23. Majhi, P.R. and Shafer, R.H. (2006) Characterization of an unusual folding pattern in a catalytically active guanine quadruplex structure. *Biopolymers*, **82**(6), 558–569.
24. Janshoff, A., Neitzert, M., Oberdorfer, Y. and Fuchs, H. (2000) Force spectroscopy of molecular systems-single molecule spectroscopy of polymers and biomolecules. *Angew Chem Int Ed*, **39**, 3212–3237.
25. Serrano-Santos, M.B., Llobet, E., Özalp, V.C. and Schäfer, T. (2012) Characterization of structural changes in aptamer films for controlled release nanodevices. *Chem Commun*, DOI:10.1039/C2CC35683J.
26. Sauerbrey, G. (1959) Verwendung von Schwingquarzen zur Wägung dünner Schichten und zur Mikrowägung. *Z Phys*, **155**(2), 206.
27. Huizenga, D.E. and Szostak, J.W. (1995) A DNA aptamer that binds adenosine and ATP. *Biochemistry*, **34**, 656–665.
28. Picuri, J.M., Frezza, B.M. and Ghadiri, M.R.J. (2009) Universal translators for nucleic acid diagnosis. *J Am Chem Soc*, **131**, 9368.
29. Stojanovic, M.N. and Kolpashchikov, D.M. (2004) Modular aptameric sensors. *J Am Chem Soc*, **126**, 9266–9270.
30. Hermann, T. and Patel, D.J. (2000) Adaptive recognition by nucleic acid aptamers. *Science*, **287**, 820–825.
31. Nonin-Lecomte, S., Lin, C.H. and Patel, D.J. (2001) Addditional hydrogen bond and base-pair kinetics in the symmetrical AMP-DNA aptamer complex. *Biophys J*, **81**, 3422–3431.
32. Tang, Z.W., Mallikaratchy, P., Yang, R.H. *et al.* (2008) Aptamer switch probe based on intramolecular displacement. *J Am Chem Soc*, **130**, 11268–11269.
33. White, R.J., Rowe, A.A. and Plaxco, K.W. (2010) Re-engineering aptamers to support reagentless, self-reporting electrochemical sensors. *Analyst (Cambridge, U.K.)*, **135**, 589–594.
34. Nielsen, L.J., Olsen, L.F. and Özalp, V.C. (2010) Aptamers embedded in polyacrylamide nanoparticles: a tool for in vivo metabolite sensing. *ACS Nano*, **4**, 4361–4370.
35. Lin, C.H. and Patel, D.J. (1997) Structural basis of DNA folding and recognition in an AMP-DNA aptamer complex: distinct architectures but common recognition motifs for DNA and RNA aptamers complexed to AMP. *Chem Biol*, **4**, 817–832.

36. Wackerbarth, H., Grubb, M., Zhang, J. *et al.* (2004) Dynamics of ordered-domain formation of DNA fragments on Au(111) with molecular resolution. *Angew Chem Int Ed*, **43**, 198–203.
37. Swann, M.J., Peel, L.L., Carrington, S. and Freeman, N.J. (2004) Dual-polarization interferometry: an analytical technique to measure changes in protein structure in real time, to determine the stoichiometry of binding events, and to differentiate between specific and nonspecific interactions. *Anal Biochem*, **329**, 190–198.
38. Deng, Q., German, I., Buchanan, D. and Kennedy, R.T. (2001) Retention and separation of adenosine and analogues by affinity chromatography with an aptamer stationary phase. *Anal Chem*, **73**, 5415–5421.
39. Lin, C.H. and Patel, D.J. (1997) Structural basis of DNA folding and recognition in an AMP-DNA aptamer complex: distinct architectures but common recognition motifs for DNA and RNA aptamers complexed to AMP. *Chem Biol*, **4**, 817–832.
40. Siwy, Z.S. and Howorka, S. (2010) Engineered voltage-responsive nanopores. *Chem Soc Rev*, **39**, 1115–1132.
41. Harrell, C.C., Kohli, P., Siwy, Z. and Martin, C.R. (2004) DNA-nanotube artificial ion channels. *J Am Chem Soc*, **126**, 15646–15647.
42. Xia, F., Guo, W., Mao, Y. *et al.* (2003) Gating of single synthetic nanopores by proton-driven DNA molecular motors. *J Am Chem Soc*, **130**, 8345–8350.
43. Sanchez-Quesada, J., Saghatelian, A., Cheley, S. *et al.* (2004) Single DNA rotaxanes of a transmembrane pore protein. *Angew Chem Int Ed*, **43**, 3063–3067.
44. Slowing, I.I., Vivero-Escoto, J.L., Trewyn, B.G. and Lin, V.S.Y. (2010) Mesoporous silica nanoparticles: structural design and applications. *J Mater Chem*, **20**, 7924–7937.
45. Lillis, B., Manning, M., Berney, H. *et al.* (2006) Dual polarisation interferometry characterisation of DNA immobilization and hybridisation detection on a silanised support. *Biosens Bioelectron*, **21**, 1459–1467.
46. Murphy, M.C., Rasnik, I., Cheng, W. *et al.* (2004) Probing single-stranded DNA conformational flexibility using fluorescence spectroscopy. *Biophys J*, **86**, 2530–2537.
47. Climent, E., Martinez-Manez, R., Sancenon, F. *et al.* (2010) Controlled delivery using oligonucleotide-capped mesoporous silica nanoparticles. *Angew Chem Int Ed*, **49**, 7281–7283.
48. Chen, C., Pu, F., Huang, Z. *et al.* (2011) Stimuli-responsive controlled-release system using quadruplex DNA-capped silica nanocontainers. *Nucleic Acids Research*, **39**, 1638–1644.
49. Ruiz-Hernandez, E., Baeza, A. and Vallet-Regi, M. (2011) Smart drug delivery through DNA/magnetic nanoparticle gates. *ACS Nano*, **5**, 1259–1266.
50. Schlossbauer, A., Warncke, S., Gramlich, P.M.E. *et al.* (2010) A programmable DNA-based molecular valve for colloidal mesoporous silica. *Angew Chem Int Ed*, **49**, 4734–4737.
51. Zhu, C.L., Lu, C.H., Song, X.Y. *et al.* (2011) Bioresponsive controlled release using mesoporous silica nanoparticles capped with aptamer-based molecular gate. *J Am Chem Soc*, **133**, 1278–1281.
52. Mairal, T., Ozalp, V.C., Lozano Sánchez, P. *et al.* (2008) Aptamers: molecular tools for analytical applications. *Anal Bioanal Chem*, **390**, 989–1007.

53. Abelow, A.E., Schepelina, O., White, R.J. *et al.* (2010) Biomimetic glass nanopores employing aptamer gates responsive to a small molecule. *Chem Commun*, **46**, 7984–7986.

54. Wandera, D., Wickramasinghe, S.R. and Husson, S.M. (2010) Stimuli-responsive membranes. *J Membrane Sci*, **357**(1–2), 6–35.

55. Özalp, V.C. and Schäfer, T. (2012) Small-molecule stimulus responsive membrane barrier with reversible gating function, submitted for publication.

56. Huizenga, D.E. and Szostak, J.W. (1995) A DNA aptamer that binds adenosine and ATP. *Biochemistry*, **34**(2), 656–665.

2

Emerging Membrane Nanomaterials – Towards Natural Selection of Functions

Mihail Barboiu

Adaptive Supramolecular Nanosystems Group, Institut Européen des Membranes –
UMR CNRS 5635, France

2.1   Introduction

Constitutional dynamic chemistry (CDC) [1] and its application dynamic combinatorial chemistry (DCC) [2, 3] are new evolutional approaches to produce chemical diversity. In contrast to the stepwise methodology of classic combinatorial chemistry, DCC allows for the simple generation of diverse interexchanging architectures from sets of building molecules interacting reversibly. With the DCC approach, the building library elements are spontaneously assembled to form all possible virtual combinations. If libraries are produced in the presence of a specific target, new ligands can be selected that resemble the naturally occurring ones. In this way, new, potentially useful affinity ligands can be generated. Compound libraries generated by DCC show special interest for a very diverse range of applications: molecular and supramolecular recognition, drug-, catalyst- and material discovery [4–10]. Kinetical or thermodynamical resolution, self-assembly followed by covalent-modification, and crystallization shed light on useful strategies of control and create convergence between self-organization and functions [1–3]. Basically, the CDC implements a dynamic reversible interface between interacting constituents, mediating the structural self-correlation of different domains of the system by virtue of their basic constitutional behaviours. The self-assembly of the molecular components allows the flow of structural information from molecular level toward nanoscale dimensions [4]. Understanding

Responsive Membranes and Materials, First Edition. Edited by D. Bhattacharyya, Thomas Schäfer, S. R. Wickramasinghe and Sylvia Daunert.
© 2013 John Wiley & Sons, Ltd. Published 2013 by John Wiley & Sons, Ltd.

and controlling such up-scale propagation of structural information might offer the potential to impose further precise order at the mesoscale and new routes to obtain highly ordered ultradense arrays over macroscopic distances. Within this context, during the last decade CDC has expressed more and more interest in dynamic interactive systems, (DIS) [10]. Networks of continuously exchanging and reorganizing reversibly connected objects (supermolecules, polymers, biomolecules, biopolymers, pores, nanoplatforms, nanotubes, surfaces, nanoparticles, liposomes, materials, cells) form the core of DIS, operating under natural selection to allow functional adaptability in response to internal constitutional or stimulant external factors. This makes DIS a great source of knowledge, highly relevant for a huge variety of direct applications. This chapter will discuss some selected examples of DIS concerning the hybrid organic/inorganic materials used for the elaboration of adaptive membrane systems. This chapter will be divided into three sections. The first part will focus on implementation of self-organization on the elaboration of ion-pair conduction pathways in liquid and hybrid membranes emphasizing the more recent developments toward biomimetic membranes. The second part describes recent work on the development of the dynamic intrapore resolution towards dynamic hybrid materials and membranes. Such systems evolve to form the fittest insidepore architecture, demonstrating flexible functionality and adaptation in confined conditions. The third part will introduce hybrid nanostructured materials with particular emphasis on self-assembly approaches used for synthetic construction of self-organized hybrid constitutional nanomaterials and dynameric systems, based on recent developments, as pursued especially in our laboratory. Finally, basic working principles of emerging membranes are provided in order to better understand the requirements in material design for the generation of functional nanostructured materials. Actual and potential applications of such emerging systems presenting combined features of structural adaptation in a specific hybrid nanospace will be presented.

2.2 Ion-Pair Conduction Pathways in Liquid and Hybrid Membranes

The structural functionality of the transporting active-channel membrane-spanning proteins is defined by very simple functional moieties (i.e. carbonyl, hydroxyl, amide, etc.) pointing toward the protein core and surrounding the transport direction. The end result when such moieties come together is a supramolecular/spatial organization of binding sites collectively contributing to the selective translocation of solutes along the protein wall [11–16]. One important aim arising from the mimicry of such highly functional devices is to develop artificial functional self-organized membrane materials using simple molecules that form patterns by collective self-assembly so as to enable efficient translocation events.

Other groups and ourselves previously reported several findings related to this aim and convergent multidimensional self-assembly strategies have been used for the synthesis of self-organized polymeric or hybrid membranes in which protons and ions are envisaged to diffuse along hydrophilic pathways, designed to mimic natural proteins at the micrometric scale.

These systems have been successfully employed to design solid dense membranes, functioning as ion-channels, and illustrate how the self-organized membranes perform interesting and potentially useful transport functions. Intermolecular interactions involving

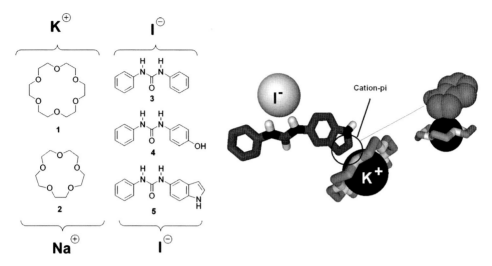

Figure 2.1 *Structures of the cation-carriers 18-crown-6, **1** and 15-crown-6, **2** and of the pheny-lureidoarene anion-carriers **3–5**; Crystal structure (stick representation) of the [**1**·K]⁺[**5**·I]⁻ complex showing clear cation-π interactions between the macrocyclic complexed K⁺ cation and indole group of the phenylureidoindole **5**. K⁺ : black and I⁻ : white spheres.*

aromatic rings are key processes in both chemical and biological recognition. Amongst these processes, cation-π interactions between positively charged species (alkali cations, ammonium, etc.) and aromatic systems with delocalized π-electrons are now recognized as important non-covalent binding forces of increasing relevance [15, 16]. The importance of interactions between alkali cations and aromatic side-chains of aromatic amino acids of proteins has been known for many years and they are of particular biological significance. Heterocomplex structures emphasizing particular $K^+ - \pi$ contacts with phenyl, phenol and indole rings were recently reported by many groups including ours [17–20].

Within this context, the selective transport of alkali cations can been performed through bulk liquid membranes in which mixtures of macrocyclic crown-ethers **1**, **2** and phenylureidoarene **3–5** derivatives are used as separate cation- and anion-carriers. As already observed in many cases, the ability of both the cation-carrier and anion-carrier are needed for transport (Figure 2.1) [19]. For separation purposes one of the most important properties is the transport selectivity of specific carriers in liquid membranes. It has been observed that the resulting flux in competitive transport experiments does not show a linear dependency with the stability constant: binding of the solute that is too strong has a negative effect on the transport selectivity. Therefore there is an optimal association constant for obtaining good transport selectivity. The differences in transport performances can be attributed to differences between the stability of the resulting [cation-carrier][anion-carrier] complexes in the membrane phase, inversely related to the hydrophilicity of the anion-carrier. In terms of performance, the extraction in the membrane phase, the transport-rate and the selectivity are higher for the K^+ cations than the Na^+ cations.

This indicates that the competitive extraction of $[\mathbf{1}^*K]^+$ by anion-carriers **3–5** is much more effective and specific than the extraction of $[\mathbf{2}^*Na]^+$. The energetic

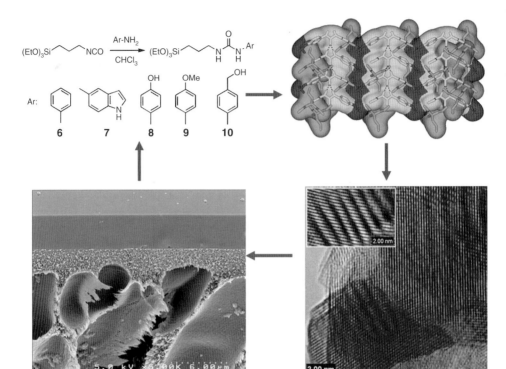

Figure 2.2   *Schematic multiscale nanostructuration of ureidoaromatic receptors in orientated aromatic cation-π and urea anion conduction pathways within thin-layer hybrid membranes [19, 20]. (Reprinted with permission from [19] Copyright (2008) Elsevier Ltd; and [20] Copyright (2008) Wiley-VCH). See plate section for colour figure.*

balance between the extraction of the KI in the membrane phase and the cation releasing step at the membrane–strip phase interface by the pairing of 18-crown-6/phenylureidoindole mixed carrier system (Figure 2.1) led to the best K^+ selectivity ($S_K{}^+/Na^+ = 12.5$) for the indole-type (tryptophane) compound [19].

These results focus us next towards designing functional amino acid ureidoaromatic precursors, **6–10** (Figure 2.2) for designing suitable molecular pathways in hybrid solid membranes. Such hybrid membrane materials are composed from nanodomains randomly ordered in the hybrid matrix, resulting from the self-assembly of simple molecular components that encode the required information for ionic assisted-diffusion within hydrophilic or hydrophobic aromatic cation-π conduction pathways.

Aromatic cation-π conduction pathways are essential in the diffusion process and in the selectivity of the transport of hydrated alkali cations. Although these pathways do not merge to cross the micrometric films, they are well defined along nanometric distances, reminiscent of the supramolecular organization of binding sites in channel-type proteins collectively contributing to the selective translocation of metabolites [11–16].

Using this knowledge, the same strategy employing a combined supramolecular self-organization and sol-gel synthetic route can be used in order to control the formation of

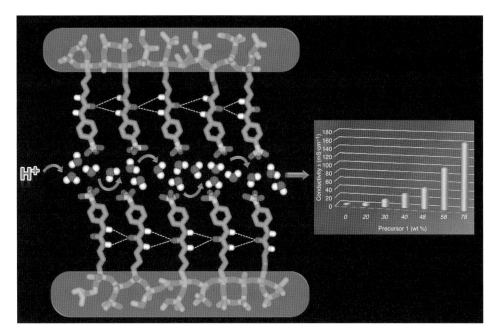

Figure 2.3 *Self-organization of supramolecular -SO$_3$H—H$_2$O proton conducting nanometric pathways and proton conductivity at 25°C and 100% relative humidity. (Reprinted with permission from [29] Copyright (2009) Royal Society of Chemistry). See plate section for colour figure.*

hydrophilic proton pathways as a straightforward approach to the design of a novel class of proton-exchange membranes – PEMs [29]. Recently, we described a PEM system in which the self-organized precursor generates directional proton-conducting superstructures in a scaffolding hydrophobic hybrid material (Figure 2.3).

The controlled generation of connected self-organized channels along hundreds of nanometers, for directional proton diffusion, result in the production of PEMs hybrid materials with high ionic conductivities up to $\sigma = 160.2$ mS cm^{-1}. This value is up to one and half times higher than the maximum Nafion 117 commercial membrane. While the water uptake constantly increases with high IEC, an even sharper rise in the conductivity with the IEC was recorded for these membranes, with an extended density and orientation of the directionally water-sulfonic group channels, hierarchically self-organized along domains of nanometers. This generates, therefore, a more efficient percolated conducting supramolecular network in "high concentrated" hybrid materials, which in turn results in correspondingly high proton conductivity. These results extend the applications of self-organized hybrid materials toward functional supramolecular devices.

The novelty of these findings with respect to previous work on self-organized membranes [21–28] using specific complexant receptors, is that simple molecules that collectively define transporting devices by self-assembly can be successfully used to transfer their overall functionality of their supramolecular self-organization in hybrid membrane materials. These results imply that control of molecular interactions can define the self-organized architectures with supramolecular functionality, such us hydrophobic and hydrophilic transporting pathways of different chemical properties. They might provide new insights into the basic features that control the design of new materials mimicking the proton channels, with applications in fuel-cell technology, chemical separations, sensors or as storage-delivery devices.

2.3   Dynamic Insidepore Resolution Towards Emergent Membrane Functions

Unlike soft hybrid materials, mesoporous materials are often able to form orientated nanopores with well-defined and controlled dimensions and geometries. Nanoscale confinement of specific receptors within integrated hybrid systems can induce collective specific properties with different functionality, going beyond the properties of any of these components acting in solution or fixed covalently in a solid matrix. Within this context, molecular libraries can organize and further evolve in such a scaffolding porous nanospace to develop adaptive "dynamic functions" related to confinement or anisotropic orientation. Thus, as an outgrowth of dynamic constitutional chemistry [1] principles, this leads to a dynamic intrapore resolution of DCLs. These phenomena might be considered an upregulation of the most adapted insidepore/super- or nanostructure under confined conditions, naturally adapting the system architecture/functionality in response to various effectors. It embodies a reorganization of the system spatial/temporal configuration, evolving an improved response in the presence of the effectors that produced this change in the first place. Operating such evolutive systems is mainly related to reversible interactions between the continually interchanging confined organic components and the inorganic porous preformed structure of specific dimensional and directional behaviours. This makes them adaptive, responding to internal constitutional, or to external, stimuli. It results in the formation of the most efficient functional superstructures, in the presence of the fittest stimulus, selected from a set of diverse less-selective possible architectures that can form by their self-assembly, within such confined conditions. This can lead to an adaptive functionality, directed by the evolution of "dynamic encodable" interactions between components self-designing the available space under confinement.

More generally, applying such consideration to hybrid constitutional systems leads to the definition of dynamic hybrid materials, in which inorganic porous matrices are reversibly connected to dynamic organic/supramolecular networks. Within this context constitutional self-instructed membranes were developed by our group and used for mimicking the adaptive structural functionality of natural ion-channel systems [30]. From the conceptual point of view and in a similar manner, columnar silica mesopores can be used as a scaffolding matrix to orientate self-organized ion-channel-type artificial systems. These membranes are based on dynamic hybrid materials in which the functional self-organized macrocycles are reversibly connected with the inorganic silica through hydrophobic non-covalent interactions (Figure 2.4).

The silica filled alumina anodisk membranes (AAMs) were prepared with the template sol-gel method to form silica mesopores orientated along the macropore walls of an AAM membrane. Afterwards, a calcination step removing the surfactant results in the formation of the MCM41-type material (40 Å pore diameter) which was reacted with

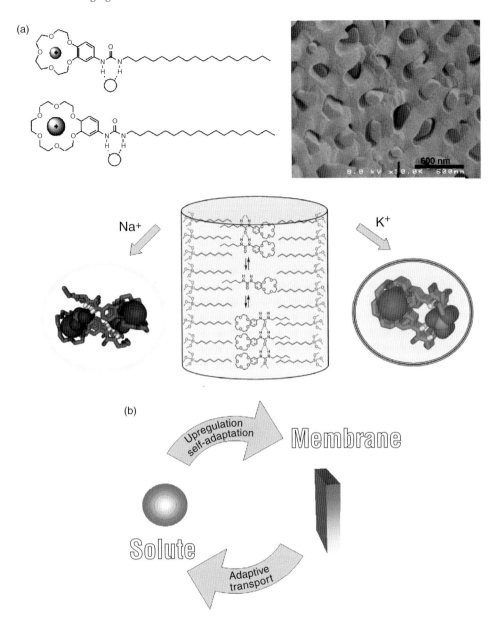

Figure 2.4 *(a) Generation of directional ion-conduction pathways which can be morphologically tuned by alkali salts templating within dynamic hybrid materials by the hydrophobic confinement of ureido-macrocyclic receptors within silica mesopores. (b) It embodies a reorganization of the membrane configuration evolving an improved response in the presence of the solute that produced this change in the first place.*

octadecyltrichlorosilane-ODS and then carefully washed, in order to obtain hydrophobic MCM41 silica materials. Then, macrocyclic receptors were physically confined within the hydrophobic mesopores, resulting in the formation of dynamic hybrid membranes. Supramolecular columnar ion-channel architectures confined within scaffolding hydrophobic silica mesopores can be structurally determined and morphologically tuned by alkali salt templating. The dynamic character lying in reversible interactions between the continually interchanging components makes them respond to external ionic stimuli and adjust to form the most efficient transporting superstructure, in the presence of the fittest cation, selected from a set of diverse less-selective possible architectures which can form by self-assembly. Evidence has been presented that such a membrane adapts and evolves its internal structure so as to improve its ion-transport properties: the dynamic non-covalent bonded macrocyclic ion-channel-type architectures can be morphologically tuned by alkali salt templating during the transport experiments or the conditioning steps. From the conceptual point of view these membranes express a synergistic adaptive behaviour: the addition of the fittest alkali ion drives a constitutional evolution of the membrane toward the selection and amplification of the specific transporting superstructures within the membrane in the presence of the cation that promoted its generation in the first place. This is a nice example of dynamic self-instructed ("trained") membranes where a solute induces the upregulation (prepares itself) of its own selective membrane. Moreover, the tradeoff in the low-permeability/high-selectivity, or vice versa, balance has disappeared (Figure 2.5) [30].

This led to the discovery of the functional supramolecular architecture evolving from a mixture of reversibly insidepore exchanging devices via ionic stimuli so as to improve membrane ion-transport properties. These phenomena might be considered an upregulation of the most adapted 3D "insidepore" superstructure, enhancing the membrane efficiency and selectivity by the binding of the ion-effectors (Figure 2.4). Finally these results extend the application of constitutional dynamic chemistry [1] from materials science to functional dynamic interactive membranes – SYSTEMS-MEMBRANES [10]. This feature

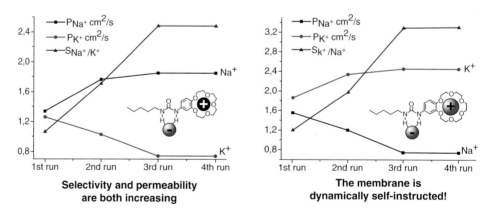

Figure 2.5 *Improved permeabilities and selectivities of the fittest cations through ion-conditioned dynamic hybrid membranes in a series of repetitive transport experiments. (Reprinted with permission from [30] Copyright (2009) PNAS).*

offers membrane science perspectives towards self-designed materials evolving their own functional superstructure so as to improve their transport performance. Prospects for the future include the development of these original methodologies towards dynamic materials, presenting a greater degree of structural complexity.

2.4 Dynameric Membranes and Materials

2.4.1 Constitutional Hybrid Materials

Convergent multidimensional self-assembly of molecular components has been used extensively over the last decades for the synthesis of functional supramolecular devices. Kinetical or thermodynamical self-assembly followed by covalent modification or crystallization shed light on useful strategies to control and create convergence between self-organization and supramolecular functions. Considerable challenges still lie ahead, and a more significant one is the "dynamic marriage" between supramolecular self-assembly and the polymerization process. Both supramolecular and polymerization processes might "correlate" in order to converge toward self-organized nanomaterials. The weak supramolecular interactions' positioning of the molecular components is typically less robust and they are usually destabilized when the cross-linked covalent polymeric network forms. The resulting hybrid architectures lead in general to amorphous or polymorphic materials. One solution to overcome these difficulties is to improve the binding efficiency of molecular components. An increased number of interactions between molecular components may improve the stability of the templating supramolecular systems within the inorganic siloxane network. Numerous successful examples, pioneered by Shinkai *et al.* concern the use of organogels acting as robust templates during the tetraetoxysilane (TEOS) sol-gel process [31]. We have recently reported nucleobase-based hybrids obtained under thermodynamic control [32, 33]. A second strategy is to use a reversible non-covalent interface between the supramolecular/ organic and the siloxane/polymeric constituents [34–37]. As the self-organization of organic/polymeric/inorganic domains has to explore the hypersurface of all structure/energy combinations, the built up of final hybrid material upon a growing/self-reparation/termination sequence might be able to select the correct spatial geometries for the supramolecular organic and inorganic entities emerging from a collection of building blocks. A dynamic reversible interface might mediate the structural self-correlation of supramolecular and inorganic domains by virtue of their basic constitutional behaviours. The resultant hybrid material might undergo continual change in its constitution, through dissociation/reconstitution of different mesophases during the sol-gel process. The reversibility of interactions between components of a hybrid material might be a crucial factor and, accordingly, a dynamic reversible hydrophobic interface might render the emergence of organic/inorganic mesophases self-adaptive, which mutually (synergistically) may adapt their 3D spatial distribution based on their own structural constitution during the simultaneous formation of hybrid self-organized domains.

In this context, we recently introduced the concept of constitutional hybrid materials prepared under thermodynamic control, containing supramolecular architectures combined with hydrophobic siloxanes as precursors for the inorganic silica matrix [34].

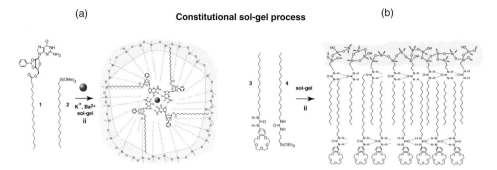

Figure 2.6 *Constitutional hybrid materials based on (a) G-quadruplex and (b) ureido-crown-ether architectures. (Reprinted with permission from [34] Copyright (2009) Royal Society of Chemistry).*

The G-quartet, the hydrogen-bonded macrocycle formed by the self-assembly of guanosine, is stabilized by alkali cations (Figure 2.6a). The role of cation templating is to stabilize by coordination to the eight carbonyl oxygens of two sandwiched G-quartets, the G-quadruplex, the columnar device formed by the vertical stacking of four G-quartets. It represents a dynamic supramolecular system, in equilibrium with linear oligomers. The dynamic convergence between supramolecular self-assembly of G-quartet architectures and the resolution process (polymerization, phase change) may "communicate". The results showed that highly self-organized G-quartet materials can be synthesized by using a dynamic communicating interface between organic and inorganic hybrid mesophases [34, 35]. The reversibility of interactions between G-quartet and polymeric components is a crucial factor and, accordingly, a reversible covalent connection might render the emergence of system mesophases self-adaptative, which may mutually adapt their 3D spatial distribution based on their own structural constitution during the simultaneous formation of mixed self-organized domains.

Another class of ion-channel forming architectures is represented by the heteroditopic ureido-crown-ethers complexing both cations and anions and self-organizing into columnar oligomeric superstructures in solution and the solid state (Figure 2.6b). Their self-organization in the liquid or bilayer membranes provided evidence for a hybrid carrier/channel transporting mechanism. Furthermore these dynamic systems have been "frozen" in a hybrid matrix by a sol-gel process, transcribing their dynamic columnar self-organization in solid dense hybrid nanomembranes [34, 36]. As observed for G-quartet systems, the same premises true hold for the ureido-crown-ether hybrid systems. These constitutional hybrids reach higher levels of organization giving access to a large number of crystalline orders in view of the higher number of diffraction peaks observed in X-ray power diffraction patterns. Moreover, while the classical hybrids are compact dense solids with a granular morphology, the constitutional hybrids consist of crystalline rods of micrometric dimension [34]. The superior long-range organization at micrometric length of both hybrid constitutional materials, compared with the classical ones showing nanometric organization, is certainly dependent of the superior emergence of self-organizing collective domains adapting along hydrophobic reversible interfaces.

On the other hand, polyoxometalates (POMs) not only exhibit beautiful topologies of nanometric dimensions, but also possess many interesting properties and applications in the fields of catalysis, optics magnetism, medicine, and biology [37]. Spherical porous anionic molybdenum capsules can be considered models for cellular cation transport, in particular with respect to the exchange of alkali metal ions between the interior of the capsules and the surrounding solution. In particular, the overall charge of internal surfaces of these nanoobjects can be finely tuned, allowing a controlled stepwise encapsulation of ions and, more importantly, the crown-ether-type pores can regulate the ion flux in a on/off supramolecular fashion. Moreover molecular dynamic simulations shed light onto the route towards embedding such nanocapsules with unprecedented molecular-scale filter properties into lipid bilayer membranes via self-assembly.

Hybrid materials combining POMs with a silica matrix have been recently developed with the hope of taking advantage of unique properties of POMs and the versatility (processability) of sol-gel materials [38]. We recently focused our work on the preparation of novel dynamic hybrid materials based on polyoxomolybdate-type nanocapsule and lypophilic silica systems using two different synthesis strategies. We found that the surfactant hybrid {DODA$_{40}$**Mo132**} architectures had a general tendency to assemble into lamellar or columnar aggregates within the silica lipophilic environment. This adaptive behaviour is attributed to the rearrangement of the DODA surfactants on the external surface of the POMs. Their dynamic character lies in reversible interactions between the continually interchanging {DODA$_{40}$**Mo132**} capsules and lipophilic ODS silica making them better able to adapt their structure from a set of diverse less-selective possible architectures which can form by self-assembly generating directional superstructures and which can organize in a scaffolding silica monolithic and mesoporous nanospace (Figure 2.7).

The dynamic hybrid materials described here are nice examples of constitutional devices mutually adapting their morphology. Finally the results obtained extend the application of constitutional dynamic chemistry from materials science to constitutional devices. This feature offers material science perspectives towards self-designed materials evolving their own functional superstructure so as to improve their transport performances. Operating such evolutive dynamic interactive systems [10] is mainly related to reversible interactions between the continually interchanging confined organic components and the inorganic porous preformed structure of specific dimensional and directional behaviours. This makes them adaptive, responding to internal constitutional or to external stimuli. It would result in the formation of the most efficient functional superstructures, in the presence of the fittest stimulus, selected from a set of diverse less-selective possible architectures which can form by their self-assembly within such confined conditions. This can lead to an adaptive functionality, directed by the evolution of "dynamic encodable" interactions between components self-designing the available space under confinement.

2.4.2 Dynameric Membranes Displaying Tunable Properties on Constitutional Exchange

Supramolecular chemistry is by nature a dynamic chemistry in view of the lability of the non-covalent interactions connecting the molecular components of a supramolecular

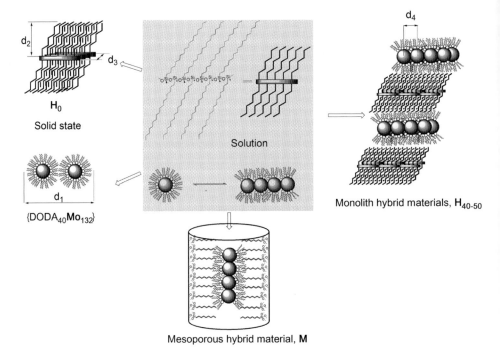

Figure 2.7 *Assembly mechanism of hybrid polyoxomolybdate {DODA$_{40}$**Mo132**} capsules condensed in the solid state under different conditions, including 2D lamellar architectures of* **ODS** *or dimers of {DODA$_{40}$**Mo132**} and layered mixed architectures {**ODS**DODA$_{40}$**Mo132**} or MCM41ODS⊂{DODA$_{40}$**Mo132**}.*

entity. Importing such dynamic features into molecular as well as supramolecular polymeric materials implies looking at dynamic polymers, "dynamers", which are dynamic by nature (supramolecular) or by design (molecular) and are capable of undergoing exchange, incorporation or decorporation of their monomeric subunits, linked together by respectively labile non-covalent interactions or reversible covalent bonds [39].

Compared to a physical mixture of classic polymers (Figure 2.8a), the mixtures of dynamers give access to new homogenous species presenting controllable modulation of their structure at molecular level in response to external stimuli or experimental conditions. They are capable of undergoing exchange, incorporation or decorporation of their monomeric subunits (Figure 2.8b), linked together by respectively labile non-covalent interactions or reversible covalent bonds. This might play an important role in the ability to more finely mutate the mechanical or functional properties of such new molecularly tunable dynamers, compared to physical mixtures of polymers [40]. In this way, specific recognition superstructures (for example the AAAA signature in Figure 2.8c) would be of much interest, giving access to a new class of functional membrane materials.

Within this context we recently reported an example of "dynameric" membrane films [40, 41]. The proposed strategy for using dynamic covalent polymer films provides a very useful methodology for easily modifying transport properties, giving access to smart and

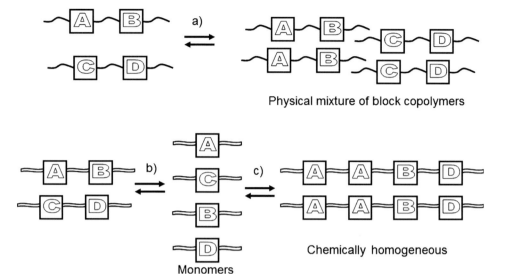

Figure 2.8   *Structural comparative features between classical mixtures of polymers (a) and the features of dynamers capable of undergoing exchange, incorporation or decorporation of their monomeric subunits, (b) resulting in the formation of specific recognition superstructures (for example the AAAA signature – (c)).*

adaptive membrane materials. We have demonstrated that the reversible covalent exchange processes between the constituent dynameric sub-components (hard or soft, complexant or not, etc.) during the membrane synthesis results in the formation of stable solid-thin layer membranes conserving their structural properties during and after the membrane transport experiments (Figure 2.9).

   Thanks to the possibility of combining the structural and functional features of different monomers, the heteropolymeric membrane materials can exhibit very different properties from their original homopolymeric components. In the developed examples, this strategy revealed itself to be a versatile way for the synthesis of new membranes presenting different permeabilities and preserving their selectivity. This feature offers membrane science perspectives towards functional materials that involve modification and control of the intrinsic structural properties of dynamic entities correlated with the dynamic features of the diffusional controlled transport.

   The dynameric membranes have been also used for the long-range amplification of the G-quadruplex self-organization into macroscopic polymeric functional films. The proposed strategy for stabilizing the G-quadruplexes in the double dynamer systems is novel. It involves the double covalent iminoboronate reversible connection between a dynamic supramolecular system (G-quartet/G-quadruplex superstructures) and a rigid polymeric network. This contributes to make functional G-quadruplexes by correlating the reversible supramolecular (G-quartets) with the polymeric self-assembly via the reversible molecular connections (iminoboronate G-macromonomer bonds) between the components (Figure 2.10). The fixed ("frozen") G-quadruplexes self-correlate with a directional order as proved by XPRD and transport experiments, to generate anisotropic mesophases

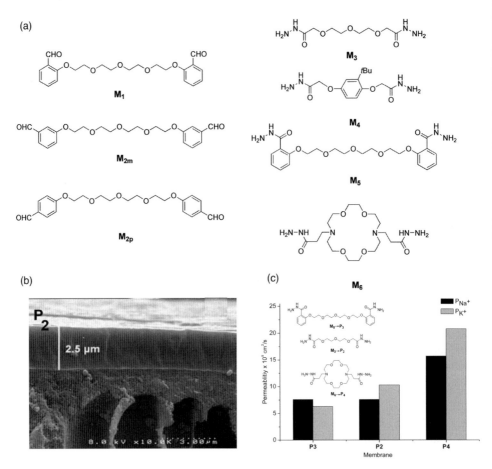

Figure 2.9 (a) Structures of the dialdehydes and of the bis-hydrazide monomers used for the synthesis of (b) dynameric membrane supported films and their (c) constitutional transporting properties on combinatorial combination of functional components.

interconnected via condensed macromonomeric hydrophobic bridges. The G-quadruplex ordered membrane films contribute to the fast electron/proton transfer by the formation of directional conduction pathways. Mixed cationic Na^+/K^+ or selective K^+ transport enable us to better understand the diffusional ion exchanges along "fixed" G-quadruplex polymeric pathways.

Within this context we saw that producing functional transporting devices like adaptive membranes or biosensing surfaces with precise structural ultradense arrays of addressable nanoscopic domains that are perfectly ordered over macroscopic length scales, might be very attractive from the standpoints of nanosynthesis but also from the real need for highly selective self-adaptive membrane systems used for industrial applications. The research in dynameric membranes is relevant to a broad section of the polymer and potentially the biomedical industries as it combines fundamental research with potential relevant

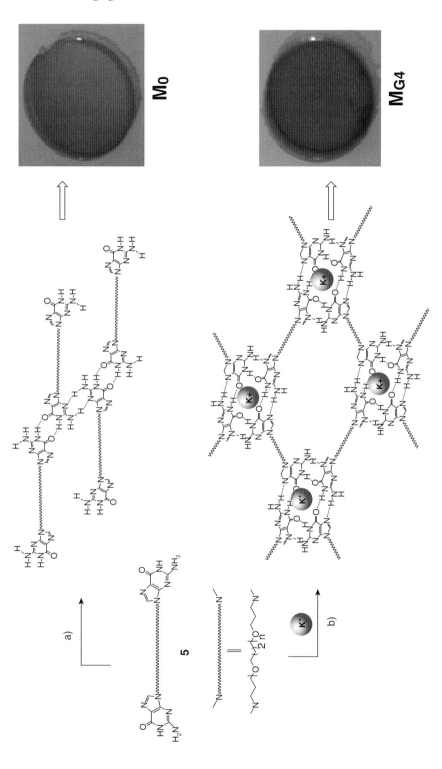

Figure 2.10 *Hierarchical cation-templated self-assembly of bis-iminoboronate-guanosine **5** macromonomer gives the G-quartet networks in the solid self-standing polymeric membrane films (a) in the absence **M$_0$** and (b) in the presence of templating K$^+$ cation, **M$_{G4}$**. See plate section for colour figure.*

applications. Opening the field of dynameric systems would represent a paradigm shift in current research efforts. Gaining a better understanding of the basic concepts of self-assembly at mesoscopic level should help to bridge the gap between synthetic chemists and biological researchers, thus providing for more immediate combined efforts in drug delivery, diagnostics and sensing between the two communities. The dynameric membrane systems may be the next modest steps towards this fascinating goal by comparing different architectures of constitutional dynameric platforms, based on selection from dynamic networks made from relatively simple molecular/macromolecular building blocks, of the fittest self-organized architectures which can undergo driven evolution under the effect of different stimuli or may present multivalent recognition properties.

2.5   Conclusion

Highly interconnected networks of chemical objects relate to a systems chemistry [5]. Lehn considered that constitutional dynamic chemistry confers the fifth dimension of the matter, that of constitution, in addition to 4D spatial/temporal chemical space [9]. The self-assembly of system components controlled by mastering molecular/ supramolecular interactions, may embody the flow of structural information from the molecular level towards nanoscale dimensions. The resulting spatio/temporal affinity is highly dependant on the nature of the scaffold as multivalent interactions have been shown to be influenced by such factors as shape, valency, orientation and flexibility. The possibility of designing and engineering nanometric multivalent platforms is at the forefront of cross-disciplinary orientated research. Thus one of the next challenges is to implement the "living" dynamic interactive systems [10] supporting natural selection and functional evolution as a viable solution to post-synthetically assembled systems. Natural systems have evolved for billions of years to accept complex evolutive structures, not synthetic generated systems. Of scientific significance is the fact that dynamic interactive systems could be extended to the vast field of scientific challenges, for example, resulting in the property (function)-driven generation of new adjustable (adaptive) system structures. They can generate dynamic "smart" membranes where a solute induces the upregulation (prepares itself) of its own selective membrane (Figure 2.11).

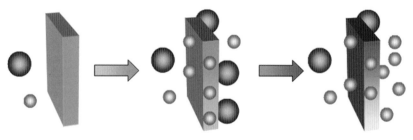

Membrane self-preparing step   **Membrane self-responding step**

Figure 2.11   *Self-reorganization of the membrane configuration evolving an adaptive response in the presence of the solute that produced the change in the first place.*

Constitutional selection of architectures from dynamic combinatorial libraries and the generation of dynamic hybrid materials and membranes led to the definition of SYSTEMS MATERIALS – SYSMAT and more specifically toward SYSTEMS MEMBRANES – SYSMEM and concern highly interconnected networks of reactional and constitutional materials and functional membrane systems respectively [10]. They should expand the fundamental understanding of nanoscale structures and functional properties as it relates to creating products and manufacturing processes. Moving dynamic systems into products and microsystems and the "functional" arena is now the main objective of major technologies. Combined supramolecular and combinatorial dynamic strategies to produce dynamic nanosystems can be effectively shared as merged marketable nanotechnology to benefit most research laboratories and nanomaterial producers. The combined features of structural adaptation in a specific hybrid nanospace and of the dynamic supramolecular selection process make the dynamic-site membranes, presented in the third section, of general interest for the development of a specific approach toward nanomembranes of increasing structural selectivity.

Acknowledgements

This work was financed as part of the Marie Curie Research Training Network – "DYNAMIC" (MRTN-CT-2005-019561), a EURYI scheme award. (www.esf.org/euryi) and ANR 2010 BLAN 717 2. I thank the past and the present members of my laboratory and the many individuals with whom we have collaborated in these studies: Adinela Cazacu, Gihane Nasr, Mathieu Michau, Remi Caraballo, Carole Arnal-Herault, Andreea Pasc-Banu, Arnauld Gilles, Arie van der Lee, Eddy Petit, Anca Meffre and Yves Marie Legrand.

References

1. Lehn, J.-M. (2007) From supramolecular chemistry towards constitutional dynamic chemistry and adaptative chemistry. *Chem. Soc. Rev.*, **36**, 151–160.
2. Corbett, P.T., Leclaire, J., Vial, L. *et al.* (2006) Dynamic combinatorial chemistry. *Chem. Rev.*, **106**, 3652–3711.
3. Lehn, J.-M. (1999) Dynamic combinatorial chemistry and virtual combinatorial libraries. *Chem. Eur. J.*, **5**, 2455–2463.
4. (2005) *Chem. Rev.*, **105**, *Special Issue on Functional Nanostructures*.
5. Kindermann, M., Stahl, I., Reimold, M. *et al.* (2005) Systems chemistry: Kinetic and computational analysis of a nearly exponential organic replicator. *Angew. Chem. Int. Ed.*, **44**, 6750–6755.
6. Müller, A. (2003) Bringing inorganic chemistry to life. *Chem. Commun.*, 803–806.
7. Prouzet, E., Ravaine, S., Sanchez, C. and Backov, R. (2008) Bio-inspired synthetic pathways and beyond: integrative chemistry. *New J. Chem.*, **32**, 1284–1299.
8. Menger, F. (1991) Chemical collectivism. *Angew. Chem. Int. Ed.*, **30**, 1086–1089.
9. Sanchez, C., Julian, B., Belleville, P. and Popall, M. (2005) Applications of hybrid organic-inorganic nanocomposites. *J. Mater. Chem.*, **15**, 3559–3592.

10. Barboiu, M. (2010) Dynamic interactive systems - dynamic selection in hybrid organic-inorganic constitutional networks. *Chem. Commun.*, **46**, 7466–7476.
11. Vaitheeeswaran, S., Yin, H., Raisaiah, J.C. and Hummer, G. (2004) Water clusters in nonpolar cavities. *Proc. Natl. Acad. Sci. U.S.A.*, **101**, 17002–17005.
12. Cukierman, S. (2000) Proton mobilities in water and in different stereoisomers of covalently linked gramicidin A channels. *Biophys. J.*, **78**, 1825–1834.
13. MacKinnon, R. (2004) Potassium channels and the atomic basis of selective ion conduction (Nobel Lecture). *Angew. Chem. Int. Ed.*, **43**, 4265–4289.
14. Agre, P. (2004) Aquaporin water channels. *Angew. Chem. Int. Ed.*, **43**, 4278–4290.
15. Dougherty, D.A. (1996) Cation-π interactions in chemistry and biology: A new view of benzene, Phe, Tyr, and Trp. *Science*, **271**, 163–168.
16. Gallivan, J.P. and Dougherty, D.A. (1999) Cation-π interactions in structural biology. *Proc. Natl. Acad. Sci. USA*, **96**, 9459–9464.
17. G. W. Gokel, A. Mukhopadhyay, (2001) Synthetic models of cation-conducting channels. *Chem. Soc. Rev.*, **30**, 274–286.
18. Arnal-Herault, C., Barboiu, M., Petit, E. *et al.* (2005) Cation-π interaction: A case for macrocycle-cation π-interaction by its ureidoarene counteranion. *New J. Chem.*, **29**, 1535–1539.
19. Michau, M., Caraballo, R., Arnal-Hérault, C. and Barboiu, M. (2008) Alkali cation-π aromatic conduction pathways in self-organized hybrid membranes. *J. Memb. Sci.*, **321**, 22–30.
20. Michau, M., Barboiu, M., Caraballo, R. *et al.* (2008) Ion-conduction pathways in self-organized ureidoarene heteropolysiloxane hybrid membranes. *Chem. Eur.J.*, **14**, 1776–1783.
21. Barboiu, M., Luca, C., Guizard, C. *et al.* (1997) Hybrid organic-inorganic fixed site dibenzo-18-crown complexant membranes. *J. Memb. Sci.*, **129**, 197–207.
22. Lachowicz, E., Rózànska, B., Teixidor, F. *et al.* (2002) Comparison of sulphur and sulphur–oxygen ligands as ionophores for liquid–liquid extraction and facilitated transport. *J. Memb. Sci.*, **210**, 279–290.
23. Barboiu, M., Guizard, C., Luca, C. *et al.* (1999) A new alternative to amino acid transport: Facilitated transport of L-Phenylalanine by hybrid siloxane membrane containing a fixed site macrocyclic complexant. *J. Memb. Sci.*, **161**, 193–206.
24. Guizard, C., Bac, A., Barboiu, M. and Hovnanaian, N. (2000) Organic-inorganic hybrid materials with specific solute and gas transport properties for membrane and sensors applications. *Mol. Cryst and Liq. Cryst.*, **354**, 91–106.
25. Barboiu, M., Guizard, C., Luca, C. *et al.* (2000) Facilitated transport of organics of biological interest II. Selective transport of organic acids by macrocyclic fixed site complexant membranes. *J. Memb. Sci.*, **174**, 277–286.
26. Barboiu, M., Guizard, C., Hovnanian, N. *et al.* (2000) Facilitated transport of organics of biological interest I. A new alternative for the amino acids separations by fixed-site crown-ether polysiloxane membranes. *J. Memb. Sci.*, **172**, 91–103.
27. Barboiu, M., Guizard, C., Hovnanian, N. and Cot, L. (2001) New molecular receptors for organics of biological interest for the facilitated transport in liquid and solid membranes. *Sep. Pur. Tech.*, **25**, 211–218.

28. Guizard, C., Bac, A., Barboiu, M. and Hovnanian, N. (2001) Hybrid organic-inorganic membranes with specific transport properties. Applications in separation and sensors technologies. *Sep. Pur. Tech.*, **25**, 167–180.
29. Michau, M. and Barboiu, M. (2009) Self-organized proton conductive layers in hybrid proton exchange membranes, exhibiting high ionic conductivity. *J. Mater. Chem.*, **19**, 6124–6131.
30. Cazacu, A., Legrand, Y.M., Pasc, A. *et al.* (2009) Dynamic hybrid materials for constitutional selective membranes. *Proc. Natl. Acad. Sci.*, **106**(20), 8117–8122.
31. van Bommel, K.J.C., Frigerri, A. and Shinkai, S. (2003) Organic templates for the generation of inorganic materials. *Angew. Chem. Int. Ed. Engl.*, **42**, 980–999.
32. Arnal-Hérault, C., Barboiu, M., Pasc, A. *et al.* (2007) Constitutional self-organization of adenine-uracil-derived hybrid materials. *Chem. Eur. J.*, **13**, 6792–6800.
33. Arnal-Herault, C., Pasc-Banu, A., Barboiu, M. *et al.* (2007) Amplification and transcription of the dynamic supramolecular chirality of the G-quadruplex. *Angew. Chem.*, **119**, 4346–4350; (2007) *Angew. Chem. Int. Ed.*, **46**, 4268–4272.
34. Mihai, S., Cazacu, A., Arnal-Herault, C. *et al.* (2009) Supramolecular self-organization in constitutional hybrid materials. *New. J. Chem*, **33**, 2335–2343.
35. Mihai, S., Le Duc, Y., Cot, D. and Barboiu, M. (2010) Sol-gel selection of hybrid G-quadruplex architectures from dynamic supramolecular guanosine libraries. *J. Mater Chem*, **20**, 9443–9448.
36. Barboiu, M., Cazacu, A., Mihai, S. *et al.* (2011) Dynamic constitutional hybrid materials-toward adaptive self-organized devices. *Microp. Mesop. Mat.*, **140**, 51–57.
37. Cronin, L. (2006) Inorganic molecular capsules: From structure to function. *Angew. Chem. Int. Ed. Engl.*, **45**, 3576–3578.
38. Cazacu, A., Mihai, S., Nasr, G. *et al.* (2010) Lipophilic polyoxomolybdate nanocapsules in constitutional dynamic hybrid materials. *Inorg. Chim. Acta*, **363**(15), 4214–4219.
39. Barboiu, M., Aimar, P. and Lehn, J.M. (2008) From simple molecules to complex membrane systems, Editorial. *J. Memb. Sci.*, **321**, 1–2.
40. Nasr, G., Barboiu, M., Ono, T. *et al.* (2008) Dynamic polymer membranes displaying tunable transport properties on constitutional exchange. *J. Memb. Sci.*, **321**, 8–14.
41. Arnal-Herault, C., Pasc-Banu, A., Michau, M. *et al.* (2007) Functional G-quartet macroscopic membrane films. *Angew. Chem.*, **119**, 8561–8565; (2007) *Angew. Chem. Int. Ed.*, **46**, 8409–8413.

3

Carbon Nanotube Membranes as an Idealized Platform for Protein Channel Mimetic Pumps

Bruce Hinds

Department of Chemical and Materials Engineering, University of Kentucky, USA

3.1 Introduction

Biological protein channels have the unmatched ability to selectively pump necessary chemicals through cell walls at rates orders of magnitude faster than simple diffusion or Newtonian fluid flow. Mimicking this function in large-area robust man-made structures can bring orders of magnitude improvement in performance for applications in water purification, chemical separations, drug delivery, and sensing. The key to Nature's protein channels are (i) selective receptor chemistry at the pore entrance, (ii) a mechanism for pumping and retaining fast transport through sub-nm scale channels and (iii) signal chemistry at the exit side of the protein to activate the channel [1–3]. The mass transport through proteins is so selective and enhanced that it seemingly defies the laws of continuum physics where, by chemisorption or kinetic scattering, the boundary condition of surface flow velocity is zero and core velocity only slowly increases by Newtonian viscous force. Nature achieves this feat by a large superstructure of folded polypeptides that through hydrogen bonding form channels of precise atomic distance as shown in Figure 3.1. In the case of aquaporin, the placement of the functional groups with either unscreened charge or hydrogen bonding precisely orientates the dipoles of molecules as small as water allowing them to flow in concerted motion without kinetic scattering. In the case of proton channels the proton shuttles down chains or hydrogen bond-ordered water with dramatic mobility. So important was the structural insight of protein channels that the 2003 Nobel Prize in

Responsive Membranes and Materials, First Edition. Edited by D. Bhattacharyya, Thomas Schäfer, S. R. Wickramasinghe and Sylvia Daunert.
© 2013 John Wiley & Sons, Ltd. Published 2013 by John Wiley & Sons, Ltd.

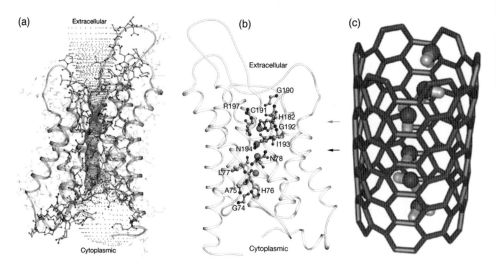

Figure 3.1 *(a) High resolution crystal structure of aquaporin 1 water channel showing supra-structure channel (blue) constriction for water purification [3]. (b) AQP1 region of smallest constriction with green spheres showing water–hydrogen bonding locations [3]. (c) Molecular dynamics simulation of hydrogen bond coupled water within the CNT core showing fast water transport equivalent to the aquaporin protein [5]. (Reprinted under the terms of the STM agreement from [3] and [5] Copyright (2001) Nature Publishing Ltd). See plate section for colour figure.*

chemistry was awarded to Peter Agre and Roderick MacKinnon. Of particular importance is the fact that this knowledge helps design new drugs to interact/bind with specific protein channels sites to modulate cellular chemistry. These structures can also inspire engineers to design a new generation of membranes for a wide variety of applications. Mimicking the complete protein superstructure is a daunting challenge and requires the use of fragile micelles to form the membrane. It would be ideal to mimic the key steps of chemical selectivity at the entrance to a rigid nm-scale pore and transmit flow down a nearly frictionless conduit to cross mechanically stable, micron-scale membranes. This "gatekeeper" would ideally be a monolayer thick and be able to actively pump chemicals through a long nearly frictionless pore.

Carbon nanotubes (CNTs) offer a unique platform to mimic proteins with a precise location for a monolayer of gatekeepers on the unsaturated carbon bonds at the cut ends of tubes and a graphitic core that has large van der Waals distances and atomic smoothness. Fluid flow through the cores of carbon nanotubes is predicted to show dramatically enhanced transport of hydrocarbon gases and water [4]. An early molecular dynamics (MD) study [5] predicted that water flow through "hydrophobic" single-walled CNTs (SWCNTs) should be not only possible but dramatically enhanced. This is because the process of water entering CNT can induce H-bond ordering in a chain of water molecules as shown in Figure 3.1c. This can make up for the energy cost of losing two of the four weak hydrogen bonds as the water molecules separate from the bulk water into the hydrophobic CNT core. In order to preserve the H-bond ordering of the water chain, it is critical to have nearly frictionless interaction with the CNT graphite sheets so as not to scatter or even rotate the quickly flowing water. The large van der Waals distance ($\sim$2 angstroms) and flat ordering of graphite sheets were predicted to allow this. Using the study's theoretical volume-rate (which is comparable to that of the protein channel aquapourin-*1*) divided by the CNT cross-sectional area, a water flow velocity of around 90 cm/s is predicted, which is 5 orders of magnitude faster than it would be for conventional materials of similar pore size (0.6 nm). High flow velocities of molecules through CNTs were also predicted due to the nearly frictionless nature of the CNT walls [6] as well as the fast diffusion rates for hydrocarbons [7, 8]. In the latter case, favourable interactions of the methane with the CNT wall predict that the molecule will "skate" down the tube wall and not scatter in random directions, as would conventionally happen in the case of Knudsen diffusion. The flow velocity of gaseous methane is predicted to be approximately 260 cm/s at 1 bar [7]. In all models, the atomically flat nature of graphite sheets inherent to the CNT microstructure makes the enhanced flow possible over long length scales. In one particularly insightful MD simulation, the effects of Lenard–Jones potential (van der Waals distance) and atomic smoothness were tested. With both reduced van der Walls distance, by using SiN bond potential profiles, and the introduction of an atomically rough step edge within CNT core, MD simulation predicts a dramatic drop in flow rates [9]. Thus CNT cores, with atomically smooth and inert surfaces, are theoretically predicted to have dramatic flow enhancements rivalling those of protein channels.

The most stringent requirement for the protein mimetic system is the ability to orientate small molecules as they pass by the "gatekeeper" for selective transport through the channel. For instance, enzymes are catalytically active due to the fact that the substrate molecule is precisely orientated as it reacts with catalytic site to break even C–C bonds at room temperature in aqueous environments. This is a feat not accomplished modern inorganic catalysis due to the trade-off between selectivity and turn over frequency. Bulky steric groups result in selectivity but block the metal site from random trajectories of the substrate molecule reducing the turnover frequency. In the case of protein channels the substrate is directed down the channel to hit the metal centre with precise orientation and kinetic energy. Mimicking this subtle natural phenomenon would be a dramatic advance for membrane reactor schemes and would be a pinnacle achievement in biomimetics. The ordering of the substrate molecule also results in dramatic flow rates due to a hopping or "fire-bucket brigade" mechanism. In proton channels hydronium ions can travel down ordered water channels with high mobility, which would have important application in hydrogen fuel cells. In the aquaporin channel there is a flip in water dipole that prevents proton transport, allowing only pure water transport. Though it is known, by the observed transport properties of proteins, that it is possible to orientate molecules in channels, MD simulations can give insight into the minimum required charge to orientate water with meaningful probability through CNT. The Fang group [10] in a controversial [11, 12] paper predicted the net pumping of water with full point charge near the entrance of the membrane and distributed charge near the middle. It appeared from this simulation, that by ordering water at the entrance (and a required second ordering point within the tube) faster bursts of coupled water transport were allowed, but this was not possible from the reverse direction, thereby giving net pumping. This simulation was made on a short time scale in order to not allow the second law of thermodynamics to be realized and there was an issue with the simulation step size to get net pumping [11]. However the work does show induced order can be transmitted down CNT cores and sets a charge length scale requirement of approximately 1.2 angstroms outside the tube to see dipole orientation effects. This short distance of charge outside the CNT wall is primarily due to the large van der Waals distance ($\sim$2 ang) within the CNT core. This predicted distance is less than even a C–F bond, hence is it is difficult to achieve the required charge on the outside of CNTs. To reach the desired distance of charge, functional chemistry will have to be placed at the entrances to CNT pores to act as ordering "gatekeepers" similar to the constructs of protein channels. Another interesting phenomena observed by MD simulations [13] is that when MeOH/H_2O mixtures flow through CNT cores the MeOH preferential segregates to the hydrophobic CNT wall. Upon exiting the tube at high flow rates, the MeOH preferentially hits the catalyst at the C–Pt bonds by necessity at the cut exit of CNTs. This in principle mimics the design of enzymes where substrate flow is directed over catalyst sites in a controlled fashion.

The experimental field of carbon nanotubes had largely focused on the superb mechanical strength and nearly ballistic electron conduction properties. However several groups [14–16] recognized early on the potential interest of fluid flow through CNTs. The Crooks group had isolated a single CNT by a microtome method and measured electro-osmotic flow rates [16], primarily using polystyrene beads blocking ionic current through the CNT via the Coulter counting method. The flow values were close to conventional expectations; however the CNT was approximately 200 nm in diameter and had an amorphous carbon coating that would interfere with the perfect slip conditions. With the large diameter, it was difficult to observe any enhanced flow rates due to the r^4 dependence on flow rate using the Haagen–Poiseuille equation of conventional Newtonian flow. Another important early study was from the Martin group where porous anodized aluminium oxide (AAO) was coated with carbon giving largely amorphous CNTs (a-CNT) membranes. Elegant elephoretic and osmotic studies [17–19] showed that the flow rates were improved but comparable to conventional expectations. However these CNTs were amorphous samples and their diameters were approximately 100 nm making it difficult to detect any flow enhancement. To observe fast fluid flow properties within CNT cores, ordered graphitic CNT membranes with small diameters (1–10 nm) were needed.

The ideal geometry of a membrane structure is to have aligned arrays of CNTs filled with a polymer barrier layer to force fluid flow through the CNT cores. Growth of arrays of the starting material has been demonstrated at the University of Kentucky [20, 21] and elsewhere [22–24]. Although the outer diameters have significant variance (30 $\pm$ 10 nm), the hollow inner core diameter is well controlled to 7 $\pm$ 1 nm. Since this is a thermal chemical vapour deposition (CVD) process using readily available xylene/ferrocene, it is an industrially scalable process with an estimated cost of \$0.60/m^2. The primary goal is to form a membrane structure taking advantage of the as-deposited alignment of multi-walled CNTs to form a well-controlled nanoporous membrane structure [25]. If the space between CNTs could be filled with a polymer barrier, then a membrane with a rigid pore structure, high porosity, and small pore size dispersion could be synthesized as shown diagrammatically in Figure 3.2.

Figure 3.3 illustrates the process to form CNT membranes. First a 50 wt.% solution of polystyrene (PS) and toluene is spun coated over the surface of the aligned CNT array. PS is known to have a high wettability with CNTs and thus can impregnate into a CNT array without disrupting the alignment of the CNT forest [26, 27]. Excess polymer on top of the structure is removed during the spin coating process due to the high viscous drag within

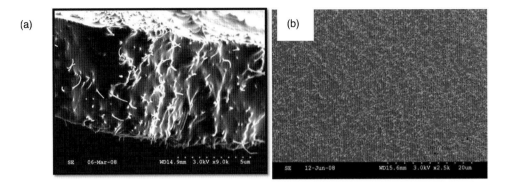

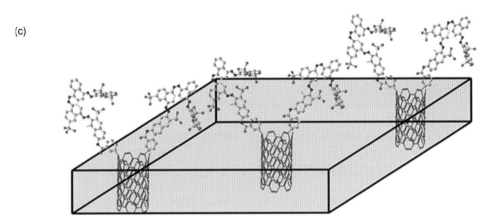

Figure 3.2 *SEM images of microtome-cut CNT membrane (a) cross-sectional view and (b) top view; (c) schematic shows the molecular structure of the anionic dye covalently functionalized on the surface of CNTs (grey: C; red: O; blue: N; yellow: S). (Reprinted with permission from [40] Copyright (2010) PNAS). See plate section for colour figure.*

the CNT array. HF acid is then used remove CNT/PS composite from the quartz substrate. To remove any thin layer of excess polymer on the top surface and open the normally closed CNT tips a H_2O plasma enhanced oxidation process [28] is used. Importantly this process removes the Fe nanocrystal catalysts from the tips of CNTs and leaves the tips of CNTs functionalized with carboxylic acid that are readily functionalized with selective receptors [29].

Recently, a simpler fabrication method was reported by Hinds' group [40] based on a microtoming method that was a modification of the early report from the Crooks group [16]. A 1–5 wt.% CNT/triton is mixed into epoxy hardener/resin by high shear velocity mixer (Thinky AR-100) and then formed into 5 mm diameter composite dies and cured. Commercially available CNT powders with inner diameters ranging from 10–1.5 nm can be easily mixed into these polymer composites. Thin 5 micron slices are made by microtome and plasma oxidation is used to remove any polymer residue and functionalize the CNT

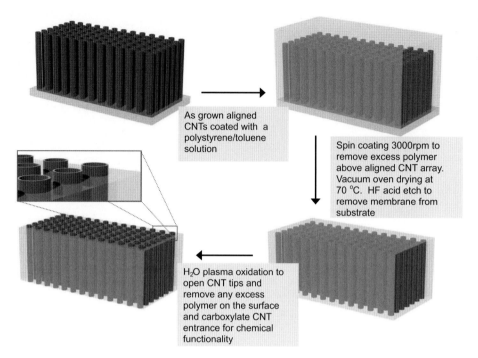

Figure 3.3   *Processing steps involved in aligned CNT membrane fabrication process for large areas [25]. Alternatively microtoming of CNT/epoxy composite can be used for laboratory scale studies. (Reprinted with permission from [40] Copyright (2010) PNAS).*

tips. Though small in area, and thus difficult for large-scale production, the technique is particularly useful for laboratory studies and can be widely adapted by the membrane research community.

3.2   Experimental Understanding of Mass Transport Through CNTs

With such dramatic theoretical predictions, it was a clear necessity to experimentally confirm the fast transport phenomena. Additionally the unique geometry of the cut CNTs allows chemistry to be placed precisely at the pore entrances, allowing for gatekeeper activity. The term "gatekeeper" is used here as a chemical layer only at the pore entrance that selectively allows chemicals to pass into and through the pores of the membrane, much how natural protein channels work. Both enhanced transport *and* gatekeeper selectivity are key elements for mimicking natural protein channels. Success with this approach would allow for fast transport and high selectivity at the performance level of natural protein channels. In particular, as a first broad understanding, it is important to examine the three critical areas of mass-transport with ionic diffusion, pressure driven gas and liquid flow.

3.2.1   Ionic Diffusion and Gatekeeper Activity

One of the key merits of CNT membranes is the ability to place chemistry at the reactive tips of cut CNTs. The plasma oxidation process to open CNTs results in carboxylate groups at the pore entrances that can be used to form covalently bonded carboimide linkages. Most any ligand with a primary amine can be covalently attached to the tips of CNTs and act as a chemically selective gatekeeper. The first systematic study [30] to prove that "gatekeeper" chemistry can be formed on CNT membranes was with a series of four chemically distinct gatekeepers: short chained alkane, long chained alkane, long polypeptide, and highly charged dye molecule. The formation of carboimide linking chemistry at the surface was later confirmed by electrochemical studies [31]. To demonstrate gatekeeper activity we looked at the ratio of small dye molecule methyl viologen^{2+} (MV) and large dye Ru(bipyr)$_3{}^{2+}$ (Ru) permeation through the membrane, where the larger dye should be more hindered. With the longest molecule (polypeptide) covalently tethered to the entrance of the 7 nm diameter pore, the ratio of small to large molecule (MV/Ru) passing through the membrane was approximately 3.6. This is significantly higher than the approximate 1.6 to ratio of bulk water diffusivity, hence showing modest gatekeeper activity. With the long and short alkanes, only modest separations were seen but the overall flux of aqueous ionic dyes decreased due to the tip becoming hydrophobic. The observed transport effects were a combination of both steric bulk and the hydrophobic/phylic nature of the CNT tip chemistry. In small pores, the hindered diffusion coefficient can be calculated by the Renkin equation based on the ratio of permeate molecule diameter to pore diameter. The observed separations were modelled with hindered diffusion at the tip gatekeeper region and with bulk diffusivity through the long length of the CNT core. This is consistent with the oxidation process that cut the CNT tips and placed functional chemistry at the pore entrances. An interesting point in this study was that the negatively charged dye molecule tethered to the CNT entrance increased the flux (five-fold) of the positively charged dyes by electrostatic attraction. The effect could be screened by spectator ions (KCl) in solutions of high ionic strength. Thus electrostatics can be a strong force in enhancing flow rates or manipulating charged gatekeepers. A more detailed study [36] showed that for a large number of ions with a variety of sizes and charges, the observed diffusion coefficients were between hindered diffusion and bulk value. However it is reasonably expected that there will not be an enhancement process for diffusion, since ions would have little interaction with hydrophobic CNT cores and interact with water in the cores.

3.2.2   Gas and Fluid Flow

Molecular dynamics simulation had predicted very rapid fluid and gas flow within CNTs due to a nearly atomically flat surface with minimal scattering or chemical attraction. The phenomenon was experimentally observed with pressure induced solvent flow [32] and gas flow [33]. In the initial report, flow through the aligned CNT membrane was measured in a syringe-pump pressure cell apparatus [32], where mass (hence volume) of the solvent passing through the membrane area is directly measured on a pan balance as a function of time. Flow data is summarized in Table 3.1. There is a remarkable 4–5 orders of magnitude increase in water flow over what would be seen in conventional nanoporous structures. For gas flow there is a 1–2 order of magnitude increase in flux over Knudsen diffusion [33] in 2 nm pores. For our larger multi-walled CNTs with 7 nm inner

Table 3.1   *Pressure driven liquid flow enhancement through MWCNT membranes as a function of solvent in order of decreasing polarity [32]. A near perfect slip boundary allows over 10 000-fold flow enhancement over atomically rough nm-scale pores. (Reprinted under the terms of the STM agreement from [32] Copyright (2005) Nature Publishing Ltd).*

Liquid	Flow velocity normalized at 1 bar (cm/s)	Calculated Newtonian flow velocity at 1 bar (cm/s)	Enhancement factor	Slip length (micron)
Water	26	5.7×10^{-4}	4.5×10^4	54
EtOH	4.5	1.4×10^{-4}	3.2×10^4	28
IPA	1.12	7.7×10^{-4}	1.5×10^4	13
Hexane	5.6	5.16×10^{-4}	1.1×10^4	9.5
Decane	0.67	1.72×10^{-4}	3.9×10^3	3.4

diameter, the gas flow enhancement is about 20. Both studies are consistent with specular reflection, where gas molecules keep their forward momentum (i.e. no back scattering). The result of such fast transport of liquids has important practical applications since less than one thousandth of the membrane area will be needed for the same amount of chemical separation from conventional membranes. The enhanced liquid flow is also one of the three critical components for mimicking protein channels.

The flow flux rate of liquids (J) through conventional porous membranes can be predicted using the well known Haagen–Poiseuille equation [34] given by:

$$J = (\varepsilon r^2 \Delta P)/(8\mu\tau L) \tag{3.1}$$

In this formula, ε is the relative porosity, r is pore radius (7 nm for our system), P is pressure applied, μ is dynamic viscosity, τ is tortuosity (1.1), and L is the length of the pore. The basic assumptions of this equation are laminar flow and "no-slip" at the boundary layer, that is the velocity of the liquid at the CNT wall is zero. This zero velocity at pore walls is the physical origin of low flow velocities in conventional membrane pores. For Newtonian liquids in pores the velocity goes from zero at the wall to a maximum "core" velocity at the pore centre based on the viscous shear from applied pressure. However in nm-scale pores this core velocity is exceedingly small. High velocities along pore walls are needed, referred to as slip conditions. Protein channels have ideal slip conditions with single file pumping of solvent or permeate. A useful convention for slip boundary is slip length that can be calculated from the equation [35]:

$$V(\lambda)/Vns = 1 + 4\lambda/r; \tag{3.2}$$

In this formula, $V(\lambda)$ is the experimentally observed flow velocity (cm/s), Vns is the "no-slip" flow velocity calculated from the Haagen–Poiseuille equation, λ is the slip length, and r is the radius of the nanotube. For water, dramatic slip lengths of 50–100 μm, compared to the CNT radius of 3.5 nm, are seen in experiments. This means that the surface velocity on the CNT wall is nearly identical to that of the pore centre, or a nearly ideal slip condition. The longest slip lengths were observed for the polar molecules (i.e. water and MeOH) that are expected to have the weakest interactions with hydrophobic CNTs. Both for the polar and non-polar liquids, the slip lengths decrease with the longer hydrocarbon chain length.

This is consistent with the concept that more interaction with the CNT wall will decrease the slip lengths and hence the flow enhancement. However, even for long chain alkanes, flow rates are still dramatically enhanced compared to conventional materials.

The broad understanding of transport through the CNT cores gives us nearly no enhancement in ionic diffusion, about a factor of 20–100 for gas flow, and a remarkable enhancement (4–5 orders of magnitude) of fluid flow. In the former case of ionic diffusion, the non-interacting CNTs offer no advantages for transport; essentially acting like a mirror for ions and molecules to bounce off. Brownian scattering of ions within the solvent inside of CNT cores would dominate and thus follow Fick's law of diffusion. Experimental measurements of ionic transport across the CNT membrane closely correlate with bulk diffusivity. However an import caveat is that selective gatekeeper chemistry at CNT tips can regulate diffusional fluxes. For gasses, scattering is only off the smooth CNT walls and the retention of forward momentum gives significant enhancement. However the rate limiting step becomes the gas molecule entering the pore entrances, which are a very low percentage of surface area. Since pore size cannot yet be controlled to sub-angstrom precision, chemical selectivity is given only by differences in gas velocity ($v \propto mass^{1/2}$). Having highly selective chemistry at pore entrances (i.e. gatekeepers) would by necessity slow the gas molecule to zero velocity and thus completely negate the fast flow phenomena of forward scattering off CNT wall. In the case of the much denser liquid flow, dramatic enhancement is observed because of the nearly frictionless interface of the liquid and the CNT. This allows for very high wall velocity ($\sim$10 cm/s at 1 atm) whereas conventional materials (atomically rough and short van der Waals distances) have zero net velocity at pore walls. However a significant intellectual puzzle emerges: How do we selectively let chemicals into CNT cores while maintaining enhanced flow rates that are based on a nearly ideal non-interaction phenomenon? Adding selective chemistry to the pore entrances or along CNT cores ruins the enhanced liquid flow effect that is based on the atomically flat non-interacting CNT wall. The flow enhancement dropped from 46 000 to 220 to less than five-fold enhancement over Haagen–Poiseuille flow with the sequential hydrophilic/bulky (direct blue-71 dye) functionalization of (i) as-made CNT, (ii) tip functionalization and (iii) core functionalization respectively [36]. This experiment on the same membrane proves that the flow phenomena are based on smooth CNT surfaces but also unfortunately shows that it is not possible to get both high selectivity and high flow rate by simple chemical functionalization at the pore entrance. What is needed is a pumping mechanism at the location of selective chemistry. The velocity or momentum developed at this region can be transmitted down the nearly frictionless CNT core as plug flow. The most promising methods to induce pumping at the CNT tip region are by molecular motion or by electric field induced electro-osmosis and electrophoresis, thereby mimicking the very essence of protein channel operation.

3.3 Electrostatic Gatekeeping and Electro-osmotic Pumping

The concept of an active "gatekeeper" that can be pulled into and out of the pore mimics the fundamental operation of many protein channels where chemical potential gradients exists. To prove the concept of an electrostatically actuated gatekeeper, a long quadra-charged dye molecule was tethered to the entrances of CNTs as shown in Figure 3.2c [37]. Importantly,

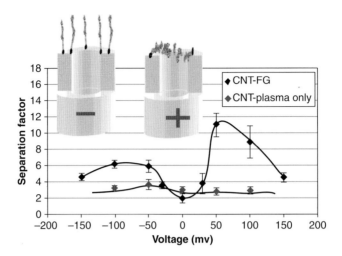

Figure 3.4 *Separation factor between small MV and large Ru(bipyr)$_3^{2+}$ cations through tip modified MWCNT membrane. On the tip of the CNT membrane is a sterically bulky anionic dye. With positive applied bias across the membrane, the anionic "gatekeeper" is drawn within the CNT core blocking the large cation, thereby demonstrating electrostatic molecular actuation. (Reprinted with permission from [37] Copyright (2007) American Chemical Society). See plate section for colour figure.*

we needed a very high density of molecules on the CNT surface to be effective gatekeepers in the pore. To achieve this, electrochemical grafting of highly reactive diazonium salt onto CNT surfaces [31,38] was performed. The diazonium salts are highly reactive and can react with the basal planes of CNT, not just the more reactive cut tips. To avoid this aggressive reaction inside the CNT, a continuous flow of inert water, at the previously mentioned remarkably high velocities, is passed through the CNT core. This prevents the diazonium reaction inside the CNT core but allows the reaction only at the CNT tips exposed to the diazonium solution. This is referred to as "flow grafting", which is a powerful tool to place chemistry only at the tips of CNTs to act as "gatekeepers". The bias dependence of transport through the membranes with anionic gatekeepers is illustrated and experimental separation factors are shown in Figure 3.4. At 0 V bias, selectivity between the two permeate molecules across the membrane is close to the ratio of bulk diffusivity ($\sim$1.6). After application of positive bias, the separation factor increases to about 10, significantly higher than what had been achieved in the earlier study [30] without bias being applied. The applied bias evidently gives enough electrostatic force to the tethered charged gatekeeper molecule in the channel to effectively block the CNT pore entrance. In the case of "flow grafted" chemistry at the top tip surface, positive bias attracts the negatively charged tethers into the pore blocking the larger permeate. With negative bias the tethers are repelled, opening the pore. However this study did not provide effective blocking for smaller molecules such as nicotine and clonidine that would be required for drug delivery applications. A more effective blocking method or active pumping mechanism is needed for drug delivery therapies, where a delivery device will require on/off ratios greater than 10.

Electro-osmotic flow (EOF) is the phenomena where a membrane pore has a covalently bound charge, usually anionic, that allows cations to flow under an electric field but rejects

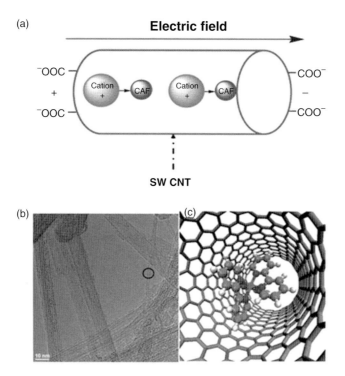

Figure 3.5 *(a) Schematic of electro-osmotic (EO) phenomena within CNT membranes. Anionic entrances to CNTs allow cations to enter CNT and be accelerated by an electric field pushing neutral molecules and solvents. Anions that would travel in the opposite direction are excluded by the anionic entrance. (b) TEM image of typical SWCNTs of about 2 nm i.d. (c) Space filling model of large Ru(bipyr)$_3{}^{2+}$ ion within CNT giving highly efficient (1:1 ratio) EO pumping due to large cation:pore diameter ratio. (Reprinted with permission from [39] Copyright (2011) Royal Society of Chemistry).*

anions travelling in the opposite direction. The moving cations induce a net flow of solvent and neutral molecules. Generally this is a very inefficient process limited by the Debye screening lengths within the first nm of the pore surface. Ideally one can have charge at CNT entrances to exclude anionic charge and allow cation to pump liquid efficiently down the fast CNT core as illustrated in Figure 3.5. This pumping can occur through regions of chemical selectivity, thereby solving the problem of how to achieve fast flux and selective chemistry in the CNT system. Electro-osmosis has been studied in the a-CNT membranes [15] showing moderately enhanced performance compared to conventional materials. However, it is expected that the fast slip boundary conditions of graphitic CNT cores can allow for more efficient EOF. EOF was recently studied [39] in CNT membranes and found to be 40–100-fold more power efficient than comparable conventional materials, as seen in Table 3.2. This power efficiency is gained by both low applied voltages and efficient pumping by ions in the small diameter CNTs. Using large cation complexes, such as Ru(bipyr)$_3{}^{2+}$ (~1.1 nm diameter), and single walled CNTs (SWCNTs), less than 1.5 nm in diameter, a 1:1 ratio of cations to neutral small molecules is achieved through EOF at

Table 3.2 *Comparison of electro-osmosis properties of MWCNT, SWCNT, a-CNT/AAO and AAO membranes. Atomically smooth CNT cores allow for highly energy efficient pumping of solvents and neutral molecules by cations under an electric field. (Reprinted with permission from [39] Copyright (2011) Royal Society of Chemistry).*

	Diameter (nm)	Veo (cm/s-V)	Power consumption (W.hr/nanomole)	Power consumption ratio	Ratio of cations to caffeine
MWCNTs	7	1.6E-01	2.5E-08	1	18
SWCNTs	1.5	1.8E-01	3.3E-08	1.3	1
AAO/a-CNT [21]	120	2.2E-03	9.9E-07	40	—
AAO membrane	20	1.1E-04	2.8E-06	112	172
AAO membrane functionalized (SO_3^-)	20	3.7E-04	6.2E-07	25	38

equal concentrations in the feed solution. This compares very well to the 200–40:1 ratio found in conventional materials. With these SWCNTs, the case of ideal electro-osmosis is nearly realized; where a single ion can pump a column of solution through the length of a CNT. The power efficiency of EOF is particularly important for biomedical devices. The EOF driven transdermal drug delivery system was recently demonstrated [40] to deliver nicotine at the fluxes used in commercially available skin patches. In particular for *in vitro* skin studies, the flux could be switched by EOF between that of high dose and low dose commercially available patches, thus allowing programmable treatment regiments after counselling. This allows a mixture of psychological counselling and chemical stimuli to optimize addiction treatment. The 40-fold improvement in EOF power efficiency of the CNT membrane system allows the device to operate for 10 days on a single standard watch battery which is critical for compact medical devices. Ongoing efforts are directed toward utilizing wireless communication to activate the patch from small interactive computer devices to program the dosing regimen with a psychology based algorithm or remote counselling.

3.3.1   Biological Gating

The primary function of protein channels is to respond to the chemical environment inside or outside of the cell and turn on or off the channel. This chemical trigger can be something as simple as pH, chemical potential, or specific ion concentration. Or the trigger can be a trace amount of a complex polypeptide biomolecule with a very specific binding site on or in the protein channel. In fact much of the small-molecule drug discovery of the pharmaceutical industry is based on drugs that interfere with cellular protein channels disrupting or modifying a complex cascade of chemical events. Making a truly protein mimetic membrane system would invariably require the ability of biochemistry to modulate pumping. With the relatively large pore diameters of 7 nm, the multi-walled CNT (MWCNT) membranes are a natural choice for examining large biological molecules as gatekeepers.

In the first report of graphitic CNT membranes [25], biotin was tethered to CNT entrances and irreversibly bound to streptavidin protein to block the pores. In a subsequent study [41] reversible desthiobiotin was covalently tethered to CNT entrances. In the presence of large streptavidin protein, the pore was blocked with a corresponding drop in flux of a dye across the membrane. With the addition of the much stronger biotin molecule to the solution, the tethered desthiobiotin is displaced from the protein by solution phase biotin, opening the membrane pores and returning to original flux values. This is an important experiment on several levels. The first is that the streptavidin protein is of a well-defined size ($\sim$4 nm) that can block only CNT pores (radius) but not large macroscopic cracks. The fact that that blockage was removed by chemistry to displace the four binding sites shows that this was not a physisorption phenomena, but due to selective surface chemistry binding. On a practical level, this system provides two sensor routes: (i) large proteins selectively blocking CNT pores or (ii) proper release chemistry opening pores. In another biological gating study, the primary hypothesis was to see if known biocatalytic (enzymatic) activity could occur at the tips of CNTs and affect the mass transport through membrane structures [42]. A peptide sequence G-R-T-G-R-R-N-S-I-NH2, specific to Protein Kinase, was covalently bound to the CNT tip. The serin was pholphated by Protein Kinase A/ATP and subsequently dephospholated by Alkaline Phosphatase. The state of the tethered peptide ligand (phosphylated or not) was detected by monoclonal anti-phosphoserine antibody binding to the tethered peptide in the phospylated state. The diffusional flux through the CNT membrane was modulated by these events showing that enzymatic catalysis (ATP cycle) could be performed at the tips of CNTs. This has important implications for drug delivery where natural biological process can open pores for drug delivery when the chemistry requires it. Sensors can also be developed by monitoring ionic diffusion to an electrode through a CNT membrane in the presence or absence of bound biochemistry at CNT tips.

3.4   CNT Membrane Applications

As with other membrane systems, numerous applications are possible ranging from water desalination, industrial chemical separation, energy separations, electrochemical conversions, pharmaceutical separations, and medical devices. In conventional membrane systems either the surface chemistry or the material processing methods, for microstructure control, are modified for the specific application. In the case of the CNT membrane, with fast transport through graphitic cores, the key lies in changing the diameter of CNT and tip chemistry for the application. Another area of modification is the composition of the matrix filler between CNTs that can render the membrane composite chemically or thermally resistant. However the uniform (or aligned) dispersion of CNTs is difficult to achieve with agglomeration and is not compatible with many common polymer systems. At present, the greatest challenge for CNT membrane applications is achieving high porosity of more than 1% where many ultrafiltration membranes are greater than 30%. To make up for the 1–3 orders of magnitude lower porosity we can take advantage of the 3–4 orders of magnitude faster flow properties within the CNT core, but the mechanism of selectivity becomes critical. Several research groups have succeeded in synthesizing graphitic CNT

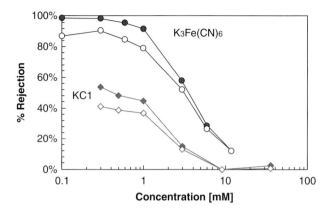

Figure 3.6   *Ionic rejection of $K_3Fe(CN)_6$ and KCl as a function of feed solution concentration through carboxyl functionalized DWCNT membranes (~1.6 nm i.d.) for desalination applications. Debye screening length at 1 mM is approximately 9 nm. (Reprinted with permission from [43] Copyright (2008) PNAS).*

membranes and demonstrating interesting flow phenomena or separation applications. A semiconductor fabrication approach to small diameter CNTs, around 1.5 nm inner diameter (i.d.) embedded in Si_3N_4 showed dramatic gas and liquid flow [33]. As seen in Figure 3.6, these membranes subsequently showed 40% ion rejection at approximately 1 mmol KCl concentrations and greater than 90% rejection for approximately 1 mmol $K_3Fe(CN)_6$. [43] Desalination is one of the most important potential applications for CNT membranes that can have an enormous impact on future water shortage needs [44, 45]. A polymer of just a few weight percent of CNTs can have the same volume rate as commercial reverse osmosis (RO) polymer membranes using $1/100^{th}$ the area giving dramatic capital cost and energy savings. However demonstration of ion exclusion at sea water concentrations (~0.6 M) remains an elusive challenge for CNT membranes. In the case of around 1.6 nm i.d. tubes with carboxylic acid functionality, larger multivalent anions ($Fe(CN)_6^{3-}$) give reasonable rejection due to the Donnan rejection mechanism of tighter rejection of multivalent species at the anionic CNT entrances [43]. However, of concern is the fact that at approximately 1 mmol ion concentration, significant screening effects are seen by a drop in rejection. At this concentration the Debye screening length is about 9.4 nm, significantly longer than CNT radius (~0.8 ± 0.2 nm), suggesting that it would be difficult to increase charge-based rejection just by further reduction in CNT diameter. A difficulty with nanofiltration [46] is the reduction of ion rejection as the flow rate increases due to mass transport increasing the screening of charge along pore walls, allowing ions to flow through the core. This kinetic screening is further exasperated in the CNT case since the charged region is only at the tips (<1 nm length), which is required for rapid flow rates, instead of the entire length of the pore as in nanofiltration membranes. MD simulations of high charge density at entrances to CNT can show significant ion rejection at moderate flows [47] and it is important to verify experimentally if the ion rejection is possible. At this time a more likely method of desalination is to use SWCNT with an inner diameter smaller than the hydrated radius of ions [48] with 8,8 CNTs being the optimal diameter predicted from MD simulation [49].

Size sieving is widely considered the mechanistic basis for RO in conventional polymer systems, but the CNTs can afford orders of magnitude improvements in flow rates that are not possible in solid state diffusion through the "free volume" of polymers. Our experience [39] is that even CNTs near 1 nm i.d. can support ionic current, so the key challenge then becomes to experimentally grow CNTs of a precise diameter near 0.8 ± 0.1 nm. Other routes to desalinization by pervaporation have also been proposed and may offer energy saving transport at temperatures below boiling point [50]. It is worth mentioning that conventional RO membrane materials have been modified with CNTs to improve mechanical strength and chloride resistance, which are meaningful parameters in large-scale application [51].

Gas separations can be an important application area of CNT membranes, since gas flow also shows enhanced transport. The Marand group reported that a simple fabrication technique commonly used to make Buckey paper [52] by filtration could be applied to CNT membranes. In this case a significant number of CNTs can be orientated down the filter and the proper amount of polymer added to form a continuous film. Gases followed the expected ($\mathrm{mass}^{-1/2}$) flow rate dependence and modest separations of CO_2 from CH_4 through the membrane were observed [53]. The Lin group has also made aligned CNT membranes on durable ceramic supports and demonstrated initial gas separations [54]. CNT membranes have also been made by taking advantage of the outer CNT walls showing oil separations [55]. The as-grown CNT arrays are relatively sparse ($\sim$5% volume) but can be compressed by capillary forces [56] resulting in dense membrane structures. These compressed membrane structures surprisingly showed dramatically fast transport [57] on the smooth outer walls of MWCNTs. By using applied bias, they were also able to move the liquid–air interface for gated ionic transport [58]. Recently a magnetic field has been able to align CNTs within Lyotropic polymers that are later cross-linked by UV radiation as a potential new CNT membrane fabrication technique [59]. CNT membranes are also ideal platforms for energy applications. With the successful development of RO membranes in desalination, the process can be reversed to generate electricity in regions where fresh water goes into the ocean. This process of pressure retarded osmosis (PRO) [60] is being vigorously pursued by the Statkraft corporation in Norway. CNTs can also act as catalyst supports [61] with high mass activity for thin uniform layers of Pt catalyst by diazonium grafting pretreatment of CNT surfaces [62] for MeOH fuel cells. The most ideal use of an expensive catalyst, from the perspective of cost efficiency, would be to have a narrow stream of MeOH flowing over an atomically thick catalyst surface at the exit or entrance of a CNT. Particularly intriguing is that MeOH in flowing aqueous solutions is seen to partition to the hydrophobic walls of CNTs [63] and to thus efficiently flow over the catalytic site at the exit of CNTs [64]. Inside 1 nm i.d. CNTs, 1 molar MeOH/H2O solutions flowing at rapid velocities of around 10 cm/s result in approximately 90% efficiency in terms of transporting MeOH to the reactive Pt catalyst site by partitioning to the sidewalls of CNTs. What is particularly promising about this geometry is that we can start to consider controlling the orientation of reactants while hitting metallic catalyst centres, which is mechanistically similar to what is done within enzyme proteins. Success in this approach would represent the pinnacle achievement of protein mimetics.

Separation applications have also been demonstrated using CNT membranes. A polymer/ MWCNT composite ultrafiltration membrane has been demonstrated by the Wu group to separate poultry products. In going from 0 to 5% MWCNT loading, significant improvement in water flux from around 200 to around 500 L/m^2-h were seen with a

94% membrane rejection rate to egg albumin [65]. Another initial biomolecule separation demonstration [66] with MWCNTs was the test case of Bovine Syrum Antibody (∼10 nm diameter) and Lysosome (∼4 nm diameter). By using electrophoretic pumping between 0 and 300 mV, the Lysosome was seen to have mobilities two orders of magnitude lower than its mobility in bulk water. However it should be noted that in conventional nanoporous material mobility is typically 5 orders of magnitude slower due to strong protein interactions with pore surfaces. The CNT interior core is relatively inert to proteins thereby allowing for chemical separation applications and conduits for intercellular injection. At higher voltages (>0.5 V) the conformation of lysosome changes allowing mobilities through CNTs to approach that of small-molecule $Ru(bipyr)_3^{2+}$. BSA is completely size-excluded to the detection limit of the system, thereby demonstrating biomolecule separations.

One of the more dramatic recent developments in CNT membranes is the report of ionic conductivity through CNT cores. Dramatic proton and K^+ mobilities through SWCNTs that are 10^8 and 10^2 faster than bulk solution respectively [67] have been recently reported. Fast proton mobilities are known to exist in protein channels and the mechanism is based on protons shuttling down ordered water chains. If verified, this discovery will have enormous application in fuel cells where selective proton conductive membranes are a critical requirement. The currently used Nafion has many technical limitations including temperature stability and humidity control. However this fast ionic mobility was indirectly measured using the analysis of signal fluctuation lifetimes through a small number of CNTs (∼10–20). Membranes, with large numbers of CNT channels, offer a method to directly measure macroscopic quantities of ions injected into the permeate solution. Table 3.3 shows observed mobilities for around 2 nm i.d. CNT membranes that are close to bulk mobility. Certainly the measurement of ion and proton conduction will be an important line of enquiry for CNT membranes and has exciting potential in fuel cell applications. Electro-osmotically induced flow will also be an important "pump" through regions of chemical selectivity for separation applications.

3.5   Conclusion and Future Prospects

CNT membranes offer an exciting opportunity to mimic natural protein channels due to (i) a mechanism of dramatically enhanced liquid flow, (ii) ability to place "gatekeeper" chemistry at the entrance to pores, and (iii) the ability to pump chemicals through regions

Table 3.3 *Electrophoretic mobility of K^+ in MWCNT membrane and Ca^{2+} in SWCNT membrane. (Reprinted with permission from [39] Copyright (2011) Royal Society of Chemistry).*

Types of membrane	Types of ions	Applied voltage (mV)	Electrophoretic mobility (m^2/V.s)	Bulk electrophoretic mobility (m^2/V.s) [38, 39]
MWCNTs	K^+	−0.3	6.4E-09	7.6E-08
SWCNTs	Ca^{2+}	−0.3	5.0E-08	6.2E-08

Note: porosities of MWCNT and SWCNT membranes are 0.015 and 0.0085%, respectively.

of chemical selectivity that would otherwise slow the flow. The structure is mechanically far more robust than lipid bilayer films and does not require expensive protein expression and separation to form the active channels. All these unique features allow for large-scale chemical separations, chemical delivery or chemical sensing based on the principles of protein channels. Initial studies show ion-rejection at low salt concentrations which is promising for water desalination or forward osmotic power generation applications. CNTs are conductive and can act as electrochemical catalyst supports. Initial biomolecule separations and oil filtrations have also been reported. Interesting ionic and proton conduction studies have been recently reported and could dramatically enhance fuel cell performance.

The transport mechanisms through CNT membranes are primarily (i) ionic diffusion near bulk expectation, (ii) gas flow enhanced 1–2 orders of magnitude primarily due to specular reflection off smooth graphite core, and (iii) liquid flow 4–5 orders of magnitude faster than conventional materials due an ideal slip–boundary interface. The transport can be modulated by "gatekeeper" chemistry at the pore entrance that can block by steric hindrance, electrostatic attraction/repulsion, or biochemical state. The density of gatekeeper chemistry can be enhanced by electrochemical grafting at just the tip region of the CNT. However, a more general intellectual puzzle emerges that the dramatic flow enhancement is negated if one places a high density of selective chemistry along the CNT walls. It is important to develop systems, such as electro-osmotic flow, that can actively pump selected chemicals through the region of chemical functionality so that they can continue down the length of CNT cores at dramatic flow rates. By designing a monolayer of active "gatekeepers" at the CNT tips that change conformation and pump desired chemicals at fast rates, we would be following the elegant examples given by natural protein channels.

Acknowledgements

The author would like to thank the students and post-doctoral fellows that contributed over the years to the research programme on CNT membranes. In particular Nitin Chopra, Mainak Majumder, Dr. Ji Wu, Dr. Xinghau Sun, Karen Gerstandt and Xin Su. Support was provided by NSF CAREER (0348544), DOE EPSCoR (DE-FG02-07ER46375), DARPA MANTRA and NIH NIDA (R01DA018822).

References

1. Hille, B. (1984) *Ionic Channels of Excitable Membranes*, Sinauer Associates Inc, Sunderland, MA.
2. Murata, K., Mitsuoka, K., Hirai, T. *et al.* (2000) Structural determinants of water permeation through aquaporin-1. *Nature*, **407**, 599–605.
3. Sui, H., Han, B.-G., Lee, J.K. *et al.* (2001) Structural basis of water-specific transport through the AQP1 water channel. *Nature*, **414**, 872–878.
4. Ohba, T., Kanoh, H. and Kaneko, K. (2005) Structures and stability of water nanoclusters in hydrophobic nanospaces. *Nano Letters*, **5**, 227–230.
5. Hummer, G., Rasaiah, J.C. and Noworyta, J.P. (2001) Water conduction through the hydrophobic channel of a carbon nanotube. *Nature*, **414**, 188–190.

6. Sokhan, V.P., Nicholson, D. and Quirke, N. (2002) Fluid flow in nanopores: Accurate boundary conditions for carbon nanotubes. *Journal of Chemical Physics*, **117**, 8531–8539.

7. Mao, Z.G. and Sinnott, S.B. (2001) Separation of organic molecular mixtures in carbon nanotubes and bundles: Molecular dynamics simulations. *Journal of Physical Chemistry B*, **105**, 6916–6924.

8. Skoulidas, A.I., Ackerman, D.M., Johnson, J.K. and Sholl, D.S. (2002) Rapid transport of gases in carbon nanotubes. *Physical Review Letters*, **89**, 185901.

9. Joseph, S. and Aluru, N.R. (2008) Why are carbon nanotubes fast transporters of water? *Nano Letters*, **8**, 452–458.

10. Gong, X.J., Li, J.Y., Lu, H.J. *et al.* (2007) A charge-driven molecular water pump. *Nature Nanotechnology*, **2**, 709–712.

11. Wong-Ekkabut, J., Miettinen, M.S., Dias, C. and Karttunen, M. (2010) Static charges cannot drive a continuous flow of water molecules through a carbon nanotube. *Nature Nanotechnology*, **5**, 555–557.

12. Hinds, B. (2007) A blueprint for a nanoscale pump. *Nature Nanotechnology*, **2**, 673–674.

13. Goldsmith, J. and Hinds, B.J. (2011) Simulation of steady state methanol flux through a model carbon nanotube catalyst support. *J. Phys. Chem. C.*, **115**(39), 19158–19164.

14. Che, G., Lakshmi, B.B., Martin, C.R. *et al.* (1998) Chemical vapor deposition based synthesis of carbon nanotubes and nanofibers using a template method. *Chemistry of Materials*, **10**, 260–267.

15. Walters, D.A., Casavant, M.J., Qin, X.C. *et al.* (2001) In-plane-aligned membranes of carbon nanotubes. *Chemical Physics Letters*, **338**, 14–20.

16. Sun, L. and Crooks, R.M. (2000) Single carbon nanotube membranes: A well-defined model for studying mass transport through nanoporous materials. *Journal of the American Chemical Society*, **122**, 12340–12345.

17. Miller, S.A., Young, V.Y. and Martin, C.R. (2001) Electroosmotic flow in template-prepared carbon nanotube membranes. *Journal of the American Chemical Society*, **123**, 12335–12342.

18. Miller, S.A. and Martin, C.R. (2002) Controlling the rate and direction of electroosmotic flow in template-prepared carbon nanotube membranes. *Journal of Electroanalytical Chemistry*, **522**, 66–69.

19. Miller, S.A. and Martin, C.R. (2004) Redox modulation of electroosmotic flow in a carbon nanotube membrane. *Journal of the American Chemical Society*, **126**, 6226–6227.

20. Andrews, R., Jacques, D., Rao, A.M. *et al.* (1999) Continuous production of aligned carbon nanotubes: a step closer to commercial realization. *Chemical Physics Letters*, **303**, 467–474.

21. Sinnott, S.B., Andrews, R., Qian, D. *et al.* (1999) Model of carbon nanotube growth through chemical vapor deposition. *Chemical Physics Letters*, **315**, 25–30.

22. Ren, Z.F., Huang, Z.P., Xu, J.W. *et al.* (1998) Synthesis of large arrays of well-aligned carbon nanotubes on glass. *Science*, **282**, 1105–1107.

23. Merkulov, V.I., Melechko, A.V., Guillorn, M.A. *et al.* (2002) Controlled alignment of carbon nanofibers in a large-scale synthesis process. *Applied Physics Letters*, **80**, 4816–4818.

24. Zhang, M., Nakayama, Y. and Pan, L.J. (2000) Synthesis of carbon tubule nanocoils in high yield using iron- coated indium tin oxide as catalyst. *Japanese Journal of Applied Physics Part 2-Letters*, **39**, L1242–L1244.

25. Hinds, B.J., Chopra, N., Rantell, T. *et al.* (2004) Aligned multiwalled carbon nanotube membranes. *Science*, **303**, 62.

26. Mitchell, C.A., Bahr, J.L., Arepalli, S. *et al.* (2002) Dispersion of functionalized carbon nanotubes in polystyrene. *Macromolecules*, **35**, 8825–8830.

27. Qian, D., Dickey, E.C., Andrews, R. and Rantell T. (2000) Load transfer and deformation mechanisms in carbon nanotube- polystyrene composites. *Applied Physics Letters*, **76**, 2868–2870.

28. Huang, S.M. and Dai, L.M. (2002) Plasma etching for purification and controlled opening of aligned carbon nanotubes. *Journal of Physical Chemistry B*, **106**, 3543–3545.

29. Wong, S.S., Woolley, A.T., Joselevich, E. *et al.* (1998) Covalently-functionalized single-walled carbon nanotube probe tips for chemical force microscopy. *Journal of the American Chemical Society*, **120**, 8557–8558.

30. Majumder, M., Chopra, N. and Hinds, B.J. (2005) Effect of tip functionalization on transport through vertically oriented carbon nanotube membranes. *Journal of the American Chemical Society*, **127**, 9062–9070.

31. Majumder, M., Keis, K., Zhan, X. *et al.* (2008) Enhanced electrostatic modulation of ionic diffusion through carbon nanotube membranes by diazonium grafting chemistry. *Journal of Membrane Science*, **316**, 89–96.

32. Majumder, M., Chopra, N., Andrews, R. and Hinds, B.J. (2005) Nanoscale hydrodynamics - Enhanced flow in carbon nanotubes. *Nature*, **438**, 44–44.

33. Holt, J.K., Park, H.G., Wang, Y.M. *et al.* (2006) Fast mass transport through sub-2-nanometer carbon nanotubes. *Science*, **312**, 1034–1037.

34. Mulder M. (1994) *Basic Principles of Membrane Technology*, Kluwer Academic Publishers.

35. Lauga, E., Brenner, M.P. and Stone, H.A. (2005) The no-slip boundary condition- A review, in *Handbook of Experimental Fluid Dynamics*, Springer.

36. Majumder, M., Chopra, N. and Hinds, B.J. (2011) Mass transport through carbon nanotube membranes in three different regimes: ionic diffusion, gas, and liquid flow. *ACS Nano*, **5**, 3867.

37. Majumder, M., Zhan, X., Andrews, R. and Hinds, B.J. (2007) Voltage gated carbon nanotube membranes. *Langmuir*, **23**, 8624–8631.

38. Bahr, J.L. and Tour, J.M. (2001) Highly functionalized carbon nanotubes using in situ generated diazonium compounds. *Chemistry of Materials*, **13**, 3823.

39. Wu, J., Gerstandt, K., Majunder, M. and Hinds, B.J. (2011) Highly efficient electro-osmotic flow through functionalized carbon nanotubes membrane, *RSC Nanoscale*, **3**, 3321

40. Wu, J., Paudel, K.S., Strasinger, C. *et al.* (2010) Programmable transdermal drug delivery of nicotine using carbon nanotube membranes. *Proceedings of the National Academy of Sciences of the United States of America*, **107**, 11698–11702.

41. Nednoor, P., Chopra, N., Gavalas, V. *et al.* (2005) Reversible biochemical switching of ionic transport through aligned carbon nanotube membranes. *Chemistry of Materials*, **17**, 3595–3599.

42. Nednoor, P., Gavalas, V.G., Chopra, N. *et al.* (2007) Carbon nanotube based biomimetic membranes: Mimicking protein channels regulated by phosphorylation. *Journal of Materials Chemistry*, **17**, 1755–1757.

43. Fornasiero, F., Park, H.G., Holt, J.K. *et al.* (2008) Ion exclusion by sub-2-nm carbon nanotube pores. *Proceedings of the National Academy of Sciences of the United States of America*, **105**, 17250–17255.

44. Shannon, M.A., Bohn, P.W., Elimelech, M. *et al.* (2008) Science and technology for water purification in the coming decades. *Nature*, **452**, 301–310.

45. Li, D. and Wang, H.T. (2010) Recent developments in reverse osmosis desalination membranes. *Journal of Materials Chemistry*, **20**, 4551–4566.

46. Hilal, N., Al-Zoubi, H., Darwish, N.A. *et al.* (2004) A comprehensive review of nanofiltration membranes: Treatment, pretreatment, modelling, and atomic force microscopy. *Desalination*, **170**, 281–308.

47. Goldsmith, J. and Martens, C.C. (2010) Molecular dynamics simulation of salt rejection in model surface-modified nanopores. *Journal of Physical Chemistry Letters*, **1**, 528–535.

48. Corry, B. (2008) Designing carbon nanotube membranes for efficient water desalination. *Journal of Physical Chemistry B*, **112**, 1427–1434.

49. Jia, Y.X., Li, H.L., Wang, M. *et al.* (2010) Carbon nanotube: Possible candidate for forward osmosis. *Separation and Purification Technology*, **75**, 55–60.

50. Lee, J. and Karnik R. (2010) Desalination of water by vapor-phase transport through hydrophobic nanopores. *Journal of Applied Physics*, 108.

51. Park, J., Choi, W., Cho, J. *et al.* (2010) Carbon nanotube-based nanocomposite desalination membranes from layer-by-layer assembly. *Desalination and Water Treatment*, **15**, 76–83.

52. Bahr, J.L., Yang, J.P., Kosynkin, D.V. *et al.* (2001) Functionalization of carbon nanotubes by electrochemical reduction of aryl diazonium salts: A bucky paper electrode. *Journal of the American Chemical Society*, **123**, 6536–6542.

53. Kim, S., Jinschek, J.R., Chen, H. *et al.* (2007) Scalable fabrication of carbon nanotube/polymer nanocomposite membranes for high flux gas transport. *Nano Letters*, **7**, 2806–2811.

54. Mi, W.L., Lin, Y.S. and Li, Y.D. (2007) Vertically aligned carbon nanotube membranes on macroporous alumina supports. *Journal of Membrane Science*, **304**, 1–7.

55. Srivastava, A., Srivastava, O.N., Talapatra, S. *et al.* (2004) Carbon nanotube filters. *Nature Materials*, **3**, 610–614.

56. Wei, B.Q., Vajtai, R., Jung, Y. *et al.* (2002) Organized assembly of carbon nanotubes - Cunning refinements help to customize the architecture of nanotube structures. *Nature*, **416**, 495–496.

57. Yu, M., Funke, H.H., Falconer, J.L. and Noble, R.D. (2009) High density, vertically-aligned carbon nanotube membranes. *Nano Letters*, **9**, 225–229.

58. Yu, M.A., Funke, H.H., Falconer, J.L. and Noble, R.D. (2010) Gated ion transport through dense carbon nanotube membranes. *Journal of the American Chemical Society*, **132**, 8285–8290.

59. Mauter, M.S., Elimelech, M. and Osuji, M.J. (2010) Nanocomposites of vertically aligned single-walled carbon nanotubes by magnetic alignment and polymerization of a lyotropic precursor. *ACS Nano*, **4**(11), 6651–6658.

60. Loeb, S. (1976) Production of energy from concentrated brines by pressure-retarded osmosis.1.1 Preliminary technical and economic correlations. *Journal of Membrane Science*, **1**, 49–63.

61. Hull, R.V., Li, L., Xing, Y.C. and Chusuei, C.C. (2006) Pt nanoparticle binding on functionalized multiwalled carbon nanotubes. *Chemistry of Materials*, **18**(7), 1780–1788.

62. Su, X., Wu, J. and Hinds, B.J. (2011) Catalytic activity of ultrathin Pt films on aligned carbon nanotube arrays. *Carbon*, **49**(4) 1145–1150.

63. Zheng, J., Lennon, E.M., Tsao, H. *et al.* (2005) *Journal of Chemical Physics*, **122**, 214702.

64. Goldsmith, J. and Hinds, B.J. (2011) Simulation of steady state methanol flux through a model carbon nanotube catalyst support. *J. Phys. Chem. C.*, **115**(39), 19158–19164.

65. Wu, H.Q., Tang, B.B. and Wu, P.Y. (2010) Novel ultrafiltration membranes prepared from a multi-walled carbon nanotubes/polymer composite. *Journal of Membrane Science*, **362**, 374–383.

66. Sun, X., Su, X., Wu, J. and Hinds, B.J. (2011) Electrophoretic transport of biomolecules through carbon nanotube membranes. *Langmuir*, **27**(6), 3150–3156.

67. Lee, C.Y., Choi, W., Han, J.H. and Strano, M.S. (2010) Coherence resonance in a single-walled carbon nanotube ion channel. *Science*, **329**, 1320–1324.

4

Synthesis Aspects in the Design of Responsive Membranes

Scott M. Husson

Department of Chemical and Biomolecular Engineering and Center for Advanced Engineering Fibers and Films, Clemson University, USA

4.1   Introduction

Responsive membranes undergo reversible changes in mass transfer and interfacial properties due to changes in their environment. Change agents include temperature, pH, solution ionic strength, light, electric and magnetic fields, and chemical cues. The need for responsive membranes in technical systems such as sensors, separation processes, and drug delivery devices is manifested by the rapid advances that have occurred in the field over the last decade. Critical to the performance of responsive membranes in these systems are the synthesis aspects used to prepare them.

This chapter provides an introduction to synthesis aspects in the design of responsive membranes. Focus is given to the use of responsive polymers as a basis for developing responsive membrane systems. After a brief introduction and background on responsive mechanisms and polymers, general methodologies are described for the design and production of responsive membranes. First, limited descriptions are presented for methods to prepare membranes by processing responsive polymers and copolymers, either alone or as additives or components of blends, as well as methods for *in situ* polymerization. Thereafter, a more in-depth coverage is given to methods for modifying membrane support materials with responsive polymers. The level of coverage given in this chapter is meant to provide a starting point for readers planning to enter the rapidly progressing and growing field of responsive membranes, as well as for experts who wish to keep up with recent developments and trends.

Responsive Membranes and Materials, First Edition. Edited by D. Bhattacharyya, Thomas Schäfer, S. R. Wickramasinghe and Sylvia Daunert.
© 2013 John Wiley & Sons, Ltd. Published 2013 by John Wiley & Sons, Ltd.

4.2   Responsive Mechanisms [1]

Stimuli-responsive membranes exploit the interplay between the pore structure and changes in the conformation/polarity/reactivity of responsive polymers or functional groups in the membrane bulk or on its surfaces. Such changes in specially tailored polymer systems have been used in many systems and devices to enable applications that demand reversibly switchable material properties. It follows that novel membranes can be designed using polymers/molecules that have been shown to undergo physicochemical changes in response to environmental cues. Responsiveness is known to occur as a two step process: (i) use of stimuli to trigger specific conformational transitions on a microscopic level and (ii) amplification of these conformational transitions into macroscopically measurable changes in membrane performance properties.

Membrane stimuli-responsive properties can be explained based on phase transition mechanisms of the membrane materials (polymers) in controlled environments. Phase transitions may be induced by solvent quality, concentration or type of ions, temperature, and other chemical or physical interactions. Polymer responsive mechanisms have been well explained in reviews by Luzinov *et al.* [2] and Minko [3]. Responsiveness generally refers to changes in polymer chain conformations. All polymers are sensitive to their immediate environments. They always respond to external stimuli to some extent by changing their conformation along the backbone, side chains, segments, or end groups. Therefore, sophisticated membrane systems with responsive properties can be designed by variation of polymer chain length, chemical composition, architecture, and topography. Most polymer responsive mechanisms are based on variations in surface energy, entropy of the polymers, and segmental interactions. Surface energy drives the surface responsive reorientation because, fundamentally, systems try to minimize the interfacial energy between the polymer surface and its immediate environment.

To understand the impact of solvent quality on responsiveness, it is instructive to examine how polymer chains behave in solution. The root-mean-square end-to-end distance of a polymer chain is normally expressed as,

$$\langle r^2 \rangle^{1/2} = \alpha \, (nC_N)^{1/2} \, l \tag{4.1}$$

where α is the chain expansion factor, which is a measure of the effect of excluded volume; n is the number of freely jointed links in a hypothetical polymer chain of equal length, l; and C_N is the characteristic ratio, which contains contributions from fixed valence angles and restricted chain rotation [4]. Another way to express the above equation is by using the unperturbed (denoted by subscript 0) root-mean-square end-to-end distance:

$$\langle r^2 \rangle^{1/2} = \alpha \langle r^2 \rangle_0^{1/2} \tag{4.2}$$

The unperturbed dimensions are those of a real polymer chain in the absence of excluded volume effects, that is, for $\alpha = 1$. In a poor solvent ($\alpha < 1$), the dimensions of the polymer chain are smaller than those in the unperturbed state ($\alpha = 1$). While in a good solvent ($\alpha > 1$), where polymer–solvent interactions are stronger than polymer–polymer or solvent–solvent interactions, the dimensions of the polymer chain are larger than those in the unperturbed state ($\alpha = 1$). So it can be said that polymers expand in good solvents and collapse in poor solvents.

4.3   Responsive Polymers

Stimuli-responsive polymers are generally used as the building blocks for designing responsive membrane systems. Responsiveness is a result of polymer transitions between two states represented by metastable energy minima for the system [5]. Energy is required for the transitions to occur; it is provided by the input stimulus. Liu and Urban [5] illustrate energy relationships for polymer transitions in solutions, surfaces and interfaces, and gels and solids. These transitions commonly involve hydrogen bonding rearrangements, protonation–deprotonation, cis–trans rearrangements, complexation–dissociation, or order–disorder transitions. Table 4.1 gives examples of polymers with stimuli-responsive transitions. This table does not consider biological systems. Wandera *et al.* [6] provide numerous examples of uses of these and many other polymers in the development of stimuli-responsive membranes.

4.3.1   Temperature-Responsive Polymers

As described earlier, changes in solvent quality lead to expansion or collapse of polymer chains relative to their unperturbed dimensions. Changes to solvent quality may be direct (e.g. addition of one or more co-solvents) or indirect (e.g. resulting from a change in system temperature). The indirect strategy is commonly used to induce stimuli-responsive transitions in polymer-solvent systems that display upper or lower critical solution temperatures. Generally, the UCST is the highest temperature on a temperature-composition diagram that a system displays with two-phase behaviour, and the LCST is the lowest such temperature;

Table 4.1 *Polymers with stimuli-responsive transitions.*

Response stimulus	Commonly used polymers
Temperature	poly(N-isopropylacrylamide) (PNIPAAm)
	poly(N,N-diethylacrylamide)
	poly(methylvinylether)
	poly(N-vinylcaprolactam)
	hydroxyethyl cellulose (HEC)
pH	poly(acrylic acid) (PAAc)
	poly(methacrylic acid)
	poly(2-(dimethylamino)ethyl methacrylate)
	poly(vinylpyridine)
	poly(L-glutamic acid)
Ionic strength	above pH-responsive polymers
	poly(styrene sulfonate)
	poly((2-(methacryloyloxy)ethyl)trimethylammonium chloride)
	poly(N,N-dimethyl-N-(2-methacryloyloxyethyl-N-(3-sulfopropyl) ammonium betaine)
Light	polymers containing photo-chromic units such as azobenzene, spiropyran, diarylethene, viologen, cinnamate
Electric field	poly(2-acrylamido-2-methyl-1-propanesulfonic acid)
Specific ions	crown ethers

although there are many systems that have UCST below LCST [7]. For systems with a UCST, an increase in temperature from T less than UCST to T greater than UCST will cause the system to merge into a single homogeneous phase, and polymer chains will swell. The opposite is true for systems that display an LCST, where an increase in temperature from T less than LCST to T more than LCST will cause the mixture to split into two liquid phases to lower its Gibbs energy. In this case, the polymer chains will collapse.

The temperatures at which transitions occur between extended (coil) and collapsed (globule) states can be controlled thermodynamically by molecular design [8]. For example, incorporation of hydrophobic comonomers increases the LCST, whereas hydrophilic comonomers have the opposite effect [9, 10]. In some cases, the LCST of temperature-responsive polymers is also affected by the average molecular weight of the polymer and its chain density, if grafted to or from a surface. Thus, synthesis offers several avenues to fine tune the temperature transition.

4.3.2   Polymers that Respond to pH, Ionic Strength, Light [1]

The polymers used to prepare responsive membranes need not be neutral. Polyelectrolytes (PELs) have ionizable groups, and their interactions are determined in part by the degree of dissociation (f) of these ionizable groups. Due to their high f, strong PELs generally are insensitive to solution pH. However, at high salt concentration when the ionic strength of the solution approaches that inside the PEL, electrostatic screening results in conformational changes. Weak PELs respond to changes in external pH and ionic strength and may undergo abrupt changes in conformation in response to these external stimuli. Weakly basic PELs expand upon a decrease in pH, while weakly acidic PELs expand upon an increase in pH. At high ionic strength, weak PELs tend to collapse due to effective screening of like charges along the PEL.

Photo-chromic units (azobenzene, spiropyran, diarylethene, viologen) undergo reversible photo-isomerization reactions on absorption of light. Reversible photo-isomerism leads to switching between two states of the photo-chromic moieties, hence leading to molecular changes in group polarity, charge, colour, and size. These molecular changes can be amplified into measurable macroscopic property changes. For example, membranes containing viologen groups have permeabilities that can be regulated reversibly by redox reactions. The viologen moieties have two different redox states [11]. On treatment with a reducing agent such as sodium hydrosulfite solution, viologens undergo reversible reduction from the dicationic state to the radical cationic state. Normally viologens in the dicationic state are highly soluble in water, but their solubility decreases in the reduced radical cation state. Therefore, in viologen grafted membranes, when the grafted viologen is in its dicationic state, the polymer chain may be expelled by the charges on the side chains and extend more in the pores leading to low permeabilities. Whereas, when the grafted viologen polymer is changed to its cationic state, the hydrophobic radical chains may be in a more entangled or collapsed state leading to higher permeabilities.

Finally, while many works cited employ one responsive mechanism, the literature contains examples of membranes modified by mixed polymers or block copolymers, where each polymer responds to a different stimulus. Mixed polymer brushes and block copolymers may impart adaptive/switching properties due to reversible microphase segregation between the different functionalities in different environmental conditions. For example, the individual polymers may change their surface energetic states upon exposure to different solvents. By imposing combinations of two or more independent stimuli, such membranes exhibit more sophisticated permeability responses than membranes modified by a single polymer type.

4.4 Preparation of Responsive Membranes

Production practices for responsive membranes naturally fall into three categories, with some overlap: (i) pre-synthesis of stimuli-responsive polymers or copolymers and processing into membranes; (ii) *in situ* polymerization to form a membrane film or interpenetrating polymer network; and (iii) surface modification of membranes by chemical and physical processes to incorporate stimuli-responsive polymers or copolymers. The remainder of this chapter describes approaches within each category. However, since the focus of this chapter is synthesis aspects in the design of responsive membranes, only very brief descriptions are given on general strategies to prepare membranes from pre-formed stimuli-responsive polymers and *in situ* polymerization methods, allowing more detailed discussion of synthetic approaches to surface modify pre-existing membrane supports with functional, responsive polymers that enhance their performance.

There is another reason that focus is given to surface modification. Polymer networks must have desirable spatial and energetic properties to enable a collective, and perhaps orchestrated, responsiveness amongst individual chains in response to stimuli [5]. Figure 4.1

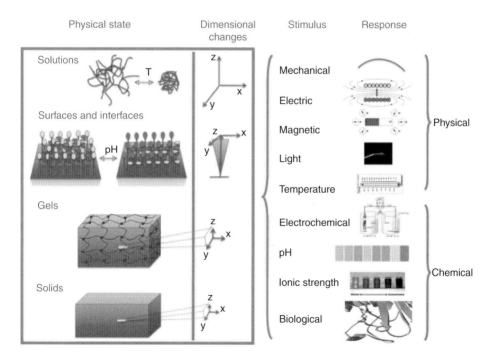

Figure 4.1 *Schematic representation of dimensional changes in polymeric solutions, at surfaces and interfaces, in polymeric gels, and polymer solids resulting from physical or chemical stimuli. (Reprinted with permission from [5] Copyright (2010) Elsevier Ltd). See plate section for colour figure.*

shows the relative dimensional changes for responsive polymers in different physical states: solutions, surfaces and interfaces, gels, and solids. Segment and chain mobilities decrease moving from solutions to surfaces and interfaces to gel and solid states, resulting in smaller displacement vectors and different energetic requirements for inducing a response [5]. Liu and Urban [5] illustrate that responsiveness is more easily attainable for polymer systems that have high solvent content and minimal energy inputs. Thus, building responsive polymer surface layers on membrane (pore) surfaces is an attractive avenue for preparing responsive membranes. Synthesis methods that further enable a high degree of control over the molecular architecture of the grafted polymer are still more attractive, since polymer structure is important for tuning responsiveness. Two such methods, surface-initiated atom transfer radical polymerization and reversible addition-fragmentation chain transfer polymerization will be covered in great detail. Both have been used to modify membranes with responsive polymers [12–20], and it is likely that both will play more important future roles in modification of porous membranes, where pore confinement demands precise control of the physical polymer layer properties.

4.5   Polymer Processing into Membranes [1]

Stimuli-responsive membranes can be prepared by mature techniques such as solvent casting and phase inversion using pre-synthesized, stimuli-responsive polymers and copolymers. These materials can be used alone or as additives in polymer blends.

4.5.1   Solvent Casting

Casting solutions of mixtures containing stimuli-responsive polymers or copolymers onto flat surfaces has been used to develop composite membranes that respond to different stimuli. Membrane preparation involves dissolving stimuli-responsive polymers or copolymers in an appropriate solvent, casting the solutions obtained on flat glass plates or laboratory dishes, and allowing the solvent to evaporate. The free-standing membranes that are formed are dried and cross-linked by annealing them [21–26].

4.5.2   Phase Inversion

Traditional membrane preparation techniques such as the wet phase-inversion process have been utilized to fabricate stimuli-responsive membranes, again by using stimuli-responsive polymers in the membrane formulation. Solutions containing stimuli-responsive polymers or copolymers are cast on flat surfaces and then immersed in an appropriate solvent such as water to enable membrane formation [27–29].

4.6   *In Situ* Polymerization [1]

4.6.1   Radiation-Based Methods

Radiation curing has been applied to develop stimuli-responsive membranes. In this method, a mixture of stimuli-sensitive and cross-linking monomers (and/or prepolymers) is coated on the pore surfaces of a porous film and the coated layer is cured with UV irradiation.

The coating formulation may also include chemical additives for controlled release applications. A variety of monomers may be used to prepare composite membranes with permeation/release profiles that respond to changes in pH, temperature, ionic strength, and so on.

4.6.2   Interpenetrating Polymer Networks (IPNs)

Interpenetrating polymer networks (IPNs) have been used as responsive membranes. The high level of cross-linking in these membranes normally leads to responsive systems with good mechanical strength. A stimuli-responsive monomer is polymerized within a physically entangled copolymer in the presence of an initiator and a cross-linker to form the stimuli-responsive IPN membrane.

4.7   Surface Modification Using Stimuli-Responsive Polymers

Membrane surface modification with stimuli-responsive polymers provides an additional degree of freedom in the synthesis of responsive membranes. When modification is done correctly, the useful properties of the base membrane are maintained, and responsive properties are introduced to the membrane surface where they interface directly with the external environment. When modification is made using controlled, surface-initiated polymerization strategies, polymer molecular architecture can be controlled precisely, which is beneficial since surface architecture plays a critical role in membrane responsiveness and performance. For example, Husson and colleagues [6] discuss the effect of chain density on the conformational responsiveness of grafted polymer chains. Figure 4.2 illustrates that, at low chain density, in the absence of strong interactions between the grafted polymer and the support surface, polymer chains respond in ways similar to free chains in solution; whereas, at high grafting density, the response is weak due to initial spatial limitations amongst grafted

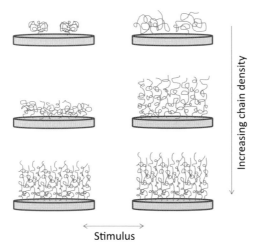

Figure 4.2   *Effect of polymer chain grafting density on stimuli responsiveness.*

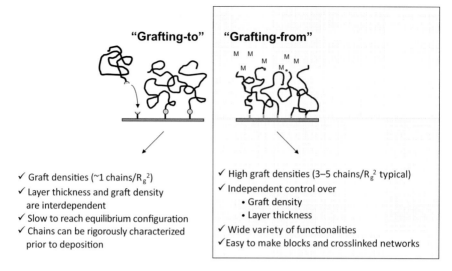

Figure 4.3 *Two common approaches to membrane surface modification with macromolecules. Shown are strategies that lead to covalently bound polymer modifiers.*

chains (i.e. chains are stretched significantly prior to application of the stimulus). Responsive surfaces require synergistic chain reorganizations that are achieved most effectively at intermediate grafting densities (near the defined onset of the polymer brush regime where the interchain distance equals twice the radius of gyration of the polymer in free solution). In the illustration, chains at these intermediate grafting densities normally stretch to the surface to avoid overlap with neighbouring chains.

Membrane modification must preserve the structural properties of the base membranes while introducing functional (responsive) moieties to the membranes. When modification is done with polymers, two distinct approaches are most common: "grafting to" and "grafting from" the membrane surface. Figure 4.3 illustrates these two approaches. The "grafting to" approach is based on the reaction of end-functionalized macromolecules with functional groups on the membrane surface. Physical adsorption to the surface is also possible but such coatings are prone to leaching (desorption of chains). One method to overcome this leaching of adsorbed macromolecules is to entrap the chains by swelling the membrane support material during the adsorption process, and then deswelling by removing the solvent [30, 31]. The "grafting from" approach uses a surface-initiated polymerization process to graft polymer chains from initiator sites on the membrane surface by monomer addition. Advantages and disadvantages of "grafting to" and "grafting from" approaches have been summarized in detail elsewhere [32]. Briefly, the polymer structures from "grafting to" techniques are generally better controlled and characterized, as the polymer can be isolated, purified, and studied prior to grafting. However, this process leads to a lower grafting density of the polymer chains, as it is limited by kinetic and thermodynamic factors. Favourable kinetic and thermodynamic factors allow the "grafting from" method to achieve higher grafting densities. In principle, "grafting from" provides a means to deposit substantially thicker polymer layers, since monomer can be added over time. Moreover, it reduces preparative steps for surface modification; no isolation and purification of the grafting material are necessary. However, characterization of the polymer is more difficult.

As indicated, one important distinction between these two approaches is that the polymer chain density and layer thickness achievable by the "grafting to" method depend on the chain molecular weight; whereas these design parameters are independent in "grafting from" strategies. Grafting density plays a major role in membrane performance. High grafting density is advantageous in applications such as membrane adsorbers and those that require the polymer coating to shield the underlying membrane support. However, once the density of chains increases beyond the onset of the polymer brush regime, increasing density further begins to limit the response to external stimuli since the chains adopt an extended configuration (Figure 4.2). For example, Biesalski and Rühe [33] showed that the degree of swelling decreases significantly with increasing graft density. The choice of modification approach will be determined by considering the ease of processing ("grafting to" advantage) and the flexibility in membrane design ("grafting from" advantage).

4.8   "Grafting to" Methods

Responsive polymers can be "grafted to" membrane surfaces by physical adsorption or chemical grafting of pre-formed polymer chains. Chemical grafting involves the reaction of functional groups on the membrane material with a reactive end group or side groups along the backbone of a functional polymer modifier. The functional groups may be inherent to the membrane material or generated (photo)chemically. The result of chemical grafting is permanent immobilization of responsive macromolecules onto the membrane surface.

4.8.1   Physical Adsorption – Non-covalent

Coating by physical adsorption is done by immersing a membrane in a solution containing the stimuli-responsive polymer. After some time, the membrane is removed from the solution and dried. Annealing the membrane at an elevated temperature may be done to improve the permanence of the coating.

4.8.2   Chemical Grafting – Covalent

This involves grafting pre-formed polymer chains or hydrogels onto a membrane surface. In one strategy, modification is achieved using photoreaction between the membrane surface material and a polymer modifier, where either the membrane or modifier can be activated photochemically. Figure 4.4 depicts the methods for photo-functionalization of non-porous and porous polymeric membranes with small molecules and macromolecules. Immobilizing stimuli-responsive polymers, hydrogels or cross-linked networks onto membrane surfaces makes them responsive to specific external stimuli. An example is the early work of Park *et al.* [34], who grafted azido-phenyl derivatized poly(NIPAAm)s in track-etched polycarbonate membranes to create composite membranes with temperature-responsive barrier properties. He *et al.* [35] provide a number of other examples based on photo-functionalization.

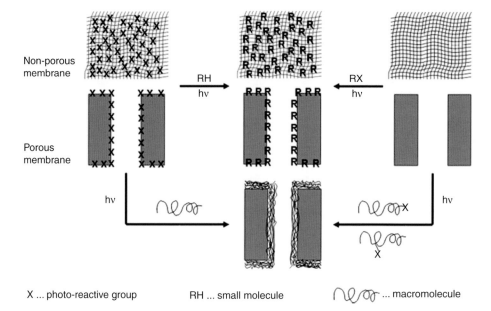

Figure 4.4 *Methods for photo-functionalization of non-porous and porous polymeric membranes with small molecules and macromolecules: via photo-reactive membrane polymer (from left) and via photo-reactive functionalization agents (from right). (Reprinted with permission from [35] Copyright (2009) Elsevier Ltd).*

4.8.3 Surface Entrapment – Non-covalent, Physically Entangled

A more robust, non-covalent "grafting to" method involves surface entrapment of adsorbed macromolecules. Figure 4.5 illustrates this surface modification technique, which involves the deposition of an amphiphilic block copolymer onto a polymeric membrane surface from a solvent that swells the membrane surface material. In the typical case where the membrane surface is hydrophobic, the hydrophobic block of the copolymer penetrates into the membrane surface layer while the hydrophobic block remains exposed at the surface [30]. After exposing the membrane to a non-solvent for the membrane polymer, the

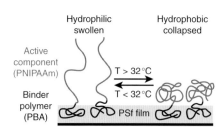

Figure 4.5 *Schematic concept of surface functionalization by entrapment of a block copolymer composed of a temperature-switchable functional block and an anchor block. (Reprinted with permission from [31] Copyright (2009) Elsevier Ltd).*

surface layer collapses, resulting in the "surface entrapment" of the block copolymer and, thus, enhanced stability over purely adsorptive surface functionalizations [31].

4.9 "Grafting from" – a.k.a. Surface-Initiated Polymerization

Membrane modification by surface-initiated polymerization is normally achieved in two steps: the first activates the membrane surface and the second initiates polymer growth by monomer addition to the activated sites. Methods for grafting from modification include radiation-initiated grafting (UV and non-UV), redox-initiated grafting, plasma-initiated grafting, thermal grafting, and controlled radical grafting methods such as atom transfer radical polymerization (ATRP) and reversible addition-fragmentation chain transfer (RAFT) polymerization.

4.9.1 Photo-Initiated Polymerization

Heterogeneous photo-initiated, radical graft-polymerization can be used to modify membranes with stimuli-responsive polymers. Ulbricht and workers [35] provide an excellent starting point for those interested in learning the fundamentals of photo-irradiation for preparing and modifying polymeric membranes. The focus here will be on methods for photo-"grafting from" as a modification strategy.

Two distinct approaches exist for photo-grafting [35]: direct generation of radicals on the membrane polymer surface by UV irradiation without added photo-initiator, and indirect generation of radicals on the membrane surface by reaction of added photo-initiator with the membrane polymer under UV irradiation. The direct method requires the use of membranes prepared from photo-reactive polymers, such as poly(arylsulfone)s, that generate radicals on exposure to UV radiation. Functionalization depends on the UV wavelength and intensity, and irradiation energy must be controlled to prevent undesirable side reactions such as homopolymerization in solution, and polymer chain scission without grafting [35]. A now well-known example of direct photo-graft modification started with the work of Crivello, Belfort, and workers [36], who used UV irradiation to modify poly(arylsulfone) membranes. Kilduff *et al.* [37] illustrate the proposed mechanism for the photo-chemical grafting of vinyl monomers from poly(arylsulfone)s, which begins with absorption of light energy by the phenoxyphenyl sulfone groups in the polymer, resulting in cleavage of the sulfone C−S bond to form radicals that may initiate polymerization in the presence of a monomer in solution. Membranes may be immersed in a monomer solution during irradiation, or, in a more preferred strategy, dipped in a monomer solution and then irradiated in an inert atmosphere [35].

There are a number of approaches for the indirect generation of radicals on membrane surfaces by reaction of added photo-initiator with the membrane polymer under UV irradiation. Type II photo-initiators such as benzophenone and its derivatives are commonly used. This initiator type abstracts a hydrogen atom from the membrane polymer, which generates a radical that can initiate graft polymerization. Figure 4.6 depicts methods for the immobilization of a "type II" photo-initiator for photo-graft modification of membrane surfaces. In the simplest case, called the simultaneous method, the initiator is dissolved in the monomer solution and the membrane is immersed in this solution and irradiated

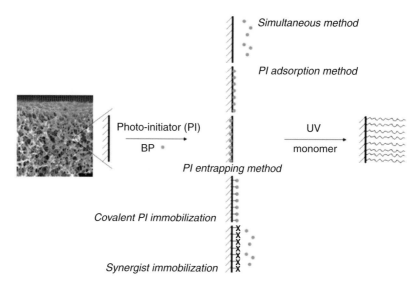

Figure 4.6   *Methods for the immobilization of a "type II" photo-initiator (e.g. benzophenone, BP) for photo-graft modification of membrane surfaces. (Reprinted with permission from [35] Copyright (2009) Elsevier Ltd).*

with UV light. This method has been used to modify membranes with temperature- and pH-responsive membranes [38–40]. While straightforward, this approach suffers from a number of limitations, including a low concentration of initiator on the membrane surface [35]. To increase the surface concentration of initiator and thereby improve photo-grafting efficiency, the photo-initiator can be adsorbed onto the membrane in a pre-coating step (Figure 4.6) [41]. By using non-solvents for the initiator during polymerization, its surface concentration is maintained at high values.

Ulbricht and coworkers [42, 43] further improved the photo-initiator adsorption method by introducing charged groups on the membrane surface and using counter-charged ben-zophenone derivatives. Incorporating ionic interactions between functional groups on the membrane surface and the photo-initiator improved grafting efficiency and control in the design of temperature- and pH-responsive poly(ethylene terephthalate) (PET) membranes.

Akin to the method of polymer surface entrapment described under "grafting to" methods, photo-initiator entrapment can also be used to improve initiator immobilization on the membrane surface [44]. He and Ulbricht cover other variants for indirect radical generation on membrane surfaces.

Controlled radical grafting methods like ATRP and RAFT have gained significant at-tention for membrane modification, as they allow precise control over chain architecture. Along these lines, the use of photo-iniferters (initiator chain-transfer termination agents) offers the possibility of controlled membrane modification by UV irradiation. Otsu [45] describes the concept of iniferters, which are initiators that induce radical polymerization that proceeds via initiation, propagation, primary radical termination, and transfer to ini-tiator. Like RAFT (see below), polymerization occurs through insertion of monomer units into the C−S iniferter bond. Well-designed iniferters yield polymers with end groups that serve as new, polymeric iniferters, and polymerization proceeds in a controlled fashion. While, surprisingly, no examples were found for the use of the photo-iniferter mediated graft polymerization to synthesize responsive membranes, Hattori *et al.* [46] immobilized benzyl N,N-diethyldithiocarbamate groups on a cellulose membrane and used these iniferter groups to carry out controlled graft polymerization from the membrane surface. He and Ulbricht [47] adopted this method for the surface-selective graft functionalization of hydrophilized polypropylene microfiltration membranes and showed that the modification was uniform and allowed good control over grafted layer thickness. As for other controlled radical polymerizations, the use of photo-iniferter mediated grafting could be used to synthesize block copolymers on membrane surfaces, since chain re-initiation efficiency can be high. This area seems ripe for exploration as it pertains to synthesis of responsive membranes.

As for all membrane modification strategies, limitations exist for the use of photo-graft polymerization. Photo-degradation must be considered to develop synthesis strategies that limit unwanted polymer degradation and loss of desirable properties of the base membrane. Another consideration for applications that require modification throughout the membrane thickness is that light intensity decays exponentially with increasing depth due to absorption or scattering by the membrane material [35]. Since the degree of modification depends on light intensity, uniform modification throughout the membrane may be hampered by light attenuation. When these limitations can be managed, photo-graft modification of membranes is a versatile technology.

4.9.2 Atom Transfer Radical Polymerization

"Grafting from" using surface-initiated atom transfer radical polymerization (ATRP) provides controlled growth of polymer chains from the membrane surface. Modification is done in two steps: immobilization of an ATRP initiator onto the membrane surface and catalyst-activated polymerization from the immobilized initiator sites. Matyjaszewski and Xia [48] provide an excellent starting point for those interested in the fundamentals of ATRP, which is perhaps the most widely studied controlled radical polymerization technique. Because ATRP is a controlled radical technique, it is extremely well suited to modification of pore surfaces in porous membranes, which requires precise control of the physical nanolayer properties. It produces homogeneous, highly uniform polymer films [49]. In the absence of chain transfer, growth occurs only from initiator sites on the surface; the absence of solution-phase polymerization avoids pore plugging. Nanolayer thickness is varied easily by adjusting polymerization time [14, 50–52]. In turn, pore dimensions can be adjusted easily and uniformly [52]. Finally, modification can be done within membrane modules since it is activated by a solution-phase catalyst and does not require the membrane housing to be optically transparent.

Figure 4.7 depicts the general mechanism of ATRP in the context of growing polymer chains from initiation sites on a membrane surface. The initiation sites often are alkyl halide (R-X) groups that have been reacted onto the surface in a first step. After initiation, propagation occurs at the polymer chain end. On the left-hand-side of the reaction equation is a dormant polymer chain that is capped by the transferable halogen atom (X). The chain is activated when a transition metal catalyst (commonly Cu^IY; $Y = Cl, Br$) abstracts the halogen atom from the dormant chain to generate a radical. The halogen atom is

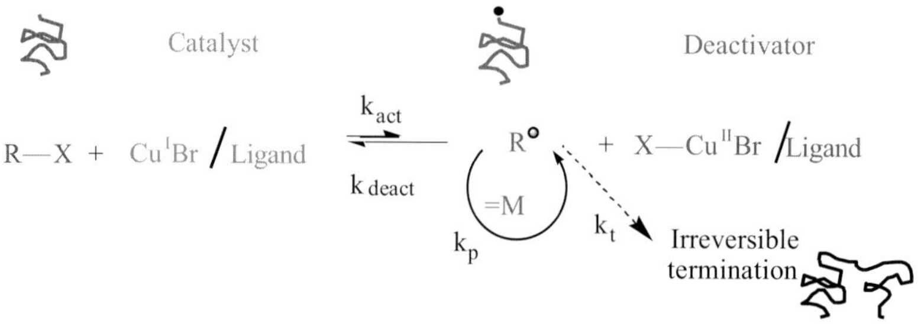

Figure 4.7   *Transition metal-catalysed atom transfer radical polymerization.*

transferred reversibly to the metal centre, which undergoes a one-electron oxidation reaction to become X-Cu^{II}Y. The polymer chain grows by addition of monomer units from the bulk. The reversible equilibrium between the Cu^I and Cu^{II} complexes ensures that after short sequences of repeat unit additions to the growing polymer chain, the chain end is capped with the halogen atom. By preserving a high percentage of polymer chains that can be reactivated, ATRP enables facile synthesis of block copolymers [15, 31, 48, 53–57].

In a well-controlled ATRP system, chain–chain termination is suppressed to a low level, and the polymer chains grow uniformly. Nevertheless, an important factor in controlled ATRP is the accumulation of sufficient Cu^{II} that occurs by termination events early in the process. This "trapped" Cu^{II} acts as a deactivator species to maintain the proper reaction equilibrium needed for controlled growth throughout the remainder of the process. Thus, one characteristic of bulk or solution-phase ATRP is that polydispersity index values decrease over time. For surface-initiated ATRP from low surface area substrates, such as polymeric membranes, insufficient accumulation of Cu^{II} occurs. To have a well-controlled ATRP system in such cases, deactivator Cu^{II} can be added to the starting polymerization formulation. The exact concentration ratio of Cu^I/Cu^{II} needed for control must be determined experimentally for each new study system.

Husson and colleagues have used ATRP to modify NF, UF, MF membranes with responsive nanolayers of poly(2-vinylpyridine) [12]; polyacrylic acid [13]; poly(dimethylaminoethyl methacrylate) [14, 16]; poly(NIPAAm) [15]; and poly(NIPAAm-b-PEG methacrylate) [15, 58]. Membrane materials in these studies have included poly(vinylidene fluoride) (PVDF), regenerated cellulose, and polyamide thin-film composites. ATRP reaction conditions are versatile, making it compatible with most polymeric membranes. One requirement is that the membrane polymer must have functional groups available for initiator attachment. These may be inherent to the membrane material or generated (photo)chemically. Cellulose membranes provide many reaction sites for surface modification without pre-treatment. Membranes made of polymers such as polycarbonate, PET, polysulfone, polyamides, and so on, have functional end groups on the polymer chains that provide reaction sites for surface modification with initiator groups or for reaction site amplification. One example of reaction site amplification is the work of Singh *et al.* [12], who coated plasma-modified PVDP membranes with poly(glycidyl methacrylate) and used the epoxy groups of the coating to incorporate an ATRP initiator. Site amplification may also be accomplished by hydrolysis of bonds to yield reactive groups. Friebe and Ulbricht have used this approach to modify PET membranes with poly(NIPAAm) [59]

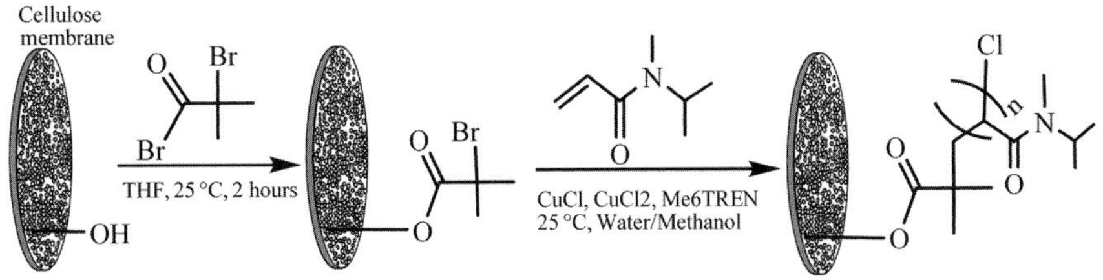

Figure 4.8 *Surface-initiated ATRP of poly(N-isopropylacrylamide) from a regenerated cellulose membrane.*

and poly(NIPAAm)-b-poly(acrylic acid) [57] using ATRP. Tomer *et al.* [15] used this approach to modify polyamide membranes with poly(NIPAAm) and poly(NIPAAm-b-PEG methacrylate).

Figure 4.8 gives an example surface-initiated ATRP protocol to modify regenerated cellulose membranes with poly(NIPAAm). Such membranes are under study for use as self-cleaning prefilters in the treatment of oily (produced) waters [15, 58]. In the first step, the membrane is activated by reaction with an ATRP initiator precursor (in this example, 2-bromoisobutyryl bromide). The density of polymer chains grafted from the membrane surface can be controlled during this step by varying the concentration ratio of ATRP initiator precursor and a non-ATRP-active analogue of this molecule, which serves as a site blocker [60, 61]. The activated membrane is modified further by surface-initiated ATRP. The reaction formulation for this step includes monomer NIPAAm; a catalyst system comprising activator CuCl, deactivator $CuCl_2$, ligand Me_6TREN, and solvent water/methanol. The ligand coordinates the metal centre, making it soluble, and adjusts its oxidation/reduction potential. The metal catalyst is susceptible to oxidation, so polymerization is often done in an inert atmosphere. As illustrated in Figure 4.8, polymerization is commonly done at ambient or near-ambient temperature, making ATRP compatible with most polymeric membrane supports.

4.9.3 Reversible Addition-Fragmentation Chain Transfer Polymerization

Surface modification using reversible addition-fragmentation chain transfer (RAFT)-mediated graft polymerization provides an alternate method for the controlled growth of polymer chains from a membrane surface. Lowe and McCormick [62] provide an excellent starting point for those interested in the fundamentals of RAFT. A notable feature of RAFT is its excellent versatility with respect to monomer choice. Nonionic, cationic, anionic, and zwitterionic monomers that yield responsive polymers can be polymerized in a controlled way using RAFT [62]. Of special note, RAFT enables the controlled polymerization of acrylamido-based monomers that historically have been difficult to polymerize in a controlled fashion using other controlled radical polymerization techniques such as ATRP. An important example is N-isopropylacrylamide, which is the most widely studied temperature-responsive acrylamido monomer. Limitations of RAFT centre on the susceptibility of the thiocarbonylthio chain transfer molecules to primary and secondary amines [63] and hydrolysis in aqueous media, particularly at high pH and temperature [64].

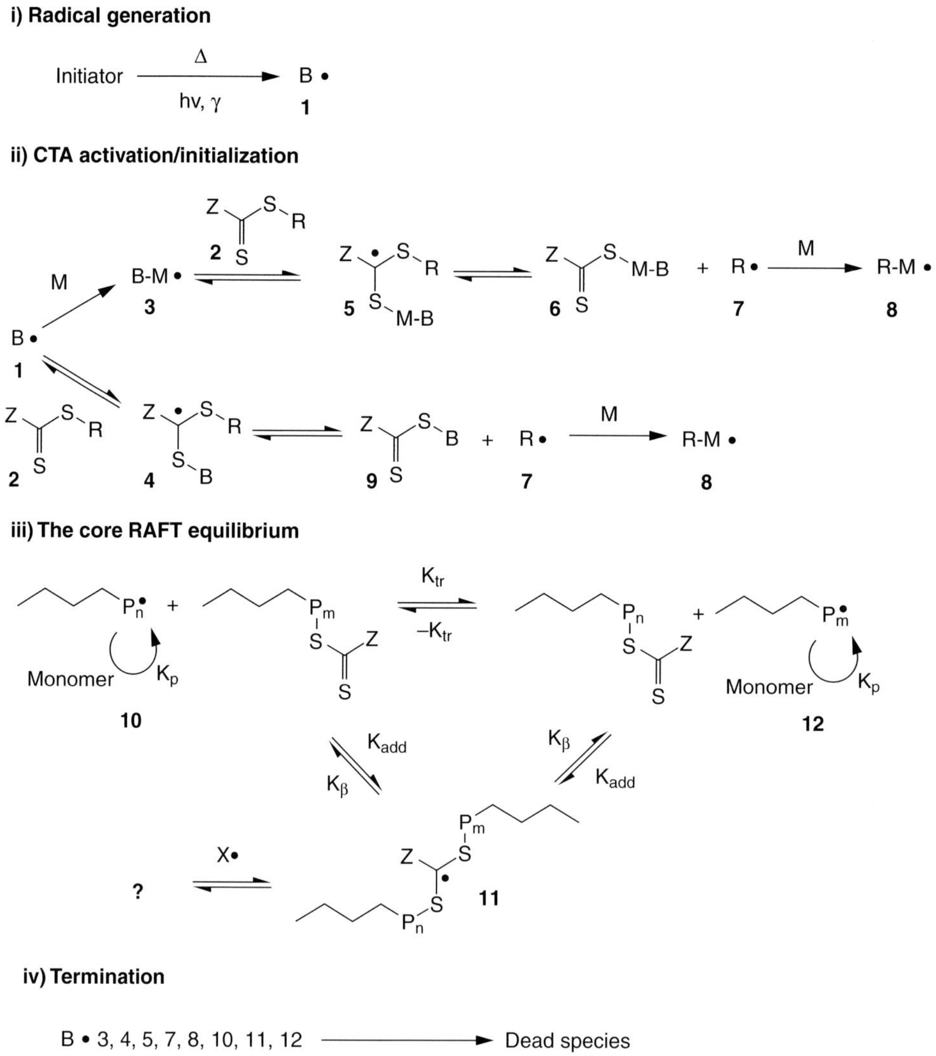

i) Radical generation

ii) CTA activation/initialization

iii) The core RAFT equilibrium

iv) Termination

B • 3, 4, 5, 7, 8, 10, 11, 12 ⟶ Dead species

Figure 4.9 *The RAFT mechanism. (Reprinted with permission from [62] Copyright (2007) Elsevier Ltd).*

Figure 4.9 depicts the RAFT mechanism. Put simply, RAFT is a conventional free radical polymerization carried out in the presence of a suitable thiocarbonylthio chain transfer agent. Figure 4.10 shows general structures of common thiocarbonylthio RAFT agents. A key to successful polymerization is the choice of a suitable RAFT agent. Guidelines for the choice of a RAFT agent can be found in reviews by Moad *et al.* [65] and Favier and Charreyre [66]. Moad *et al.* [65] also discuss approaches for RAFT agent synthesis, as their commercial availability is somewhat limited. Important structural features of RAFT agents are the Z and R groups. The structure of the Z group defines the "activity" of the agent by

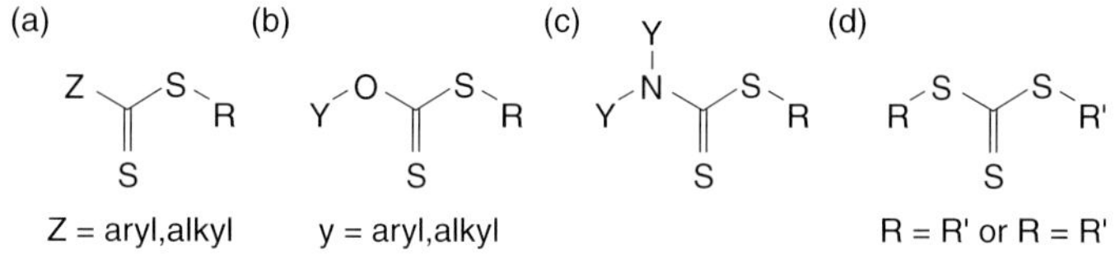

Figure 4.10 *General structures of common thiocarbonylthio RAFT agents. (Reprinted with permission from [62] Copyright (2007) Elsevier Ltd).*

adjusting the reactivity of the C=S bond toward radical addition; whereas, the R group fine tunes the reactivity to enable controlled polymerization [62].

The first step (i) for RAFT polymerization (Figure 4.9) is radical **1** generation, using standard radical initiator compounds like 2,2′-azobis(2-methylpropionitrile) (AIBN), photo-initiators, gamma radiation, thermal initiation, and so on. Next, this radical initiates step (ii), the so-called RAFT pre-equilibrium, where chain transfer agents are activated to oligomeric-type RAFT agents [62]. In this step, radical **1** may propagate by addition of monomer **M** (top sequence in Figure 4.9, step (ii)) or add to the RAFT agent **2** (bottom sequence) in a reversible reaction to form an intermediate radical **4**. Continuing to follow the bottom sequence of reactions in step (ii), reversible fragmentation of **4** to yield a new RAFT agent **9** and radical **7** is favoured with proper choice of RAFT agent. This radical **7** may add a monomer to generate an oligomeric radical **8**, which may add back to the new RAFT agent **9**. Following this pre-equilibrium stage is the main RAFT equilibrium stage, shown as step (iii) in Figure 4.9, which involves degenerative chain transfer of the thiocarbonylthio species between polymer chains **10** and **12** via radical intermediate **11** [62]. RAFT achieves controlled polymer growth (i.e. limits the extent of step (iv) in Figure 4.9) by maintaining a low number of active radicals, relative to the number of polymer chains.

Figure 4.11 gives an example of photo-induced RAFT-mediated protocol to modify polypropylene membranes with poly(acrylic acid)-*block*-poly(NIPAAm). The permeability of aqueous solutions through the PPMM-g-PAAc-b-PNIPAAm membranes is both pH- and temperature-sensitive. In the first step, benzophenone (BP) was immobilized on the membrane surface by UV irradiation. Next, acrylic acid (AAc) was grafted from the BP functionalized membrane surface by surface-initiated RAFT under UV irradiation using dibenzyltrithiocarbonate (DBTTC) as a chain transfer agent. In the final step, the AAc-modified membrane surface containing dormant end groups (i.e. the membrane is a macro-chain transfer agent), was further functionalized by surface-initiated block copolymerization with NIPAAm in the presence of free radical initiator, AIBN.

In a similar example illustrated in Figure 4.12, Neoh and coworkers [19] used RAFT-mediated copolymerization of AAc with ozone-pretreated PVDF and cast the copolymers into pH-responsive microfiltration membranes using phase inversion in aqueous solution. Because the pore surfaces were enriched in PAAc, and since some fraction of the PAAc chains remained end-capped with thiocarbonylthio groups, they could be re-activated for surface-initiated, RAFT-mediated graft polymerization of PNIPAAm. The final PVDF-g-PAAc-b-PNIPAAm membranes exhibited pH- and temperature-dependent permeability to aqueous media.

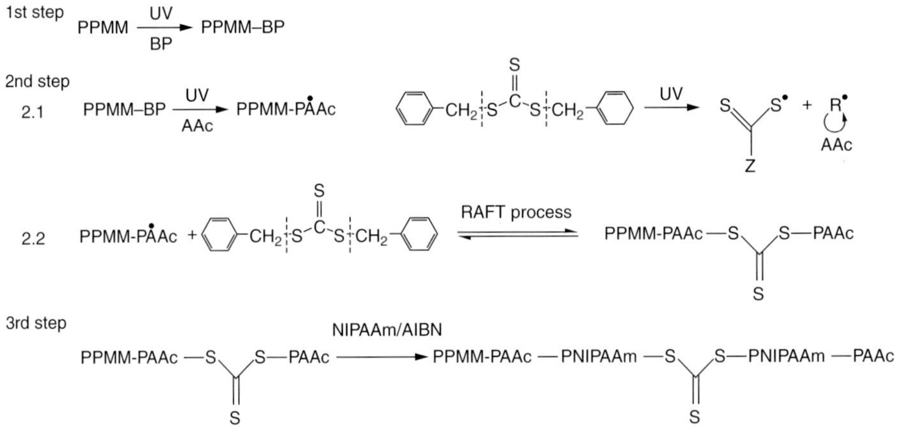

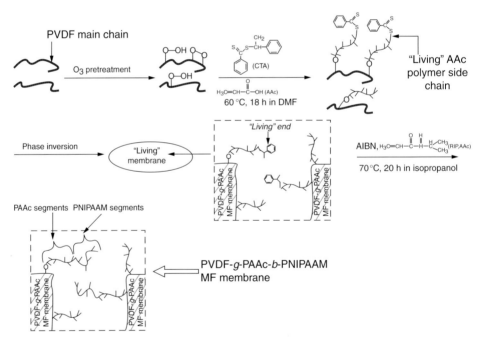

1st step   PPMM $\xrightarrow[\text{BP}]{\text{UV}}$ PPMM–BP

2nd step

2.1   PPMM–BP $\xrightarrow[\text{AAc}]{\text{UV}}$ PPMM-PȦAc

2.2   PPMM-PȦAc + $\xrightarrow{\text{RAFT process}}$ PPMM-PAAc—S S—PAAc

3rd step

PPMM-PAAc —S S—PAAc $\xrightarrow{\text{NIPAAm/AIBN}}$ PPMM-PAAc — PNIPAAm —S S—PNIPAAm —PAAc

Figure 4.11   *Schematic representation of the photoinduced RAFT-mediated graft copolymerization of acrylic acid (AAc) and N-isopropylacrylamide (NIPAAm) from a polypropylene microfiltration membrane (PPMM). (Reprinted with permission from [20] Copyright (2009) Elsevier Ltd).*

Figure 4.12   *Schematic illustration of the synthesis of PVDF-g-PAAc copolymer by RAFT-mediated graft polymerization; the preparation of the PVDF-g-PAAc membrane with dormant, thiocarbonylthio end-capped PAAc chains; and the preparation of the pH- and temperature-sensitive PVDF-g-PAAc-b-PNIPAAM microfiltration membrane via the surface-initiated, RAFT-mediated block copolymerization. (Reprinted with permission from [19] Copyright (2004) American Chemical Society).*

4.9.4   Other Grafting Methods [1]

Redox-initiated polymerization: Chemicals such as Fenton's reagent (Fe^{2+}-H_2O_2), persulfate salts, and cerium ammonium nitrate in nitric acid are used to produce free radicals from which graft polymerization is carried out in a nitrogen atmosphere. This method can be used to graft different stimuli-responsive polymers onto membranes leading to membranes that can respond to pH, temperature, oxidoreduction, ionic strength, light, and so on.

Radiation-induced polymerization: Radiation-induced graft polymerization can be used to modify polymeric membranes with stimuli-responsive polymers to develop novel responsive membranes. In one method, membranes are immersed in aqueous solutions of the monomer with various concentrations of hydrated copper(II) sulfate, bubbled with pure nitrogen, and irradiated with ^{60}Co γ-ray radiation.

Plasma-graft-filling polymerization: Plasma-graft-filling polymerization has been used to modify membranes and introduce stimuli-responsiveness to them. Membranes are irradiated with argon plasma to form initiator radicals, and then polymerization is carried out from these initiator sites.

4.9.5   Summary of "Grafting from" Methods

In closing this section, it is worth reiterating that controlled radical grafting methods such as ATRP, RAFT, and photo-iniferter methods yield polymer chains with low polydispersity. The result is that modification layers can be highly uniform. Furthermore, these synthetic methods allow facile control over molecular structure and, as stated earlier, will play important future roles in the design of porous, responsive membranes, where pore confinement demands precise control of the physical polymer layer properties.

4.10   Future Directions [1]

Responsive membrane systems with changing barrier properties and highly adaptive surfaces have been created using different approaches in recent decades, but most of the work has focused on how the different responsive interactions within the membranes can be tuned and monitored in controlled environments. The next generation of responsive membranes will move towards advanced functions, that is beyond barrier functions, in less well-defined environments. Innovations will lead to the design and synthesis of more complex membrane systems capable of mimicking functions of living systems.

Stimulation of responsive membranes will shift from non-specific triggers such as temperature and pH to specific, affinity type triggers. Molecular imprinting is one platform that may see growth for developing membranes that respond to such chemical cues. Imprinted polymers made to mimic natural receptors may find applications in modelling cellular transmembrane transport.

A less well developed research area that is ripe for exploration is the fabrication of membranes that respond to biochemical stimuli, such as enzyme substrates and antibodies/antigens. Membrane synthesis might involve immobilization of a biomolecule (enzyme/antibody/antigen) onto the membrane surface. Interaction of the immobilized agent with a substrate or biochemical in solution would elicit a conformation change directly or indirectly by generating a reaction product that elicits the change. Such membranes could be used to regulate the transport and release of chemicals, for example, drugs, by receiving and responding to biochemical cues.

Stimuli-responsive membranes have strong potential for future applications in tissue engineering, bioseparations, antifouling surfaces, and drug delivery amongst others. Reversible changes to surface composition, surface energy, adhesion, and wettability of stimuli-responsive membranes will provide ways of fabricating membranes with new functions, such as self-cleaning and self-refreshing abilities. Switchable membrane surface properties will improve the efficiency of many technological processes.

References

1. Excerpt reprinted from *J. Membr. Sci*, **357**, Wandera, D., Wickramasinghe, S.R. and Husson, S.M. Stimuli-responsive membranes, 6–35, Copyright (2010), with permission from Elsevier.
2. Luzinov, I., Minko, S. and Tsukruk, V.V. (2004) Adaptive and responsive surfaces through controlled reorganization of interfacial polymer layers. *Prog. Polym. Sci.*, **29**, 635–698.
3. Minko, S. (2006) Responsive polymer brushes. *J. Macromol. Sci., Part C: Polymer Reviews*, **46**, 397–420.
4. Fried, J.R. (2003) *Polymer Science and Technology*, 2nd edn, Prentice Hall, New Jersey.
5. Liu, F. and Urban, M.W. (2010) Recent advances and challenges in designing stimuli-responsive polymers. *Prog. Polym. Sci.*, **35**, 3–23.
6. Wandera, D., Wickramasinghe, S.R. and Husson, S.M. (2010) Stimuli-responsive membranes. *J. Membr. Sci.*, **357**, 6–35.
7. Siow, K.S., Delmas, G. and Patterson, D. (1972) Cloud-point curves in polymer solutions with adjacent upper and lower critical solution temperatures. *Macromolecules*, **5**, 29–34.
8. Kujawa, P. and Winnik, F.M. (2001) Volumetric studies of aqueous polymer solutions using pressure perturbation calorimetry: a new look at the temperature-induced phase transition of poly(N-isopropylacrylamide) in water and D_2O. *Macromolecules*, **34**, 4130–4135.
9. Kuckling, D., Adler, H.J.P., Arndt, K.F. *et al.* (2000) Temperature and pH dependent solubility of novel poly(N-isopropylacrylamide) copolymers. *Macromol. Chem. Phys.*, **201**, 273–280.
10. Principi, T., Goh, C.C.E., Liu, R.C.W. and Winnik, F.M. (2000) Solution properties of hydrophobically modified copolymers of N-isopropylacrylamide and N-glycine acrylamide: A study by microcalorimetry and fluorescence spectroscopy. *Macromolecules*, **33**, 2958–2966.
11. Liu, X., Neoh, K.G. and Kang, E.T. (2003) Redox-sensitive microporous membranes prepared from poly(vinylidene fluoride) grafted with viologen-containing polymer side chains. *Macromolecules*, **36**, 8361–8367.
12. Singh, N., Husson, S.M., Zdyrko, B. and Luzinov, I. (2005) Surface modification of microporous PVDF membranes by ATRP. *J. Membr. Sci.*, **262**, 81–90.

13. Singh, N., Wang, J., Ulbricht, M. *et al.* (2008) Surface-initiated atom transfer radical polymerization: A new method for preparation of polymeric membrane adsorbers. *J. Membr. Sci.*, **309**, 64–72.

14. Bhut, B.V., Wickramasinghe, S.R. and Husson, S.M. (2008) Preparation of high-capacity, weak anion-exchange membranes for protein separations using surface-initiated atom transfer radical polymerization. *J. Membr. Sci.*, **325**, 176–183.

15. Tomer, N., Mondal, S., Wandera, D. *et al.* (2009) Modification of nanofiltration membranes by surface-initiated atom transfer radical polymerization for produced water filtration. *Sep. Sci. Technol.*, **44**, 3346–3368.

16. Bhut, B.V. and Husson, S.M. (2009) Dramatic performance improvement of weak anion-exchange membranes for chromatographic bioseparations. *J. Membr. Sci.*, **337**, 215–223.

17. Wan, L.S., Yang, Y.F., Tian, J. *et al.* (2009) Construction of comb-like poly(N-isopropylacrylamide) layers on microporous polypropylene membrane by surface-initiated atom transfer radical polymerization. *J. Membr. Sci.*, **327**, 174–181.

18. Li, P.F., Xie, R., Jiang, J.C. *et al.* (2009) Thermo-responsive gating membranes with controllable length and density of poly(N-isopropylacrylamide) chains grafted by ATRP method. *J. Membr. Sci.*, **337**, 310–317.

19. Ying, L., Yu, W.H., Kang, E.T. and Neoh, K.G. (2004) Functional and surface-active membranes from poly(vinylidene fluoride)-graft-poly(acrylic acid) prepared via RAFT-mediated graft copolymerization. *Langmuir*, **20**, 6032–6040.

20. Yu, H.-Y., Li, W., Zhou, J. *et al.* (2009) Thermo- and pH-responsive polypropylene microporous membrane prepared by the photoinduced RAFT-mediated graft copolymerization. *J. Membr. Sci.*, **343**, 82–89.

21. Nonaka, T., Ogata, T. and Kurihara, S. (1994) Preparation of poly(vinyl alcohol)-*graft*-N-isopropylacrylamide copolymer membranes and permeation of solutes through the membranes. *J. Appl. Polym. Sci.*, **52**, 951–957.

22. Ogata, T., Nonaka, T. and Kurihara, S. (1995) Permeation of solutes with different molecular size and hydrophobicity through the poly(vinyl alcohol)-*graft*-N-isopropylacrylamide copolymer membrane. *J. Membr. Sci.*, **103**, 159–165.

23. Tokarev, I., Orlov, M. and Minko, S. (2006) Responsive polyelectrolyte gel membranes. *Adv. Mater.*, **18**, 2458–2460.

24. Orlov, M., Tokarev, I., Doran, A. and Minko, S. (2007) pH-responsive thin film membranes from poly(2-vinylpyridine): Water vapor-induced formation of a microporous structure. *Macromolecules*, **40**, 2086–2091.

25. Tokarev, I., Orlov, M., Katz, E. and Minko, S. (2007) An electrochemical gate based on a stimuli-responsive membrane associated with an electrode surface. *J. Phys. Chem. B*, **111**, 12141–12145.

26. Darkow, R., Yoshikawa, M., Kitao, T. *et al.* (1994) Photomodification of a poly(acrylonitrile-*co*-butadiene-*co*-styrene) containing diaryltetrazolyl groups. *J. Polym. Sci. A: Polym. Chem.*, **32**, 1657–1664.

27. Ying, L., Kang, E.T., Neoh, K.G. *et al.* (2004) Drug permeation through temperature-sensitive membranes prepared from poly(vinylidene fluoride) with grafted poly(N-isopropylacrylamide) chains. *J. Membr. Sci.*, **243**, 253–262.

28. Oak, M.S., Kobayashi, T., Wang, H.Y. *et al.* (1997) pH effect on molecular size exclusion of polyacrylonitrile ultrafiltration membranes having carboxylic acid groups. *J. Membr. Sci.*, **123**, 185–195.

29. Xue, J., Chen, L., Wang, H.L. *et al.* (2008) Stimuli-responsive multifunctional membranes of controllable morphology from poly(vinylidene fluoride)-*graft*-poly[2-(N,N-dimethylamino)ethyl methacrylate] prepared by atom transfer radical polymerization. *Langmuir*, **24**, 14151–14158.

30. Ruckenstein, E. and Chung, D.B. (1988) Surface modification by a two-liquid process deposition of A-B block copolymers. *J. Colloid. Interface Sci.*, **123**, 170–185.

31. Berndt, E. and Ulbricht, M. (2009) Synthesis of block copolymers for surface functionalization with stimuli-responsive macromolecules. *Polymer*, **50**, 5181–5191.

32. Zhao, B. and Brittain, W.J. (2000) Polymer brushes: Surface immobilized macromolecules. *Prog. Polym. Sci.*, **25**, 677–710.

33. Biesalski, M. and Rühe, J. (2002) Scaling laws for the swelling of neutral and charged polymer brushes in good solvents. *Macromolecules*, **35**, 499–507.

34. Park, Y.S., Ito, Y. and Imanishi, Y. (1998) Permeation control through porous membranes immobilized with thermosensitive polymer. *Langmuir*, **14**, 910–914.

35. He, D., Susanto, H. and Ulbricht, M. (2009) Photo-irradiation for preparation, modification and stimulation of polymeric membranes. *Prog. Polym. Sci.*, **34**, 62–98.

36. Yamagishi, H., Crivello, J.V. and Belfort, G. (1995) Development of a novel photochemical technique for modifying poly(arylsulfone) ultrafiltration membranes. *J. Membr. Sci.*, **105**, 237–247.

37. Kilduff, J.E., Mattaraj, S., Zhou, M. and Belfort, G. (2005) Kinetics of membrane flux decline: the role of natural colloids and mitigation via membrane surface modification. *J. Nanopart. Res.*, **7**, 525–544.

38. Yang, B. and Yang, W. (2003) Thermo-sensitive switching membranes regulated by pore-covering polymer brushes. *J. Membr. Sci.*, **218**, 247–255.

39. Yang, B. and Yang, W. (2005) Novel pore-covering membrane as a full open/close valve. *J. Membr. Sci.*, **258**, 133–139.

40. Wu, G., Li, Y., Han, M. and Liu, X. (2006) Novel thermo-sensitive membranes prepared by rapid bulk photo-grafting polymerization of N,N-diethylacrylamide onto the microfiltration membranes Nylon. *J. Membr. Sci.*, **283**, 13–20.

41. Ulbricht, M. (1996) Photograft-polymer-modified microporous membranes with environment-sensitive permeabilities. *React. Funct. Polym.*, **31**, 165–177.

42. Geismann, C. and Ulbricht, M. (2005) Photoreactive functionalization of poly(ethylene terephthalate) track-etched pore surfaces with "smart" polymer systems. *Macromol. Chem. Phys.*, **206**, 268–281.

43. Geismann, C., Yaroshchuk, A. and Ulbricht, M. (2007) Permeability and electrokinetic characterization of poly(ethylene terephthalate) capillary pore membranes with grafted temperature-responsive polymers. *Langmuir*, **23**, 76–83.

44. Ulbricht, M. and Yang, H. (2005) Porous polypropylene membranes with different carboxyl polymer brush layers for reversible protein binding via surface-initiated graft copolymerization. *Chem. Mater.*, **17**, 2622–2631.

45. Otsu, T. (2000) Iniferter concept and living radical polymerization. *J. Polym. Sci. A: Polym. Chem.*, **38**, 2121–2136.

46. Hattori, K., Hiwatari, M., Iiyama, C. *et al.* (2004) Gate effect of theophylline-imprinted polymers grafted to the cellulose by living radical polymerization. *J. Membr. Sci.*, **233**, 169–173.

47. He, D. and Ulbricht, M. (2009) Tailored "grafting-from" functionalization of microfiltration membrane surface photo-initiated by immobilized iniferter. *Macromol. Chem. Phys.*, **210**, 1149–1158.

48. Matyjaszewski, K. and Xia, J. (2001) Atom transfer radical polymerization. *Chem. Rev.*, **101**, 2921–2990.

49. Li, X., Wei, X. and Husson, S.M. (2004) Thermodynamic studies on the adsorption of fibronectin adhesion-promoting peptide on nanothin films of poly(2-vinylpyridine) by SPR. *Biomacromolecules*, **5**, 869–876.

50. Singh, N., Cui, X., Boland, T. and Husson, S.M. (2007) The role of independently variable grafting densities and layer thicknesses of polymer nanolayers on peptide adsorption and cell adhesion. *Biomaterials*, **28**, 763–771.

51. Sankhe, A.Y., Husson, S.M. and Kilbey, S.M., II. (2007) Direct polymerization of surface-tethered electrolytes in aqueous solution via surface-confined atom transfer radical polymerization. *J. Polym. Sci A: Polym. Chem.*, **45**, 566–575.

52. Singh, N., Chen, Z., Tomer, N. *et al.* (2008) Modification of regenerated cellulose ultrafiltration membranes by surface-initiated atom transfer radical polymerization. *J. Membr. Sci.*, **311**, 225–234.

53. Matyjaszewski, K., Miller, P.J., Shukla, N. *et al.* (1999) Polymers at interfaces: Using atom transfer radical polymerization in the controlled growth of homopolymers and block copolymers from silicon surfaces in the absence of untethered sacrificial initiator. *Macromolecules*, **32**, 8716–8724.

54. Davis, K.A. and Matyjaszewski, K. (2001) ABC triblock copolymers prepared using atom transfer radical polymerization techniques. *Macromolecules*, **34**, 2101–2107.

55. Eastwood, E.A. and Dadmun, M.D. (2001) A method to synthesize multiblock copolymers of methyl methacrylate and styrene regardless of monomer sequence. *Macromolecules*, **34**, 740–747.

56. Kim, J., Huang, W., Bruening, M.L. and Baker, G.L. (2002) Synthesis of triblock copolymer brushes by surface-initiated atom transfer radical polymerization. *Macromolecules*, **35**, 5410–5416.

57. Friebe, A. and Ulbricht, M. (2009) Cylindrical pores responding to two different stimuli via surface-initiated atom transfer radical polymerization for synthesis of grafted diblock copolymers. *Macromolecules*, **42**, 1838–1848.

58. Wandera, D., Wickramasinghe, S.R. and Husson, S.M. (2011) Modification and characterization of ultrafiltration membranes for treatment of produced water. *J. Membr. Sci.*, **373**, 178–188.

59. Friebe, A. and Ulbricht, M. (2007) Controlled pore functionalization of poly(ethylene terephthalate) track-etched membranes via surface-initiated atom transfer radical polymerization. *Langmuir*, **23**, 10316–10322.

60. Bhut, B.V., Weaver, J., Carter, A.R. *et al.* (2011) The role of polymer nanolayer architecture on the separation performance of anion-exchange membrane adsorbers. Part I: Protein separations. *Biotechnol. Bioeng.*, **108**, 2645–2653.

61. Bhut, B.V., Weaver, J., Carter, A.R. *et al.* (2011), The role of polymer nanolayer architecture on the separation performance of anion-exchange membrane adsorbers. Part II: DNA and virus separations. *Biotechnol. Bioeng.*, **108**, 2654–2660.

62. Lowe, A.B. and McCormick, C.L. (2007) Reversible addition–fragmentation chain transfer (RAFT) radical polymerization and the synthesis of water-soluble (co)polymers under homogeneous conditions in organic and aqueous media. *Prog. Polym. Sci.*, **32**, 283–351.

63. Delêtre, M. and Levesque, G. (1990) Kinetics and mechanism of polythioamidation in solution. 1. Reaction of mono- and bis(dithioester)s with excess amine. *Macromolecules*, **23**, 4733–4741.

64. Thomas, D.B., Convertine, A.J., Hester, R.D. *et al.* (2004) Hydrolytic susceptibility of dithioester chain transfer agents and implications in aqueous RAFT polymerizations. *Macromolecules*, **37**, 1735–1741.

65. Moad, G., Rizzardo, E. and Thang, S.H. (2005) Living radical polymerization by the RAFT process. *Aust. J. Chem.*, **58**, 379–410.

66. Favier, A. and Charreyre, M.-T. (2006) Experimental requirements for an efficient control of free-radical polymerizations via the reversible addition-fragmentation chain transfer (RAFT) process. *Macromol. Rapid Commun.*, **27**, 653–692.

5

Tunable Separations, Reactions, and Nanoparticle Synthesis in Functionalized Membranes

Scott R. Lewis, Vasile Smuleac, Li Xiao and D. Bhattacharyya
Department of Chemical and Materials Engineering, University of Kentucky, USA

5.1 Introduction

Membrane-based separations and reactions have wide applications ranging from clean water production to selective separations, chemical synthesis, and biotechnology. The development of nanocomposite membranes with tunable and responsive properties provides opportunities in flux modulations, and separation/reaction selectivity control for various applications. Depending on the physico-chemical properties of the functional groups, these membranes could be used in applications ranging from metal (or oxyanions) separation to protein purification, and enzymatic catalysis to water detoxification by iron nanoparticles. The functionalization of microfiltration (MF) membrane pores with macromolecules (such as, polypeptides or polyelectrolytes) provides nano-scale interactions in a confined domain.

If the selected macromolecule is a polypeptide that undergoes conformational changes (for example, poly-glutamic or poly-aspartic acid), then in addition to creating a highly charged field in the membrane pores, tunable nanofiltration type separations or high capacity metal capture can be conducted with these macroporous membranes at low pressures. Various pore functionalization approaches show that by linking macromolecules, biomolecules, or by synthesizing nanoparticles in membrane pores one can go far beyond conventional ion exchange, reverse osmosis, or microfiltration applications. This chapter deals with functionalized, responsive membranes with applications ranging from metal capture to nanoparticles synthesis and layer-by-layer assembly in membrane pores for enzyme catalysis.

Responsive Membranes and Materials, First Edition. Edited by D. Bhattacharyya, Thomas Schäfer, S. R. Wickramasinghe and Sylvia Daunert.
© 2013 John Wiley & Sons, Ltd. Published 2013 by John Wiley & Sons, Ltd.

5.2   Membrane Functionalization

Membrane functionalization provides significant advances in various non-desalination applications. There are a variety of reasons for functionalizing the membrane surface: flux enhancement, reduction of fouling and concentration polarization (hydrophilization), or attachment of ligands which can offer separations based on specific interactions with the target molecules (affinity membranes). In addition, depending on the attached ligand, the functionalized membranes can be used not only for separations, but also for manufacturing of advanced materials, biocatalysis, or biosensing. The most common ligands used for membrane functionalization [1, 2] are:

- hydrophilic polymers to enhance water transport;
- amino/polyamino acids;
- metal affinity ligands;
- ion exchange ligands;
- antigen and antibodies;
- enzymes for catalysis and biosensing.

The most common methods of functionalization (Figure 5.1) of a base membrane are chemical modification, plasma treatment, graft polymerization, pore filling, and layer by layer (LbL) assembly.

5.2.1   Chemical Modification

A key parameter for chemical modification of any base membrane is the availability of active groups (such as -COOH, -SH, -OH, epoxy, amine etc.). In the absence of any reactive groups, the functionalization involves two steps: membrane activation and ligand coupling to the activated membrane. There are a many chemistries which can be implemented for chemical activation, and the most common are summarized in Table 5.1, and briefly described below.

5.2.1.1   Periodate Oxidation

Sodium periodate reacts with vicinal hydroxyl groups on agarose, cellulose, or polyvinyl alcohol, yielding aldehyde groups. These can be easily reacted with amine-terminated ligands, producing Schiff bases. Subsequent reduction with sodium borohydride or

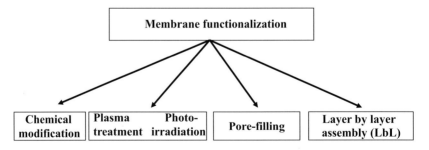

Figure 5.1   *Various approaches for membrane functionalization.*

Table 5.1 *The most common agents used for membrane activation.*

Activation reagent	Functional groups on substrates	Reactive groups
Glutaric aldehyde	$-NH_2$, $-OH$, $-SH$	$-CHO$
1,4-butanediol diglycidyl ether	$-NH_2$, $-OH$	epoxy
Epichlorohydrin	$-OH$	epoxy
Tri -chloro, -ethoxy, -methoxy silane	$-NH_2$, $-OH$, $-COOH$, $-SH$, epoxy, etc.	$-NH_2$, $-OH$, $-COOH$, $-SH$, epoxy, etc.
Sulfonyl chloride	$-NH_2$, $-SH$	-SOCl
Diglycolic anhydride	$-NH_2$, $-OH$	$-COOH$
Vinylsulfone	$-OH$, $-SH$	(structure: sulfone with vinyl group)
2,4-Disulfonic acid chloride	$-NH_2$, $-OH$	$-SO_2Cl$
2,2,2-Trifluoroethansulfonyl chloride	$-NH_2$, $-OH$, $-SH$	$-CF_3$
2,2,2-Trifluoroethansulfonyl chloride	$-NH_2$, $-OH$, $-SH$	$-CF_3$
P,p-Difluoro-m,m-nitrophenyl sulfone	$-NH_2$, $-OH$	(structure: difluoro-nitrophenyl sulfone)

cyanoborohydride produces a secondary amine; reaction with dihydrazides produces hydrazones. The major advantage is that this method uses non-toxic reagents. However the oxidation has to be carefully controlled, as the base membrane can undergo serious modifications due to the disruption of C–C bonds, which can lead to major damage of the matrix [3, 4].

5.2.1.2 Epoxy Activation

Epoxy rings react with such groups like -OH or $-NH_2$; the most commonly described in literature are 1,4-butanediol diglycidyl ether [5, 6] or epichlorohydrin [7]. Ring opening occurs under both acidic and basic conditions, and due to its simplicity it plays an increasing role in the membrane activation. The major advantage is that once a membrane is epoxy activated, it can be used to attach a variety of ligands containing amino, thiol, hydroxyl, or carboxyl groups.

5.2.1.3 Carbodiimide Coupling

Carbodiimide condensation is one of the most common methods of formation of peptide bonds. The reaction can occur in organic media [8] as well as in aqueous conditions [9]. The carboxyl activated membrane condenses with an amino-terminated ligand to form a peptide bond and urea, which is removed from the membrane surface.

5.2.1.4   Triazine (2,4,6-trichlorotriazine) Activation

This technique is extensively used to bind colouring compounds on cellulosic materials, and consists of activation of the membrane with 2-amino-4,6-trichlorotriazine [10]. During the activation step, one chlorine from 2,4,6-trichlorotriazine is replaced by the solubilization groups. In the next step, another chlorine undergoes a nucleophilic substitution and the ligand is bound covalently via the amino group. This reaction can be carried out at room temperature.

5.2.1.5   Substituted Sulfonyl Chloride

In these reactions, nucleophiles react with alkyl and aryl sulfonates and lead to direct bonding between the ester group of the sulfonates and nucleophile. Various substituents (such as tosyl, tresyl, or mesyl) on the sulfonyl group facilitate these reactions, making the displaced alkyl group electron deficient [11]. Displacement of the sulfonyl group takes place because of the greater nucleophilicity of amine, Br^-, CN^-, or OH^-.

5.2.1.6   Carbonylation

This is a very simple and effective method in which reactive carbonyl diimidazole (CDI) can be attached to amino or hydroxyl groups on the substrates, the most common being glass and silica [12]. The major inconveniences of this technique are the tendency to form internal cross-linking (reduces the number of ligands available for further coupling reactions) and the need of dry solvent (CDI is moisture sensitive) during the activation step.

5.2.1.7   Silylation of Inorganic (Ceramic) Materials

Silylation is the most common and versatile method of modifying inorganic materials (membranes), such as silica, alumina, titania, and zirconia. The reaction can be achieved using a variety of di- or trifunctional sylilating agents (the most common ones are tri -chloro, -ethoxy, or methoxy silanes), and can take place in aqueous or anhydrous phases.

In aqueous phase silylation, silanes undergo hydrolysis and condensation into polysilane networks prior to the deposition onto the surface. However, the deposition and adsorption of bulk-formed polysilanes can make portions of the surface inaccessible for further silylation, resulting in an incomplete and non-uniform surface coverage. A more effective silylation occurs in anhydrous conditions, resulting in a controlled degree of surface silylation coverage. Thus, with multifunctional silanes, anhydrous silylation occurs by hydrolysis of one or more of the chloro or alkoxy groups, followed by a condensation reaction between surface hydroxyl groups and hydrolysed silane molecules [13]. One to three bonds have been reported to occur in this reaction [14, 15]. The hydrolysis reaction is a requirement for condensation; direct condensation between chloro- or alkoxysilanes and surface hydroxyls is not observed [13, 16]. The water for hydrolysis can be provided by water molecules adsorbed on the support surface or trace amounts of water in the organic solvent [17].

Since a variety of silanes can be used, virtually any functional group can be introduced on the surface, with a possibility of further derivatization. This approach has been applied for surface modification of ceramic membranes. Thus, a glycidoxypropyl trimethoxysilane-modified silica [18] or alumina [19] membranes were used as supports to immobilize polypeptides, by reacting the epoxy group with the terminal amine on the polypeptide.

5.2.1.8   Thiol-Gold Chemistry

Thiol-based self-assembled monolayers (SAMs) on gold coated surfaces have been used in a variety of areas, such as: molecular recognition, model substrates, and biomembrane mimetics in studies of selective binding of enzymes to surfaces, corrosion protection, molecular crystal growth, alignment of liquid crystals, pH-sensing devices, electrically conducting molecular wires, and so on. There are several reasons for using thiol SAM in fundamental studies. There is a strong interaction between thiols and the gold substrate ($\sim$44 kcal/mol), and thiols usually form well-ordered, dense structures. This technique has the advantage of being simple and versatile; virtually any functional group can be introduced in the SAM. Gold-coated particles, electrodes, films, or membranes can be used as supports.

Gold coated membranes are ideal systems to study the role of various parameters, such as pore size, charge, ionic strength, and pH on solute transport. For this purpose, membranes with a uniform structure (narrow pore size distribution, straight pores) such as track-etched polycarbonate or polyethylene terephtalate membranes, or aluminium oxide membranes, prepared by anodic oxidation, are normally used. Depending on the nature of the thiol used, the SAM can induce chemical [20–22], ion transport [23–25] or protein [26–28] selectivity. Functionalization with charged polymers can be useful to design stimuli-responsive materials [29, 30].

5.2.2   Surface Initiated Membrane Modification

In the case of membranes which are chemically inert, with no reactive groups, such as polyethylene or polypropylene, alternative techniques for surface modification are plasma treatment, radiation grafting, or atom transfer radical polymerization.

The plasma treatment technique involves the use of plasma (a low pressure gas containing electron, photons, ions, and other charged particles) in the presence of oxygen to form peroxides on the surface of the membrane [31, 32]. The peroxides then decompose to form oxygen-containing radical groups such as hydroxyls, carbonyls, or carboxyls. Nitrogen-, H_2O-, or Ar-based plasma systems can also be used [33, 34]. The actual composition of the plasma used in surface modification can significantly affect the membrane properties. Generally, polyelectrolytes are attached to the modified surface, in order to make the membrane hydrophilic. This procedure has two benefits: there is an increased flux through the modified membrane and a lowered degree of fouling.

The radiation grafting methods are also used mostly on chemically inert membranes; ultraviolet, visible, and infrared radiation is used to form active sites (radicals) on the surface of the membrane [35]. Polymer chains or monomers (with subsequent polymerization) can then be grafted to these active sites. Similar to plasma treatment methods, various polymers are usually attached, such as polyelectrolytes [36–38] or copolymers [39], with a broad area of application, ranging from fouling reduction to improvement of the membrane wetting properties and separation characteristics.

Both plasma treatment and radiation grafting may also be used to pretreat the membranes for a subsequent technique known as surface graft polymerization. Surface graft polymerization involves the attachment of synthetic monomers to the active sites established by the plasma or radiation treatment. The attached monomers then polymerize on the surface of the membrane [40, 41].

5.2.3   Cross-Linked Hydrogel (Pore Filled) Membranes

Pore-filled hydrogel membranes are formed via *in situ* polymerization and concurrent cross-linking resulting in a high degree of ionizable groups in the pore structure. The promise of a specific gel as the active component of a membrane is governed both by its charge density and by its effective hydrodynamic (Darcy) permeability, k_m, which can be described by Darcy's Law

$$J_v = \frac{k_m}{\eta \Delta x} \Delta P = A \Delta P \qquad (5.1)$$

where J_v is the superficial solvent flux, η is the solvent viscosity, Δx is the membrane thickness, ΔP is the applied pressure gradient, and A is the observed hydraulic permeability. For thin films of homogeneous gel, a plot of volume flux versus applied pressure is often non-linear (k_m decreases at enhanced pressure). This is due to compression of the polymer chains and results in a loss of the desired morphology. In contrast to bulk gels, whose phase transition can reduce the polymer volume by several hundred percent, pore-filled hydrogel networks are forced to maintain their total volume due to steric restrictions resulting from intermeshing with the rigid polymer support [42]. Therefore, the relatively high charge concentrations associated with these gel networks can be utilized to separate electrolytes using composite supports with adequate mechanical strength and reduced sensitivity to osmotic forces [42]. In terms of permeability valves, the cross-linked poly(4-vinylpyridine)-filled polyethylene membranes showed excellent performance. Mika and Childs have linked this stimuli-responsive behaviour to a local microphase separation in the immobilized gel. In terms of ion exclusion, the pore-enmeshed hydrogel membranes exhibited solute rejection values greater than 90% for divalent co-ions. However, under these conditions the polyelectrolyte gel filled membrane permeances ranged from $0.2–1.0 \times 10^{-4}$ cm^3/(cm^{2}s bar). These values do not offer a significant increase when compared to thin-film nanofiltration composite membranes available commercially. This is perhaps due to enhanced polyelectrolyte loading (mass gains up to 190% [42]) and the nature of cross-linked polymer networks. Although cross-linking provides mechanical stability and reduces solute leakage effects, it also severely restricts chain mobility and thus limits possible gains in membrane permeability over high performance nanofiltration. Tunable membrane performances in terms of permeability and separation capabilities (for both charged and neutral molecules) can be achieved via pore filling with a pH-sensitive hydrogel, such as poly(acrylic acid) [43–46]. Again, the degree of cross-linking plays a very important role in gel compression/expansion inside the pores, as a response to changes in pH.

5.2.3.1   Temperature Responsive Hydrogels

While hydrogels can be easily synthesized within and on the porous structure of a support membrane, they are more commonly synthesized and used without a supporting structure. Responsive hydrogels are a class of materials which exhibit changes in network structure, mechanical properties, or swelling behaviour when exposed to a change in stimulus such as light, pH, temperature, particular analyte, or ionic strength [47, 48]. Because of their highly tunable properties, these responsive polymers have been used extensively in the areas of drug delivery, sensors, bioseparation, and optical transduction of chemical signals [49, 50].

Varying temperature in order to induce a volume phase change in temperature responsive hydrogels has been used to achieve controlled and tunable release of a variety of compounds [51–54]. Poly(N-isopropylacrylamide) (pNIPAAm) is a temperature sensitive polymer which undergoes a sharp transition at its lower critical solution temperature (LCST) of approximately 32°C [55]. Below the LCST, a pNIPAAm hydrogel swells with water due to the formation of hydrogen bonds between its amide groups and water. Above the LCST, the hydrogel becomes hydrophobic and collapses [56]. Hydrogels based on p(NIPAAm) have been used for applications including drug delivery [57], enzyme immobilization [58], bioseparation [59], and protein adsorption [60]. Other temperature responsive hydrogels include those based on poly(N-vinylcaprolactam), hydroxypropylcellulose, poly(N,N-diethylacrylamide), and poly(methylvinylether) [61].

5.2.4   Layer by Layer Assemblies

Layer by layer (LbL) assembly technique, most commonly conducted by intercalation of positive and negative polyelectrolytes [62, 63], is a powerful, versatile, and simple method for assembling supramolecular structures. These structures exhibit negative and positive charges, which allow for incorporation of a variety of materials: dyes, crystals, particles, or biomolecules. For this reason, the LbL assembly technique has found applications in a multitude of areas: non-linear optical materials formation [64, 65], patterning [66], separations [67–72], biosensing [73, 74], catalytic reactions [75], biocatalysis, and others.

Using membranes as a support, layer by layer assemblies are usually built by alternate polyanion/polycation deposition on the external surface, via a soaking mode. Figure 5.2 shows this principle for making freely suspended, thin PAH-PSS multilayers, using a sacrificial film of polyvinyl acetate [76]. The deposition leads to ultra-thin films of polyelectrolyte multilayers deposited on the external surface of a porous support (usually an ultrafiltration membrane). The alternate deposition of polyelectrolyte layers changes the membrane charge from positive to negative, as determined from zeta potential measurements, shown in Figure 5.3. Depending on the deposited layers, a variety of surfaces with different

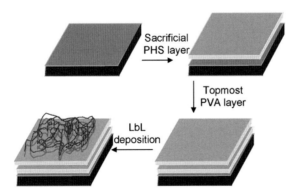

Figure 5.2   *General scheme for the deposition of PHS (polyhydroxystyrene) sacrificial and topmost hydrophobic PVA (polyvinyl acetate) films for the fabrication of perforated (PAH-PSS)$_n$ LbL assembly. (Reprinted with permission from [76] Copyright (2008) American Chemical Society).*

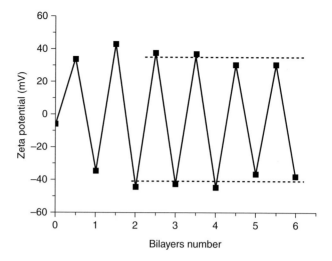

Figure 5.3 *Zeta potential of [PAH/PSS]$_n$ films on PES membranes deduced from streaming potential measurements. (Reprinted with permission from [156] Copyright (2010) Elsevier Ltd).*

applications can be obtained, including anti-adhesive, biocide leaching, or microorganism contact killing properties (Figure 5.4). The deposition on the external membrane surface may lead to pore blockage (due to polyelectrolyte adsorption on the external membrane surface and pore mouth), increasing the pressure drop. On the other hand, too thin films are more susceptible to defects. Another approach is to form the polyelectrolyte multilayer within the membrane pore domain under convective conditions. The support (usually a more open, microfiltration membrane) has a more open structure and the polyelectrolyte assemblies are created mainly inside the pores. Ultra-low pressures are required during operation, thus saving the energy. A schematic of the two approaches, polyelectrolyte adsorption/pore assembly formation, with subsequent protein immobilization is presented in Figure 5.5. The main difference is that in the pore assembled multilayers of polyelectrolytes the first layer is covalently attached to the pore wall, providing a more stable structure to changes in operational conditions such as pH, ionic strength, and so on.

5.3 Applications

5.3.1 Water Flux Tunability

Various polelectrolytes/polypeptides can be used to functionalize the membrane pores; membranes responsive to a variety of stimuli, such as temperature, pH, electric or magnetic fields, light, and so on, have been reported in the literature [77]. If the selected macromolecule is a polypeptide (such as polyglutamic or polyaspartic acid), the incorporation into the membrane domain provides both a highly charged field in the membrane pores and conformational changes. Poly-L-glutamic acid (PLGA), is a molecule that undergoes conformational changes as a function of pH. At low pH (<5.5), PLGA (pK$_a$ 4.38) is in a helix conformation, in which the polymer chains shorten and the carboxylic acid groups

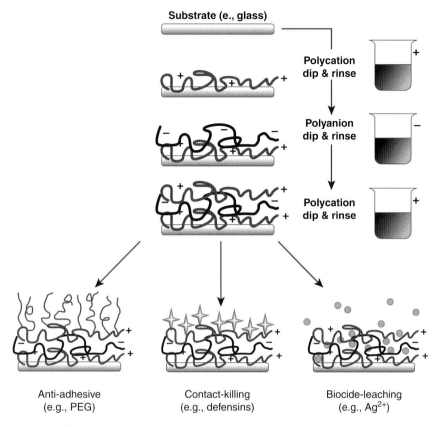

Substrate (e., glass)

Polycation
dip & rinse

Polyanion
dip & rinse

Polycation
dip & rinse

Anti-adhesive
(e.g., PEG)

Contact-killing
(e.g., defensins)

Biocide-leaching
(e.g., Ag^{2+})

Figure 5.4 *Polyelectrolyte multilayers (LbL approach) to create various surface coatings. (Reprinted with permission from [157] Copyright (2011) Materials Research Society). See plate section for colour figure.*

remain uncharged. At pH >5.5 the polyamino acid is in a random coil conformation, in which the chains are extended and the carboxylic acid groups are charged (Figure 5.6).

The variation of flux for the polymer-modified membranes will be greatly influenced by the polymer functional group. Figure 5.7 shows the expected trend for rejection and flux, as a function of pH. For polymers which are charged over the entire pH range, such as polyarginine (PArg), polylysine (PLL) the flux would show no variation. In contrast, for polyglutamic acid (PLGA) or polyaspartic acid (PAsp) the flux is highly influenced by pH; high flux when these are neutral and lower flux and high rejection at a pH above pK$_a$. The actual data for the flux is shown in Figure 5.8. It can be observed that for PArg immobilized on a polyethylene/silica (PE/Si) composite membrane support there is no change in flux or rejection with pH, whereas for PLGA (immobilized on PE/Si) and polyvinyl pyridine (PVP, immobilized on PE) pH has a significant effect. It can also be observed that the trend is the opposite as PVP is charged (and swollen) at low pH and PLGA at high pH. Tunable flux as a response to changes in pH for PLGA immobilized on other membrane supports (such as gold-coated polycarbonate, PC) was also reported by Ito *et al.* [29, 79]. The polymer

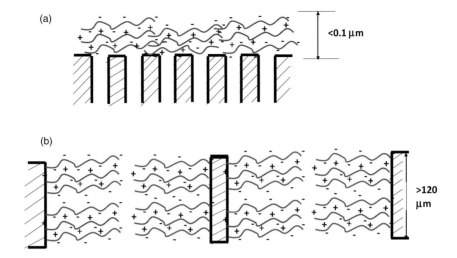

Figure 5.5 *Polyelectrolyte multilayer assemblies: (a) thin films of polyanion and polycation, obtained by alternate adsorption on the membrane surface; and (b) pore assembled polyelectrolyte multilayers.*

chain length:pore radius ratio as well as polymer chain density play a significant role in flux modulation. Besides PLGA, flux modulation by changing pH was demonstrated for other polymers immobilized on membranes, such as polymethacrylic acid or polyacrylic acid [78, 79].

5.3.1.1 Modelling the pH-Responsive Behaviour of Poly(Vinylidene Fluoride) (PVDF)-PAA Membranes

In contrast to the helix-coil transitions undergone by the poly(amino acids), polyelectrolyte gels can undergo shrinking/swelling that corresponds to a decrease/increase in water content of the gel. Several studies have demonstrated the controlled opening and closing of

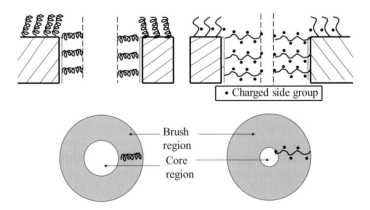

Figure 5.6 *Schematic of helix-coil transitions for an anionic polypeptide (PLGA, PAsp), immobilized in a microfiltration membrane pore.*

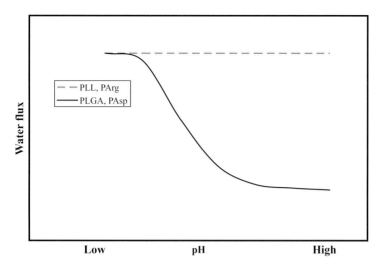

Figure 5.7 *Expected behaviour of water flux variation vs. pH of the feed solution for membranes immobilized with positively (PLL, PArg), and negatively (PLGA, PAsp) charged polymers.*

membrane pores modified with PAA hydrogels via pH modulation [43, 80, 81]. The pH range in which the most rapid change in membrane flux occurs can be altered by changing the pK_a of the PAA chains/network and can be achieved by varying the concentration of monomer, concentration and type of initiator, type of solvent, and the conditions under which the polymerization takes place. As pH increases, more of the carboxylate groups become ionized, causing the PAA network to expand, thereby reducing R_1 (core radius, see Figure 5.9a). Datta *et al.* [152] studied the effects of pH variation on water flux through

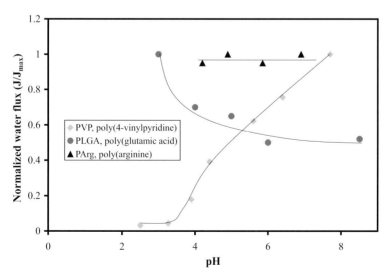

Figure 5.8 *Water flux variation as a function of the pH of the feed solution, for membranes functionalized with PArg, PLGA (data from [82]) and PVP (data from [42]).*

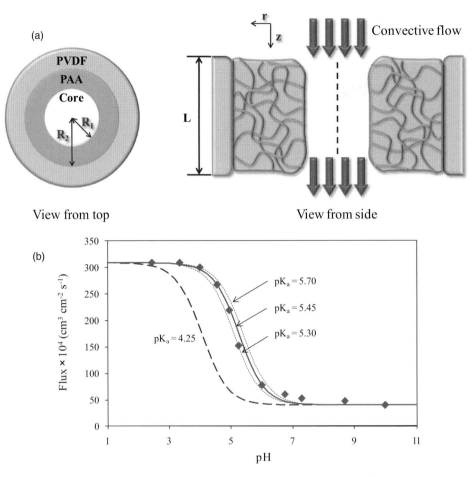

Figure 5.9 *(a) Schematic of PVDF-PAA membrane pore for modelling of pH-responsive behaviour. (b) Experimental (diamonds) and predicted solution flux (lines) as a function of pH through a PVDF-PAA. The flux for $pK_a = 5.45$ (solid line) is surrounded by the flux for pK_a 5.3 and 5.7 (thin dashed lines). $\Delta P = 1.38$ bar (data taken from [152]).*

PVDF-PAA membranes over a broad pH range. They found that by increasing the pH from 2.4 to 10.0, the volumetric solution flux (J_V) at 1.38 bar decreased from 309×10^{-4} cm^3 cm^{-2} s^{-1} ($J_{V,max}$) to 40×10^{-4} cm^3 cm^{-2} s^{-1} ($J_{V,min}$).

$$J_V = \frac{Q}{A_{membrane}} \tag{5.2}$$

where Q is the volume flow rate (cm^3 s^{-1}), $A_{membrane}$ (m^2) is the cross-sectional area of the membrane. Assuming the volume occupied by the PAA gel, V_{PAA}, expands linearly with the molar fraction of ionized carboxylate groups, this data can then be used to predict J_V

as a function of pH.

$$V_{PAA} = \pi L \times \left(R_2^2 - R_1^2\right) \tag{5.3}$$

$$V_{PAA} = V_{PAA,\min} + \frac{[COO^-]}{[COOH] + [COO^-]}(V_{PAA,\max} - V_{PAA,\min}) \tag{5.4}$$

where $V_{PAA,\min}$ is the minimum volume occupied by the PAA gel and $V_{PAA,\max}$ is the maximum volume occupied by the PAA gel, L is membrane thickness, and R_1, R_2 are the radii of the core and membrane pore, respectively. Combining Equations (5.3) and (5.4) and assuming flow through the pores follows the Hagen–Pouiselle equation, $R_1 \propto J_V^{1/4}$:

$$J_V = \left(J_{V,\max}^{1/2} - \frac{[COO^-]}{[COOH] + [COO^-]}\left(J_{V,\max}^{1/2} - J_{V,\min}^{1/2}\right)\right)^2 \tag{5.5}$$

where $[COO^-]$ is the concentration of ionized carboxylate groups and $[COOH]$ is the concentration of non-ionized carboxylate groups. The molar fraction of ionized to non-ionized carboxylate groups was calculated using Equation (5.6).

$$\frac{[COO^-]}{[COOH] + [COO^-]} = \frac{1}{1 + 10^{pK_a - pH}} \tag{5.6}$$

For the case where the pK_a of the PAA network is the same as that of acrylic acid, 4.25, the flux at 1.38 bar as a function of pH is shown in Figure 5.9b. When compared to the experimental data obtained by Datta *et al.* [152], this pK_a value incorrectly predicted J_V whereas $pK_a = 5.45$ resulted in the best prediction of the experimental data. This pK_a is higher than the previously reported values for PAA gels, but may be due to differences in the aforementioned variables (cross-linker, etc.) associated with this PAA network. Similarly, J_V predictions for $pK_a = 5.3$ and 5.7 are shown in Figure 5.9b. This PAA with increased pK_a is advantageous for use at circumneutral pH since it allows for higher permeability than similar PAA coatings with lower pK_a values. Similar pK_a shift behaviour has been modelled and reported for single point attachment (covalent) of PLGA on a membrane pore surface [82].

5.3.2   Tunable Separation of Salts

As shown previously in Figure 5.6, at pH greater than 5.5 the polyamino acid (PLGA) is in a random coil conformation, in which the chains are extended and the carboxylic acid groups are charged. This can be utilized in nanofiltration type separations with tunable selectivity in response to stimuli such as pH or ionic strength [30,83]. As an example, Figure 5.10 shows the rejection of a charged solute (SO_4^{2-}) on two different membranes, 30 and 100 nm pore diameter membranes, functionalized with PLGA. At high pH the chains are fully charged and extended from the pore wall and the rejection is more than 80%; at pH values below pK_a, the polypeptide chains are neutral and shrunk and the rejection is negligible ($\sim 0\%$). This is reversible, the rejection increases to the original value as the pH of the feed solution is above pK_a [30].

The variation of separation capabilities of the polymer-modified membranes will be greatly influenced by the polymer functional group. Figure 5.11 shows the expected trend

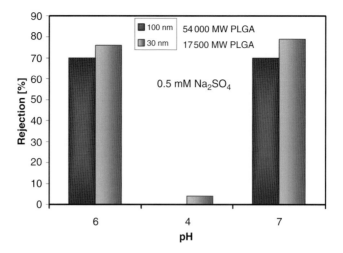

Figure 5.10 *Tunable separation of charged solutes in a PLGA-functionalized PCTE membrane, as a function of the pH of the feed solution. (Reprinted with permission from [30] Copyright (2004) American Chemical Society).*

for rejection as a function of pH. For polymers which are charged over the entire pH range, such as polyarginine (PArg) or polylysine (PLL), the rejection would show no variation. In contrast, for polyglutamic acid (PLGA) or polyaspartic acid (PAsp) the rejection is greatly influenced by pH; zero rejection when these are neutral and lower flux and high rejection at a pH above pK_a. The actual data for flux and rejection is shown in Figure 5.12. It can be observed that for PArg there is no change in rejection as a function of pH, whereas for PLGA and polyvinyl pyridine (PVP) pH has a significant effect. It can also be observed that the trend is opposite, as PVP is charged (and swollen) at low pH and PLGA at high pH.

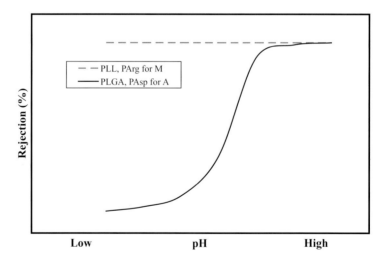

Figure 5.11 *Expected behaviour for separation of charged solutes in polymer functionalized membranes, as a function of the pH of the feed solution (M: metal cation; A: anion).*

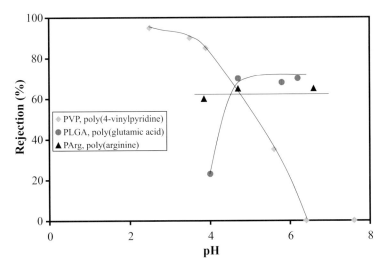

Figure 5.12　*Tunable separation of charged solutes in polymer functionalized membranes, as a function of the pH of the feed solution (data taken from [42] and [82]).*

5.3.2.1　*Modelling the Ion Transport Through Polypeptide (PLGA)-Functionalized Membranes*

The transport of electrolytes in charged media has been a well-discussed topic in membrane research; the usual assumption is that the flux of each species through the membrane is directly proportional to the driving forces acting on the system. However, critical properties of the membrane itself, such as the electrostatic potential, are not taken into account. Salt transport is attributed solely to concentration gradients and/or convective coupling to the overall volume flux. The effects of electrostatic interaction between the fixed membrane charge (Q) and the co-ion in solution are essentially neglected. To address the limitations associated with irreversible thermodynamics theory, a number of research groups have focused primarily on the development of models based on the extended Nernst–Planck equations. These models encompass all three modes of electrolyte transport, namely diffusion, convection, and electromigration. The last one is fundamental to the study of salt transport in charged media.

The separation behaviour of these poly(amino acid) functionalized membranes was modelled assuming a cylindrical pore geometry with polypeptide chains uniformly bound along the pore wall [82]. The effects of pore size distribution and irregular pore geometry on solute retention were ignored for simplicity. This idealized pore, shown in Figure 5.13a suggests two distinctly different regions of ionic transport.

Charged solute transport through the semi-permeable brush region ($R_c < r \leq R_p$) can be determined using the well known Debye–Brinkman equation,

$$\frac{1}{r}\frac{d}{dr}\left(r\frac{du^B}{dr}\right) - \frac{u^B}{\kappa} + \frac{1}{\eta}\frac{dP}{dx} = 0 \qquad (5.7)$$

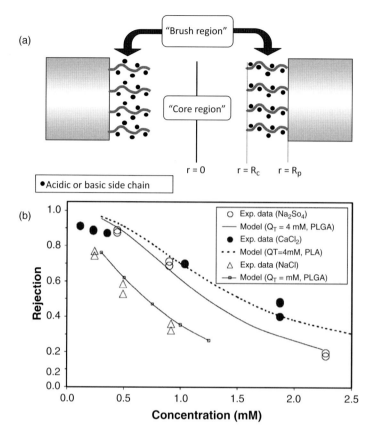

Figure 5.13   *(a) Schematic of the polypeptide (PLGA)-modified membrane pore (b) Comparison of experimentally determined solute rejection with model calculations. (Reprinted with permission from [82] Copyright (2004) Elsevier Ltd).*

where $u^B(r)$ is the PAA region velocity profile and κ is the solvent-specific hydraulic permeability coefficient associated with this polypeptide containing region [84]. The coupled velocity distribution, $u(r)$, was subject to the boundary conditions,

$$u^B(R_P) = 0$$

$$\frac{du^c}{dr}(0) = 0$$

$$u^c(R_c) = u^B(R_c)$$

$$\frac{du^c}{dr}(R_c) = \frac{du^B}{dr}(R_c)$$

Donnan potential is induced to maintain electrochemical equilibrium. Given the conditions of electroneutrality in the membrane and solution phases, the co-ion distribution for a negatively charged membrane in contact with a dilute single salt solution is described by

$$\frac{c_2^m}{c_2} = \left[\frac{|z_2| c_2}{|z_2| c_2^m + |z_x| Q} \right]^{\left(\frac{|z_2|}{|z_1|} \right)}$$

(5.8)

where z_i is the ion valency (1 = cation, 2 = anion) and z_x is the valency associated with the fixed membrane charge, Q (superscript m denotes membrane phase). In relatively dense charged membranes (pore < 3 nm), ion transport is generally well described through a one-dimensional approach that assumes a uniform distribution of electric potential and solute concentration in the radial direction [85]. However, this assumption is not valid for membranes with pore dimensions beyond a few nanometers. For two-dimensional ion transport at steady state, the flux of ion i in the axial ($j_{x,i}$) and radial ($j_{r,i}$) direction can be described using the extended Nernst–Planck equations,

$$j_{x,i} = u(r)c_i^m - D_i \left(\frac{\partial c_i^m}{\partial x} + \frac{z_i c_i^m F}{RT} \frac{\partial \Phi}{\partial x} \right)$$

(5.9)

$$j_{r,i} = -D_i \left(\frac{\partial c_i^m}{\partial r} + \frac{z_i c_i^m F}{RT} \frac{\partial \Phi}{\partial r} \right)$$

(5.10)

where D_i is the ion diffusivity in the membrane phase, R is the gas constant, F is the Faraday constant, T is the absolute temperature, and Φ is the total electric potential. Figure 5.13b shows a good correlation between experimental and calculated ion rejection vs. feed concentration. It also allows for evaluation of the average pK_a associated with the attached polypeptide. The estimated pK_a shift for the immobilized PLGA was found to be 0.5 [82], consistent with solution-phase experiments found in the literature.

5.3.3   Charged-Polymer Multilayer Assemblies for Environmental Applications

One of the major disadvantages of the polypeptide-functionalized microfiltration membranes is that the chains are not extended throughout the pore. The central part of the pore, called the "core" is not functionalized and causes a leakage, thus lowering the overall rejection. In order to ensure a better pore coverage and minimize this "core leakage", multilayer assemblies of charged polymers can be formed inside the pore as shown in Figure 5.5, and as was described in Section 5.2.4. Thus, electrostatic assembly based on layer by layer (LbL) deposition of polyelectrolytes leads to an enhanced volume density of charged groups allowing for Donnan exclusion of ionic species. This provides a simple and versatile method of preparing thin charged films characterized by a hydraulic permeability far beyond conventional nanofiltration membranes. This is presented in Figure 5.14, the multilayer assemblies (PLGA-PAH-PSS or PLGA-PLL-PLGA) showed more than 90% rejection of toxic anions (AsO_4^{2-}) in the presence of non toxic species NaCl and Na_2SO_4, at remarkably low pressures of less than 3 bars [86].

The membrane can be used successfully for environmental applications (Figure 5.15), the selectivity increases with feed concentration for divalent ions such as AsO_4^{2-} vs. common non-toxic ions that are usually present in ground water (such as Cl^-). Tunable selectivity can also be achieved by increasing the flux, monovalent ions being more susceptible to "leakage" through the core region of the membrane than divalent ones. This constitutes

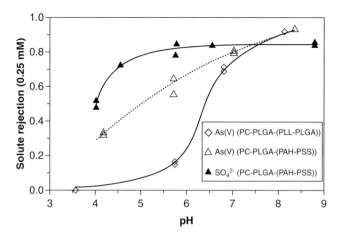

Figure 5.14   *Effect of solution pH on the ion separation As(V) (Na$_2$HAsO$_4$) or Na$_2$SO$_4$ solutions using a PC–PLGA functionalized membrane containing either a PLL/PLGA or a PAH/PSS bilayer. (Reprinted with permission from [86] Copyright (2004) American Chemical Society).*

a unique characteristic of these functionalized membranes, not found in other types of separations such as classical nanofiltration.

The selectivity can be further tuned as a function of the charged groups on the polymer chains which are forming the assembly. Thus, strong acid groups (such as for PSS) will be fully charged for the entire pH working range (greater than 2), whereas weak acid groups (such as carboxyl in PAA, PLGA, PAsp etc.), will be neutral below pH 5 and charged above this value.

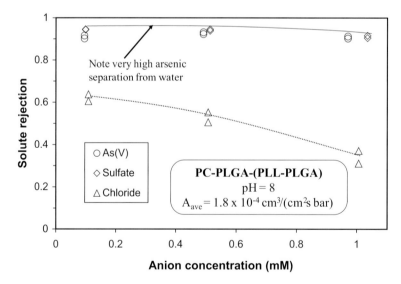

Figure 5.15   *Dependence of charged solute rejection on the feed concentration (equimolar mixture) for a multilayer membrane prepared in 0.5 M NaCl. (Reprinted with permission from [86] Copyright (2004) American Chemical Society).*

5.3.3.1   Ultra-High Capacity Metal Capture

Functionalization of membrane materials also allows for the development of high capacity heavy metal (Cd^{2+}, Pb^{2+}, Cu^{2+} etc.) adsorbents [4, 87, 88]. Unlike charge-based exclusion, where separation occurs due to the electrostatic repulsive force between fixed ionizable groups and permeating co-ions in solution, adsorption mechanisms (i.e. ion exchange, chelation etc.) are due to the attraction of the counterion to the immobilized polyelectrolyte. Ion exchange is driven by electroneutrality, such that a divalent counterion requires two binding sites for metal capture. Chelation, on the other hand, is based on ion-ligand complex formation. This mechanism allows for enhanced selectivity and a maximum solute to ligand binding ratio of 1:1. Sorbent materials, therefore, have a finite capacity and require regeneration for multiuse applications. Microporous membrane supports (cellulose, silica, etc.) are typically used for the development of these functionalized materials due to a relatively high surface area and low resistance to mass transfer. This allows for maximizing both capacity and throughput. An additional advantage of these membrane-based adsorbents is that functional groups can be grafted within their porous structure as polymeric ligands with multiple metal binding sites, rather than as monomeric surface functionalities. Hence, interactions between these groups and the heavy metal ions are very rapid, as sorption proceeds under convective flow and ion-to-ligand transfer resistance is minimized. Also, concerns regarding pressure-drop and channelling inherent to ion-exchange columns are minimized using convective flow membrane operation.

Glycidyl methacrylate (GMP) is often used as a monomer for the preparation of function-alized sorbent membranes because its epoxide functionality is easily mutable to iminodiac-etate (IDA) [89], sulfonic acid (SO_3-H) [90], phosphoric acid (PO_3-H), and triethylamine moieties [87]. Supports containing a high density of IDA groups, for instance, are particu-larly attractive for solutions having high levels of hardness because they offer much greater selectivity over non-toxic Ca^{2+} and Mg^{2+} ions [91]. Bhattacharyya and colleagues have developed high-capacity sorbent materials based on the tethering of specific polypeptide homopolymers (poly(L-glutamic acid), poly(aspartic acid)) within microporous supports [91, 92]. These materials have shown toxic metal adsorption capacities exceeding 1 gram per gram of dry membrane sorbent. They have attributed these anomalously high capacities to the additional capture mechanism of counterion condensation. This capture mechanism results from the superposition of the electrostatic fields generated by regularly-spaced (<0.7 nm separation) charged functionalities [91]. Although providing increased capacity, counterion condensation is non-specific and thus results in reduced overall selectivity.

5.4   Responsive Membranes and Materials for Catalysis and Reactions

In addition to using functionalized membranes for advanced separation processes, their tunable properties make them suitable platforms for conducting a wide range of reactions. Since many catalysts are derived from heavy metals, the same responsive materials used to capture heavy metal ions can be used for their synthesis and immobilization. One important aspect of catalyst immobilization is to eliminate the costs associated with separating them from reaction mixtures. Support materials with a high surface area to volume ratio are desirable since large amounts of catalyst can be loaded onto a relatively small volume of material. Although materials such as ion exchange resins do have a high surface area to volume ratio, reactants must diffuse through the bead's porous network in order to access all of the immobilized reactants. On the other hand, functionalized membranes containing immobilized catalysts can be operated under convective flow, thereby significantly reducing the mass transfer limitations associated with the previously-mentioned materials.

In some instances, the membrane can be directly functionalized with the catalytic material. For example, Shah *et al.* [159] grafted sulfonated polystyrene chains to commercial polyethersulfone (PES) microfiltration membranes for use as an esterification catalyst. The authors demonstrated that the catalytic activity of these membranes was similar to that of commercially available ion exchange resins in batch reaction. However, when operating the membrane under convective flow, the same reactant conversion was obtained in 20 s residence time as in 11 h of batch reaction, demonstrating greatly improved kinetics [93]. Additionally, the authors showed how they could control the number of repeat units per graft and hence the number of catalytic sites per unit of membrane area.

The properties of such functionalized materials can also be used to control the exposure of catalysts, reactants, or other moieties imbedded within the responsive layer. For instance, in order to demonstrate controllable activity of catalytic nanoparticles, Lu *et al.* [95] synthesized silver nanoparticles within a thermoresponsive polymer layer supported by a larger particle [94, 95]. By varying the system's temperature, they were able to control the nanoparticles' exposure to the surrounding solution and hence the catalytic activity of the composite system. In order to obtain a tunable response to various nitro-aromatic compounds, Dotzauer *et al.* [75, 162] used versatile LbL assemblies to immobilize catalytic Au nanoparticles within multiple types of synthetic membranes. Commercially available alumina, polycarbonate, and nylon membranes were modified and operated under convective flow with varying residence time and temperature to alter the distribution of products from the catalytic reduction of the nitro compounds used.

5.4.1   Iron-Functionalized Responsive Membranes

A particular non-toxic heavy metal used in a variety of environmentally important reactions is iron. Depending on its form, iron can be used as a catalyst for either advanced oxidative or reductive reactions and can also be easily immobilized within membranes functionalized with negatively charged polyelectrolytes. Membrane-immobilized iron ions (Fe(II)/Fe(III)) and/or iron oxide nanoparticles with hydrogen peroxide (H_2O_2) have been used to degrade toxic organic compounds via oxidation whereas membrane-immobilized iron-based nanoparticles (Fe^0, Fe/Pd bimetallic, etc.) have been used to degrade similar chemicals via reduction. An overview of how a versatile PVDF-PAA pore filled membrane (PVDF-PAA) can be used for these technologies is presented in Figure 5.16.

5.4.1.1   Membrane-Immobilized, Iron-Catalysed Free Radical Reactions

The addition of H_2O_2 to a solution containing iron ions initiates a series of well-established free radical generating reactions [96]. These reactions, also known as iron-catalysed free radical reactions, have been used to oxidize and/or degrade a variety of non-toxic and toxic organic compounds. The main species of interest, the hydroxyl radical (OH•), is formed

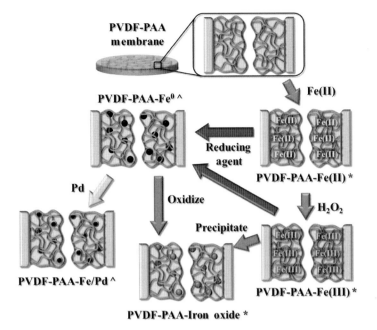

Figure 5.16 *Schematic of PVDF-PAA membranes for use as platforms in both advanced oxidative (*) and reductive (ˆ) reactions. Note that the oxidative reactions require the addition of H_2O_2. PVDF-PAA membranes can also be used as a platform for electrostatic enzyme immobilization (see Figure 5.26). See plate section for colour figure.*

via Equation (5.11) [97] and reacts with organic contaminants present (Equation 5.12).

$$Fe^{2+} + H_2O_2 \rightarrow Fe^{3+} + OH^- + OH\bullet \tag{5.11}$$

$$Organic \xrightarrow{\ OH\bullet\ } Intermediates \xrightarrow{\ OH\bullet\ } Products \tag{5.12}$$

Due to iron speciation, these reactions take place most efficiently at low pH; however, most environmental contaminants are present in waters near neutral pH. For this and other reasons, several investigators have modified these reactions through the use of iron chelates or by using iron oxide particles which can also generate free radicals in the presence of H_2O_2 [98–101]. Nonetheless, these catalysts, both homogeneous and heterogeneous, must be separated from solution if treatment of wastewater or drinking water is desired. By immobilizing these iron-based catalysts on appropriate support materials, this potentially costly separation step can be avoided.

The materials required for catalyst immobilization should be robust and provide a large surface area in order to maximize catalyst exposure. For example, various porous polymer resins have been used as high-capacity supports for reagents and catalysts with specific surface areas ranging from 50 to 1000 m^2/g [102–104]. Many of these polymers are easily functionalized, enabling their use for an array of applications. Using iron ions immobilized within strong-acid ion exchange resins and adding H_2O_2, several investigators have degraded a variety of organic compounds [105, 106]. Similarly, iron ions immobilized within zeolites have been shown to react with H_2O_2 to form free radicals and can be used to degrade organic contaminants such as reactive azo dyes [107, 108]. Although these materials are high-capacity, the mobile reactants, H_2O_2 and the organic compounds, must diffuse through the material's pores in order to access the immobilized reactive iron species.

Previous work has demonstrated how functionalized iron-loaded membranes can be used to mimic the catalytic properties of materials found in nature. Parton *et al.* [158] oxidized alkanes at rates similar to that of the enzyme cytochrome P-450 at room temperature by using iron phthalocyanine complexes immobilized in zeolite Y crystals which were embedded in polydimethylsiloxane membranes. Similarly, iron ion-loaded membranes have been used to generate free radicals in the presence of H_2O_2 for water treatment applications [109, 110]. Fernandez *et al.* [111] demonstrated how iron ions immobilized by negatively-charged functional groups, SO_3^- and COO^-, in Nafion membranes and carboxylated polyethylene mats, respectively, can be used as photocatalysts for the free radical mediated degradation of azo-dyes and/or 4-chlorophenol in the presence of H_2O_2 [111–113].

5.4.1.1.1   Operation Under Convective Flow (Residence Time Variation).   These studies illustrate membranes' effectiveness to be used as platforms for immobilized, iron-based free radical reactions. The membranes distinct advantage over other high-surface area materials is the ability to operate them under convective flow conditions. This provides reactants in solution with rapid access to the immobilized catalysts/reactants thereby reducing, if not eliminating, the mass transfer limitations of diffusion. Lewis *et al.* [44] used a PVDF-PAA membrane with immobilized Fe(II)/Fe(III) for free radical reactions by permeating a solution containing H_2O_2 through the membrane pores. Since Fe(II) is oxidized to Fe(III) (Equation 5.11) much faster than Fe(III) is reduced to Fe(II) in these reactions (Equation 5.13), a majority of the iron species present in the membrane domain were Fe(III) after prolonged reaction with H_2O_2.

$$Fe^{3+} + H_2O_2 \rightarrow Fe^{2+} + H^+ + HO_2 \bullet \qquad (5.13)$$

In a PVDF-PAA membrane containing mostly Fe(III) (PVDF-PAA-Fe(III)), the authors demonstrated how the conversion of H_2O_2 could be easily controlled by changing the residence time of the permeating solution. Figure 5.17 shows the conversion of H_2O_2 as a function of residence time and the predicted conversion assuming Equation (5.13) is the rate-limiting step in the decomposition of H_2O_2. These same PVDF-PAA-Fe(III) membranes were then used to degrade pentachlorophenol, a toxic organic compound, at near neutral pH and at low residence times (<1 min). The pH-responsive behaviour of the membrane also permits loading/unloading of the iron ions, present as Fe(II) and/or Fe(III), when desired [114]. Since direct Fe(III) loading in PAA (COOH groups) functionalized pores is not possible at a pH of less than 5, thus Fe(II) capture followed by oxidation to Fe(III) is a new approach.

Responsive membrane platforms have also been used for the controlled synthesis of iron oxide and/or ferrihydrite nanoparticles within the membrane domain. Iron oxide nanoparticles have been used with H_2O_2 for the selective oxidation and synthesis of organic compounds [115]. In order to demonstrate the control with which these nanoparticles can be synthesized, Winnik *et al.* [116] utilized microporous polypropylene membranes (200 nm pore diameter) with PAA immobilized within the pores via photo-initiated grafting. By introducing a ferrous iron solution to the membrane and subjecting it to oxidation with either

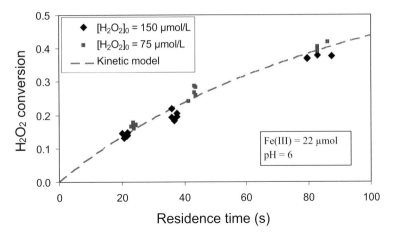

Figure 5.17 *Experimental and model-predicted H_2O_2 conversion as a function of residence time for PVDF-PAA-Fe(III) membrane. External membrane area = 33.2 cm^2, thickness = 125 μm.*

O_2 or H_2O_2 at 70°C and pH 14, they formed superparamagnetic ferrihydrite nanocrystals of varying shapes. In the presence of H_2O_2, the authors noted the formation of disc-like particles 3–4 nm in diameter and needle-shaped particles (4 nm × 50 nm); however, in the presence of O_2, only the disk-like particles formed [116]. Using a PVDF membrane containing a cross-linked PAA gel, Lewis *et al.* [114] synthesized ferrihydrite/iron oxide nanoparticles from iron ions which were immobilized in the PAA gel. These membranes were then used with H_2O_2 to degrade trichloroethylene (TCE), a common groundwater pollutant. The use of such pH-responsive membranes can aid in controlling nanoparticle agglomeration and prevent pore clogging. By synthesizing the particles at a higher pH, the PAA gel swells and occupies more of the membrane pore volume, thereby preventing the formation of larger particle agglomerates. Subsequent pH reduction causes the PAA gel to collapse, increasing the permeability of the membrane. With these or similar responsive materials, one could envision making various iron oxide nanostructures within the membrane pores by manipulating the conditions under which iron oxidation occurs.

5.4.1.2 Tunable Fe^0 Nanoparticle Synthesis for Treatment of Contaminated Water

Functionalized membranes containing heavy metal ions can also be used for nanoparticle synthesis and subsequent reactions. For instance, the aforementioned PVDF-PAA membranes containing immobilized iron ions have been used to synthesize membrane-immobilized, zero-valent iron (Fe^0) nanoparticles. Bulk Fe^0 particles have been used to treat a wide variety of water contaminants including organic solvents, pesticides, fertilizers, organic dyes, heavy metal ions, and inorganic anions [117]. By synthesizing these particles on the nanoscale, not only do they have a much higher specific surface area than their bulk counterparts (i.e. <1 m^2/g for bulk vs 35 m^2/g for nanoparticles), but they have also been shown to have higher surface area normalized reactivity [118]. There are several advantages associated with immobilizing reactive nanoparticles within a membrane-supported negatively-charged polyelectrolyte matrix. In addition to preventing nanoparticle loss, the need for separation of the particles from the treated solution is eliminated. Whereas other materials such as zeolites, metal oxides, and activated carbon, have been used to support similar nanoparticles, they are generally very dense and can inhibit contaminant transport [119–121]. Membrane-based platforms can provide contaminants with easier access to the reactive nanoparticles and they are also ideal for nanoparticle synthesis because they help prevent nanoparticle agglomeration, a problem commonly encountered during conventional solution-phase synthesis due to the magnetic properties of Fe^0-based materials [122]. A stimuli-responsive polyelectrolyte matrix provides the added benefit of tunable residence time and iron loading, both of which are important parameters in determining the reactivity of the nanoparticle-membrane composite material. Xu and Bhattacharyya [128] demonstrated how nanoparticle loading and size could be controlled by varying the ratio of Fe^{2+}:COOH in PVDF-PAA membranes as shown in Figure 5.18.

5.4.1.2.1 Iron Nanoparticles for the Reductive Transformation of Chlorinated Organic Compounds. The majority of Fe^0 nanoparticles used for contaminant removal are synthesized using a powerful reducing agent such as sodium borohydride. The overall reaction for the complete reductive dechlorination of a chlorinated organic compound, RCl_n, is shown in Equation (5.14).

$$nFe^0 + RCl_n + nH^+ \rightarrow nFe^{2+} + RH_n + nCl^- \tag{5.14}$$

While these nanoparticles have been used directly, doping a small amount of a catalytic metal such as Pd, Ni, or Cu onto the surface of the nanoparticles (as little as 0.1wt%) has been shown to increase their reactivity with chlorinated organic compounds by up to 2 orders of magnitude [118, 123, 124], reduce the rate of oxidation of the Fe^0 core thereby increasing the particle's reactive lifetime, and increase the formation of completely dechlorinated products [125]. However, some catalytic metals, such as Ni, can be hazardous if released into the environment, making control of the nanoparticle transport desirable. Immobilization of the nanoparticles within a negatively-charged polyelectrolyte matrix alleviates this problem by capturing the heavy metal ions if the nanoparticles do dissolve.

Bimetallic Fe^0-based core/shell nanoparticles immobilized within a PVDF-PAA membrane have been used to rapidly dechlorinate toxic organic compounds including polychlorinated biphenyls (PCBs), TCE, and trichloroacetic acid (TCAA) [44, 46, 126–129]. A typical procedure for the functionalization of such membranes is shown in Figure 5.19.

5.4.1.2.2 Bimetallic Fe/Pd Nanoparticle Synthesis. The flowchart for membrane functionalization and nanoparticle synthesis is shown in Figure 5.19. PVDF membranes were coated with poly(acrylic acid) by *in situ* polymerization of acrylic acid. The polymerization reaction was carried out in aqueous [44, 45] and organic phases [46], thus cross-linked PAA-coated PVDF membranes were obtained. Prior to Fe^{2+} ion exchange, PAA-coated PVDF membranes were immersed in NaCl (5 to 10% wt) solution at pH 10 for 10 h to convert the -COOH to COONa form. The membrane was immersed in $FeCl_2$ solution, the volume and concentration were adjusted according to the desired amount of Fe^{2+} on the membrane. Nitrogen gas was bubbled to minimize Fe^{2+} oxidation. The reduction with either $NaBH_4$ or tea extract [45, 46, 130] ensured Fe nanoparticle formation and the membranes were stored in ethanol to prevent oxidation. The secondary metal, Pd, was

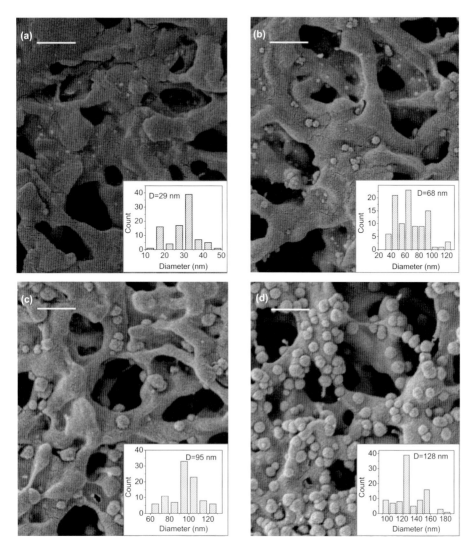

Figure 5.18 *SEM images and particle size distributions of Fe/Pd nanoparticles immobilized in PVDF-PAA membranes. Nanoparticle size and loading were varied by changing the molar ratio of PAA (monomer unit) to Fe(II) ions as follows: (a) 100, (b) 50, (c) 8 and (d) 4. For all cases, NaBH$_4$ concentration was 0.5 M. Scale bars represent 500 nm. (Reprinted with permission from [128] Copyright (2005) AICHE).*

deposited on the Fe nanoparticles by immersing the membrane in a K$_2$PdCl$_4$ solution via the well known redox reaction:

$$Pd^{2+} + Fe^0 \rightarrow Fe^{2+} + Pd^0$$

The concentration of the K$_2$PdCl$_4$ solution varied, depending on the amount of Pd desired to be deposited on the Fe-modified membrane (typically in the range of 1 to 5% wt. compared

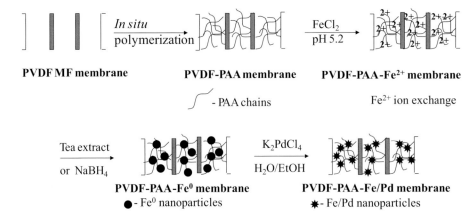

Figure 5.19   Schematic for in situ *polymerization of PAA and Fe/Pd nanoparticle synthesis in PVDF membranes. (Reprinted with permission from [130] Copyright (2011) Elsevier Ltd).*

to Fe); this created core-shell Fe/Pd bimetallic nanoparticles [45, 46]. In a Fe/Pd bimetallic system, the role of Fe (reactant) in zero-valent form is to generate hydrogen (near neutral pH operation) via a corrosion reaction. Pd acts as a hydrogenation catalyst, thus providing the active sites in the bimetallic system.

5.4.1.2.3   Contaminant Degradation in Batch Reactions.   Xu *et al.* [129] utilized Fe/Pd and Fe/Ni nanoparticles immobilized in PAA-functionalized microfiltration membranes to achieve complete TCE dechlorination in batch studies in 1 h with ethane as the main product. The Fe/Pd nanoparticles immobilized in a PVDF-PAA membrane (PVDF-PAA-Fe/Pd) were also used to completely dechlorinate PCBs, which are very difficult to dechlorinate through the use of Fe^0 alone, in 2 h batch reactions. Although these studies demonstrate the high reactivity of the modified membranes, they were used in batch reactions where contaminant transport was mainly due to diffusion into the membrane.

In order to illustrate the effectiveness of operating the PVDF-PAA-Fe/Pd membranes under convective flow, Smuleac *et al.* [45] permeated a solution containing 2,2'-dichlorobiphenyl (DiCB) through the membrane pores. By varying the residence time, the authors demonstrated they could control biphenyl (the final product) formation, as shown in Figure 5.20. As expected, the biphenyl concentration in the permeate (and thus the DiCB conversion) is a strong function of the residence time, and increases from 14 to 43% as the residence time is varied from 23 to 37 s. Another tunable aspect of these systems compared to similar batch reactions is the ability to control intermediate and product formation by varying residence time.

In contrast to using conventional reducing agents, highly reactive nanoparticles have been synthesized in the homogeneous phase using "green" reducing agents. These are non-toxic, biodegradable materials such as ascorbic acid (vitamin C) [131, 132], coffee [133], tea extracts [133, 134], grape pomace [135], and some polymers [136]. In addition to eliminating the need for harsher reducing agents, some of these "green" reducing agents can also be used to tune nanoparticle reactivity by varying their coating of the nanoparticle surface.

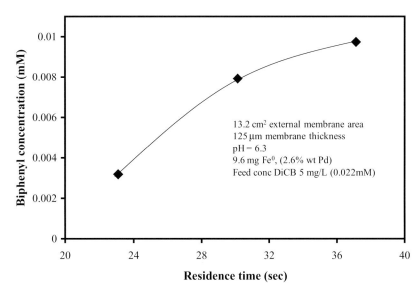

Figure 5.20 *Permeate concentration of biphenyl, the final product, as a function of residence time for convective mode dechlorination of 2,2′-Dichlorobiphenyl (DiCB) in a PVDF-PAA-Fe/Pd membrane. (Reprinted with permission from [45] Copyright (2010) Elsevier Ltd).*

Recent studies illustrate how such reducing agents can be used to synthesize reactive Fe^0, Fe/Pd, and Pd/Fe nanoparticles within a PVDF-PAA membrane domain. Smuleac *et al.* [130] synthesized Fe^0 and Fe/Pd nanoparticles (base nanoparticle diameter of 20–30 nm) immobilized within PVDF-PAA membranes using only green tea extract. These membranes were then used to dechlorinate TCE in both batch reactions and under convective flow conditions. One additional advantage the Fe^0 nanoparticles synthesized using green tea extract offered over those synthesized with $NaBH_4$ was increased reactive lifetime. After four cycles of PVDF-PAA-Fe^0 use for TCE dechlorination, those synthesized using $NaBH_4$ retained only approximately 15% of their original activity whereas those synthesized using green tea extract retained over 80% of their original stability. Corresponding to the change in reactivity, the Fe^0 nanoparticles coated with green tea extract and/or Pd did not show obvious signs of surface oxidation (Figure 5.21b–d), indicated by the presence of an orange rust colour (Figure 5.21a). As a demonstration of the usefulness of the "green", immobilized Fe^0 nanoparticles in real world applications, they were used to effectively dechlorinate TCE in contaminated groundwater from a Superfund site.

5.4.1.2.4   Model Development for Catalytic Dechlorination of PCB's using Fe/Pd Nanoparticles Immobilized in Tunable Membranes Operated Under Convective Flow. While measuring the reactive properties of these responsive materials is of the utmost importance, transport models of such systems are convenient and important tools for predicting a membrane's behaviour through changes in operating and design conditions, which include tunable membrane properties, solution flux modulation, and catalyst loading and composition. By developing a two-dimensional reactor model based on the convective

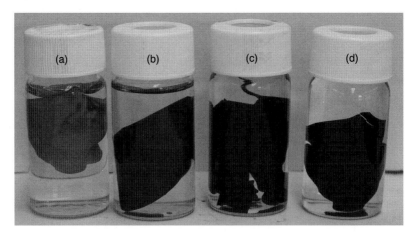

Figure 5.21 *PVDF-PAA membranes with immobilized nanoparticles after use for TCE dechlorination. The types of nanoparticles, reducing agent, and reaction times are given as follows: (a) Fe⁰, sodium borohydride, 5 h; (b) Fe/Pd, sodium borohydride, 2 h; (c) Fe⁰, tea extract, 23 h; (d) Fe/Pd, tea extract, 23 h. (Reprinted with permission from [130] Copyright (2011) Elsevier Ltd). See plate section for colour figure.*

flow through the membrane pores, Xu and Bhattacharyya [127] were able to predict PCB dechlorination using PVDF-PAA-Fe/Pd membranes operated in convective flow mode.

By controlling the residence time, through pressure modulation and PVDF/PAA pore diameter (Figure 5.22a), and tuning the catalytic properties of the Fe/Pd nanoparticles (Figure 5.22b), the authors were able to adjust the rate of DiCB dechlorination.

The key assumptions used for this model (see Figure 5.9a) include: the uniform distribution of Fe/Pd nanoparticles throughout the PAA domain, laminar flow in the core of the membrane pore, uniform and cylindrical pores. For the transport of PCBs in the two domains: (i) core domain – diffusion and convection in axial direction, no convection in the radial direction and (ii) PAA domain – no convective transport. As the PCB solution permeates through the membrane pores, the PCBs will diffuse into the PAA hydrogel layer containing the Fe/Pd nanoparticles and react with the Fe/Pd nanoparticles.

Operating under these assumptions, the governing equations for this model consist of balances on PCBs in the open "core" portion of the membrane pore (Equation 5.15) as well as in the PAA gel (Equation 5.16).

$$-U(r)\frac{\partial[PCB]}{\partial z} + D_{PCB,aq}\left(\frac{1}{r}\frac{\partial}{\partial r}\left(r\frac{\partial[PCB]}{\partial r}\right) + \frac{\partial^2[PCB]}{\partial z^2}\right) = 0 \qquad (5.15)$$

$$D_{PCB,PAA}\left(\frac{\partial^2[PCB]}{\partial^2 z} + \frac{1}{r}\frac{\partial}{\partial r}\left(r \times \frac{\partial[PCB]}{\partial r}\right)\right) + r_{PCB} = 0 \qquad (5.16)$$

where $U(r)$ (m s^{-1}) is the velocity profile of the solution permeating through the membrane pore, $[PCB]$ (mol m^{-3}) is the concentration of PCBs, $D_{PCB,aq}$ is the diffusion coefficient of PCBs in water and $D_{PCB,PAA}$ is the diffusion coefficient of PCBs in the hydrated PAA gel, and r_{PCB} is the rate law governing the decomposition of PCBs by the Fe/Pd nanoparticles.

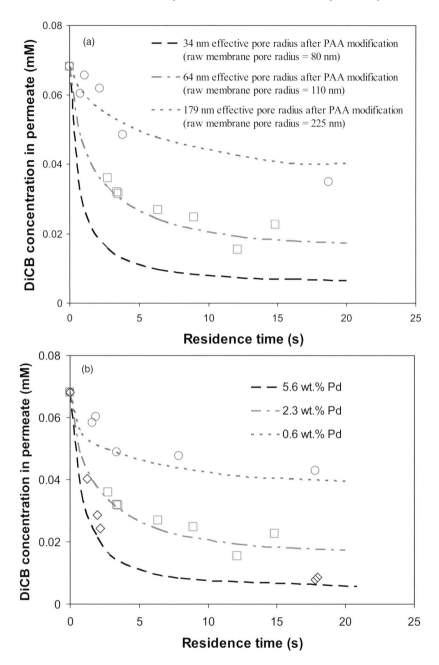

Figure 5.22 *Dechlorination of DiCB using PVDF-PAA-Fe/Pd membrane operated under convective flow with varying residence times. Symbols represent experimental data and lines represent model-predicted concentrations. Fe/Pd nanoparticle loading in the PAA layer was 0.15 g cm^{-3}. (a) Effect of tuning pore radius on DiCB conversion with 2.3%wt. Pd Fe/Pd nanoparticles. (b) Effect of tuning the Pd loading on the Fe/Pd nanoparticles in membranes with 64 nm effective pore radius. (Reprinted with permission from [127] Copyright (2008) American Chemical Society).*

This reaction term can be expressed using Equation (5.17).

$$r_{PCB} = -k_{in}\rho a_S[PCB] \tag{5.17}$$

where k_{in} is the surface area normalized reaction rate constant (L m^{-2} s^{-1}), ρ is the density of the Fe/Pd nanoparticles in the PAA layer (g L^{-1}), and a_s specific surface area of the Fe/Pd nanoparticles (m^2 g^{-1}).

U(r) is a function of the volumetric solution membrane flux, J_V (m^3 m^{-2} s^{-1}), which can be controlled by varying transmembrane pressure or pore dimensions through responsive behaviour, and was calculated using Equation (5.18) as follows:

$$U(r) = 2 \times U_{avg}\left(1 - \left(\frac{r}{R_1}\right)^2\right) = 2\frac{J_V}{\varepsilon}\left(1 - \left(\frac{r}{R_1}\right)^2\right) \tag{5.18}$$

where U_{avg} is the average fluid velocity through the membrane pore (m s^{-1}), ε is the membrane porosity (dimensionless), and R_1 is the average pore radius of the PVDF-PAA membrane (m). R_1 was calculated using the Ferry–Faxen equation and the observed rejection of dextran [127, 137, 138].

The appropriate boundary conditions were developed for Equations (5.15) and (5.16) including that at $r = R_1$,

$$D_{PCB,aq}\frac{\partial[PCB]}{\partial r} = k_{PCB}([PCB]_2 - H[PCB]_1)$$

where $[PCB]_1$ is the PCB concentration in the pore at R_1 (mM), $[PCB]_2$ is the PCB concentration in the PAA gel (mM), and H is the PCB partitioning coefficient (dimensionless, determined experimentally). k_{PCB} is the mass transfer coefficient determined for membrane filtration at laminar flow using Sherwood number correlations [127, 139, 140–142].

Equations (5.15) and (5.16) were combined and solved with the boundary conditions using multiphysics modelling and simulation software to obtain the PCB and biphenyl concentration profiles as a function of distance along the z-axis. Formation of the intermediate, 2-chlorobiphenyl, was ignored since it only appeared in trace quantities. Through this model development, the authors were able to characterize the convective flow membrane reactor by the membrane permeability, pore diameter, thickness, and porosity, the nanoparticle properties and composition, and the bulk solution and PCB properties.

5.4.1.2.5 Thermosensitive Hydrogel Networks for Metallic Nanoparticle Catalysis. Although most of this section has focused on the use of PAA hydrogels synthesized within membrane supports, unsupported hydrogels have also been used for the immobilization and stabilization of metal nanoparticles. A variety of ligands and coatings have been used to reduce the agglomeration of metallic nanoparticles; however, many of these surface coatings may reduce nanoparticle reactivity. The customizable properties and phase change behaviour of hydrogels could help maintain nanoparticle reactivity while preventing agglomeration. Also, by incorporating the catalytic particles into the hydrogel structure, they are easier to separate from the reaction media and can even be used as nanoreactors [143]. For instance, Pich *et al.* [163] synthesized Ag nanoparticles with tunable size within a thermosensitive poly[vinylcaprolactam-co-(acetoacetoxyethyl methacrylate)] hydrogel network for the reduction of 4-nitrophenol. When compared to traditional aqueous systems, the hydrogels offered the advantages of tunability and an increased reaction rate.

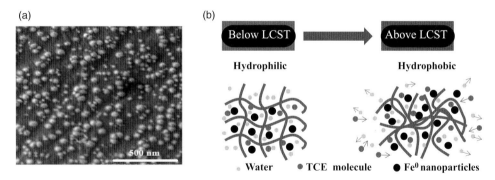

Figure 5.23 *(a) Fe/Pd nanoparticles (~40 nm diameter) formed via in situ synthesis with borohydride in poly(NIPAAm) hydrogel. (b) Schematic of water and TCE (a hydrophobic water contaminant) partitioning above and below the LCST of poly(NIPAAm). See plate section for colour figure.*

Poly(NIPAAm) has been used to immobilize Fe/Pd nanoparticles via either encapsulation or *in situ* synthesis (Figure 5.23a) and provides a unique environment for the degradation of toxic organic compounds such as TCE. Below the LCST of poly(NIPAAm), the hydrogel is in its swollen, hydrophilic state, making the Fe/Pd nanoparticles completely accessible to TCE, as shown in Figure 5.23b. Above the LCST, the hydrogel is in its collapsed, hydrophobic state, which, although it decreases the porosity of the hydrogel, creates a more favourable environment for the hydrophobic TCE. Table 5.2 shows the surface area normalized rate constants for the dechlorination of TCE above and below the LCST of poly(NIPAAm) using the hydrogel-Fe/Pd systems and conventional Fe/Pd nanoparticles stabilized in solution. The nanoparticle immobilized hydrogel system indicated three times increase in reactivity from 30°C to 34°C (compared to only two times increase with Fe/Pd in solution phase for the same temperature change) through control of the swelling and collapse of the hydrogel by thermal transition (below and above LCST). This demonstrates the importance of the increase of pollutant partitioning on reactivity. The development of reactive polymer hydrogel with nanoparticles and controlled partitioning should lead to applications ranging from organic synthesis to pollution control.

5.4.2   Responsive Membranes for Enzymatic Catalysis

In addition to using stimuli-responsive materials for heavy metal capture and subsequent reactions, membranes with a net charge have been used to electrostatically immobilize enzymes for conducting catalytic reactions [144]. Although enzyme immobilization

Table 5.2 *Surface area normalized rate constants (k_{SA}) for hydrogel-immobilized and conventional solution phase Fe/Pd nanoparticles below and above the LCST of p(NIPAAm).*

Reaction system	K_{SA} (L h^{-1} m^{-2})	R^2
Nano Fe/Pd in solution phase at 30°C	0.0275	0.9783
Hydrogel immobilized Fe/Pd nanoparticles at 30°C	0.0156	0.9799
Nano Fe/Pd with CMC in solution phase at 34°C	0.0568	0.9437
Hydrogel immobilized Fe/Pd nanoparticles at 34°C	0.0411	0.956

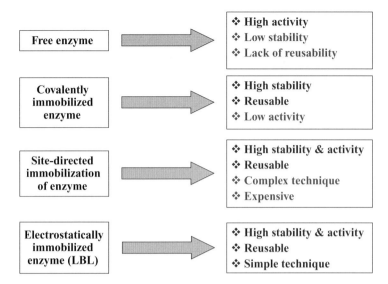

Figure 5.24   *Advantages and disadvantages of enzyme catalysis using free and immobilized enzymes.*

generally lowers activity, the use of free enzymes in the homogeneous phase is restricted due to lower stability and product inhibition [145]. Methods for enzyme immobilization include covalent, site-directed, and electrostatic attachment; a summary of the advantages and disadvantages of each is shown in Figure 5.24 [146, 147].

The chosen method of immobilization can have a significant impact on the sensitivity and behaviour of many types of responsive materials, an important class of which is biosensors. These materials rely on the integration of a molecular recognition element (usually a biofunctional membrane) and a transducer into one device in order to detect a target analyte in the surrounding environment [148]. The molecular recognition element can be either catalytic or affinity-based, with enzymes constituting a large portion of the former [148]. Glucose biosensors are a classic example of such responsive materials, first described by Clark and Lyons in 1962 [160], which utilize glucose oxidase (GOx) coupled with a potentiometric measurement to determine blood glucose levels. A recent and major advancement of this technology is the ability to immobilize both an enzyme and redox mediator in close proximity on thin films, providing the opportunity to use such sensors *in vivo* [149]. Electrostatic LbL deposition provides a very thin and flexible platform capable of incorporating a variety of responsive materials such as polyelectrolytes, biomolecules, and carbon nanotubes for improved sensor performance [150, 151].

Membranes have been used to support enzyme immobilization on both their exterior surfaces and throughout their porous structure [67]. LbL assembly of polyelectrolytes for enzyme immobilization is a very simple technique that provides increased enzyme stability similar to that of covalent attachment, but with less loss of activity [152]. For instance, the activity of GOx covalently attached to a membrane was 31–34% of free enzyme whereas the activity of GOx electrostatically immobilized on a membrane was 65–76% (Figure 5.25) [153, 154].

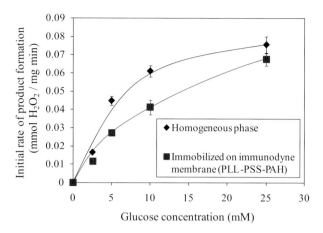

Figure 5.25 *Activity of homogeneous and membrane immobilized GOx measured by H_2O_2 generation at pH 6.9. An immunodyne membrane with covalently attached PLL and LbL assembly of PSS/PAH layers was used to electrostatically immobilize 0.699 mg of GOx within the membrane pores (data adapted from [153]).*

Though many studies construct a LbL assembly to support one layer of enzymes, Lu and Hsieh [161] demonstrated how membranes containing multiple layers of enzymes can be formed to increase loading while maintaining significant enzymatic activity. Alternating layers of Cibacron Blue F3GA (CB) dye and lipase were electrostatically immobilized on the surface of an ultra-fine cellulose fibrous membrane in a series of bilayers, each of which was only 10–12 nm thick. The lipase activity as a percentage of free enzyme activity was 82, 74, 70, 63, and 45% for 1, 2, 3, 4, and 5 bilayers, respectively. Despite this drop in activity, using multiple bilayers resulted in increases in total lipase loading and total catalytic activity.

Though a LbL assembly consisting of multiple layers of polyelectrolytes is an excellent platform for enzyme immobilization, covalent attachment of the first layer may be necessary in order to impart greater stability to the assembly (Figure 5.26a). However, covalent attachment is far from trivial for many inexpensive and stable support membranes which may be desirable for use in these systems. Creating a polyelectrolyte gel via *in situ* polymerization is a flexible and simple technique that can be used to immobilize a variety of functional groups within the existing structure of the support membrane (as with the previously mentioned PVDF-PAA membranes). This eliminates the need for specific functional groups on the surface of the support membrane and subsequent polyelectrolyte layers can be added as desired (Figure 5.26b).

5.4.2.1 Tunable Stacked Membrane Systems

As mentioned earlier, a distinct advantage of using responsive membranes over other conventional materials with high specific surface areas is the ability to operate them under convective flow. Additionally, the versatility of membrane systems allows for the use of multiple responsive/catalytic membranes in series. Lewis *et al.* [114] and Smuleac *et al.*

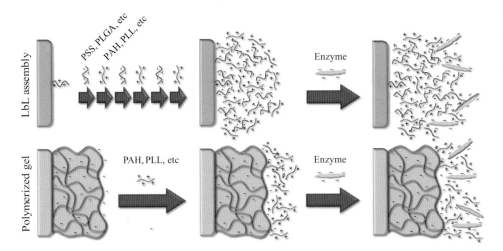

Figure 5.26 *Schematic of two easily-synthesized, charged membrane platforms for the electrostatic immobilization of a negatively charged enzyme. (a) LbL assembly of multiple positively and negatively charged polyelectrolytes on a base membrane with covalently attached positively charged polyelectrolyte. (b) Negatively charged gel polymerized within the membrane pore with a layer of positively charged polyelectrolytes.*

[155] developed nanostructured, tunable stacked membrane systems for the degradation of toxic organic compounds and greener chemical synthesis through the addition of an inexpensive and abundant substrate, glucose.

One can easily synthesize bifunctional, two-stack catalytic membranes to obtain high selectivity and product yield with minimal separation needs. This principle was described in the literature [155], with two model enzymes: glucose oxidase is immobilized on the top membrane and catalase on the bottom one. For this particular system glucose solution and air was supplied in the feed and the product resulting from the reaction in the first membrane (H_2O_2) becomes the reactant in the second one. Thus, a high value chemical: gluconic acid, commonly used as Ca or Fe gluconate (nutrition supplement) and pure oxygen are obtained without additional separation steps. This synthesis protocol of creating highly enriched oxygen for medical and other energy-related applications (under low energy consumption), provides a dramatically different approach compared to conventional air separation processes.

For the organic degradation case (Figure 5.27), the first membrane contains the enzyme GOx electrostatically immobilized in a polycation/polyanion LbL assembly (Figure 5.27a) which generates H_2O_2 from the added glucose. Pressure-induced convective flow then passes the solution to the second membrane (Figure 5.27b), PVDF-PAA, which contains some form of reactive iron species (ions and/or iron oxide nanoparticles) immobilized within the pH-responsive PAA layer. Here, H_2O_2 reacts with the iron species to generate free radicals which react with and degrade the target contaminants. In addition to pressure modulation, the responsive behaviour of the PAA gel provides another route for hydrodynamic control of residence time.

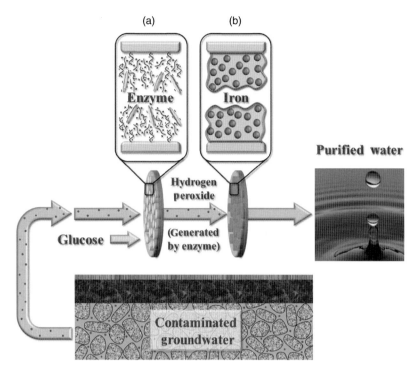

Figure 5.27 *Schematic of tunable, reactive nanostructured stacked membrane system for the degradation of toxic organic contaminants from water. The contaminated water is convectively permeated through the first membrane (a) with added glucose, where the immobilized enzyme, GOx generates H_2O_2 from the glucose. This solution then passes to the second membrane (b), a PVDF-PAA membrane, which contains immobilized iron ions and/or iron oxide nanoparticles that react with H_2O_2 to form free radicals for contaminant degradation.*

In order to demonstrate the effectiveness of this system for contaminant degradation, the authors permeated an oxygen-saturated solution containing trichlorophenol (TCP) and glucose through the membrane stack (Figure 5.28a). The first membrane consisted of a base membrane of regenerated cellulose (RC) with covalently attached poly(L-lysine-hydrochloride) (PLL), three alternating layers of poly(styrene sulfonate) (PSS) and PAH (RC-PLL-(PSS-PAH)$_3$ or RC-LbL), and electrostatically-immobilized GOx. The second membrane was PVDF-PAA with immobilized Fe(II) ions. Initially, the conversion of TCP was 100%, but decreased to 55–70% after 30 min at only 2 s residence time (Figure 5.28b). This decrease was expected since the Fe(II) was being converted to Fe(III), which does not react as quickly with H_2O_2, resulting in a decrease in the rate of free radical production and hence TCP conversion [114]. Since TCP contains three chlorine atoms, they will be released as chloride ions as the degradation of the parent compound proceeds. Though the conversion of TCP decreased with time, the mol of Cl$^-$ formed per mol of TCP degraded was approximately 2 for the entire study, with unaccounted Cl$^-$ most likely present as chlorinated intermediates. Additionally, the conversion of TCP can be adjusted by changing the residence time, iron loading, and/or rate of H_2O_2 generation.

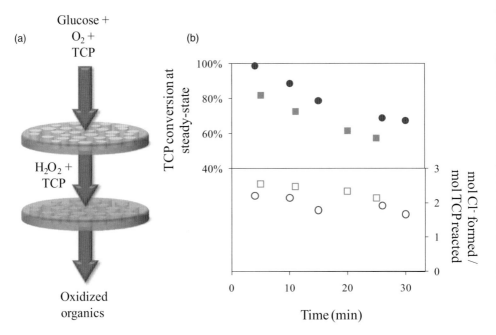

Figure 5.28   *Use of a tunable, nanostructured membrane stack for the degradation of a toxic organic compound, TCP, with the addition of glucose. (a) Schematic of membrane stack operated under convective flow and consisting of a RC-LbL-GOx membrane on a PVDF-PAA-Fe(II) membrane. (b) Steady-state TCP conversion and Cl⁻ formation as a function of clock time. Residence time in first and second membranes = 2.7 and 2 s, respectively. Squares: 0.142 mmol/L TCP, [H₂O₂] = 0.13 mM; circles: 0.076 mmol/L TCP, [H₂O₂] = 0.10 mM. Fe loading = 0.09 mmol, pH 5.5. (Reprinted with permission from [114] Copyright (2011) PNAS).*

Acknowledgements

This work was funded by NIEHS-SRP (Grant number: P42ES007380), DOE-KRCEE, US EPA, and NSF-IGERT programmes. The authors would like to acknowledge the contributions of Drs. S. Sikdar, A. Hollman, J. Xu, and S. Datta.

References

1. Klein, E. (2000) Affinity membranes: a 10-year review. *Journal of Membrane Science*, **179**(1–2), 1–27.
2. Ulbricht, M. (2006) Advanced functional polymer membranes. *Polymer*, **47**(7), 2217–2262.
3. Derwinska, K., Sauer, U., and Preininger, C. (2008) Adsorption versus covalent, statistically oriented and covalent, site-specific IgG immobilization on poly(vinyl alcohol)-based surfaces. *Talanta*, **77**(2), 652–658.

4. Hestekin, J.A., Bachas, L.G., and Bhattacharyya, D. (2001) Poly(amino acid)-functionalized cellulosic membranes: Metal sorption mechanisms and results. *Industrial & Engineering Chemistry Research*, **40**(12), 2668–2678.

5. Boi, C., Cattoli, F., Facchini, R. *et al.* (2006) Adsorption of lectins on affinity membranes. *Journal of Membrane Science*, **273**(1–2), 12–19.

6. Shi, W., Zhang, F.B., and Zhang, G.L. (2005) Adsorption of bilirubin with polylysine carrying chitosan-coated nylon affinity membranes. *Journal of Chromatography B-Analytical Technologies in the Biomedical and Life Sciences*, **819**(2), 301–306.

7. Liu, Y.C. Suen, S.Y., Huang, C.W., *et al.* (2005) Effects of spacer arm on penicillin G acylase purification using immobilized metal affinity membranes. *Journal of Membrane Science*, **251**(1–2), 201–207.

8. Rebek, J. and Feitler, D. (1974) Mechanism of carbodiimide reaction. 2. Peptide-synthesis on solid-phase. *Journal of the American Chemical Society*, **96**(5), 1606–1607.

9. Ulbricht, M. and Papra, A. (1997) Polyacrylonitrile enzyme ultrafiltration membranes prepared by adsorption, cross-linking, and covalent binding. *Enzyme and Microbial Technology*, **20**(1), 61–68.

10. Neame, P.J. and Parikh, I. (1982) Sepharose-immobilized triazine dyes as adsorbants for human-lymphoblastoid interferon purification. *Applied Biochemistry and Biotechnology*, **7**(4), 295–305.

11. Mcconway, M.G. and Chapman, R.S. (1986) Application of solid-phase antibodies to radioimmunoassay - evaluation of 2 polymeric microparticles, dynospheres and nylon, activated by carbonyldiimidazole or tresyl chloride. *Journal of Immunological Methods*, **95**(2), 259–266.

12. Klein, E. (1991) *Affinity Membranes: Their Chemistry and Performance in Adsorptive Separation Processes*, Wiley, New York, x, 152 p.

13. Blitz, J.P., Murthy, R.S.S., and Leyden, D.E. (1987) Ammonia-catalyzed silylation reactions of Cab-O-Sil with methoxymethylsilanes. *Journal of the American Chemical Society*, **109**(23), 7141–7145.

14. Sindorf, D.W. and Maciel, G.E. (1983) Solid-state Nmr-studies of the reactions of silica surfaces with polyfunctional chloromethylsilanes and ethoxymethylsilanes. *Journal of the American Chemical Society*, **105**(12), 3767–3776.

15. Zeigler, R.C. and Maciel, G.E. (1991) A study of the structure and dynamics of dimethyloctadecylsilyl-modified silica using wide-line H-2-Nmr techniques. *Journal of the American Chemical Society*, **113**(17), 6349–6358.

16. Tripp, C.P. and Hair, M.L. (1992) An infrared study of the reaction of octadecyl-trichlorosilane with silica. *Langmuir*, **8**(4), 1120–1126.

17. Silberzan, P. Léger, D., Ausserré, J. *et al.* (1991) Silanation of silica surfaces - a new method of constructing pure or mixed monolayers. *Langmuir*, **7**(8), 1647–1651.

18. Ritchie, S.M.C. Kissick, K.E., Bachas, L.G., *et al.* (2001) Polycysteine and other polyamino acid functionalized microfiltration membranes for heavy metal capture. *Environmental Science & Technology*, **35**(15), 3252–3258.

19. Smuleac, V., Buttereld, D.A., Sikdar, S.K. *et al.* (2005) Polythiol-functionalized alumina membranes for mercury capture. *Journal of Membrane Science*, **251**(1–2), 169–178.

20. Hulteen, J.C., Jirage, K.B., and Martin, C.R. (1998) Introducing chemical transport selectivity into gold nanotubule membranes. *Journal of the American Chemical Society*, **120**(26), 6603–6604.

21. Jirage, K.B., Hulteen, J.C., and Martin, C.R. (1999) Effect of thiol chemisorption on the transport properties of gold nanotubule membranes. *Analytical Chemistry*, **71**(21), 4913–4918.

22. Martin, C.R. and Gasparac, R. (2001) Investigations of the transport properties of gold nanotubule membranes. *Journal of Physical Chemistry B*, **105**(10), 1925–1934.

23. Chun, K.Y. and Stroeve, P. (2001) External control of ion transport in nanoporous membranes with surfaces modified with self-assembled monolayers. *Langmuir*, **17**(17), 5271–5275.

24. Hou, Z.Z., Abbott, N.L., and Stroeve, P. (2000) Self-assembled monolayers on electroless gold impart pH-responsive transport of ions in porous membranes. *Langmuir*, **16**(5), 2401–2404.

25. Kang, M.S. and Martin, C.R. (2001) Investigations of potential-dependent fluxes of ionic permeates in gold nanotubule membranes prepared via the template method. *Langmuir*, **17**(9), 2753–2759.

26. Chun, K.Y. and Stroeve, P. (2002) Protein transport in nanoporous membranes modified with self-assembled monolayers of functionalized thiols. *Langmuir*, **18**(12), 4653–4658.

27. Ku, J.R. and Stroeve, P. (2004) Protein diffusion in charged nanotubes: "On-Off" behavior of molecular transport. *Langmuir*, **20**(5), 2030–2032.

28. Yu, S.F. Sang Bok, L., Kang, M. *et al.* (2001) Size-based protein separations in poly(ethylene glycol)-derivatized gold nanotubule membranes. *Nano Letters*, **1**(9), 495–498.

29. Ito, Y., Park, Y.S., and Imanishi, Y. (2000) Nanometer-sized channel gating by a self-assembled polypeptide brush. *Langmuir*, **16**(12), 5376–5381.

30. Smuleac, V., Butterfield, D.A., and Bhattacharyya, D. (2004) Permeability and separation characteristics of polypeptide-functionalized polycarbonate track-etched membranes. *Chemistry of Materials*, **16**(14), 2762–2771.

31. Alvarez, S., Garcia, A., Manolache, S. *et al.* (2005) Plasma-enhanced modification of the pore size of ceramic membranes. *Desalination*, **184**(1–3), 99–104.

32. Greene, G. and Tannenbaum, R. (2004) Adsorbtion of polyelectrolyte multilayers on plasma-modified porous polyethylene. *Applied Surface Science*, **238**(1–4), 101–107.

33. Kull, K.R., Steen, M.L., and Fisher, E.R. (2005) Surface modification with nitrogen-containing plasmas to produce hydrophilic, low-fouling membranes. *Journal of Membrane Science*, **246**(2), 203–215.

34. Steen, M.L., Hymas, L., Havey, E.D. *et al.* (2001) Low temperature plasma treatment of asymmetric polysulfone membranes for permanent hydrophilic surface modification. *Journal of Membrane Science*, **188**(1), 97–114.

35. He, D.M., Susanto, H., and Ulbricht, M. (2009) Photo-irradiation for preparation, modification and stimulation of polymeric membranes. *Progress in Polymer Science*, **34**(1), 62–98.

36. Freger, V., Gilron, J., and Belfer, S. (2002) TFC polyamide membranes modified by grafting of hydrophilic polymers: an FT-IR/AFM/TEM study. *Journal of Membrane Science*, **209**(1), 283–292.

37. Himstedt, H.H., Marshall, K.M., and Wickramasinghe, S.R. (2011) pH-responsive nanofiltration membranes by surface modification. *Journal of Membrane Science*, **366**(1–2), 373–381.
38. Meier-Haack, J., Booker, N.A., and Carroll, T. (2003) A permeability-controlled microfiltration membrane for reduced fouling in drinking water treatment. *Water Research*, **37**(3), 585–588.
39. Ulbricht, M. and Yang, H. (2005) Porous polypropylene membranes with different carboxyl polymer brush layers for reversible protein binding via surface-initiated graft copolymerization. *Chemistry of Materials*, **17**(10), 2622–2631.
40. Higa, M. and Yamakawa, T. (2004) Design and preparation of a novel temperature-responsive ionic gel. 1. A fast and reversible temperature response in the charge density. *Journal of Physical Chemistry B*, **108**(43), 16703–16707.
41. Higa, M. and Yamakawa, T. (2005) Design and preparation of a novel temperature-responsive ionic gel. 2. Concentration modulation of specific ions in response to temperature changes. *Journal of Physical Chemistry B*, **109**(22), 11373–11378.
42. Mika, A.M., Childs, R.F., and Dickson, J.M. (2002) Salt separation and hydrodynamic permeability of a porous membrane filled with pH-sensitive gel. *Journal of Membrane Science*, **206**(1–2), 19–30.
43. Hu, K. and Dickson, J.M. (2007) Development and characterization of poly(vinylidene fluoride)-poly(acrylic acid) pore-filled pH-sensitive membranes. *Journal of Membrane Science*, **301**(1–2), 19–28.
44. Lewis, S., Smuleac, V., Montague, A. *et al.* (2009) Iron-functionalized membranes for nanoparticle synthesis and reactions. *Separation Science and Technology*, **44**(14), 3289–3311.
45. Smuleac, V., Bachas, L., and Bhattacharyya, D. (2010) Aqueous-phase synthesis of PAA in PVDF membrane pores for nanoparticle synthesis and dichlorobiphenyl degradation. *Journal of Membrane Science*, **346**(2), 310–317.
46. Xu, J. and Bhattacharyya, D. (2007) Fe/Pd nanoparticle immobilization in microfiltration membrane pores: Synthesis, characterization, and application in the dechlorination of polychlorinated biphenyls. *Industrial & Engineering Chemistry Research*, **46**(8), 2348–2359.
47. Ilmain, F., Tanaka, T., and Kokufuta, E. (1991), Volume transition in a gel driven by hydrogen-bonding. *Nature*, **349**(6308), 400–401.
48. Lahann, J. and Yoshida, M. (2008) Smart nanomaterials. *Acs Nano*, **2**(6), 1101–1107.
49. Fernandez-Barbero, A. *et al.* (2009) Gels and microgels for nanotechnological applications. *Advances in Colloid and Interface Science*, **147–148**, 88–108.
50. Stuart, M.A.C., Huck, W., Genzer, J. *et al.* (2010) Emerging applications of stimuli-responsive polymer materials. *Nature Materials*, **9**(2), 101–113.
51. Kuckling, D. (2009) Responsive hydrogel layers-from synthesis to applications. *Colloid and Polymer Science*, **287**(8), 881–891.
52. Schueneman, S.M. and Chen, W. (2002), Environmentally responsive hydrogels. *Journal of Chemical Education*, **79**(7), 860–862.
53. West, J.L. and Strong, L.E. (2011) Thermally responsive polymer-nanoparticle composites for biomedical applications. *Wiley Interdisciplinary Reviews-Nanomedicine and Nanobiotechnology*, **3**(3), 307–317.

54. Milasinovic, N., Kalagasidis Krušić, M., Knežević-Jugović, Z. *et al.* (2010) Hydrogels of N-isopropylacrylamide copolymers with controlled release of a model protein. *International Journal of Pharmaceutics*, **383**(1–2), 53–61.

55. Schild, H.G. (1992) Poly (N-isopropylacrylamide) - experiment, theory and application. *Progress in Polymer Science*, **17**(2), 163–249.

56. Makino, K., Hiyoshi, J., and Ohshima, H. (2000) Kinetics of swelling and shrinking of poly (N-isopropylacrylamide) hydrogels at different temperatures. *Colloids and Surfaces B-Biointerfaces*, **19**(2), 197–204.

57. Gupta, P., Vermani, K., and Garg, S. (2002) Hydrogels: from controlled release to pH-responsive drug delivery. *Drug Discovery Today*, **7**(10), 569–579.

58. Dong, L.C. and Hoffman, A.S. (1986) Thermally reversible hydrogels: III. Immobilization of enzymes for feedback reaction control. *Journal of Controlled Release*, **4**(3), 223–227.

59. Kanazawa, H. and Okano, T. (2011) Temperature-responsive chromatography for the separation of biomolecules. *Journal of Chromatography A*, **1218**, (49), 8738–8747.

60. Gan, D.J. and Lyon, L.A. (2002) Synthesis and protein adsorption resistance of PEG-modified poly(N-isopropylacrylamide) core/shell microgels. *Macromolecules*, **35**(26), 9634–9639.

61. Qiu, Y. and Park, K. (2001) Environment-sensitive hydrogels for drug delivery. *Advanced Drug Delivery Reviews*, **53**(3), 321–339.

62. Decher, G. (1997) Fuzzy nanoassemblies: Toward layered polymeric multicomposites. *Science*, **277**(5330), 1232–1237.

63. Decher, G., Hong, J.D., and Schmitt, J. (1992) Buildup of ultrathin multilayer films by a self-assembly process. 3. Consecutively alternating adsorption of anionic and cationic polyelectrolytes on charged surfaces. *Thin Solid Films*, **210**(1–2), 831–835.

64. Lvov, Y., Yamada, S., and Kunitake, T. (1997) Non-linear optical effects in layer-by-layer alternate films of polycations and an azobenzene-containing polyanion. *Thin Solid Films*, **300**(1–2), 107–112.

65. Shimazaki, Y., Ito, S., and Tsutsumi, N. (2000) Adsorption-induced second harmonic generation from the layer-by-layer deposited ultrathin film based on the charge-transfer interaction. *Langmuir*, **16**(24), 9478–9482.

66. Zheng, H.P., Lee, I., Rubner, M.F *et al.* (2002) Two component particle arrays on patterned polyelectrolyte multilayer templates. *Advanced Materials*, **14**(8), 569–572.

67. Bruening, M.L., Dotzauer, D. M., Jain, P. *et al.* (2008) Creation of functional membranes using polyelectrolyte multilayers and polymer brushes. *Langmuir*, **24**(15), 7663–7673.

68. Krasemann, L. and Tieke, B. (1998) Ultrathin self-assembled polyelectrolyte membranes for pervaporation. *Journal of Membrane Science*, **150**(1), 23–30.

69. Liu, X.Y. and Bruening, M.L. (2004) Size-selective transport of uncharged solutes through multilayer polyelectrolyte membranes. *Chemistry of Materials*, **16**(2), 351–357.

70. Malaisamy, R. and Bruening, M.L. (2005) High-flux nanofiltration membranes prepared by adsorption of multilayer polyelectrolyte membranes on polymeric supports. *Langmuir*, **21**(23), 10587–10592.

71. Jin, W.Q., Toutianoush, A. and Tieke, B. (2003) Use of polyelectrolyte layer-by-layer assemblies as nanofiltration and reverse osmosis membranes. *Langmuir*, **19**(7), 2550–2553.

72. Tarabara, V.V.,Wang, F., Davis, T.E. *et al.* (2010) Polyelectrolyte multilayer films as backflushable nanofiltration membranes with tunable hydrophilicity and surface charge. *Journal of Membrane Science*, **349**(1–2), 268–278.

73. Wu, Z.Y., Guan, L., Shen, G. *et al.* (2002) Renewable urea sensor based on a self-assembled polyelectrolyte layer. *Analyst*, **127**(3), 391–395.

74. Forzani, E.S., Solis, V.M., and Calvo, E.J. (2000) Electrochemical behavior of polyphenol oxidase immobilized in self-assembled structures layer by layer with cationic polyallylamine. *Analytical Chemistry*, **72**(21), 5300–5307.

75. Dotzauer, D.M., Abusaloua, A., Miachon, S. *et al.* (2009) Wet air oxidation with tubular ceramic membranes modified with polyelectrolyte/Pt nanoparticle films. *Applied Catalysis B: Environmental*, **91**(1–2), 180–188.

76. Zimnitsky, D., Shevchenko, V.V., and Tsukruk, V.V. (2008) Perforated, freely suspended layer-by-layer nanoscale membranes. *Langmuir*, **24**(12), 5996–6006.

77. Wandera, D., Wickramasinghe, S.R., and Husson, S.M. (2010) Stimuli-responsive membranes. *Journal of Membrane Science*, **357**(1–2), 6–35.

78. Zhang, H.J. and Ito, Y. (2001) pH control of transport through a porous membrane self-assembled with a poly(acrylic acid) loop brush. *Langmuir*, **17**(26), 8336–8340.

79. Ito, Y., Park, Y.S., and Imanishi, Y. (1997) Visualization of critical pH-controlled gating of a porous membrane grafted with polyelectrolyte brushes. *Journal of the American Chemical Society*, **119**(11), 2739–2740.

80. Hu, K. and Dickson, J.M. (2009) In vitro investigation of potential application of pH-sensitive poly(vinylidene fluoride)-poly(acrylic acid) pore-filled membranes for controlled drug release in ruminant animals. *Journal of Membrane Science*, **337**(1–2), 9–16.

81. Ito, Y., Kotera, S., Inaba, M. *et al.* (1990) Control of pore-size of polycarbonate membrane with straight pores by poly(acrylic acid) grafts. *Polymer*, **31**(11), 2157–2161.

82. Hollman, A.M., Scherrer, N.T., Cammers-Goodwin, A. *et al.* (2004) Separation of dilute electrolytes in poly(amino acid) functionalized microporous membranes: model evaluation and experimental results. *Journal of Membrane Science*, **239**(1), 65–79.

83. Hollman, A.M. and Bhattacharyya, D. (2002) Controlled permeability and ion exclusion in microporous membranes functionalized with poly(L-glutamic acid). *Langmuir*, **18**(15), 5946–5952.

84. Castro, R.P., Monbouquette, H.G., and Cohen, Y. (2000) Shear-induced permeability changes in a polymer grafted silica membrane. *Journal of Membrane Science*, **179**(1–2), 207–220.

85. Bowen, W.R. and Mukhtar, H. (1996) Characterisation and prediction of separation performance of nanofiltration membranes. *Journal of Membrane Science*, **112**(2), 263–274.

86. Hollman, A.M. and Bhattacharyya, D. (2004) Pore-assembled multilayers of charged polypeptides in microporous membranes for ion separation. *Langmuir*, **20**(13), 5418–5424.

87. Choi, S.H., Han Jeong, Y., Jeong Ryoo, J. *et al.* (2001) Desalination by electrodialysis with the ion-exchange membrane prepared by radiation-induced graft polymerization. *Radiation Physics and Chemistry*, **60**(4–5), 503–511.

88. Ritchie, S.M.C., Bhattacharyya, D., Bachas, L.G. *et al.* (1999) Surface modification of silica- and cellulose-based microfiltration membranes with functional polyamino acids for heavy metal sorption. *Langmuir*, **15**(19), 6346–6357.

89. Konishi, S., Saito, K., Furusaki S. *et al.* (1992) Sorption kinetics of cobalt in chelating porous membrane. *Industrial & Engineering Chemistry Research*, **31**(12), 2722–2727.

90. Choi, S.H. and Nho, Y.C. (1999) Adsorption of Co2+ by stylene-g-polyethylene membrane bearing sulfonic acid groups modified by radiation-induced graft copolymerization. *Journal of Applied Polymer Science*, **71**(13), 2227–2235.

91. Ritchie, S.M.C. and Bhattacharyya, D. (2001) Polymeric ligand-based functionalized materials and membranes for ion exchange, in *Ion Exchange and Solvent Extraction: A Series of Advances* (eds A.K. SenGupta, J.A. Marinsky and Y. Marcus), Marcel Dekker, Inc., New York, pp. 81–118.

92. Bhattacharyya, D., Hestekin, P., Brushaber, L. *et al.* (1998) Novel poly-glutamic acid functionalized microfiltration membranes for sorption of heavy metals at high capacity. *Journal of Membrane Science*, **141**(1), 121–135.

93. Shah, T.N. and Ritchie, S.M.C. (2005) Esterification catalysis using functionalized membranes. *Applied Catalysis a-General*, **296**(1), 12–20.

94. Lu, Y. Mei, Y., Drechsler, M. *et al.* (2006) Thermosensitive core-shell particles as carriers for Ag nanoparticles: Modulating the catalytic activity by a phase transition in networks. *Angewandte Chemie-International Edition*, **45**(5), 813–816.

95. Lu, Y., Proch S., Schrinner, M. *et al.* (2009) Thermosensitive core-shell microgel as a "nanoreactor" for catalytic active metal nanoparticles. *Journal of Materials Chemistry*, **19**(23), 3955–3961.

96. Haber, F. and Weiss, J. (1934) The catalytic decomposition of hydrogern peroxide by iron salts. *Proceedings of the Royal Society*, **A147**, 332–351.

97. Barb, W.G., Baxendale, J. H., George, Philip, *et al.* (1949) Reactions of ferrous and ferric ions with hydrogen peroxide. *Nature*, **163**(4148), 692–694.

98. Watts, R.J. and Teel, A.L. (2005) Chemistry of modified Fenton's reagent (catalyzed H2O2 propagations-CHP) for in situ soil and groundwater remediation. *Journal of Environmental Engineering-ASCE*, **131**(4), 612–622.

99. Lewis, S., Lynch, A., Bachas, L. *et al.* (2009) Chelate-modified Fenton reaction for the degradation of trichloroethylene in aqueous and two-phase systems. *Environmental Engineering Science*, **26**(4), 849–859.

100. Li, Y.C., Bachas, L.G., and Bhattacharyya, D. (2007) Selected chloro-organic detoxifications by polychelate (Poly(acrylic acid)) and citrate-based Fenton reaction at neutral pH environment. *Industrial & Engineering Chemistry Research*, **46**(24), 7984–7992.

101. Lin, S.S. and Gurol, M.D. (1998) Catalytic decomposition of hydrogen peroxide on iron oxide: Kinetics, mechanism, and implications. *Environmental Science & Technology*, **32**(10), 1417–1423.

102. Benaglia, M., Puglisi, A., and Cozzi, F. (2003) Polymer-supported organic catalysts. *Chemical Reviews*, **103**(9), 3401–3429.

103. Haag, R. and Roller, S. (2004) Polymeric supports for the immobilisation of catalysts. *Immobilized Catalysts*, **242**, 1–42.

104. Kirschning, A., Monenschein, H., and Wittenberg, R. (2001) Functionalized polymers – Emerging versatile tools for solution-phase chemistry and automated parallel synthesis. *Angewandte Chemie-International Edition*, **40**(4), 650–679.

105. Feng, J.Y., Hu, X.J., and Yue, P.L. (2004) Degradation of salicylic acid by photo-assisted Fenton reaction using Fe ions on strongly acidic ion exchange resin as catalyst. *Chemical Engineering Journal*, **100**(1–3), 159–165.

106. Ma, W.H., Huang, Y., Li, J., *et al.* (2003) An efficient approach for the photodegradation of organic pollutants by immobilized iron ions at neutral pHs. *Chemical Communications*, (13), 1582–1583.

107. Ruda, T.A. and Dutta, P.K. (2005) Fenton chemistry of Fe-III-exchanged zeolitic minerals treated with antioxidants. *Environmental Science & Technology*, **39**(16), 6147–6152.

108. Neamtu, M., Zaharia, C., Catrinescu, C. *et al.* (2004) Fe-exchanged Y zeolite as catalyst for wet peroxide oxidation of reactive azo dye Procion Marine H-EXL. *Applied Catalysis B-Environmental*, **48**(4), 287–294.

109. Gonzalez-Bahamon, L.F., García, H., Gómez-García, C.J. *et al.* (2011) Photo-Fenton degradation of resorcinol mediated by catalysts based on iron species supported on polymers. *Journal of Photochemistry and Photobiology a-Chemistry*, **217**(1), 201–206.

110. Ramirez, J., Godínez, L., Méndez, M. *et al.* (2010) Heterogeneous photo-electro-Fenton process using different iron supporting materials. *Journal of Applied Electrochemistry*, **40**(10), 1729–1736.

111. Fernandez, J., Nadtochenko V., Bozzi A. *et al.* (2003) Testing and performance of immobilized Fenton photoreactions via membranes, mats, and modified copolymers. *International Journal of Photoenergy*, **5**(2), 107–113.

112. Parra, S., Henao L., Mielczarski E. *et al.* (2004) Synthesis, testing, and characterization of a novel Nafion membrane with superior performance in photoassisted immobilized Fenton catalysis. *Langmuir*, **20**(13), 5621–5629.

113. Fernandez, J., Bandara , J., Kiwi, J. *et al.* (1998) Efficient photo-assisted Fenton catalysis mediated by Fe ions on Nafion membranes active in the abatement of non-biodegradable azo-dye. *Chemical Communciations*, (14), 1493–1494.

114. Lewis, S.R., Datta, S., Gui, M. *et al.* (2011) Reactive nanostructured membranes for water purification. *Proceedings of the National Academy of Sciences of the United States of America*, **108**(21), 8577–8582.

115. Shi, F., Tse, M., Pohl, M-M. *et al.* (2008) Nano-iron oxide-catalyzed selective oxidations of alcohols and olefins with hydrogen peroxide. *Journal of Molecular Catalysis a-Chemical*, **292**(1–2), 28–35.

116. Winnik, F.M., Morneau, A., Mika, A.M. *et al.* (1998) Polyacrylic acid pore-filled microporous membranes and their use in membrane-mediated synthesis of nanocrystalline ferrihydrite. *Canadian Journal of Chemistry*, **76**(1), 10–17.

117. Zhang, W.X. (2003) Nanoscale iron particles for environmental remediation: An overview. *Journal of Nanoparticle Research*, **5**(3–4), 323–332.

118. Li, X.Q., Elliott, D.W., and Zhang, W.X. (2006) Zero-valent iron nanoparticles for abatement of environmental pollutants: Materials and engineering aspects. *Critical Reviews in Solid State and Materials Sciences*, **31**(4), 111–122.

119. Liu, Z.L., Ling, X.Y., Su, X. *et al.* (2004) Carbon-supported Pt and PtRu nanoparticles as catalysts for a direct methanol fuel cell. *Journal of Physical Chemistry B*, **108**(24), 8234–8240.

120. Mallick, K. and Scurrell, M.S. (2003) CO oxidation over gold nanoparticles supported on TiO2 and TiO2-ZnO: catalytic activity effects due to surface modification of TiO2 with ZnO. *Applied Catalysis a-General*, **253**(2), 527–536.

121. Sun, C.L., Peltre, M-J., Briend, M. *et al.* (2003) Catalysts for aromatics hydrogenation in presence of sulfur: reactivities of nanoparticles of ruthenium metal and sulfide dispersed in acidic Y zeolites. *Applied Catalysis a-General*, **245**(2), 245–255.

122. Lu, A.H., Salabas, E.L., and Schuth, F. (2007) Magnetic nanoparticles: Synthesis, protection, functionalization, and application. *Angewandte Chemie-International Edition*, **46**(8), 1222–1244.

123. He, F. and Zhao, D.Y. (2005) Preparation and characterization of a new class of starch-stabilized bimetallic nanoparticles for degradation of chlorinated hydrocarbons in water. *Environmental Science & Technology*, **39**(9), 3314–3320.

124. Meyer, D.E., Hampson, S., Ormsbee., L. *et al.* (2009) A study of groundwater matrix effects for the destruction of trichloroethylene using Fe/Pd nanoaggregates. *Environmental Progress & Sustainable Energy*, **28**(4), 507–518.

125. Liu, Y.Q., Majetich, S. A., Tilton, R. D. *et al.* (2005) TCE dechlorination rates, pathways, and efficiency of nanoscale iron particles with different properties. *Environmental Science & Technology*, **39**(5), 1338–1345.

126. Wang, X.Y., Li, J., Liu, H. *et al.* (2008) Preparation and characterization of PAA/PVDF membrane-immobilized Pd/Fe nanoparticles for dechlorination of trichloroacetic acid. *Water Research*, **42**(18), 4656–4664.

127. Xu, J. and Bhattacharyya, D. (2008) Modeling of Fe/Pd nanoparticle-based functionalized membrane reactor for PCB dechlorination at room temperature. *Journal of Physical Chemistry C*, **112**(25), 9133–9144.

128. Xu, J. and Bhattacharyya, D. (2005) Membrane-based bimetallic nanoparticles for environmental remediation: Synthesis and reactive properties. *Environmental Progress*, **24**(4), 358–366.

129. Xu, J., Dozier, A., and Bhattacharyya, D. (2005) Synthesis of nanoscale bimetallic particles in polyelectrolyte membrane matrix for reductive transformation of halogenated organic compounds. *Journal of Nanoparticle Research*, **7**(4–5), 449–467.

130. Smuleac, V., Varma, R., Sikdar, S. *et al.* (2011) Green synthesis of Fe and Fe/Pd bimetallic nanoparticles in membranes for reductive degradation of chlorinated organics. *Journal of Membrane Science*, **379**, (1–2), 131–137.

131. He, F., Liu, J., Roberts, C.B. *et al.* (2009) One-step "green" synthesis of Pd nanoparticles of controlled size and their catalytic activity for trichloroethene hydrodechlorination. *Industrial & Engineering Chemistry Research*, **48**(14), 6550–6557.

132. Nadagouda, M.N. and Varma, R.S. (2007) A greener synthesis of core (Fe, Cu)-shell (An, Pt, Pd, and Ag) nanocrystals using aqueous vitamin C. *Crystal Growth & Design*, **7**(12), 2582–2587.

133. Nadagouda, M.N. and Varma, R.S. (2008) Green synthesis of silver and palladium nanoparticles at room temperature using coffee and tea extract. *Green Chemistry*, **10**(8), 859–862.

134. Nadagouda, M.N. *et al.* (2010) In vitro biocompatibility of nanoscale zerovalent iron particles (NZVI) synthesized using tea polyphenols. *Green Chemistry*, **12**(1), 114–122.

135. Baruwati, B. and Varma, R.S. (2009) High value products from waste: Grape pomace extract-A three-in-one package for the synthesis of metal nanoparticles. *Chemsuschem*, **2**(11), 1041–1044.
136. Virkutyte, J. and Varma, R.S. (2011) Green synthesis of metal nanoparticles: Biodegradable polymers and enzymes in stabilization and surface functionalization. *Chemical Science*, **2**(5), 837–846.
137. Ferry, J.D. (1936) Statistical evaluation of sieve constants in ultrafiltration. *Journal of General Physiology*, **20**(1), 95–104.
138. Lindau, J., Jonsson, A.S., and Bottino, A. (1998) Flux reduction of ultrafiltration membranes with different cut-off due to adsorption of a low-molecular-weight hydrophobic solute-correlation between flux decline and pore size. *Journal of Membrane Science*, **149**(1), 11–20.
139. van den Berg, G.B., Racz, I.G., and Smolders, C.A. (1989) Mass-transfer coefficients in cross-flow ultrafiltration. *Journal of Membrane Science*, **47**(1–2), 25–51.
140. Tu, S.C., Ravindran, V., Den W., *et al.* (2001) Predictive membrane transport model for nanofiltration precesses in water treatment. *Aiche Journal*, **47**(6), 1346–1362.
141. Tu, S.C., Ravindran, V., and Pirbazari, M. (2005) A pore diffusion transport model for forecasting the performance of membrane processes. *Journal of Membrane Science*, **265**(1–2), 29–50.
142. Wiley, D.E., Fell, C.J.D., and Fane, A.G. (1985) Optimization of membrane module design for brackish water desalination. *Desalination*, **52**(3), 249–265.
143. Lu, Y., Spyra, P., Mei, Y. *et al.* (2007) Composite hydrogels: Robust carriers for catalytic nanoparticles. *Macromolecular Chemistry and Physics*, **208**(3), 254–261.
144. Butterfield, D.A. (2008) An overview of biofunctional membranes for tunable separations, metal-ion capture, and enzyme catalysis based on research from the laboratories of Allan Butterfield and Dibakar Bhattacharyya. *Separation Science and Technology*, **43**(16), 3942–3954.
145. Woodward, J. (1985) *Immobilised Cells and Enzymes: A Practical Approach, Practical Approach Series*, IRL Press, Oxford, England, Washington, DC, xiv, 177 p.
146. Cheryan, M. and Mehaia, M.A. (1986) Membrane bioreactors, in *Membrane Separations in Biotechnology* (ed. W.S. McGregor), Marcel Dekker, New York, pp. 255–295.
147. Liu, J.L.,Wang, J., Bachas, L.G. *et al.* (2001) Activity studies of immobilized subtilisin on functionalized pure cellulose-based membranes. *Biotechnology Progress*, **17**(5), 866–871.
148. Nakamura, H. and Karube, I. (2003) Current research activity in biosensors. *Analytical and Bioanalytical Chemistry*, **377**(3), 446–468.
149. Calvo, E.J., Danilowicz, C., Lagier, C. *et al.* (2004) Characterization of self-assembled redox polymer and antibody molecules on thiolated gold electrodes. *Biosensors & Bioelectronics*, **19**(10), 1219–1228.
150. Harper, A. and Anderson, M.R. (2010) Electrochemical glucose sensors-developments using electrostatic assembly and carbon nanotubes for biosensor construction. *Sensors*, **10**(9), 8248–8274.
151. Yan, X.B., Chen, X.J., Tay, B.K. *et al.* (2007) Transparent and flexible glucose biosensor via layer-by-layer assembly of multi-wall carbon nanotubes and glucose oxidase. *Electrochemistry Communications*, **9**(6), 1269–1275.

152. Datta, S., Cecil, C., and Bhattacharyya, D. (2008) Functionalized membranes by layer-by-layer assembly of polyelectrolytes and in situ polymerization of acrylic acid for applications in enzymatic catalysis. *Industrial & Engineering Chemistry Research*, **47**(14), 4586–4597.

153. Smuleac, V., Butterfield, D.A., and Bhattacharyya, D. (2006) Layer-by-layer-assembled microfiltration membranes for biomolecule immobilization and enzymatic catalysis. *Langmuir*, **22**(24), 10118–10124.

154. Ying, L., Kang, E.T., and Neoh, K.G. (2002) Covalent immobilization of glucose oxidase on microporous membranes prepared from poly(vinylidene fluoride) with grafted poly(acrylic acid) side chains. *Journal of Membrane Science*, **208**(1–2), 361–374.

155. Smuleac, V., Varma, R., Baruwati, B. *et al.* (2011) Nanostructured membranes for enzyme catalysis and green synthesis of nanoparticles, *Chemsuschem*, **4**(12), 1773–1777.

156. Egueh, A.N.D., Lakard, B., Fievet, P. *et al.* (2010) Charge properties of membranes modified by multilayer polyelectrolyte adsorption. *Journal of Colloid and Interface Science*, **344**(1), 221–227.

157. Khoo, X. and Grinstaff, M.W. (2011) Novel infection-resistant surface coatings: A bioengineering approach. *MRS Bulletin*, **36**(05), 357–366.

158. Parton, R.F., Vankelecom, I.F.J., Casselman, M.J.A. et al. (1994) An efficient mimic of cytochrome-P-450 from a zeolite encaged iron complex in a polymer membrane. *Nature*, **370**(6490), 541–544.

159. Shah, T.N., Goodwin, J.C., and Ritchie, S.M.C. (2005) Development and characterization of a microfiltration membrane catalyst containing sulfonated polystyrene grafts. *Journal of Membrane Science*, **251**(1–2), 81–89.

160. Clark, L.C. and Lyons, C. (1962) Electrode systems for continuous monitoring in cardiovascular surgery. *Annals of the New York Academy of Sciences*, **102**(1), 29–40.

161. Lu, P. and Hsieh, Y.L. (2010) Layer-by-layer self-assembly of Cibacron Blue F3GA and lipase on ultra-fine cellulose fibrous membrane. *Journal of Membrane Science*, **348**(1–2), 21–27.

162. Dotzauer, D.A., Bhattacharjee, S., Wen Y. *et al.* (2009) Nanoparticle-containing membranes for the catalytic reduction of nitroaromatic compounds. *Langmuir*, **25**(3), 1865–1871.

163. Pich, A., Karak, A., Lu, Y. *et al.* (2006) Preparation of hybrid microgels functionalized by silver nanoparticles. *Macromolecular Rapid Communications*, **27**(5), 344–350.

6

Responsive Membranes for Water Treatment

Qian Yang[1] and S. R. Wickramasinghe[1,2]
[1]Lehrstuhl für Technische Chemie II, Universität Duisburg-Essen, Germany
[2]Ralph E. Martin Department of Chemical Engineering,
University of Arkansas, USA

6.1 Introduction

Membranes are frequently used for water treatment. While reverse osmosis (RO) [1], for the desalination of sea water, is perhaps the best known membrane separation process, other pressure driven filtration processes such as nanofiltration (NF) [2], ultrafiltration (UF) [3], and microfiltration [4] are growing in importance. Both ceramic and polymeric membranes are used [5]. In addition, non-pressure driven processes such as osmotically driven processes, for example, forward osmosis and pressure retarded osmosis, and thermally driven processes, for example, membrane distillation, are of interest as potential future membrane separation processes for water treatment [6].

Stimuli responsive membranes have been developed for many applications, some of the original being for controlled release of drugs [7, 8]. Stimuli responsive membranes change their physical properties in response to changes in environmental conditions such as pH, ionic strength, and temperature or due to changes in external fields due to photo-irradiation or changing electric and magnetic fields [9–13]. Changes in the physical properties of the membrane in response to changed environmental conditions can lead to changes in the mass transfer and interfacial properties of the membrane [14–17].

Membrane fouling often limits the use of membrane based processes for water treatment [18]. Commonly used membrane materials are cellulose acetate, polyamide, and polyethersulfone (PES) [19]. While these materials have good mechanical, thermal, and

Responsive Membranes and Materials, First Edition. Edited by D. Bhattacharyya, Thomas Schäfer, S. R. Wickramasinghe and Sylvia Daunert.
© 2013 John Wiley & Sons, Ltd. Published 2013 by John Wiley & Sons, Ltd.

chemical stability they are hydrophobic in character [20] and are susceptible to fouling. Various cleaning methods, such as backflushing, pulsing of the feed, backpulsing of the permeate, and air sparging, have been proposed and are frequently used. In addition membrane cleaning by pumping cleaning solutions as well as an acid and base is also used industrially. However, these methods always add to the operating cost of the system and may reduce the useful life of the membrane [21]. Moreover, the membrane process has to be interrupted to incorporate such cleaning procedures. Development of fouling resistant membrane surfaces that are in contact with the feed stream offer a potentially more efficient way to suppress membrane fouling and reduce the frequency and severity of cleaning protocols. Numerous studies have indicated that reducing membrane hydrophobicity leads to reduced fouling [22–25]. Given the critical role membrane hydrophobicity plays in determining the level of membrane fouling during operation, numerous studies have focused on producing more hydrophilic membrane surfaces [26].

In general, fouling may be suppressed by altering the membrane surface chemistry (e.g. creating more hydrophilic membrane surfaces), inducing mixing at low Reynolds number near the membrane–fluid interface (e.g. Taylor and Dean vortices, inserts) and changing the properties of the feed solution (e.g. flocculation) [27]. In recent years, researchers have investigated the use of responsive membranes for enhanced fouling resistance during water treatment. To date, most of these studies have focused on the development of pH, ionic strength, and temperature responsive membranes. By changing the pH, ionic strength, or temperature the responsive groups, typically grafted to the membrane surface, change their conformation. This change in conformation between a more hydrophilic and hydrophobic state can lead to desorption of adsorbed solutes. In a very recent study, Himstedt *et al.* [28] grafted magnetically responsive brushes to the surface of commercially available nanofiltration membranes. These investigations indicate that in an oscillating magnetic field, the grafted polymer chains will move, creating mixing at the membrane–fluid interface which suppresses fouling. Thus, unlike more traditional surface modifications that aim to suppress fouling by increasing membrane hydrophilicity, these authors show that a responsive layer could be designed to induce low Reynolds number mixing at the membrane–fluid interface.

The development of responsive membranes for the suppression of fouling in water treatment applications is relatively new. Several different strategies are available to produce responsive membranes for water treatment. In this chapter previous attempts to develop responsive membranes are described in terms of the strategy used to produce the responsive membrane. The chapter covers reverse osmosis, nanofiltration, ultrafiltration, and microfiltration membranes since these are the membrane separation processes used most frequently for water treatment. It also focuses on responsive membranes which result in changes in permeate flux and suppression of membrane fouling. Finally some of the possible future developments in responsive membranes for water treatment are discussed.

6.2   Fabrication of Responsive Membranes

Three methods have been described for the development of responsive membranes for water treatment applications: incubation in liquids to produce responsive groups, incorporation of responsive monomers in the base membrane material, and surface modification to add response groups to the membrane surface. The latter surface modification method has the potential to graft responsive groups only on the membrane surface that is in contact with the feed stream. Each modification is described below as well as its application to membranes for water treatment. While responsive membranes are reported using all three strategies it is the latter surface modification method that is the most popular for development of responsive membranes for control of membrane fouling.

6.2.1 Functionalization by Incubation in Liquids

Polyacrylonitrile (PAN) membranes may be hydrolysed using aqueous NaOH solutions [29]. Wang has shown that some of the PAN molecules are converted into poly(acrylic acid) (PAA). Yang and Tong [30] hydrolysed PAN hollow fibres using aqueous NaOH to produce a thin PAA layer on the surface of the membrane. At pH values above the pKa of the PAA molecules, the carboxylic groups are deportonated and swell. Yang and Tong observed that at pH values between 5 and 7 the swelling of the PAA layer led to a decrease in permeability. Lohokare *et al.* [31] surface modified PAN membranes by incubation with ethanolamine, triethylamine, sodium hydroxide, and potassium hydroxide. They also observed that the membranes swell at higher pH values.

At pH values above the pKa of the PAA chains present at the membrane surface, a high density of COO^- ions will be present which leads to swelling. Conversely at pH values below the pKa of the PAA chains, the chains will become more hydrophobic and collapse. Switching between the expanded and collapsed conformations of PAA could be used to desorb adsorbed proteins [26]. Incubation in liquid solutions is easy to implement. Furthermore, membranes that have been already put in a module can be modified. However the major disadvantage of this method is a lack of control. Neither can it be used to add new functionalities to the membrane surface. In addition, one must be careful as it is easy to adversely alter membrane rejection and transport properties by over-incubation of the membrane.

6.2.2 Functionalization by Incorporation of Responsive Groups in the Base Membrane

Functionalization by incorporation of responsive groups into the base membrane may be classified as either *in situ* or post-synthesis. In the former method, the group is part of one of the monomers used to form the membrane. In the latter method, the membrane polymer is synthesized and then a reactive group present is functionalized prior to membrane casting.

Using the *in situ* method, the monomer containing the responsive group and other co-monomers are copolymerized and the copolymers are fabricated into the desired membrane. Schacher *et al.* synthesized polystyrene-block-poly(N, N-dimethylaminoethyl methacrylate) (PS-*b*-PDMAEMA) copolymer via sequential living anionic polymerization and fabricated ultrafiltration membranes from this copolymer by a non-solvent-induced phase separation (NIPS) process [32]. The degree of polymerization of the two blocks was 607 for polystyrene and 92 for PDMAEMA and the molecular weight distribution was 1.04. In the presence of the PDMAEMA block, the membrane showed both pH- and temperature-responsive behaviour.

As shown in Figure 6.1a, at pH 2 and room temperature the water flux was very low and no flux increase was detected upon increasing the temperature to 65°C. This is due to the

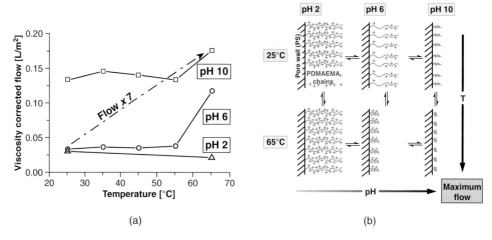

Figure 6.1   *Water flux of PS-b-PDMAEMA membrane (a) and schematic depiction of the differ-*
ent states of the inner part of the membrane pores (b) at different pH values and temperatures.
(Reprinted under the terms of the STM agreement from [32] Copyright (2009) Wiley-VCH).
See plate section for colour figure.

fact that PDMAEMA segments are protonated leading to a more extended conformation
(see Figure 6.1b, left). A low water flux is observed at pH 6 due to partial protonation of
the chains (Figure 6.1b upper middle). No significant change in flux was detected upon
increasing the temperature to 55 °C. Above 55°C the low charge density on PDMAEMA
segments allowed the chains to collapse at temperatures higher than the lower critical
solution temperature (LCST) and the flux increased four-fold at 65°C. Further increasing
the pH to 10 resulted in full deprotonation of the PDMAEMA chains (Figure 6.1b right)
and even higher fluxes could be achieved. By combining the high pH and temperature, up
to a seven-fold increase in flux was obtained compared to low pH and temperature. While
the focus of these model studies was to use pH and temperature responsive behaviour to
modify permeate flux, other researchers (using surface grafting methods) have shown that
by oscillating between the swollen and collapsed conformations of the polymer, adsorbed
foulants may be released (see Section 6.2).

   As an example of a post-synthesis strategy, Shi and coworkers [33] grafted poly
(methacrylic acid) (PMAA) onto PES using benzoyl peroxide (BPO) as the initiator. The
resulting copolymer, PES-*g*-PMAA, was cast into asymmetric ultrafiltration membranes via
the phase inversion process. The PMAA component rendered the membrane pH-responsive
property. As shown in Figure 6.2, the water flux of the PES membrane was pH-independent.
However, the membrane cast from PES-*g*-PMAA showed a pH-dependent permeability.
The aqueous solution flux through the membrane showed a drastic decrease between pH 4.0
and 6.0. Interestingly, this decrease turned out to be less the higher the degree of grafting
(DG). As can be seen from Table 6.1, the increase of DG led to an increased coverage
of PMAA on the membrane surface and a decreased membrane pore size. This is due to
the relatively high hydrophilicity of PMAA which slowed down the solvent/non-solvent
exchange rate during membrane casting. Therefore, despite the fact that membranes had

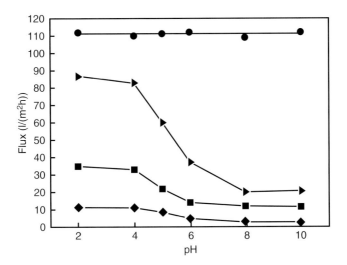

Figure 6.2 *pH dependence of water permeation through an unmodified PES membrane (solid circles), PES-g-PMAA membrane with a DG of 3.7% (solid triangles), 8.2% (solid squares) and 17.0% (solid diamonds). (Reprinted with permission from [33] Copyright (2010) Elsevier Ltd).*

higher surface coverage of pH-sensitive PMAA, the PES-*g*-PMAA membrane with higher DG was apparently less sensitive to pH change.

Similarly, Hester *et al*. made a PVDF-*g*-PMAA copolymer and blended it with PVDF to cast a pH-sensitive membrane [34]. The surface segregation of the PVDF-*g*-PMAA during the coagulation step resulted in a PMAA coverage of 29wt.% at the membrane surface and a bulk composition of only 5 wt.%. The membrane permeability underwent a pH-responsive change of 17.5-fold between pH 2 and 8. These investigators estimated the pH-dependent conformational changes of the PMAA surface layer by atomic force microscopy. As can be seen from Figure 6.3, the width of the transition region of the force–distance curve (between the vertical dashed lines) provides a rough measure of the PMAA layer height. Qualitatively, this width is clearly much larger at pH 8 (68.5 nm) than at pH 2 (9.2 nm).

The examples given above indicate the feasibility of producing responsive membranes by incorporation of a responsive group into the bulk membrane polymer. Again while

Table 6.1 *Surface coverage of PMAA and the pore size of PES and PES-g-PMAA membranes. (Reprinted with permission from [33] Copyright (2010) Elsevier Ltd).*

Graft yield (%)	Surface coverage of PMAA	Pore-size range (nm)
0	0	6–25
3.7	20.6	5–17
8.2	28.0	3–14
17.0	43.2	3–8

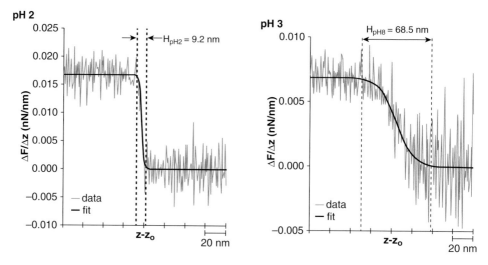

Figure 6.3 *Numerical derivatives of AFM force–distance curves for blend membranes immersed in pH 2 and pH 8 buffers. ΔF and ΔZ represent increments of the force and decrements of the distance respectively. (Reprinted with permission from [34] Copyright (2002) Elsevier Ltd).*

these membranes were developed in order to adjust flux as a function of pH, switching between the collapsed and expanded conformations of the responsive groups could lead to detachment of adsorbed foulants as described in Section 6.2.3. A disadvantage of this method is that the new membranes could display compromised mechanical, chemical, and thermal stability compared to the original membranes that did not incorporate responsive groups.

6.2.3   Surface Modification of Existing Membranes

The method of preference for imparting responsive groups on membranes for water treatment is surface modification of existing/commercial membranes. This can be achieved in a highly surface selective way and, consequently, minimizes the influence on membrane bulk properties like pore structure and porosity. Generally, membrane surface modification with tethered polymer brushes can be achieved by either chemical binding (covalent attachment), including "grafting to" and "grafting from", or physical interaction, including physical adsorption, interfacial cross-linking, and pore-filling.

6.2.3.1   "Grafting to"

The "grafting to" approach refers to preformed polymer chains with end-functional or pendant reactive groups that are covalently coupled to the membrane surface [35]. Polymers with a designed molecular structure can be used to form tethered polymer brushes and the molecular weight can be precisely controlled and characterized in the synthesis step. Nevertheless the macromolecular chains must diffuse to the membrane surface. Thus, compared to "grafting from" (see Section 6.2.3.2), the grafting density is limited as further grafting is hindered by the polymer chains already tethered on the surface [35–37].

$$R_{CA}^{-OH} + CH_2 = CHSO_2CH = CH_2 + CH - R_{HPC}$$
$$\rightarrow R_{CA} - OCH_2CH_2SO_2CH_2CH_2 - O - R_{HPC}$$

Figure 6.4 *Reaction for immobilization of HPC to CA ultrafiltration membrane. (Reprinted under the terms of the STM agreement from [38] Copyright (2008) Taylor and Francis).*

Gorey *et al.* produced a fouling-resistant cellulose acetate (CA) ultrafiltration membrane by attaching a stimuli-responsive polymer film on the membrane surface [38]. A thermally responsive polymer, hydroxypropyl cellulose (HPC), was grafted onto the membrane surface by the reaction showed in Figure 6.4, in which divinyl sulfone (DVS) was used as a coupling agent. The thermal responsive property of the grafted HPC was directly determined by monitoring the roughness of the membranes in solutions of different temperatures with AFM. As shown in Figure 6.5, the unmodified membrane displayed a negligible difference in roughness at 25 and 60°C indicating that temperature had no effect on the unmodified membrane. On the other hand, the HPC modified membrane showed a roughness of 8.398 nm at 25°C and 0.915 nm at 60°C. The change in roughness was attributed to the extension and collapse of the surface attached HPC layer below and above LCST (46°C).

Continuously triggering the phase transition by temperature oscillation, that is, increasing the temperature to collapse the film and then immediately decreasing the temperature to expand it, could achieve a sweeping motion at the molecular (nanometre) level and offer better protection of the surface from foulants. Interestingly, antibodies were immobilized at the end of surface grafted HPC chains and used for detection of bacteria (*Mycobacteria*)

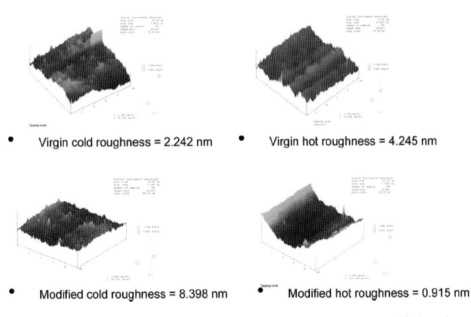

• Virgin cold roughness = 2.242 nm   • Virgin hot roughness = 4.245 nm

• Modified cold roughness = 8.398 nm   • Modified hot roughness = 0.915 nm

Figure 6.5 *AFM images used to determine the roughness of the virgin (unmodified) and HPC modified membranes at low (left) and high (right) temperatures. (Reprinted under the terms of the STM agreement from [38] Copyright (2008) Taylor and Francis).*

in water. This modified membrane could be a promising anti-fouling and bacteria detection platform in membrane-based drinking water systems.

6.2.3.2   "Grafting-from"

Using the "grafting from" method, much higher polymer surface densities are possible. Numerous surface initiated "grafting from" techniques have been described such as: UV-induced graft polymerization [39], plasma treatment [40], ozone treatment [41], and γ-Ray irradiation [42]. Amongst them, UV-induced graft polymerization has been frequently studied due to its low cost of operation, mild reaction conditions and ease of use [43]. However, all of these traditional "grafting from" techniques suffer from a lack of control. Consequently it is often difficult to achieve reproducible surface structures. In particular, the grafting density and grafted chain length are often kinetically limited and difficult to control. Fortunately, progress in polymerization technology gives us a chance to solve the problem of lack of control. It is possible to prepare well-defined polymer brushes with controlled chain length and chain density by living/controlled polymerization techniques such as cationic polymerization [44], anionic polymerization [45], ring-opening metathesis polymerization (ROMP) [46], nitroxide mediated polymerization (NMP) [47], reversible addition-fragmentation chain-transfer polymerization (RAFT) [48] and atom transfer radical polymerization (ATRP) [49]. Amongst these controlled polymerization methods, ATRP is especially attractive and has gained much attention because of its tolerance to impurities and mild polymerization conditions [50]. Below, three recent examples of surface modification of commercial membranes to impart responsive groups are described.

Himstedt *et al.* modified nanofiltration membranes by growing PAA nanobrushes from the surface of the membrane via UV-induced graft polymerization [26]. These nanobrushes were grown from the membrane surface with minimal impact on permeate flux. The flux data of unmodified and PAA grafted membranes under different pH conditions are shown in Figure 6.6. For the unmodified membrane, no significant difference was observed between the flux at the two different pH values tested. However membranes with grafted PAA nanobrushes exhibited significant changes in the filtrate flux as a function of pH. The flux was higher at higher pH values compared to lower pH. This could be ascribed to the deprotonated state of carboxyl groups and the consequent stretched structure of the grafted nanobrushes caused by the static repulsion between negatively charged chains at pH values above the pKa of the grafted PAA chains.

Hinmstedt *et al.* indicate that the behaviour of grafted PAA chains is often complicated and difficult to predict based on free PAA in solution. A theoretical study of the pKa of oligo-methacrylic acids demonstrates that the pKa values of methacrylic acid increase with the DG to a plateau value of about 7.5 for polymerization degrees greater than 12 [51]. This result also suggests that in order to achieve a sharp transition between the swollen and collapsed states of PAA, the polydispersity of the grafted nanobrushes should be minimized. Moreover, careful tuning of polymer structures (chain length, chain density etc.) is needed for construction of responsive membranes by tethering polymer brushes. For example, due to steric hindrance effects the grafted brushes should have optimized density to ensure enough space for the swelling and deswelling transition and to simultaneously realize sufficient coverage on the membrane surface.

Every year, trillions of barrels of produced water (PW) from oil and gas exploration are produced and need to be disposed of with minimum impact on the environment. A

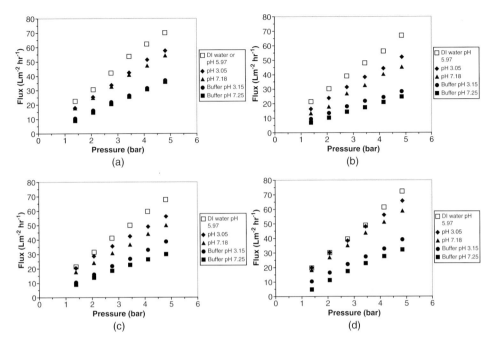

Figure 6.6 *Variation of filtrate flux with pH for unmodified membranes (a) and membranes modified with 1% AA, 15 min UV irradiation (b); 2% AA, 10 min (c) and 15 min (d) UV irradiation. (Reprinted with permission from [26] Copyright (2011) Elsevier Ltd).*

number of studies have demonstrated successful treatment of PW by the membrane process [52, 53, 54] though membrane fouling is often a serious concern. Wandera and coworkers modified regenerated cellulose UF membranes with PNIPAAm and PNIPAAm-*b*-PPEGMA copolymer grafted by surface-initiated ATRP [55] for treatment of PW.

As shown in Figure 6.7, the ATRP initiator was first immobilized at the membrane surface and then surface initiated ATRP of NIPAAm was carried out. Since ATRP is a living polymerization, the grafted PNIPAAm chains were reinitiated from the end to obtain the second PPEGMA block. Afterwards, the performance of unmodified, PNIPAAm and PNIPAAm-*b*-PPEGMA copolymer grafted membranes was evaluated. Cross-flow filtration experiments revealed that the total volume of permeate processed through the modified membrane was 13.8% higher than the unmodified membrane after 40 h of operation (see Figure 6.8a). This result also indicates that longer operation times prior to cleaning and relatively lower operation costs can be expected for the modified membrane.

Figure 6.8b also indicates that grafting a responsive layer from the surface of the membrane leads to an increase in resistance to permeate flow and hence lower permeate fluxes compared to the unmodified membrane. This additional resistance to flow caused by the dense grafted layer indicates the importance of controlling the layer thickness and chain density. With the thermo-responsive property of PNIPAAm, the flux recovery was better for the modified membranes after a cold (below LCST) water rinse. The foulants attached to the surface at temperatures above 32°C detached during the cold water rinse as a result

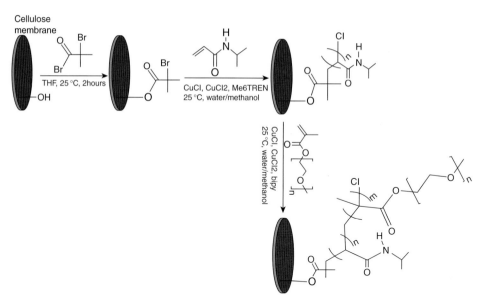

Figure 6.7 *Surface-initiated ATRP of PNIPAAm-b-PPEGMA from cellulose UF membranes. (Reprinted with permission from [55] Copyright (2011) Elsevier Ltd).*

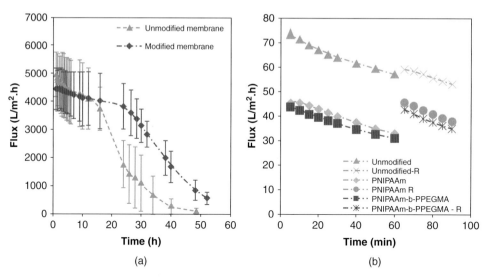

Figure 6.8 *Synthetically produced water flux measurements by cross-flow ((a) experiments were carried out at a temperature of 50°C and a TMP of 414 kPa for unmodified and PNIPAAm-b-PPEGMA-modified membranes) and dead-end ((b) a second filtration run was carried out for each of these membranes after a cold water (15°C) rinse, indicated by the letter R in the legend. A constant pressure of 207 kPa was used for all of the experiments) filtration for unmodified and modified cellulose UF membranes. (Reprinted with permission from [55] Copyright (2011) Elsevier Ltd). See plate section for colour figure.*

of phase change (swelling at higher temperature). Modified membranes achieved 100% flux recovery, while only about 81% of the initial flux was recovered for the unmodified membrane.

Living polymerizaton protocols, like ATRP, not only offer the poential to independently control chain length and density but also to selectively functionalize chain ends. Himstedt *et al.* grafted polymer chains from the surface of NF270 membranes and then immobilized magnetically responsive nanoparticles to the chain ends [28]. By using an oscillating magnetic field, the magnetic nanoparticles acted like micromixers and induced mixing directly above the membrane surface (see Figure 6.9a). They further showed that this movement of the polymer brushes reduced concentration polarization and, potentially, also fouling by colloids.

As shown in Figure 6.9b, after immobilization of the initiator surface, initiated ATRP was conducted to grow poly(2-hydroxyethyl methacrylate) (PHEMA) chains from the membrane surface. Next the bromine at the chain ends was converted into an amine group by Gabriel synthesis. After that, iron oxide magnetite superparamagnetic nanoparticles with carboxyl groups on the surface were immobilized at the end of PHEMA brushes. The magnetically actuated movement was observed by using a particle image velocimetry (PIV) system.

The PIV system can monitor the velocity profile of water above the membrane and can supply time-resolved flow patterns. Thus, the movement of the grafted polymer brushes and the macroscopic fluid mixing behaviour were observed in the presence of a magnetic field at various oscillating frequencies. As can be seen from Figure 6.10, at all tested oscillating frequencies the fluid behaviour above the unmodified membrane was relatively ordered, that is the red field lines were orientated in the same direction and roughly parallel, indicating no mixing occured. However the pattern above the modified membrane at 9 and 22 Hz exhibited obvious changes in velocity vectors of varying direction and/or length as well as field lines with chaotic (non-parallel) pathways. These results significantly suggested fluid mixing at the membrane surface. The mixing at a lower oscillating frequency (9 Hz) was more pronounced than that at 22 Hz and the mixing almost vanished when the oscillating frequency increased to 30 Hz. This could be explained by the fact that at higher oscillating frequency there is the less time for the chains to move in a given direction. In addition, chain length and chain density could affect the mixing and should be taken into consideration when constructing such responsive membrane systems.

The rejection of $CaCl_2$ and $MgSO_4$ was also improved. Compared with the result for unmodified membranes, the modified membranes showed identical salt rejection without a magnetic field. However, higher salt rejection was then observed under an oscillating magnetic field which provides further evidence that the concentration polorization is being supressed by the movement of magnetically responsive brushes at the membrane surface. Importantly, the membrane was activated through an external field which makes it highly practical for industrial applications where changes in the bulk feed temperature or pH are not required.

6.2.3.3   *Physical Adsorption, Interfacial Cross-Linking, and Pore-Filling*

Physical adsorption is a reversible process. It is always achieved by the self-assembly of polymers and/or copolymers on the membrane surface in which the polymers are acting more like a surfactant [56]. This method is very easy to carry out and the adsorption process

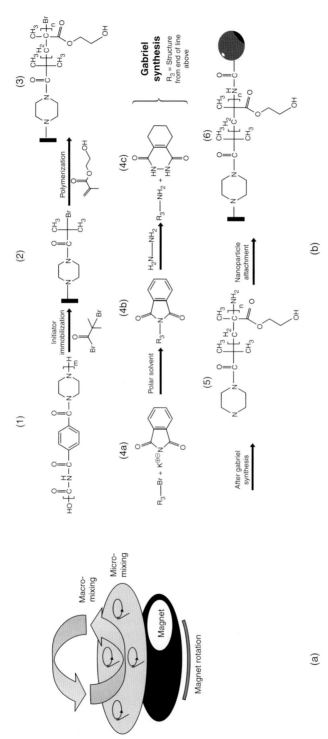

Figure 6.9 *Schematic depiction of macromixing caused by micromixers on the membrane surface activated by an oscillating magnetic field (a) and the reaction sequence for immobilization of magnetic NPs (b). (Reprinted with permission from [28] Copyright (2011) American Chemical Society).*

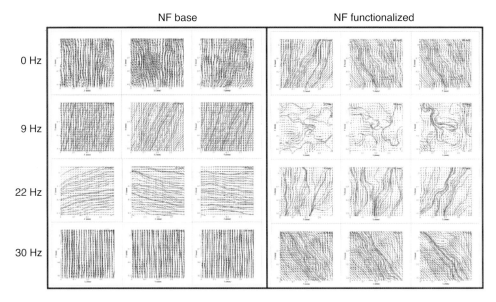

Figure 6.10 *Series of 4 PIV vector diagrams for magnet rotation frequencies of 0, 9, 22 and 30 Hz. Each vector diagram is averaged over 1 ms of time. (Reprinted with permission from [28] Copyright (2011) American Chemical Society). See plate section for colour figure.*

with respect to the resulting modification parameters, for example, surface grafting density, are controlled by thermodynamic equilibrium [57]. However, the interaction between the polymer and the membrane surface is always weak and adsorbed polymers can easily desorb from the membrane surface.

Interfacial cross-linking involves the formation of functional polymer networks on the surface or in the pores which also penetrate into the bulk membrane. Pore-filling is a relatively broad term [37], here it refers to the *in situ* polymerization in the membrane pores to give a hydrogel padding without covalent coupling to the membrane pore wall. Once the barrier pores of a membrane are completely filled with a hydrogel, the transport of water and solutes through the membrane will be controlled by the hydrogel [58].

Yu *et al.* modified a thin-film composite polyamide RO membrane by depositing poly(N-isopropylacrylamide-*co*-acrylamide) (PNIPAAm-*co*-PAAm) copolymer on the membrane surface via an "*in situ*" method [59]. They placed the membrane in a cross-flow module and the thermally responsive copolymer was deposited on the membrane surface from the feed stream by hydrogen bonding. As shown in Figure 6.11, the PNIPAAm-*co*-PAAm coating layer improved the fouling resistance of the membrane, especially at higher feed foulant concentrations.

Importantly, surface modification enhanced membrane cleaning. A sharp improvement in cleaning efficiency around the LCST of the deposited copolymer (42.5°C) was observed. The flux recovery of modified membranes cleaned at 45°C (above LCST) was about 88.5%, which was much higher than the value (about 68.5%) obtained with the membrane cleaned at 40°C (below LCST, see Figure 6.12). This was due to the phase transition of the deposited PNIPAAm-*co*-PAAm) copolymer layer on the membrane surface which

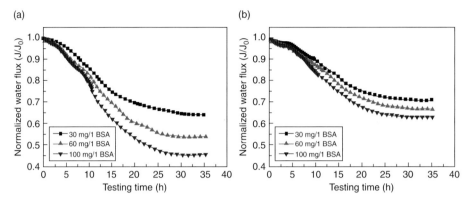

Figure 6.11 *Normalized flux with different feed BSA concentrations of 30 mg/l (■), 60 mg/l (▲) and 100 mg/l (▼) of the unmodified (a) and modified (b) membranes. Test conditions employed were: ionic strength = 50 mM NaCl, pH = 6.8, initial permeate flux = 40 l/m² h, temperature = 25°C and cross-flow velocity = 0.092 m/s. Membrane modification conditions: PNIPAAm-co-PAAm) concentration = 200 mg/l and deposition time = 10 h. (Reprinted with permission from [59] Copyright (2011) Elsevier Ltd).*

loosened the deposited protein layer and pushed it away from the membrane surface, thereby facilitating the removal of the foulant by water flushing. Moreover, this surface modification approach would be of particular interest for commercial/industrial applications as the membranes can be treated in their original module assembly and the anti-fouling properties can be improved without sacrificing permeation properties (flux and salt rejection).

Petrov *et al.* modified PAN based UF membranes by cross-linking temperature sensitive poly(vinylalcohol-*co*-vinylacetal) (PVA-ac, LCST 34.5°C) to form an interpenetrating

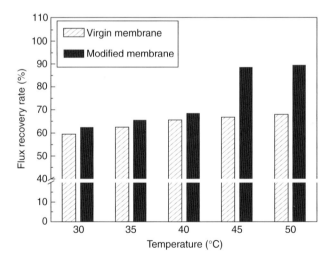

Figure 6.12 *Flux recovery rates of the virgin and modified membranes cleaned with deionized water at different temperatures. (Reprinted with permission from [59] Copyright (2011) Elsevier Ltd).*

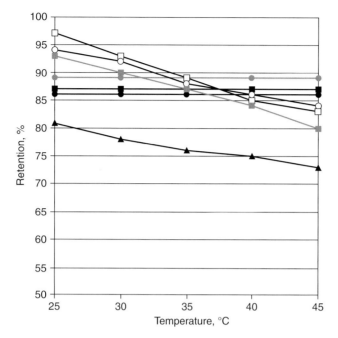

Figure 6.13 *Retention of the modified PAN UF membranes at different temperatures for calibrant albumin: (▲) basic membrane; (■) modified on the selective layer at PVA-ac concentration 25%; (▪) modified on the selective layer at PVA-ac concentration 12.5%; (□) modified on the selective layer at PVA-ac concentration 6.25%; (●) modified on the matt side at PVA-ac concentration 25%; (◦) modified on the matt side at PVA-ac concentration 12.5%; (○) modified on the matt side at PVA-ac concentration 6.2%. (Reprinted with permission from [60] Copyright (2005) Elsevier Ltd).*

polymer network (IPN) on the membrane surface and in the pores [60]. They found that with a high concentration of PVA-ac, a dense IPN layer formed mainly on the membrane surface and no temperature sensitivity was observed. However, decreased PVA-ac concentration led to a thin IPN layer mainly in the membrane pores and significant temperature response was detected. As shown in Figure 6.13, the retention of the unmodified membrane decreased with the increase of temperature due to the pore widening at high temperature. On the other hand, all modified membranes had higher retentions than the unmodified one. With the IPN layer on the surface, the membranes modified with higher concentration solutions of PVA-ac were not responsive to the temperature while the retention of the membranes prepared with diluted solutions of the PVA-ac polymer decreased significantly with the increase of temperature.

Latukippe *et al.* fabricated a pH-responsive ultrafiltration membrane by filling the pores of either microfiltration membranes or non-woven supports with cross-linked PAA-*co*-PAAm [61]. An interesting salt concentration dependence on permeability was observed by measuring the change of transmembrane pressure (TMP) at constant flux under different pH and salt concentration conditions. The TMP increased with an increase of pH (see Figure 6.14a) due to the swollen structure of the pore filling gel caused by deprotonation of carboxyl

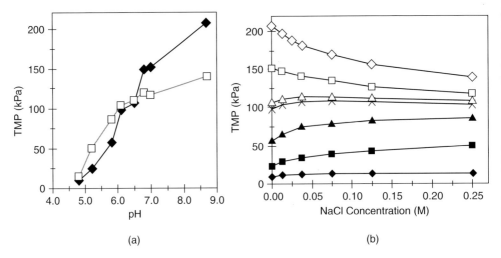

Figure 6.14   *(a) Effect of pH on TMP for PAA-co-PAAm modified membranes at a permeate flux of 9.21 × 10⁻⁵ m/s at NaCl concentrations of 0 mM (♦) and 250 mM (□); (b) effect of salt concentration on TMP for PAA-co-PAAm modified membranes at a permeate flux of 9.21 × 10⁻⁵ m/s at solution pH of 4.8 (♦), 5.2 (■), 5.8 (▲), 6.1 (×), 6.5 (△), 7.0 (□) and 8.7(◊). (Reprinted with permission from [61] Copyright (2009) Elsevier Ltd).*

groups. This pH-responsive property was influenced by adding NaCl. As can be seen from Figure 6.14b, at higher pH, TMP decreased with the increase of salt concentration. This can be ascribed to the fact that the salt screens charge repulsion effects between polymer chains and rearrange the gel network to a heterogeneous structure which subsequently leads to a reduced hydraulic resistance. On the other hand, at low pH, the effect of salt reversed and increasing salt concentration resulted in enhanced TMP. Both the authors and early research from Elbrahmi *et al.* confirmed that an increase in ionic strength resulted in an increase of the polymer ionization of acrylamide-acrylic acid copolymers [62]. The increased charge density on the PAA-*co*-PAAm chains then caused the gel network to move to a swollen configuration which has higher hydraulic resistance.

6.3   Outlook

Pressure driven membrane-based separation processes are finding an increasing number of applications in water treatment. In addition, several alternative membrane based separation processes, such as forward osmosis and membrane distillation, are being actively developed. Thus the demand for advanced membranes that show desired separation properties and fouling resistance will increase. Frequently the membrane surface and surface properties are critical for the separation. Further membrane surface contamination by fouling leads to compromised performance.

Numerous methods have been described for modification of membrane surface properties [63]. Surface modification was initially used as a method to develop more hydrophilic membrane surfaces for suppression of protein fouling. Later, these methods were used to develop hydrophilic membranes for water treatment that also showed enhanced fouling resistance [18]. Modification of membrane surfaces to increase their hydrophilicity or reduce surface roughness does not require much control over the polymer chain molecular weight.

Recently the development of controlled polymerization procedures such as ATRP has allowed for independent control of polymer chain length and density. In addition, using a living polymerization procedure, the possibility exists of growing polymer chains containing multiple blocks from the membrane surface. This has enabled the development of responsive membranes where control of both chain density and molecular weight are often essential to ensure a sharp response to external stimuli. It is likely that many new advanced multifunctional membranes that not only respond to external stimuli, but could also catalyse chemical reactions, will find future applications in water treatment.

References

1. Malaeb, L. and Ayoub, G.M. (2011) Reverse osmosis technology for water treatment: State of the art review. *Desalination*, **267**, 1–8.
2. Goncharuk, V.V., Kavitskaya, A.A. and Skil'skaya, M.D. (2011) Nanofiltration in drinking water supply. *J. Water Chem. Technol.*, **33**, 37–54.
3. Davey, J. and Schafer, A.I. (2009) Ultrafiltration to supply drinking water in international development: A review of opportunities, in *Appropriate Technologies for the Environmental Protection in the Developing World* (ed. E.K. Yanful), Springer Science + Business Media B. V., pp. 151–168.
4. Huang, H., Schwab, K. and Jacangelo, J.G. (2009) Pretreatment for low pressure membranes in water treatment: a review. *Environ. Sci. Technol.*, **43**, 3011–3019.
5. Zaidi, A., Simms, K. and Kok, S. (1992) The use of micro/ultrafiltration for removal of oil and suspended solids from oilfield brines. *Wat. Sci. Tech.*, **24**, 163–176.
6. Cathy, T.Y. (2010) Osmotically and thermally driven membrane processes for enhancement of water recovery in desalination processes. *Desalin. Water Treat.*, **15**, 279–286.
7. Shaikh, R.P., Pillay, V., Choonara, Y.E. *et al.* (2010) A review of multi-responsive membrane systems for rate-modulated drug delivery. *AAPS Pharm. Sci. Tech.*, **11**, 441–459.
8. Hoare, T., Santamaria, J., Goya, G.F. *et al.* (2009) A magnetically triggered composite membrane for on-demand drug delivery. *Nano Lett.*, **9**, 3651–3657.
9. Wandera, D., Wickramasinghe, S.R. and Huson, S.M. (2010) Stimuli-responsive membranes. *J. Membr. Sci.*, **357**, 6–35.
10. Lee, Y.M., Ihmt, S.Y., Shim, J.K. and Kim, J.H. (1995) Preparation of surface-modified stimuli-responsive polymeric membranes by plasma and ultraviolet grafting methods and their riboflavin permeation. *Polymer*, **36**, 81–85.
11. Park, Y.S., Ito, Y. and Imanishi, Y. (1998) Photocontrolled gating by polymer brushes grafted on porous glass filter. *Macromolecules*, **31**, 2606–2610.
12. Liang, L., Feng, X., Peurrung, L. and Viswanathan, V. (1999) Temperature-sensitive membranes prepared by UV photopolymerization of N-isopropylacrylamide on a surface of porous hydrophilic polypropylene membranes. *J. Membr. Sci.*, **162**, 235–246.

13. Chung, D.-J., Ito, Y. and Imanishi, Y. (1994) Preparation of porous membranes grafted with poly(spiropyran-containing methacrylate) and photocontrol of permeability. *J. Appl. Polym. Sci.*, **51**, 2027–2033.

14. Oak, M.S., Kobayashi, T., Wang, H.Y. *et al.* (1997) pH effect on molecular size exclusion of polyacrylonitrile ultrafiltration membranes having carboxylic acid groups. *J. Membr. Sci.*, **123**, 185–195.

15. Ito, Y., Park, Y.S. and Imanishi, Y. (1997) Visualization of critical pH-controlled gating of a porous membrane grafted with polyelectrolyte brushes. *J. Am. Chem. Soc.*, **119**, 2739–2740.

16. Zhang, H. and Ito, Y. (2001) pH Control of transport through a porous membrane self-assembled with a poly(acrylic acid) loop brush. *Langmuir*, **17**, 8336–8340.

17. Ito, Y. and Park, Y.S. (2000) Signal-responsive gating of porous membranes by polymer brushes. *Polym. Adv. Technol.*, **11**, 136–144.

18. Kilduff, J.E., Mattaraj, S., Zhou, M. and Belfort, G. (2005) Kinetics of membrane flux decline: the role of natural colloids and mitigation via membrane surface modification. *J. Nanoparticle Res.*, **7**, 525–544.

19. Vankelecon, I.F.J., De Smet, K., Gevers, L.E.M. and Jacobs, P.A. (2003) in *Nanofiltratoin: Principles and Applications* (eds A.I. Schäfer, A.G. Fane and T.D. Waite), Elsevier, Oxford, UK, pp. 33–65.

20. van der Bruggen, B. (2009) Chemical modification of polyethersulfone nanofiltration membranes: a REVIEW. *J. Appl. Polym. Sci.*, **114**, 630–642.

21. Akhtar, S., Hawes, C., Dudley, L. *et al.* (1995) Coatings reduce the fouling of micro-filtration membranes. *J. Membr. Sci.*, **107**, 209–218.

22. Kim, M., Saito, K., Furusaki, S. *et al.* (1991) Water flux and protein adsorption of a hollow fiber modified with hydroxyl groups. *J. Membr. Sci.*, **56**, 289–302.

23. Chang, I.S., Bag, S.O. and Lee, C.H. (2001) Effects of membrane fouling on solute rejection durign membrane filtration of activated sludge. *Process Biochem.*, **36**, 855–860.

24. Freger, V., Gilron, J. and Belfer, S. (2002) TFC polyamide membranes modified by grafting of hydrophilic polymers: an FTIR?AFM/TEM study. *J. Membr. Sci.*, **209**, 283–292.

25. Braeken, L., Boussu, K., van der Bruggen, B. and Vandecasteele, C. (2005) Modeling of the adsorption of organic compounds on polymeric nanofiltration membranes in solutions containing two compounds. *ChemPhysChem*, **6**, 1606–1612.

26. Himstedt, H.H., Marshall, K.M. and Wickramasinghe, S.R. (2011) pH responsive nanofiltration membranes by surface modification. *J. Membr. Sci.*, **366**, 373–381.

27. Belfort, G., Davis, T.H. and Zydney, A.L. (1994) The behavior of suspensions and macromolecular solutions in crossflow microfiltration. *J. Membr. Sci.*, **96**, 1–58.

28. Himstedt, H.H., Yang, Q., Dasi, L.P. *et al.* (2011) Magnetically activated micromixers for separation membranes. *Langmuir*, **27**, 5574–5581.

29. Wang, X.-P. (2000) Preparation of crosslinked alginate composite membrane for dehydration of ethanol-water mixtures. *J. Appl. Polym. Sci.*, **77**, 3054–3061.

30. Yang, M.-C. and Tong, J.-H. (1997) Loose ultrafiltration of proteins using hydrolyzed polyacrylonitrile hollow fiber. *J. Membr. Sci.*, **132**, 63–71.

31. Lohorakre, H.R., Kumbharkar, S.C., Bhole, Y.S. and Kharul, U.K. (2006) Surface modification of polyacrylonitrile based ultrafiltration membrane. *J. Appl. Polym. Sci.*, **101**, 4378–4385.

32. Schacher, F., Ulbricht, M. and Muller, A.H.E. (2009) Self-Supporting, double stimuli-responsive porous membranes from polystyrene-block-poly(N,N-dimethylaminoethyl methacrylate) diblock copolymers. *Adv. Funct. Mater.*, **19**, 1040–1045.

33. Shi, Q., Su, Y., Ning, X. *et al.* (2010) Graft polymerization of methacrylic acid onto polyethersulfone for potential pH-responsive membrane materials. *J. Membr. Sci.*, **347**, 62–68.

34. Hester, J.F., Olugebefola, S.C. and Mayes, A.M. (2002) Preparation of pH-responsive polymer membranes by self-organization. *J. Membr. Sci.*, **208**, 375–388.

35. Kato, K., Uchida, E., Kang, E.-T. *et al.* (2003) Polymer surface with graft chains. *Prog. Polym. Sci.*, **28**, 209–259.

36. Zhao, B. and Brittain, W.J. (2000) Polymer brushes: surface-immobilized macro-molecules. *Prog. Polym. Sci.*, **25**, 677–710.

37. Yang, Q., Adrus, N., Tomicki, F. and Ulbricht, M. (2011) Composites of functional polymeric hydrogels and porous membranes. *J. Mater. Chem.*, **21**, 2783–2811.

38. Gorey, C., Escobar, I.C., Gruden, C. *et al.* (2008) Development of smart membrane filters for microbial sensing. *Sep. Sci. Technol.*, **43**, 4056–4074.

39. Yang, Q., Xu, Z.K., Dai, Z.W. *et al.* (2005) Surface modification of polypropylene microporous membranes with a novel glycopolymer. *Chem. Mater.*, **17**, 3050–3058.

40. Wavhal, D.S. and Fisher, E.R. (2003) Membrane surface modification by plasma-induced polymerization of acrylamide for improved surface properties and reduced protein fouling. *Langmuir*, **19**, 79–85.

41. Wang, Y., Kim, J.H., Choo, K.H. *et al.* (2000) Hydrophilic modification of polypropylene microfiltration membranes by ozone-induced graft polymerization. *J. Membr. Sci.*, **169**, 269–276.

42. Kang, J.S., Shim, J.K., Huh, H. and Lee, Y.M. (2001) Colloidal adsorption of bovine serum albumin on porous polypropylene-g-poly(2-hydroxyethyl methacrylate) membrane. *Langmuir*, **17**, 4352–4359.

43. Yang, Q., Xu, Z.K. and Dai, Z.W. (2006) Modulate protein adsorption and enzyme activity by surface modifications on polypropylene microporous membranes, in *Proteins at the Solid/Liquid Interfaces* (ed. P. Dejardin), Springer-Verlag, Germany, pp. 271–298.

44. Jordan, R., West, N., Ulman, A. *et al.* (2001) Nanocomposites by surface-initiated living cationic polymerization of 2-oxazolines on functionalized gold nanoparticles. *Macromolecules*, **34**, 1606–1611.

45. Zheng, L., Xie, A.F. and Lean, J.T. (2004) Polystyrene Nanoparticles with anionically polymerized polybutadiene brushes. *Macromolecules*, **37**, 9954–9962.

46. Kim, N.Y., Jeon, N.L., Choi, I.S. *et al.* (2000) Surface-initiated ring-opening metathesis polymerization on Si/SiO2. *Macromolecules*, **33**, 2793–2795.

47. Blomberg, S., Ostberg, S., Harth, E. *et al.* (2002) Production of crosslinked, hollow nanoparticles by surface-initiated living free-radical polymerization. *J. Polym. Sci., Part A: Polym. Chem.*, **40**, 1309–1320.

48. Ying, L., Yu, W.H., Kang, E.T. and Neoh, K.G. (2004) Functional and surface-active membranes from poly(vinylidene fluoride)-graft-poly(acrylic acid) prepared via RAFT-mediated graft copolymerization. *Langmuir*, **20**, 6032–6040.

49. Yang, Q. and Ulbricht, M. (2011) Cylindrical membrane pores with well-defined grafted linear and comblike glycopolymer layers for lectin binding. *Macromolecules*, **44**, 1303–1310.

50. Edmondson, S., Osborne, V.L. and Huck, W.T.S. (2004) Polymer brushes via surface-initiated polymerizations. *Chem. Soc. Rev.*, **33**, 14–22.

51. Dong, H., Du, H. and Qian, X. (2009) Prediction of pKa values for oligo-methacrylic acids using combined classical and quantum approaches. *J. Phys. Chem. B*, **113**, 12857–12859.

52. Mondal, S. and Wickramasinghe, S.R. (2008) Produced water treatment by nanofiltration and reverse osmosis membranes. *J. Membr. Sci.*, **322**, 162–170.

53. Sagle, A.C., Van Wagner, E.M., Ju, H. *et al.* (2009) PEG-coated reverse osmosis membranes: desalination properties and fouling resistance. *J. Membr. Sci.*, **340**, 92–108.

54. Szep, A. and Kohlheb, R. (2010) Water treatment technology for produced water. *Water Sci. Technol.*, **62**, 2372–2380.

55. Wandera, D., Wickramasinghe, S.R. and Husson, S.M. (2011) Modification and characterization of ultrafiltration membranes for treatment of produced water. *J. Membr. Sci.*, **373**, 178–188.

56. Bug, A.L.R., Cates, M.E., Safran, S.A. and Witten, T.A. (1987) Theory of size distribution of associating polymer aggregates. I. Spherical aggregates. *J. Chem. Phys.*, **87**, 1824–1833.

57. Halperin, A., Tirrell, M. and Lodge, T.P. (1992) Tethered chains in polymer microstructures. *Adv. Polym. Sci.*, **100**, 31–71.

58. Ulbricht, M. (2006) Advanced functional membranes. *Polymer*, **47**, 2217–2262.

59. Yu, S., Liu, X., Liu, J. *et al.* (2011) Surface modification of thin-film composite polyamide reverse osmosis membranes with thermo-responsive polymer (TRP) for improved fouling resistance and cleaning efficiency. *Sep. Purif. Technol.*, **76**, 283–291.

60. Petrov, S., Ivanova, T., Christova, D. and Ivanova, S. (2005) Modification of polyacrylonitrile membranes with temperature sensitive poly(vinylalcohol-co-vinylacetal). *J. Membr. Sci.*, **261**, 1–6.

61. Latulippe, D.R., Mika, A.M., Childs, R.F. *et al.* (2009) Flux performance and macrosolute sieving behavior of environment responsive formed-in-place ultrafiltration membranes. *J. Membr. Sci.*, **342**, 227–235.

62. Elbrahmi, K., Rawiso, M. and Francois, J. (1993) Potentiometric titration of acrylamide acrylic-acid copolymers-influence of the concentration. *Eur. Polym. J.*, **29**, 1531–1537.

63. Khulbe, K.C., Feng, C. and Matsuura, T. (2010) The art of surface modification of synthetic polymeric membranes. *J. Appl. Polym. Sci.*, **111**, 855–895.

7

Functionalization of Polymeric Membranes and Feed Spacers for Fouling Control in Drinking Water Treatment Applications

Colleen Gorey, Richard Hausman and Isabel C. Escobar
Department of Chemical and Environmental Engineering,
The University of Toledo, USA

7.1 Membrane Filtration

A membrane is a thin layer of polymer or inorganic film that is capable of separating materials as a function of their physical and chemical properties when a driving force is applied across the membrane [1]. Membranes are becoming exceedingly popular because they can be designed to meet stringent standards while requiring significantly less space and time than other separation processes. Further, many conventional systems (e.g. activated sludge) are unable to remediate wastewater to the quality levels needed for discharge. The separation abilities of membranes make membrane filtration a possible candidate for achieving these goals. Not only could membranes remediate wastewater effluents to meet permitted values, but they may also be able to remediate it to the extent that the effluent meets drinking water standards, making reuse safer. The potential for cost-effective use of membranes in water treatment has been widely investigated. Application areas that have been considered include colour removal [2], removal of trihalomethanes and other disinfection by products [3, 4], suspended solids/bacterial removal, and iron removal. Furthermore, membranes, specifically reverse osmosis (RO), are one of the leading technologies for desalination.

Responsive Membranes and Materials, First Edition. Edited by D. Bhattacharyya, Thomas Schäfer, S. R. Wickramasinghe and Sylvia Daunert.
© 2013 John Wiley & Sons, Ltd. Published 2013 by John Wiley & Sons, Ltd.

Membranes for water treatment are typically thin films made up of solid materials, such as cellulose acetate (CA), cellulose diacetate (CDA), cellulose triacetate, polyamide (PA), sulfonated polysulfone, and aromatic polyamide polymers. Membranes may also be made up of a combination of these materials, such membranes are appropriately called thin film composite (TFC) membranes. The material used to make the membrane determines its properties, such as allowable chlorine concentrations and allowable pH ranges. Bhattacharyya [5] provided information on the acceptable levels of chlorine and pH ranges for the different membrane types. Cellulose acetate-based membranes were able to tolerate moderate amounts of chlorine (0.3–1.0 mg/L), but were vulnerable to biological attack, hydrolysis, and chemical reaction with feed waters to form cellulose and acetic acid. In order to minimize these vulnerabilities, it was suggested that the feed water pH should be kept between 4 and 6. Linear polyamide membranes were not tolerant to chlorine, and thus levels should be kept below 0.05 mg/L. Linear polyamide membranes typically showed allowable pH ranges between 4 and 11.

Polyamide or polyuria membranes have zero tolerance to chlorine, and the feed water should be dechlorinated prior to membrane filtration [6]. Aromatic polyamide membranes have minimal resistance to chlorine with allowable concentrations of around 0.05 mg/L. Finally, membranes made up of sulfonated polysulfone are chlorine resistance and can be operated in water with 1.0 mg/L chlorine.

There are three general structures that are used for membranes: homogeneous, asymmetric, and composite. Homogeneous membranes are made from a single polymer and the pore size is essentially uniform throughout the membrane. Asymmetric membranes consist of a uniform substrate support that is covered by a very thin film with slightly smaller pores. A composite membrane is similar to an asymmetric membrane, but the thin layer that covers the substrate is made from a different material [7]. Despite the configuration, the membrane surface characteristics directly contribute to the separation abilities of the membrane. Depending on the polymer, the resistance to chemical attack, biological stresses, and operational pH range vary.

Figure 7.1 is an exaggerated cross-section of a typical composite membrane. The porous support and fibrous support layers only provide mechanical strength to the membrane, they possess no separation capabilities. The thin selective layer provides the characteristics of separation that are inherent in a membrane. Within the selective layer, small pores or channels allow the solvent (i.e. water) to pass through the membrane, while solutes (i.e. contaminants) are rejected back to the bulk phase.

Figure 7.1   *Cross-section of a typical composite membrane.*

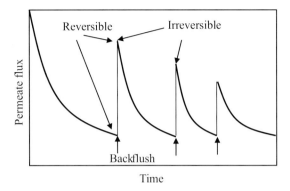

Figure 7.2   *Reduction in permeate flux over time. In this case, periodic cleaning of the membrane by hydraulic backflushing restores a portion of the membrane permeability, but with decreasing efficacy when irreversible fouling occurs. Chemical cleaning may restore a portion of the permeate flux lost to irreversible fouling.*

7.2   Fouling

As materials accumulate near, on, and within the membrane, they may reduce the permeability of the membrane by blocking or constricting pores and by forming a layer of additional resistance to flow across the membrane. Reductions in permeate flux over time may be substantial and represent a loss in the capacity of a membrane facility (Figure 7.2). The characteristics and location of the deposited materials can play an important role in determining the extent and reversibility of permeate flux decline. Cleaning the membrane, either hydraulically or chemically, may remove some of the accumulated materials and partially restore permeate flux. A reduction in permeate flux that cannot be reversed is referred to as membrane fouling. Frequently, both reversible and irreversible decreases in permeate flux decline are referred to as membrane fouling, and materials in the water that produce reductions in permeate flux are collectively referred to as foulants.

The distinction between reversible and irreversible reductions in permeate flux is entirely dependent on the context in which membranes are operated and cleaned [8], that is to say that the process of permeate flux decline is extremely path dependent. The degree of irreversible fouling tends to reflect the memory of the membrane for extreme conditions it has been exposed to during operation, such as the highest transmembrane pressure (TMP) or the worst feed water quality. A different order of addition of chemical cleaning agents (e.g. acid wash followed by base versus base followed by acid) usually produces different degrees of permeate flux recovery [9]. Also, hydraulic-based measures to reverse permeate flux decline, such as backflushing, tend to become less effective over time, to the extent that all of the permeate flux decline would be considered irreversible by this operation [10]. Under these conditions, chemical cleaning may be required to restore permeate flux. Thus, it is probably more practical to consider an irreversible loss in permeate flux to be the difference between the permeate flux of the newly installed membrane and the permeate flux observed after applying the most rigorous cleaning procedure envisioned for a given membrane system. A loss in permeate flux that is truly irreversible, usually requiring replacement of the membrane, is sometimes termed membrane poisoning [8].

Flux losses due to the formation of a removable layer of particles (i.e. reversible abiotic fouling) is called concentration polarization and is caused by the convective transport of charged solute particles toward the surface of the membrane. A typical membrane carries a negative charge while many solute constituents carry a positive charge (i.e. calcium ions), thus a natural attraction occurs between the particles and the membrane. When the concentration polarization layer reaches a critical concentration, the particles may bridge together on the surface of the membrane, may be transported through the membrane, or may flow tangentially across the membrane when shear rates are of proper magnitude.

To control the formation of concentration polarization, membrane cleaning, cross-flow filtration, or backpulsing are often used. Membrane cleaning utilizes a chemical solution (chemical cleaning) or water (aqueous cleaning) at increased shear rates to remove particles from the surface. Using an alkaline chelatant cleaning agent to clean a membrane fouled by conventionally treated surface water, Liikanen [11] was able to double the flux value of the virgin membrane. However, this increase was at the expense of ion retention, indicating that rejection of the membrane was affected. An acidic cleaning process repaired the selectivity, but reduced permeability. Lee [12] discovered that cleaning efficiency was dependant upon the characteristics of the natural organic matter (NOM) and of the cleaning agent. Hydrophobic NOM was cleaned more effectively by acid and caustic cleanings, whereas hydrophilic NOM was effectively removed by high ionic strength cleaning (0.1 M NaCl). Also, increased cross-flow velocity and longer cleaning times were observed to increase the efficiency.

Cleaning procedures are expensive, provide only temporary relief to flux losses, and are capable of damaging the membrane with application [13]. A membrane that is highly resistant to fouling requires infrequent cleanings, reduces operating and disposal costs, increases the operational life, and provides consistent permeate quality.

Backpulsing is described as reversal of the TMP for a short period of time (less than one second) every few seconds [14]. This reverses the flow of permeate through the membrane and flushes accumulated solutes away from the surface of the membrane. Ma [14] applied backpulsing to the filtration of 0.14 g/L *E. coli* and increased permeate volume collection by 1.7 times when compared to permeate volume collected without the use of backpulsing.

Any material that is not removed by these particular techniques can adsorb to the surface of the membrane and is termed an irreversible membrane foulant. Fouling can occur on the surface of the membrane, at pore openings, and within the pores, and leads to permanent flux losses and shortened membrane life. There are two major forms of membrane fouling: abiotic fouling and biofouling. Abiotic fouling is the formation of a "cake layer" or "gel layer", consisting of rejected material, while biofouling is the accumulation of microorganisms on the membrane surface and within the pores of the membrane. It is widely known that NOM is the single largest contributor to abiotic fouling [15]. Factors such as chemical conditions of the feed water (e.g. ionic strength, pH, and presence of divalent cations), configuration of NOM in the feed water (e.g. compact or linear), and initial flux values, intimately interacted and significantly varied the degree of fouling. High ionic strength, low pH, and the presence of divalent cations along with compact NOM led to a dense layer of NOM on the surface of the membrane, and resulted in severe flux decline. However, feed waters with low ionic strength, high pH, and no divalent cations with linear configured

NOM formed a loose, thin layer on the surface, leading to trivial flux decline. At very high initial permeate flux values, significant fouling and flux decline occurred despite feed water composition. Results from a subsequent study by Seidel and Elimelech [15] supported the optimum operating conditions for membrane processes as low initial flux, high cross-flow velocities, and low concentrations of calcium ions.

The rate of initial fouling of a membrane surface was investigated by Vrijenhoek [16]. Four aromatic polyamide thin-film composite membranes were characterized for physical surface morphology, surface chemical properties, surface zeta potential, and specific surface chemical structure. Initial flux, cross-flow velocity, and feed water ionic strength were varied. As previously discussed, ionic strength and cross-flow velocity affected the severity of abiotic fouling, but not the order in which each membrane was fouled. Conclusions drawn from the characterization data suggested that the rate and extent of colloidal fouling were most significantly influenced by the surface roughness of the membrane, regardless of the chemical or physical operating conditions.

A membrane that exhibits high surface roughness inherently has distinct peaks and valleys. The valleys provide paths of least resistance, therefore a majority of permeate is transported through the membrane via these valleys. However, during operation, the valleys easily become clogged, which initiates membrane fouling and leads to flux decline, frequent cleanings, and eventually, complete blockage of the membrane pores.

Another form of fouling that leads to flux decline and membrane degradation is biofouling. Biofouling occurs due to the accumulation of microorganisms on the surface. As microorganisms colonize and proliferate, a biofilm is formed. The biofilm consists of an extracellular polymeric substances (EPS), which is produced by the microorganisms themselves, and provides an ideal environment for bacteria to proliferate. Baker and Dudley [17] determined that biofilms typically consisted of moisture (>90%), organic matter (50% of dried content), humic substances (40% of organic matter), minimal inorganic components, and high microbiological counts ($>10^6$ cfu/cm^2). The biofilm was also capable of capturing nutrients from the surrounding environment to supply the microorganisms with a food source [18].

The typical method of controlling biofouling is through the use of biocides. Biocides are able to kill bacteria, but removal of the biofilm, which is directly responsible for flux decline, is difficult. Further, bacteria and viruses have been known to become resilient to biocides, and biofilms can become "conditioned" or "hardened" by frequent cleaning applications [17]. In addition, oxidizing biocides, such as chlorine, bromine, chloramines, chlorine dioxide, hydrogen dioxide, peroxyacetic acid, and ozone, can convert large molecular weight organic molecules into smaller, easily consumable organic components, which can be used by microorganisms as a food source [17, 19, 20]. These easily consumable components, termed assimilable organic carbon (AOC) compounds, are related to the biostability of the distribution system. The level of AOC in finished water to prevent increases in heterotrophic bacteria growth has been correlated to 10 µg/L for non-chlorinated systems [21]. For chlorinated systems, AOC concentrations of less than 50–100 µg/L were needed to prevent microbial growth [22]. AOC removal from feed waters typically can be achieved by biological treatment methods.

Oxidizing biocides, used to prevent biofilm formation, attack a membrane's amide groups resulting in N-chloro derivatives [23, 24]. Aromatic rings can also be chlorinated via electrophilic substitution on the ring or Orton rearrangement to form N-chloroamide [24].

Interactions with oxidants pose a significant problem since low quality raw waters require pretreatment, including biocide addition, to reduce biofouling. Furthermore, oxidizing biocides cannot be used in conjunction with potable water systems, are not compatible with most membrane polymers, and increase the handling and disposal costs.

7.3   Improving Membrane Performance

In order to maintain membrane processes as economical alternatives to conventional water treatment technologies, they must produce a high quality permeate at a fast rate, and be able to maintain that production for an extended period of time. However, the relationship between flux and selectivity along with fouling introduce a challenge that must be addressed for all membrane applications. There are three main areas of interest when it comes to improving membrane performance: (i) the synthesis process, (ii) the application process, and (iii) post-synthesis modification. Synthesis process improvement involves using the techniques, methods, and materials of the manufacturing processes to produce a high performance membrane. The application process involves the specific operating parameters for a membrane system. These include selecting the raw water characteristics, operating pressure, and cleaning intervals to allow the system to operate at maximum efficiency. Post-synthesis modification involves modifying the membrane after the initial manufacturing process is complete.

There are several main objectives of post-synthesis modification. These include, but are not limited to, decreasing the hydrophobicity of the membrane (i.e. making the membrane more hydrophilic), reducing the surface roughness of the membrane, increasing the selectivity, and altering the surface charge of the membrane. Many membranes are naturally hydrophobic due to the polymers that are used in the manufacturing process. A drawback of a membrane exhibiting a hydrophobic nature is that it increases the required TMP that must be utilized for operation, since the solvent (i.e. water) is repelled by the surface of the membrane. However, if a membrane can be rendered hydrophilic, the solvent is attracted to and transported through the membrane at a much faster rate, which reduces the required TMP for operation. In addition, many solutes found in feed waters are hydrophobic, so utilizing a hydrophilic membrane could increase the membrane's selectivity and fouling resistance.

The severity of abiotic fouling is directly linked to membrane's surface roughness [16]. Membranes with higher surface roughness have distinct peaks and valleys, and are more susceptible to fouling, and with a reduction of roughness, foulants are less likely to adhere to the membrane. Post-synthesis modifications may ultimately lead to a higher flux and selectivity and improved fouling resistance. There have been numerous reports on post-synthesis modification of membranes.

7.3.1   Plasma Treatment

Modification of polymer surfaces can be rapidly and cleanly achieved by plasma treatment due to the possibility of forming various active species on the surface (Figure 7.3). The results of several studies that have investigated this technique have concluded that plasma exposure resulted in a more hydrophilic membrane surface [25–29].

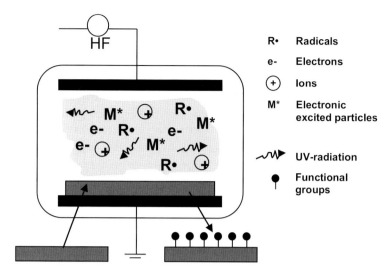

Figure 7.3   *Schematic diagram of the plasma treatment process.*

Kim *et al.* [27] exposed polysulfone UF membranes, which were inherently hydrophobic, to oxygen plasma. The exposure led to the formation of hydrophilic functional groups, such as hydroxyl, carbonyl, and carboxyl groups on the membrane surface. This caused the membrane to become more hydrophilic, and an increase in the permeability of the membrane was observed.

Another major advantage of plasma treatment is the increased fouling resistance of modified membranes. Previous studies have shown that the reduction in fouling potential was mainly a result from the increase in charge repulsion between the membrane and the solute [27]. As charge repulsion became more effective at preventing flux decline, flux recovery in turn was achievable to a greater degree. Gancarz *et al.* [26] revealed that a polysulfone membrane modified by carbon dioxide plasma had a 100% flux recovery after cleaning. A membrane with a high flux recovery is highly regarded in industrial applications. However, prolonged exposure to the carbon dioxide plasma revealed that significant membrane damage occurred in the form of pore enlargement and widening of the pores' size distribution, which could lead to a decrease in membrane selectivity [26].

Wu *et al.* [29] used oxygen and argon plasma to modify aromatic polyamide membranes. Water permeability of the membranes treated by oxygen plasma increased with increasing exposure times. This was a result of the introduction of carboxyl groups on the membrane surface, increasing the hydrophilicity. However, a slight decrease in permeability was observed with argon plasma, since carboxyl groups were not introduced by the argon treatment process. Conversely, argon plasma treatment significantly increased the chlorine resistance of the membrane with increasing treatment times, which was a result of increased cross-linking of the polyamide polymer. Oxygen plasma only moderately increased chlorine resistance, which indicated that oxygen plasma induced minimal cross-linking of the polymer. A membrane with a high resistance to chlorine requires shorter cleaning intervals and a reduced intensity of necessary pretreatment to remove chlorine from the feed water.

Wavhal and Fisher [28] rendered a naturally hydrophobic polyether sulfone water treatment membrane hydrophilic using argon plasma treatment. However, the changes induced by the plasma treatment were not permanent since the induced hydrophilicity declined during a period of two months. In order to prevent the degradation, graft polymerization of polyacrylic acid (PAA) onto the plasma modified membranes was required. All flux values measured (i.e. initial flux, flux after protein fouling, flux after aqueous cleaning, and flux after caustic cleaning) for the plasma modified and PAA grafted membrane were higher when compared to the pristine membrane. Argon plasma treatment combined with PAA grafting produced a hydrophilic membrane with a significantly higher flux and a resistance to protein adsorption fouling, which was easily cleaned by hydrodynamic or caustic methods.

Ulbricht *et al.* [30] modified the polyacrylonitrile (PAN) and polysulfone (PS) ultra-filtration membranes with H_2O plasma and with He plasma resulting in a drastic and permanent increase in the surface hydrophilicity of PS membranes. However, in contrast to the behaviour of PAN UF membranes, the PS surface pore structure was changed as indicated by altered water permeability and reduced protein retentions. Belfort *et al.* [31] also showed the increased hydrophilicity with the application of low temperature helium plasma-induced surface grafting of polyether sulfone (PES) membranes with N-vinyl-2-pyrrolidone (NVP). *Plasma* treatment roughened the membrane surface. When using a filtration protocol to simulate protein fouling and cleaning potential, the surface modified membranes were notably less susceptible to BSA fouling than the virgin PES membrane. In addition, the modified membranes were easier to clean and required little caustic to recover the permeation flux. Morphological transformation of polypropylene by a cold nitrogen plasma treatment resulted in formation of macroradicals onto the polypropylene surface [32, 33]. Plasma treatment has limitations such as slow and expensive processing and has found limited applications.

7.3.2   Ultraviolet (UV) Irradiation

Ultraviolet (UV) radiation is defined as that portion of the electromagnetic spectrum between X rays and visible light, that is, between 40 and 400 nm (30–3 eV). Ultraviolet energy has been extensively applied to assist surface graft polymerization of polymers in the presence of a photoinitiators or photosensitizers, such as benzophenone (BP) [34]. Photochemical graft polymerization utilizes monomers and/or polymers, which are photochemically sensitive. Thus, when the monomer or polymer is distributed onto the membrane, it is subsequently exposed to ultraviolet (UV) radiation at various wavelengths. The exposure to UV induces the photochemically sensitive polymers to graft onto the membrane surface, effectively becoming part of the modified membrane. This technique, as with simple graft polymerization, typically utilizes monomers and polymers that can increase the hydrophilicity of the membrane, which, in turn, can increase flux and reduce fouling.

Kaczmarek *et al.* [35] modified a commercial, purified poly(vinyl chloride) (PVC) by low temperature air-plasma and short-wavelength UV-irradiation resulting in the oxidation of PVC films, which was connected to the formation of functional groups enhancing polymer wettability. This process was very fast and efficient in air-plasma but slower during PVC exposure to *UV*. They also reported the possibility of conformational changes associated with PVC modified by plasma method during storage. In the case of UV-irradiated PVC, secondary, dark reactions were observed during storage at ambient conditions. Kim *et al.* [36] also observed improved surface wettability by grafting NVP to the PVC membrane through UV irradiation assisted grafting. Kowal *et al.* [37] also observed improved hydrophilicity of polysulfone membranes with UV irradiation and hydrogen peroxide plasma treatment. Zhengmao *et al.* [38] reported increased surface roughness above a fluence of 16 J/cm^2 indicating onset of etching. Topographical changes were not observed below a fluence of 16 J/cm^2. Below this fluence value, the poly(ethylene terepthalate) (PET) surface chemistry was characterized by an increasing loss of C=O moiety and carboxylic acid production with modification. UV irradiation of PET in the presence of bi-functional media (i.e. molecules that have the potential for ensuring cross-linking between functional groups, in this case, diallyl compounds) resulted in an increase of surface modulus, lower expansion, and lower soft/melt temperatures indicating surface cross-linking and a decrease in crystallinity [39]. A significant increase in the alkali resistance was also reported after irradiation, indicating wettability and cross-linking effects on the polymer. These studies indicated the ease and longevity of the polymer surfaces resulting from UV assisted modification. However, slight increases in the fluence dosage or exposure times can lead to deteriorating effects on the thin polymeric selective layer of the membranes.

Ma [14] investigated the modification of polypropylene membranes with photo-induced grafting of neutral poly(ethylene glycol 200)monomethacrylate (PEG200MA), positively charged dimethyl aninoethyl methacrylate (DMAEMA) or negatively charged acrylic acid (AA). Using a 0.14 g/L *E. coli* solution in a cross-flow filtration apparatus, very little change in permeate flux was observed for modified membranes. In fact, backpulsing had to be employed in order to obtain an increase of permeate flux.

Photo-induced membrane modification has consistently been studied as an effective tool to increase the hydrophilicity of membranes. The degree of grafting can be controlled by monomer solution concentration, exposure time, and UV intensity. However, a significant drawback to this procedure is the proven fact that exposure to UV radiation can increase the pore size of the membrane [40]. This leads to higher flux values and increased fouling resistance, but at the same time, the selectivity of the membrane is severely impacted. Photochemical modification of membranes is most likely limited to a select few membranes as well as a limited number of membrane applications.

7.3.3   Membrane Modification by Graft Polymerization

Another common technique that is used to modify water treatment membranes is to physically graft an additional polymer onto and/or within the surface of the membrane (Figure 7.4). The additional polymer is typically hydrophilic, which causes a given

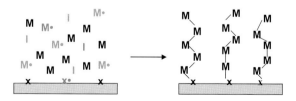

Figure 7.4   *Schematic diagram of a graft polymerization process.*

membrane that is hydrophobic to become more hydrophilic. The goal of this modification technique is to create a membrane with a highly hydrophilic surface while maintaining the flux and separation characteristics. An increase in hydrophilicity can increase the mass transfer coefficient, providing a more efficient membrane.

Over many years, graft polymerization has been studied in various fields of industrial applications, using different innovative techniques including different kinds of chemical and physical approaches. Modification via graft polymerization has many advantages over other methods, such as its ease of use and controllable introduction of graft chains to the surface with the bulk properties unchanged. Grafting of side chains can be performed in two ways, "grafting from" and "grafting to" methods [34]. In the former method, the membrane surface consists of reactive radicals, while in the latter case, grafting chains carry a functional group that may react with complementary reactive sites on the surface. The procedures involved in grafting to surfaces are not straightforward enough for industrial applications, although the precise control of chain structure is in principle easier than with other methods [34]. Grafting can be achieved using different chemical techniques such as chemical oxidation, plasma discharge method and UV irradiation, and physical methods, such as radiation polymerization [34], as previously discussed.

Graft polymerization of hydrophilic polymers has been extensively studied and has provided promising results. However, drawbacks to this procedure can be substantial. Pore size reduction is a major consideration since it is directly capable of greatly reducing membrane productivity. Modification techniques that do not require the use of additional polymers are an area of great interest since they are less apt to constrict pore size.

7.3.3.1   Free Radical Grafting Reactions

A radical is an atomic or molecular species having one or more unpaired electrons. Though they are electron deficient they are uncharged, resulting in different chemistries compared to those of ionic or even electron species. Apart from a very few stable radicals such as nitric oxide (NO•), most radicals are very unstable and reactive making them suitable for initiating chemical reactions with typical stable polymeric surfaces. The amount of extra stabilization that adjacent lone pairs, π bonds, and σ bonds provide to radicals is not as great as that which they provide to carbocations. The reason is that the interaction of a filled atomic molecular orbital with an empty atomic orbital (as in carbocations) puts two electrons in a molecular orbital with reduced energy. In contrast, the interaction of a filled atomic or molecular orbital with a half-filled atomic orbital (free radicals) puts two electrons in a molecular orbital with reduced energy and one electron in a molecular orbital with increased energy [41].

Pairs of electrically neutral free radicals can be formed via homolytic bond breakage that can be achieved by heating in non-polar solvents or the vapour phase. All molecular species dissociate into radicals at elevated temperatures. Neutral radicals can also be produced in polar environments through oxidation processes [41]. Once the free radicals are formed on the membrane surface, due to their extreme reactivity, polymerization proceeds by coupling monomer chains to the membrane surface [34]. Other polymerization reactions that are widely used in industry are polycondensation reactions. Condensation polymers are formed from monomers bearing two or more reactive groups having a character such that they may condense intermolecularly with elimination of a by-product, usually water. Unlike addition reactions, the molecular formula of the polymer in this case is not an integral multiple of the formula of the monomer [41]. Condensation polymerization offers the possibility of constructing high polymer molecules of accurately known structure, and have the advantage of achieving precise average molecular weight control through varying the degree of condensation [42, 43].

7.3.3.2 Green Chemistry

Although chemical methods have been widely used to achieve membrane surface modifications, the involvement of free radicals and strong solvents make them barely sustainable processes [44]. Green chemical processes demand an increased level of attention [44, 45] because they are more benign towards the environment. Biodegradable polymers, biocatalysis, sustainable technologies, and renewable resource utilization are amongst the important components of environmentally-friendly synthesis and processing, that is, green chemistry.

Biocatalysis consists of using enzymes, microbes, and higher organisms to carry out chemical reactions. It has many attractive features in the context of green chemistry such as mild reaction conditions (physiological pH and temperature), an environmentally compatible catalyst (an enzyme), combined with high activity and impressive chemo-, regio-, and stereo-selectivity in multifunctional molecules. Furthermore, the use of enzymes generally circumvents the need for functional group activation and avoids the protection and deprotection steps required in traditional syntheses [46]. This leads to processes that generate less chemical waste and are, therefore, both environmentally and economically more attractive than conventional routes. Rapid progress in the biotechnology industry has also led to the increasing popularity of biocatalysis. It has the great advantage of being a more sustainable process as it utilizes naturally-available enzymes to promote grafting through condensation reactions rather than radical-generating chemicals, which may have a larger impact on the environment. It is well established that enzyme-catalysed reactions in organic media have the advantage of achieving high selectivity under mild reaction conditions [44].

Hydrophilic flexible chains of PEG can be attached to the membrane surface through an enzymatic polycondensation reaction. One major advantage of using condensation polymerization for surface modification is its ability to reduce the degree of homopolymerization, which is common in free radical polymerization. In addition, the primary activation of a monomer in free radical reactions is followed by the addition of other monomers in rapid succession until the growing chain is eventually deactivated. Once the chain formation begins, it comes to an end very quickly through chain propagation. So during any instant of the reaction, the reaction mixture contains just monomer and converted polymer [43]. In free radical reactions, where monomer, membrane, and initiator react together, a large homopolymer formation might occur as liquid monomer has equal or higher access to the initiator than initiator access to the polymeric membrane surface. Conversely, in condensation polymerization, reaction occurs between the two functional groups present in the monomers and membrane surface or between two monomers. Reaction proceeds through the formation of various chains of intermediate lengths. At a given time in the reaction, dimers, trimers, and tetramers are present, and all these molecules have a chance to attach to the membrane surface depending on the available reaction time.

7.3.3.3 Graft Polymer Examples

7.3.3.3.1 Hydrophilic Enhancer: Poly(Ethylene Glycol).
Poly(ethylene glycol)-lipid (PEG-lipid) conjugates are widely used in the field of nanoparticulate drug delivery. They

are used to provide a protective cloud of polymer around liposome thereby increasing the longevity and stability of the nanoparticle in circulation by reducing disruptive interaction with different solutes such as plasma proteins [47, 48]. PEG is also known as polyethylene oxide (PEO) or poly oxyethylene (POE) depending on the definition of its monomer molecule as ethylene glycol or ethylene oxide or oxyethylene respectively. The success of PEG layers in nanoparticulate drug delivery is attributed to its hydrophilicity, which prevents the penetration of incompatible proteins to the liposome surface by disrupting their interactions. The high flexibility of PEG chains also plays a vital role in denying incoherent proteins to the surface. Hydrophilic polymers with rigid chains may not provide a sufficient protective layer for liposomes as noticed for liposome-grafted dextran [49, 50]. PEG also possesses excellent characteristics such as very low toxicity [51] and non-biodegradability [52], which make it an ideal choice for use in the water purification industry. These distinguishing properties of PEG have been explained by its chain's high mobility associated with conformational flexibility and water binding ability [49, 50, 53].

Hydrophilic polymer chains, such as PEG, are widely used as grafting polymers for water treatment membranes due to their ability to disrupt hydrophobic interactions between NOM or protein foulants and the membrane selective layer [32, 54]. PEG has a high exclusion volume in water making it a suitable material for water-based applications [55]. Apart from these properties, uncharged PEG has excellent hydrophilicity owing to its double free electron pair on the oxygen atom, and flexibility owing to its linear structure containing C, O, and H [48, 55]. These two properties also make PEG an excellent choice for the application in membrane surface modification processes. Another advantage of flexible polymers is that they more easily form a dense conformational protective cloud as graft chains than many rigid polymers [48]. PEG has been the most investigated of the polymers that are capable of reducing interactions with proteins [56]. Its hydrophilicity plays a crucial role in reducing hydrophobic interactions that are responsible for organic fouling of the membrane [32] and its flexibility eliminates incidence of steric hindrance during the grafting process.

7.3.3.3.2   Stimuli-Responsive Polymers.   Stimuli-sensitive polymers have been gaining attention in recent years due to interest in the unique property that these polymers change their conformation from a coiled state to a globular one in the presence of a stimulus [57]. The stimuli can be pH, temperature, ionic strength, and electric and magnetic fields [58]. Which stimulus works best depends on the polymer. Some polymers are sensitive to multiple stimuli. The idea that the polymer chain can expand and collapse and has led to the idea of "gateways" and triggerable particle releases such as drug delivery [58]. When temperature is the stimulus, the transition from coiled to globular occurs at a critical solution temperature and depending on the polymer and its solutions it can be an upper critical solution temperature (UCST) or a lower critical solution temperature (LCST). The UCST occurs when a polymer goes from a globular to coiled state and the LCST occurs when the polymer goes from a coiled state to a globular one, both with temperature increasing. The transition of state only occurs in the polymerized form, that is, monomer units of a chemical are only stimuli-sensitive after they have been polymerized.

An example of a temperature responsive polymer is N-isopropylacrylamide (NIPAAM). It has an LCST of 34°C in water, which is close to the human body temperature of 37°C and has been gaining interest in biomedical applications because of this feature [57]. However, the value of the LCST can vary between 30 and 35°C depending on the detailed microstructure of the polymer. The LCST can be increased or decreased depending on the desired application by adding either hydrophilic or hydrophobic monomers. Hydrophilic co-monomers increase the LCST while hydrophobic co-monomers decrease it. The unusual thermal behaviour of polyNIPAAM is the abrupt transition (linear to coiled) from a hydrophilic to a hydrophobic peripheral structure at the LCST. A gel of polyNIPAAM expels a large volume of free water as the balance between the hydrophilic (–CONH–) and hydrophobic (–CH(CH$_3$)$_2$) moieties is shifted. Below the LCST water is a good solvent for the polymer because the polymer–solvent interactions are stronger than the polymer–polymer interactions. Hydrogen bonds are bound to the hydrophilic moieties and the hydrophobic moieties are such that the polymer is in an extended coil conformation. The H-bonds between the water molecules and amide groups make water molecules orientate neatly around isopropyl groups. These structured water molecules act cooperatively to form a cage-like structure, which results in the polyNIPAAM chains becoming hydrated and soluble in the surrounding water. Above the LCST, the polymer–polymer interactions increase due to the hydrophobic interactions and at this point water is a poor solvent for the polymer. The polyNIPAAM chains precipitate from water due to the dissociation of the hydrating water molecules from polymer. The result is the aggregation and formation of compact globules as the hydrogen bonds are broken and water is expelled from the coils. Also note that H-bonding between amide–water (hydration) and amide–amide (dehydration) has a potential role in this phase transition behaviour as depicted in Figure 7.5.

Another example is hydroxypropyl cellulose (HPC), which is a commercially important cellulose ether. It is a water soluble and non-toxic polymer. As a cellulose derivative, it has the typical feature of reversed solution phase behaviour. It precipitates at high temperatures. Aqueous solutions of HPC possess a lower critical solution temperature (LCST) value of 41°C, dense polymer solutions at 43°C, and gels at 46°C. In general, HPC is an effective polymer to use. The availability of hydroxyl groups makes polymerization simple, via divinyl sulfone (DVS), and since it is a cellulose derivative it is safe to use.

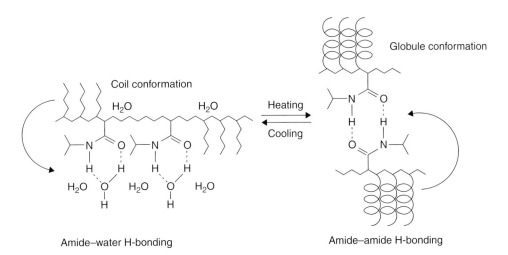

Figure 7.5 *Illustration of the coil to globule conformation change of polyNIPAAM. (Reprinted under the terms of the STM agreement from [89] Copyright (2007) Elsevier Ltd).*

In recent studies, NIPAAM and HPC were grafted to a cellulose acetate ultrafiltration membrane surface [59, 60]. While surface activation was observed at the nano-scale level, no improvements in fouling control were observed during operation under a temperature cycle. The NIPAAM-modified membranes only showed improvements with respect to constant cold operation when compared to unmodified membranes. The possible reason for this is the higher hydrophilicity at cold temperatures of the NIPAAM-modified membranes. On the other hand, under all conditions, the HPC-modified membrane showed a lower flux decline than the unmodified membrane, and supported the advantage of the cycling of hot/cold temperatures in maintaining higher flux values. Therefore, HPC-modified membranes would be more successful than NIPAAM-modified membranes in controlling flux.

7.3.4   Ion Beam Irradiation

Ion beam irradiation has long been recognized as an effective method for the synthesis and modification of diverse materials, including polymers [61]. Ion beam irradiation is the bombardment of a substance with energetic ions. When the ions penetrate through the surface of a membrane, they may eliminate the tall peaks and deep valleys, resulting in an overall reduction in surface roughness [62]. As the ions penetrate the membrane, they lose energy to their surroundings (membrane structure) by two main processes: interacting with target nuclei (nuclear stopping) and interacting with target electrons (electronic stopping, [63]).

Nuclear stopping energy losses arise from collisions between energetic particles and target nuclei [59, 63]. Atomic displacement occurs when the colliding ion imparts energy greater than certain displacement threshold energy. If the energy is not great enough for displacement, the energy dissipates as atomic vibrations known as phonons. The threshold energy is the energy that a recoil requires to overcome binding forces and to move more than one atomic spacing away from its original site. The interaction of an ion with a target nucleus is treated as the scattering of two screened particles since the nuclear collision occurs between two atoms with electrons around protons and neutrons. Nuclear stopping varies with ion velocity as well as the charges of two colliding atoms, as it is derived with consideration of the momentum transfer from ion to target atom and the inter-atomic potential between two atoms.

Electronic stopping energy losses arise from electromagnetic interaction between the positively charged ions and the target electrons [63]. There are two mechanisms, one that of glancing collisions (inelastic scattering, distant resonant collisions with small momentum transfer), which are quite frequent but each collision involves a small energy loss ($<100\,\mathrm{eV}$). The other is knock-on collisions (elastic scattering, close collisions with large momentum transfer), which are very infrequent but each collision imparts a large energy to a target electron (>100 eV). Both collisions transfer energy in two ways: electronic excitation and ionization. Excitation is the process in which an electron jumps to a higher energy level, while in ionization an orbital electron is ejected from the atom.

7.3.4.1   Ion Beam Induced Property Changes

Various gaseous molecular species are released during irradiation. The most prominent emission is hydrogen, followed by less abundant heavier molecular species, which are scission products from the pendant side groups and chain-end segments and their reaction

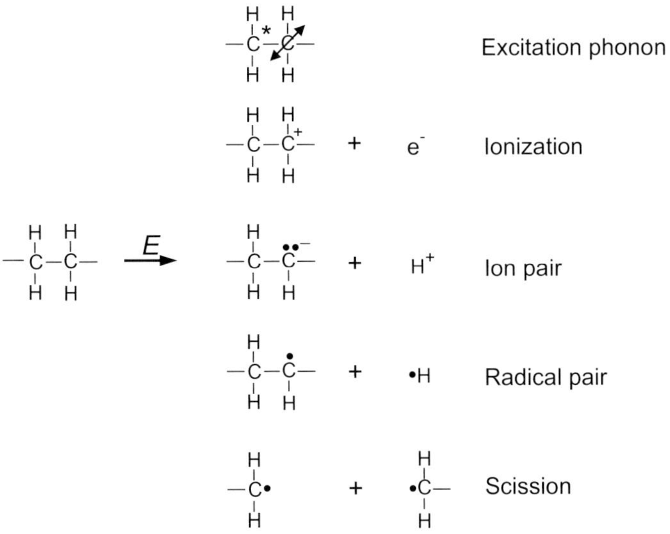

Figure 7.6   *Typical consequences induced by ion irradiation which include electronic excitation, phonons, ionization, ion pair formation, radical formation, and chain scission.*

products [63]. Figure 7.6 illustrates various functional chemical entities created by irradiation. Cross-linking occurs when two free dangling ion or radical pairs on neighbouring chains combine, whereas double or triple bonds are formed if two neighbouring radicals in the same chain combine. It has been well established that mechanical, physical, and chemical property changes in polymers are determined by the magnitude of cross-linking and scission, and that cross-linking enhances mechanical stability while scission degrades mechanical strength [64].

Cross-linking generally increases hardness and slows diffusion, improves wear and scratch resistance, and decreases solubility in chemical solvents. Electrical conductivity and optical density increase due to the formation of cross-links and conjugated double and triple bonds by irradiation. The delocalized π-electrons in the conjugated bonds are loosely bound and thus more mobile than the covalent σ-bond electrons. Furthermore, charge carrier mobility increases by cross-links, which facilitate the transport of charge carriers across the chains. Otherwise, charge carriers must hop across the chains for conduction. The loosely bound π-electrons can be excited by the energies of visible light, and thus colour changes occur because light is absorbed when these electrons are excited. Radiation induced defects, such as anions and radicals (donors) and cations (acceptors), form a broadened band in the band gap and result in the absorption of light as well. Energetic blue light is absorbed first and the colour changes from pale yellow to reddish brown and eventually to a dark colour with increasing irradiation dose. At very high doses, a metallic lustre appears because light is scattered by the abundant π-electrons in a manner similar to the effect of free electrons in metals. On the other hand, scission causes bond breakage and increases dissolution of polymers in solvents. This feature has been used for lithography with positive-resists in the electronic industry [65].

There have been a number of studies examining the effects of ion beam irradiation on gas separation membranes [62, 66–68]. These studies exposed polyimide gas separation membranes to different ion irradiation fluences and energies. Ion irradiation fluences refer to the number of ions implanted into a unit of area of the membrane. Ion energy is directly related to the depth the ions will penetrate into the membrane. As discussed earlier, ion beam irradiation provides energy to the electrons and nuclei of the membrane. The intensive energy deposition in polymers can lead to the following: (i) formation of volatile molecules and free radicals which leave defects in the polymer matrix, (ii) creation of additional cross-linking between polymer chains, (iii) formation of new chemical bonds, and (iv) chemical reactions with chemical atmosphere such as oxidation [63]. These four events can, in turn, lead to membrane microstructure alterations.

There are three regions of modification resulting from ion irradiation, low, medium, and high [69, 70]. The result of low dose modification includes small changes in chemical structure and morphology, medium dose modification includes significant cross-linking and modification, while high dose modification of the polymer causes formation of graphite like material. Even though there are defined modification ranges, there are no set doses for the ranges due to the fact that the ion dose is only one factor that determines the amount of modification. The extent of modification depends on several key parameters, such as ion energy, ion type, dose, virgin polymer structure, and current density. The impact of these parameters on the energy transfer to the polymer can be estimated using the Stopping and Range of Ions in Matter (SRIM) Monte Carlo simulation.

7.3.4.2   Ion Irradiation of Polymeric Membranes

Previous studies [66–68, 71, 72] found that ion beam irradiation resulted in gas separation membranes with both increased permeability and selectivity, two characteristics which naturally have a trade-off relationship toward each other. The improvements in membrane performance were believed to be a result of microstructure modifications, which were proved by a narrow but intensive free volume distribution. AFM studies of ion beam irradiated polyimide films showed that ion beam irradiation eliminated deep valleys and tall peaks on the surface of the polyimide films even at very low doses of irradiation and that a very smooth surface was observed after ion beam irradiation [63].

Lastly, Chennamsetty *et al.* [73, 74] irradiated commercial water treatment sulfonated polysulfone membranes. The results showed that some sulfonic and C-H bonds were broken and new C-S bonds were formed after irradiation. AFM analysis showed that the roughness of the membranes decreased after irradiation, and the decrease in surface roughness was proportional to the increase in irradiation fluence. An increase in flux after ion beam irradiation was also observed along with a smaller flux decline during operation. Hydrophobicity, pore size distribution, and membrane rejection efficiencies were not affected by ion beam irradiation. Overall, irradiation led to an improvement in membrane performance.

7.4   Design and Surface Modifications of Feed Spacers for Biofouling Control

Much work has been done showing that biofouling is essentially a feed spacer problem since biofouling on the feed spacer rather than the membrane itself causes most of the feed channel pressure drop increase, initially cellular attachment occurs at the membrane/spacer

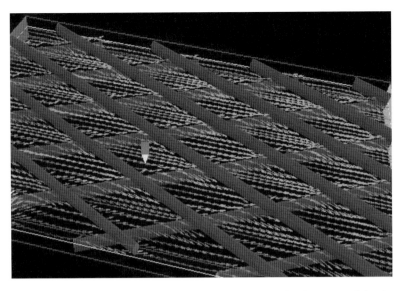

Figure 7.7   *A typical feed spacer configuration. (Reprinted under the terms of the STM agreement from [76] Copyright (2007) Elsevier Ltd).*

interface, and restricted biomass accumulation on the feed spacer already has a strong impact on the flow velocity profile [75, 76]. Feed spacers, which usually have the form of non-woven crossed cylinders, serve to separate adjacent membrane leaves and create flow passages, but also promote flow unsteadiness and enhance mass transport. Figure 7.7 is an example of a typical membrane feed spacer configuration.

Research that has been conducted on feed spacers has mostly focused on feed spacer geometry, generally to reduce the pressure drop and increase permeate flux [76, 77]. These works have shown that altering the feed spacer geometry can alter wall shear stresses and that such stresses have maxima significantly higher than those corresponding to empty channels. The non-uniformity of shear stresses was also shown to have the ability to be manipulated, which may have implications for membrane fouling. It is apparent that anti-fouling properties were not the primary focus of these studies and biofouling resistance, specifically, was not even mentioned. The only studies found regarding feed spacer surface modifications consisted of methods for bulk modifications of the entire membrane module/membrane/spacer rather than specifically the feed spacer [78, 79].

Surface modifications of feed spacers have also been attempted. Modification to the feed spacer, as compared to that of the membrane, has its advantages due to it generally being constructed from polypropylene (PP). Polypropylene is a commercial polymer that is ever-present in many fields such as textiles, medical devices, automobiles, food packaging, membrane filtration, and so on [80–82]. Chemical resistance, low cost, and versatility have made PP an attractive polymer in many applications [81, 82]. While polypropylene is known for its inherent chemical stability, surface modifications can be performed to make this polymer more attractive.

Such modifications that could increase the anti-biofouling properties of polypropylene would be of interest to the field of membrane science, as well as many industries where

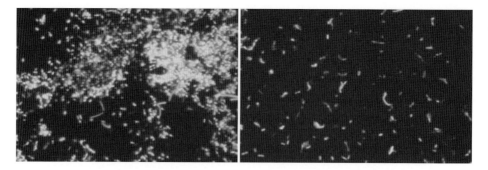

Figure 7.8 *Cells detached from membranes during filtration with* Pseudomonas fluorescens *stained with either PI or pico-green and imaged using a fluorescent microscope using regular polypropylene feed spacers (left) and copper-charged polypropylene feed spacers (right). (Reprinted under the terms of the STM agreement from [86] Copyright (2010) Elsevier Ltd). See plate section for colour figure.*

sanitation is of great significance. Grafting of unsaturated vinyl monomers onto PP is a convenient route to develop new polymeric materials with synergistic properties [83]. Polymer–metal complexes have been extensively studied and successfully employed in several fields [84]. As in low-molecular-weight compounds, a polymer ligand must donate unshared electrons to the metal ion to form metal–ligand bonds. Amongst the multidentate ligands, iminodiacetic acid (IDA) possesses one aminopolycarboxylate and provides a reactive secondary amine hydrogen to react with alternate functional groups [84]. Hence, IDA can be more easily introduced to the side chain of a polymer or vinyl monomer via an epoxy group reaction of glycidyl methacrylate (GMA) and IDA. In studies by Hausman *et al.* [85, 86], the IDA was chelated with copper (Cu) to make low biofouling Cu-charged PP. Copper was chosen since many studies have been conducted on the use of copper ions to disinfect water against microbial biofilms with effective dosages of a few tenths of 1 mg/L [87]. Copper is thought to be cytotoxic by causing changes in the plasma membrane permeability or efflux of intra-cellular K^+ during the entry of Cu^{2+} ions [88]. It can also participate in Fenton-like reactions generating reactive hydroxyl radicals, which can cause cellular damage imparted via oxidative stress [88]. Furthermore, it is believed to interfere with the enzymes involved in cellular respiration and binds to DNA at specific sites [87]. Results showed that the antimicrobial property of the Cu-charged feed spacers aided in hindering cell adhesion and, consequently, biofilm formation and biofouling, as shown in Figures 7.8 (live and dead cells on the membrane surface) and 7.9 (scanning electron microscopy (SEM) images of biofouled membranes) [86]. Therefore, the use of copper-charged feed spacers showed the potential to increase membrane life and decrease chemical cleanings associated with detrimental biofouling of membranes.

7.5 Conclusion

Developing a membrane with high flux and selectivity along with low fouling is one of the "holy grails" of membrane research. There are physical and chemical limitations, however, that cannot be disregarded. For microfiltration/ultrafiltration (MF/UF) systems higher flux

Figure 7.9 *SEM images of membranes fouled with an unmodified feed spacer (left) and a Cu-charged feed spacer (right).*

operation requires energy input, in the form of backflushes, air scouring, physical agitation, and so on, to reduce fouling. Membrane chemical and physical properties effect how a membrane fouls. Altered surface properties allow a membrane to be functionalized for a specific purpose, such as reducing fouling. Taking existing membranes and making morphological or chemical adjustments influences how the feed solution interacts with the surface. A way to change the properties of a membrane is through post-synthesis modification. Post-synthesis modification is currently an intensive area of research. It has the advantage of imparting the desired properties to the surface of interest while keeping the properties of bulk intact. Different innovative physical and chemical surface modification techniques have been used here. These techniques included ultraviolet (UV), plasma, chemical induced free radical graft polymerization, and surface irradiation. These techniques are also frequently used to functionalize wide arrays of commercial polymeric membranes that are available in the market for liquid treatment processes. Feed spacer geometry design and surface modifications have also been successfully used to minimize or prevent fouling of membranes.

Acknowledgements

Much of the work presented here was funded by the National Science Foundation grants NSF 0331778, 0610624, 0714539, 0754387, and 1037842, and by grants from the Department of Interior US Bureau of Reclamation and US Geological Survey.

References

1. Mallevialle, J., Suffet, I.H. and Chan, U.S. (1992) *Influence and Removal of Organics in Drinking Water*, Lewis Publishers, Boca Raton.
2. Tan, L. and Sudak, R.G. (1992) Removing colour from a groundwater source. *Journal of the American Water Works Association*, January, **84**(1), 79–87.

3. Blau, T.J., Taylor, J.S., Morris, K.E. and Mulford, L.A. (1992) DBP control by nanofiltration: cost and performance. *Journal of the American Water Works Association*, December, **84**(12), 104–116.

4. Taylor, J.S., Mulford, L.A., Duranceau, S.J. and Barrett, W.M. (1989) Cost and performance of a membrane pilot plant. *Journal of the American Water Works Association*, November, **81**(11), 52–60.

5. Bhattacharyya, D., Williams, M.E., Ray, R.J. and McCray, S.B. (1992) *Membrane Handbook*, Van Nostrand Reinhold, New York.

6. Soice, N.P., Greenberg, A.R., Krantz, W.B. and Norman, A.D. (2004) Studies of oxidative degradation in polyamide RO membrane barrier layers using pendant drop mechanical analysis. *Journal of Membrane Science*, **243**, 345–355.

7. Harrison, R.G., Todd, P., Rudge, S.R. and Petrides, D.P. (2003) *Bioseparations Science and Engineering*, Oxford University Press, New York.

8. Mallevialle, J., Odendaal, P.E. and Wiesner, M.R. (1996) *Water Treatment Membrane Processes*, Mc-Graw-Hill.

9. Madaeni, S., Mohamamdi, S. and Mansour, T.K.M. (2001) *Chemical* cleaning *of reverse osmosis* membranes. *Desalination*, **134**, 77–82.

10. Nabe, A., Staude, E. and Belfort, G. (1997) Surface modification of polysulfone ultrafiltration *membranes* and fouling by BSA solutions. *Journal of Membrane Science*, **133**, 57–72.

11. Liikanen, R., Yli-Kuivila, J. and Laukkanen, R. (2002) Efficiency of various chemical cleanings for nanofiltration membrane fouled by conventionally-treated surface water. *Journal of Membrane*, **195**, 265–276.

12. Lee, H., Amy, G., Cho, J. *et al.* (2001) Cleaning strategies for flux recovery of an ultrafiltration membrane fouled by natural organic matter. *Water Research*, **35**(14), 3301–3308.

13. Chen, J., Kim, S.L. and Ting, Y.P. (2003) Optimization of membrane physical and chemical cleaning by a statistically designed approach. *Journal of Membrane Science*, **219**, 27–45.

14. Ma, H., Bowman, C.N. and Davis, R.H. (2000) Membrane fouling reduction by back-pulsing and surface modification. *Journal of Membrane Science*, **173**, 191–200.

15. Seidel, A. and Elimelech, M. (2002) Coupling between chemical and physical interactions in natural organic matter (NOM) fouling of nanofiltration membranes: implications for fouling control. *Journal of Membrane Science*, **203**, 245–255.

16. Vrijenhoek, E.M., Hong, S. and Elimelech, M. (2001) Influence of membrane surface properties on initial rate of colloidal fouling of reverse osmosis and nanofiltration membranes. *Journal of Membrane Science*, **188**, 115–128.

17. Baker, J.S. and Dudley, L.Y. (1998) Biofouling in membrane systems – A review. *Desalination*, **118**, 81–90.

18. Flemming, H.C. (2002) Biofouling in water systems – cases, causes and countermeasures. *Applied Microbiology and Biotechnology*, **59**, 629–640.

19. Escobar, I.C., Randall, A.A. and Taylor, J.S. (2001) Bacterial growth in distribution systems: Effect of assimilable organic carbon and biodegradable organic carbon. *Environmental Science & Technology*, **35**, 3442–3447.

20. Escobar, I.C. and Randall, A.A. (2001) Case study: Ozonation and distribution system biostability. *Journal of the American Water Works Association*, **93**, 77–89.

21. van der Kooij, D. (1992) Assimilable organic carbon as an indicator of bacterial regrowth. *Journal of the American Water Works Association*, **84**, 57–65.
22. LeChevallier, M.W., Badcock, T.M. and Lee, R.G. (1987) Examination and characterization of distribution system biofilms. *Applied and Environmental Microbiology*, **53**, 2714–2724.
23. Glater, J., Hong, S.K. and Elimelech, M. (1994) The search for a chlorine-resistant reverse osmosis membrane. *Desalination*, **95**, 325–345.
24. Jayarani, M.M., Rajmohanan, P.R., Kulkarni, S.S. and Kharul, U.K. (2000) Synthesis of model diamide, diester and esteramine adducts and studies on their chlorine tolerance. *Desalination*, **130**, 1–16.
25. Dmitriev, S.N., Kravats, L.I., Sleptsov, V.V. and Elinson, V.M. (2002) Water permeability of poly(ethylene) terephthalate track membranes modified in plasma. *Desalination*, **146**, 279–286.
26. Ganzarz, I., Pozniak, G. and Bryjak, M. (1999) Modification of polysulfone membranes 1. CO2 plasma treatment. *European Polymer Journal*, **35**, 1419–1428.
27. Kim, K.S., Lee, K.H., Cho, K. and Park, C.E. (2002) Surface modification of polysulfone ultrafiltration membrane modified by oxygen plasma treatment. *Journal of Membrane Science*, **199**, 135–145.
28. Wavhal, D. and Fisher, E. (2002) Hydrophilic modification of polyether sulfone membranes by low temperature plasma-induced graft polymerization. *Journal of Membrane Science*, **209**, 255–269.
29. Wu, S., Xing, J., Zheng, C. *et al.* (1997) Plasma modification of aromatic polyamide reverse osmosis composite membrane surface. *Journal of Applied Polymer Science*, **64**, 1923–1926.
30. Ulbricht, M. and Belfort, G. (1996) Surface modification of ultrafiltration membranes by low temperature plasma II. Graft polymerization onto polyacrylonitrile and polysulfone. *Journal of Membrane Science*, **111**, 193–215.
31. Chen, H. and Belfort, G. (1999) Surface modification of poly (ether sulfone) ultrafiltration membranes by low –temperature plasma-induced graft polymerization. *Journal of Applied Polymer Science*, **72**, 1699–1711.
32. Gullinkala, T. and Escobar, I.C. (2008) Study of a hydrophilic-enhanced ultrafiltration membrane. *Environmental Progress*, **27**(2), 210–217.
33. Gullinkala T. and Escobar, I.C. (2010) A green membrane functionalization method to decrease natural organic matter fouling. *Journal of Membrane Science*, **360**, 155–164.
34. Kato, K., Uchida, E., Kang, E.-T. *et al.* (2003) Polymer surface with graft chains. *Progress in Polymer Science*, **28**, 209–259.
35. Kaczmarek, H., Kowalonek, J., Szalla, A. and Sionkowska, A. (2002) Surface modification of thin polymeric films by air plasma or UV-irradiation. *Surface Science*, **507**, 883–888.
36. Kim, D.S., Kang, J.S., Kim, K.Y. and Lee, Y.M. (2002) Surface modification of a poly (vinyl chloride) membrane by UV irradiation for reducing sludge adsorption. *Desalination*, **146**, 301–305.
37. Kowal, J., Czakowska, B., Bulwan, E. *et al.* (2004) Modification of polysulfone by means of UV irradiation and H2O2 plasma treatment. *European Cells and Materials*, **7**, 59–59.

38. Zhengmao, Z. and Kelly, M.J. (2004) Poly (ethylene terepthalate) surface modification by deep UV (172 nm) irradiation. *Applied Surface Science*, **236**, 416–425.
39. Gao, S.L., Hassler, R., Mader, E. *et al.* (2005) Photochemical surface modification of PET by excimer UV lamp irradiation. *Applied Physics B*, **81**, 680–690.
40. Kilduff, J.E., Mattaraj, S., Pieracci, J.P. and Belfort, G. (2000) Photochemical modification of poly(ether sulfone) and sulfonated poly(sulfone) nanofiltration membranes for control of fouling by natural organic matter. *Desalination*, **132**, 133–142.
41. Jones, M. (2005) *Organic Chemistry*, 3rd edn, Norton & Co, New York, NY.
42. Flory, P.J. (1946) Fundamental principles of condensation polymerization. *Chemical Reviews*, **39**, 137–197.
43. Carothers, W.H. (1936) Polymers and polyfunctionality. *Transactions of Faraday Society*, **32**, 39–53.
44. DeSimone, J.M. (2002) Practical approach to green solvents. *Science*, **297**, 799–803.
45. Kumar, R., Tyagi, R., Parmar, V.S. *et al.* (2004) Biocatalytic "green" synthesis of PEG-based aromatic polyesters: Optimization of the substrate and reaction conditions. *Green Chemistry*, **6**, 516–520.
46. Cheng, H.N. and Gross, R.A. (2005) *Polymer Biocatalysis and Biomaterials*, ACS Publishing, Washington, DC.
47. Song, L.Y., Ahkong, Q.F., Rong, Q. *et al.* (2002) Characterization of the inhibitory effect of PEG-lipid conjugates on the intracellular delivery of plasmid and antisense DNA mediated by cationic lipid liposomes. *Biochimica et Biophysica Acta*, **1558**, 1–13.
48. Torchilian, V.P. and Trubetskoy, V.S. (1995) Which polymers can make nanoparticulate drug carriers long-circulating? *Advanced Drug Delivery Reviews*, **16**, 141–155.
49. Blume, G. and Ceve, G. (1993) Molecular mechanism of the lipid vescile longevity in vivo. *Biochimica et Biophysica Acta*, **1146**, 157–168.
50. Torchilian, V.P., Omelyanenko, V.G., Popisov, I.M. *et al.* (1994) Poly (ethylene glycol) on the liposome surface: On the mechanism of polymer- coated liposome longevity. *Biochimica et Biophysica Acta*, **1195**, 11–20.
51. Pang, S.N.J. (1993) Final report on the safety assessment of polyethylene glycols (PEGs)-6, -8, -32, -75, -150, -14M, -20M. *Journal of the American College of Toxicology*, **12**, 429–456.
52. Zalipsky, S. (1995) Chemistry of polyethylene glycol conjugates with biologically active molecules. *Advanced Drug Delivery Reviews*, **16**, 157–182.
53. Needham, D., Melntosh, T.J. and Lasic, D.D. (1992) Repulsive interactions and mechanical stability of polymer-grafted lipid membranes. *Biochimica et Biophysica Acta*, **1108**, 40–48.
54. Morão A., de Amorim, M.T.P., Lopes, A. *et al.* (2008) Characterization of ultrafiltration and nanofiltration membranes from rejections of neutral reference solutes using a model of asymmetric pores. *Journal of Membrane Science*, **319**, 64–75.
55. Harris, J.M. (1992) Introduction to biotechnical and biomedical applications of poly (ethylene glycol), in *Poly(ethylene glycol) Chemistry* (ed. J.M. Harris), Plenum Press, New York, NY, pp. 1–14.
56. Mosqueira, V.C.F., Legrand, P., Gulik, A. *et al.* (2001) Relationship between complement activation, cellular uptake and surface physiochemical aspects of novel PEG-modified nanocapsules. *Biomaterials*, **22**, 2967–2979.

57. Masci, G., Giacomelli, L. and Crescenzi, V. (2004) Atom transfer radical polymerization of N-isopropylacrylamide. *Macromolecular Rapid Communications*, **25**, 559–564.

58. Chen, H. and Hsieh, Y.L. (2004) Dual temperature- and pH-sensitive hydrogels from interpenetrating networks and copolymerization of N-isopropylacrylamide and sodium acrylate. *Journal of Polymer Science*, **42**, 3293–3301.

59. Gorey C., Escobar, I.C., Gruden, C. *et al.* (2008) Development of smart membrane filters for microbial sensing. *Separation Science & Technology*, **43**(16), 4056–4074.

60. Gorey, C. and Escobar, I.C. (2011) N-isopropylacrylamide (NIPAAM) modified cellulose acetate ultrafiltration membranes. *Journal of Membrane Science*, **383**, 272–279.

61. Xu, X.L., Dolveck, J.Y., Boiteux, G. *et al.* (1995) Ion beam irradiation effect on gas permeatin properties of polyimid films. *Journal of Applied Polymer Science*, **55**, 99–107.

62. Xu, X.L. and Coleman, M.R. (1997) Atomic force microscopy images of ion-implanted 6FDA-pMDA polyimide films. *Journal of Applied Polymer Science*, **66**, 459–469.

63. Lee, E.H. (1999) Ion-beam modification of polymeric materials: Fundamental principles and applications. *Nuclear Instruments and Methods in Physics Research B*, **151**, 29–41.

64. Lee, E.H., Rao, G.R., Mansur, L.K. (1996) Super-hard-surfaced polymers by high-energy ion-beam irradiation. *Trends in Polymer Science*, **4**, 229–237.

65. Hall, T.M., Wagner, A. and Thompson, L.F. (1982) Ion-beam exposure characteristics of resists – experimental results. *Journal of Applied Physics*, **53**, 3997–4010.

66. Xu, X.L. and Coleman, M.R. (1999) Ion beam irradiation: An efficient method to modify the sub-nanometer scale microstructure of polymers in a controlled way. *Materials Research Society Symposium Proceedings*, **540**, 255–260.

67. Xu, X.L. and Coleman, M.R. (1999) Preliminary investigation of gas transport mechanism in a H$^+$ irradiated polyamide-ceramic composite membrane. *Nuclear Instruments and Methods in Physics Research B*, **152**, 325–334.

68. Xu, X.L., Coleman, M.R., Myler, U. and Simpson, P.J. (2000) Post-synthesis method for development of membranes using ion beam irradiation of polimide thin films, in *Membrane Formation and Modification* (eds I. Pinnau and B.D. Freeman), Oxford University Press, Washington, D.C., pp. 205–227.

69. Xu, X.L. , Coleman M.R, Myler, U. And Simpson, P.J. (2000) Post-Synthesis method for development of membranes using ion beam irradiation of polyimide thin films, in *Membrane Formation and Modification* (eds I. Pinnau and B.D. Freeman) Oxford University Press, Washington D.C., 205–227.

70. Won, J., Kim, M.H., Kang, Y.S. *et al.* (2000) Surface modification of polyimide and polysulfone membranes by ion beam for gas separation. *Journal of Applied Polymer Science*, **75**, 1554–1560.

71. Guenther, M., Gerlach, G., Suchaneck, G., *et al.* (2004) Physical properties and structure of thin ion-beam modified polymer films. *Nuclear Instruments and Methods in Physics Research Section B-Beam Interactions with Materials and Atoms*, **216**, 143–148.

72. Saha, A., Chakraborty, V. and Chintalapudi, S.N. (2000) Chemical modification of polypropylene induced by high energy carbon ions. *Nuclear Instruments and Methods in Physics Research B*, **168**, 245–251.

73. Chennamsetty R. and Escobar, I.C. (2008) Evolution of a polysulfone nanofiltration membrane following ion beam irradiation. *Langmuir*, **24**, 5569–5579.
74. Chennamsetty R. and Escobar, I.C. (2008) Effect of ion beam irradiation on two nanofiltration water treatment membranes. *Separation Science & Technology*, **43**(16), 4009–4029.
75. Vrouwenvelder, J.S. *et al.* (2009) Biofouling of spiral-wound nanofiltration and reverse osmosis membranes: A feed spacer problem. *Water Research*, **43**, 583–594.
76. Koutsou, C.P., Yiantsios, S.G. and Karabelas, A.J. (2007) Direct numerical simulation of flow in spacer-filled channels: Effect of spacer geometrical characteristics. *Journal of Membrane Science*, **291**, 53–69.
77. Dendukuri, D., Karode, S.K. and Kumar, A. (2005) Flow visualization through spacer filled channels by computational fluid dynamics-II: improved feed spacer designs. *Journal of Membrane Science*, **249**, 41–49.
78. Freeman, B.D. (2011) *Polymer Deposition and Modification of Membranes for Fouling Resistance*, Board of Regents, The University of Texas System, USA.
79. Yang, H.-L., Lin, J.C.-T. and Huang, C. (2009) Application of nanosilver surface modification to RO membrane and spacer for mitigating biofouling in seawater desalination. *Water Research*, **43**, 3777–3786.
80. Pan, Y., Ruan, J. and Zhou, D. (1997) Solid-phase grafting of glycidyl methacrylate onto polypropylene. *Journal of Applied Polymer Science*, **65**, 1905–1912.
81. Badrossamy, M. and Sun, G. (2008) Preparation of rechargeable biocidal polypropylene by reactive extrusion with diallylamino triazine. *European Polymer Journal*, **44**, 733–742.
82. Cornelissen, E.R., Vrouwenvelder, J.S., Heijman, S.G.J. *et al.* (2007) Periodic air/water cleaning for control of biofouling in spiral wound membrane elements. *Journal of Membrane Science*, **287**, 94–101.
83. Picchioni, F., Goossens, J. and Duln, M. (2001) Solid-state modification of polypropylene (PP): grafting of styrene on atactic PP. *MacroMolecular Symposium*, **176**, 245–263.
84. Chen, C.-Y. and Chen, C.-Y. (2002) Stability constants of polymer-bound iminodiacetate-type chelating agents with some transition-metal ions. *Journal of Applied Polymer Science*, **86**, 1986–1994.
85. Hausman, R., Gullinkala, T. and Escobar, I.C. (2009) Development of low-biofouling polypropylene feedspacers for reverse osmosis. *Journal of Applied Polymer Science*, **114**(5), 3068–3073.
86. Hausman R., Gullinkala, T. and Escobar, I.C. (2010) Development of copper-charged polypropylene feedspacers for biofouling control. *Journal of Membrane Science*, **358**(11), 114–121.
87. Kim, B., Anderson, J., Mueller, S. *et al.* (2002) Literature review – efficacy of various disinfectants against Legionella in water systems. *Water Research*, **36**, 4433–4444.
88. Selvaraj, S., Saha, K., Chakraborty, A. *et al.* (2009) Toxicity of free and various aminocarboxylic ligands sequestered copper(II) ions to Escherichia coli. *Journal of Hazardous Materials*, **166**, 1403–1409.
89. Rzaev, Z.M.O., Dinçer, S. and Pişkin, E. (2007) Functional copolymers of N-isopropylacrylamide for bioengineering applications. *Progress in Polymer Science*, **32**, 534–595.

8

Pore-Filled Membranes as Responsive Release Devices

Kang Hu[1] and James Dickson[2]
[1]Research and Development, Land O'Lakes, Inc., USA
[2]Department of Chemical Engineering, McMaster University, Canada

8.1 Introduction

Polymeric membranes can be fabricated for use in releasing a component, such as a drug, in a variety of applications. The component can either be encapsulated or imbibed into the polymer and the polymeric membrane affects the rate of delivery of the drug. A "responsive release device" is one in which the rate of delivery is dependent on the environmental conditions: when more drug is needed then more is released. For instance, for insulin delivery, a device that automatically adjusted the delivery rate of insulin depending on the blood glucose would belong in this category and would have great potential. With an encapsulated delivery device, for example, release of the drug through the membrane occurs through the drug partitioning into the membrane, diffusing through the membrane, and dissolving into the surrounding solution before being carried away. When the membrane permeability is a function of environmental conditions, such as temperature, ionic strength, or pH, the device can release drugs in a controlled-responsive manner.

A new type of environmental responsive membranes, pore-filled responsive membranes, has attracted research attention recently. Pore-filling technology has been applied by the Membrane Research Group at McMaster University to fabricate a new type of charged membrane that consists of a host neutral substrate and incorporated functional polyelectrolytes [1]. The host substrate, usually microporous and physically and chemically stable, provides mechanical strength for the fabricated membranes. Due to the porous structure

Responsive Membranes and Materials, First Edition. Edited by D. Bhattacharyya, Thomas Schäfer, S. R. Wickramasinghe and Sylvia Daunert.
© 2013 John Wiley & Sons, Ltd. Published 2013 by John Wiley & Sons, Ltd.

of the host substrate, various polyelectrolytes can be pore-filled in a controlled manner, resulting in dramatically different behaviour from the host membranes.

Pore-filled membranes can be environmentally responsive, owing to the fact that the pore-filling polyelectrolytes are capable of changing their conformation, and thus membrane structure, as a function of surrounding environmental conditions. In this chapter we will first review the research done on pore-filled membranes as responsive controlled release devices. Then we describe the development and characterization of a poly(vinylidene fluoride) – poly(acrylic acid) (PVDF-PAA) pore-filled pH-sensitive membrane. Finally, we show preliminary results for the release of a test drug, Aspirin, as applied to the potential application of pH responsive drug delivery in ruminant animals.

8.2   Responsive Pore-Filled Membranes

The earliest work on pore filled membranes was that of Childs and coworkers [1, 2] where the response behaviour of the membranes was discovered and originally investigated for polyvinylpyridine filled polypropylene and polyethylene microporous membranes; as summarized below. Since then several other researchers have investigated related systems, as summarized here. Depending on the properties of the pore-filling polyelectrolyte, the pore-filled membranes can be responsive to environmental ionic strength, temperature, or pH. Yu *et al.* [3] developed a pore-filled responsive membrane by the incorporation of a mixture of dextran and lectin Concanavalin A (ConA) into a Millipore glass fibre pre-filter. The membrane developed was characterized by micrograph and diffusion of a tracer as environmental glucose concentration changed. Microscope images showed that gels were formed in a glass fibre matrix. The diffusion rate of the tracer through the pore-filled membrane, indicated by the rate of UV absorbance change, increased with the glucose concentration in environment solution (to a certain concentration), as shown in Figure 8.1.

Zhang and coworkers [4, 5] developed a thermo-sensitive pore-filled membrane by synthesizing a copolymer, carboxyl methyl dextran, and poly(N-isopropylacrylamide)-NH$_2$,

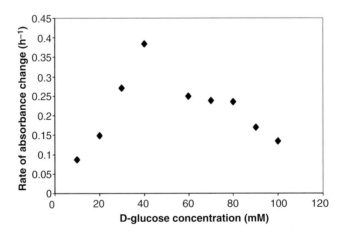

Figure 8.1 *Effect of D-glucose concentration on the diffusion rate of a tracer through a ConA dextran pore-filled membrane. (Reprinted under the terms of the STM agreement from [3] Copyright (2009) Elsevier Ltd).*

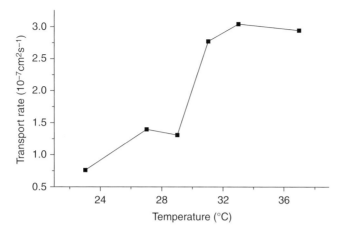

Figure 8.2 *Transport of reactive red 120 though the pore filled hydrogel membrane at different temperatures. (Reprinted under the terms of the STM agreement from [4] Copyright (2005) Elsevier Ltd).*

within Millipore glass filter discs. The membrane developed showed clear temperature responsiveness after the thermosensitive hydrogel was formed inside the matrix. As presented in Figure 8.2, the transport rate of a model compound significantly increased when the environment temperature was higher than 32°C, below which the effect could be neglected.

pH-sensitive membranes, filled with poly(4-vinylpyridine), have been fabricated by Childs and coworkers [1, 2]. Various mass gains were obtained by controlling the polymerization initiation process and the concentrations of the monomer and the cross-linker in the polymerization solution. These membranes exhibited a pH-valve effect indicated by permeability changing dramatically over a very narrow pH range. As shown in Figure 8.3, at neutral pH, the pH-valve opened as the anchored poly(4-vinylpyridine)

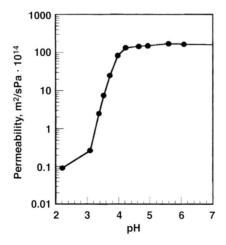

Figure 8.3 *pH valve recorded with membrane 4 under pressure of 100 kPa. (Reprinted with permission from [2] Copyright (1999) Elsevier Ltd).*

was mainly in unionized form. At low pH, the valve closed since the polyelectrolyte was protonated and formed positively charged pyridinium ions, resulting in expansion of the polymer by electrostatic repulsion forces. From open to closed the permeability dropped by about three orders of magnitude. For the pore-filled membranes anchored with some weak polyacids, the pH-valve worked in the opposite direction to the above [6]. In other words, the valve opened at lower pH and closed at higher pH.

pH-sensitive membranes have potential applications in the preparation of specific drug delivery systems, such as development of a pH-controlled drug release device, even though the reports on this are rather limited. Akerman *et al.* [7] studied the performance of PAA grafted pH-sensitive membranes and found that the diffusion flux through the membranes was influenced by the drug charge and the ionic strength. The drug release through membrane bags as a function of environmental pH was investigated *in vitro* [8] and *in vivo* [9]. In these studies, the difference in drug flux at pH acidic from that at pH neutral suggested that the PAA grafted pH-sensitive membrane might be suitable for stomach specific delivery.

In the most recent literature there are now many groups working on improving and modifying the above early work. For instance, Zhao and coworkers [10, 11] fabricated a series of hollow fibre membranes by modification of polyethersulfone using various pH-sensitive copolymers. They found the membranes showed pH valve behaviour with water flux while the pH value depended on the copolymer. Bhattacharyya and coworkers [12] incorporated bimetallic (Fe/Pd) nanoparticles inside the PAA-coated PVDF membrane pores for catalytic dechlorination. The substrate membrane was used as a platform for nanoparticle synthesis to improve reactivity.

8.3   Development and Characterization of PVDF-PAA Pore-Filled pH-Sensitive Membranes

In this section, we describe the development and characterization of a poly(vinylidene fluoride) – poly(acrylic acid) (PVDF-PAA) pore-filled pH-sensitive membrane. This includes extensive investigation, under nanofiltration (NF) pressure driven experimental conditions, on membrane gel incorporation, charge property, morphology, and permeability change as a function of pH and gel incorporation. Then, the membrane pore size at both pH acidic and pH neutral are calculated and estimated.

Pore-filling materials can be incorporated into the substrate membranes by various methods. Dafinov *et al.* [13] modified ceramic membranes by physical adsorption of alcohol to increase the membrane selectivity. However, they found that the physically adsorbed alcohol layer was not stable and could be easily removed by environmental change. Yamaguchi *et al.* [14] prepared a flat sheet pore-filled membrane by means of controlled plasma-graft polymerization. Childs and coworkers [1, 2, 15] prepared a series of pore-filled membranes by grafting 4-vinylpyridine (4VP) onto polyethylene or polypropylene microfiltration membranes, using a photochemical polymerization process. The drawback of this grafting method is the difficulty in scaling up photochemical polymerization processes to make commercial membrane modules.

Compared to the graft polymerization, cross-linking has the advantage of anchoring materials inside the porous structure. Gabriel and Gillberg [16] modified microporous polypropylene Celgard 2500 membranes by means of *in situ* polymerization of acrylic acid inside the membrane pores and achieved a permanent hydrophilicity. Kapur *et al.* [17] found that when cross-linking of polyacrylamide was conducted in a porous substrate membrane, grafting was not required to incorporate the polymer gel. The three-dimensional cross-linked polymer gel was anchored within membranes by means of physical entanglement.

In this section, the pore-filled PVDF-PAA pH-sensitive membrane is developed by incorporation of acrylic acid into PVDF using thermally initiated free radical cross-linking polymerization with N, N′-methylenebis(acrylamide) (NNMA) as the cross-linker [18]. A hydrophobic microporous PVDF membrane was chosen as the substrate membrane due to the broad chemical compatibility, superior thermal and mechanical properties, and well-controlled porosity and pore size distribution. PAA acted as the pore-filling polyelectrolyte due to the sensitivity to the environmental pH.

8.3.1   Membrane Gel Incorporation (Mass Gain)

Various membrane mass gains were obtained depending on the monomer concentration in the polymerization solution and the cross-linking degree (CLD), similar to the results of the fabrication of other pore-filled membranes [19, 20]. To illustrate the monomer concentration effect, the membrane mass gain vs. the monomer concentration, at cross-linking degree 8 mol% and in the absence of the cross-linker (CLD 0 mol%), is plotted in Figure 8.4a. At cross-linking degree 8 mol%, the mass gain increases with the acrylic acid concentration, and reaches about 90% as the monomer concentration is 7 mol/L. However, without cross-linking, almost no PAA gel was incorporated over the monomer concentration from 3 to 6 mol/L. This result suggests that the gels were anchored inside the substrate by simple gel network entanglement, as expected.

The mass gain was also found to increase linearly with the cross-linking degree at a given monomer concentration as shown in Figure 8.4b. A drastic increase in mass gain is observed with the introduction of even a small amount of cross-linker, similar to the earlier work [19].

8.3.2   Membrane pH Reversibility

The membrane pH reversibility was evaluated by the buffer solution flux at pH 2.5 and 7.4, using the apparatus described in [18]. The permeate flux of membrane M013 (see Table 8.1), as the feed buffer solution was changed over the time, is presented in Figure 8.5. Each experimental run involved 2 min equilibration in the buffer solution flow followed by 3 min sample collecting. From Figure 8.5, the flux is fully reversible between 52×10^{-6} and 0.7×10^{-6} m^3/m^{2}s as the buffer was alternated, comparable to the result obtained by Hester *et al.* [21] for the membrane fabricated by grafting PMAA on PVDF substrate membranes.

The flux response to the buffer solution was very quick (in seconds) with the response from pH 7.4 to pH 2.5 being several times faster than exchanging from pH 2.5 to pH 7.4, as expected.

8.3.3   Membrane Water Flux as pH Varied from 2 to 7.5

The effect of changing pH over the range of 2 to 7.5 on water flux through two pore-filled membranes (M013 and M015, Table 8.1), with the same cross-linking degree but different

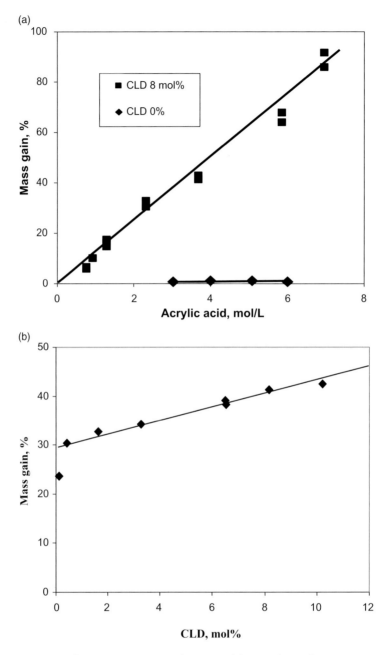

Figure 8.4 *(a): Membrane mass gain as a function of the acrylic acid concentration at cross-linking degree 8 mol% and 0 mol%, respectively. The solid lines are best-fit trend lines. (b): Membrane mass gain as a function of the cross-linking degree. Acrylic acid concentrations were 3.6 mol/L in all experiments. The solid line shows the trend after the initial increase in mass gain at a low cross-linking degree. (Reprinted with permission from [18] Copyright (2007) Elsevier Ltd).*

Table 8.1 *Membrane properties and the estimated pure water permeabilities with 95% confidence intervals of the tested membranes.*[a]

Membrane	CLD mol%	Mass gain, %	$L_P \times 10^9$ m^3/m^{2}s kPa pH 7.4	$L_P \times 10^9$ m^3/m^{2}s kPa pH 2.5	Valve ratio
M009	4	3.53	4.03 ± 0.52	33.0 ± 3.4	8.2
M011	4	8.46	1.99 ± 0.16	7.04 ± 0.60	3.5
M012	4	11.6	1.40 ± 0.13	2.68 ± 0.23	1.9
M013	8	2.47	14.0 ± 0.56	866 ± 46	62
M014	8	4.95	2.65 ± 0.78	138 ± 0.64	52
M016	8	7.73	1.55 ± 0.23	24.1 ± 3.0	16
M017	12	5.43	3.31 ± 0.31	372 ± 11	112
M018	12	6.39	2.01 ± 0.21	103 ± 2.7	51
M019	12	8.51	7.59 ± 0.27	30.8 ± 8.3	4.1
M020	12	12.8	0.56 ± 0.15	4.61 ± 3.5	8.2

[a]Reprinted with permission from [18] Copyright (2007) Elsevier Ltd.

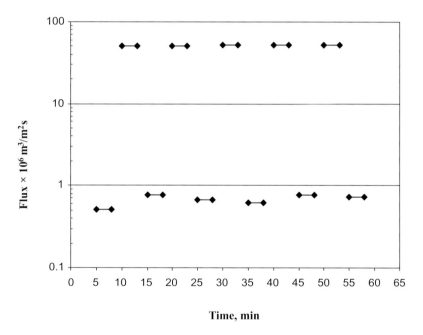

Figure 8.5 *Buffer solution permeate flux through M013 at 120 kPa as the feed was exchanged between pH 2.5 and pH 7.4. (Reprinted with permission from [18] Copyright (2007) Elsevier Ltd).*

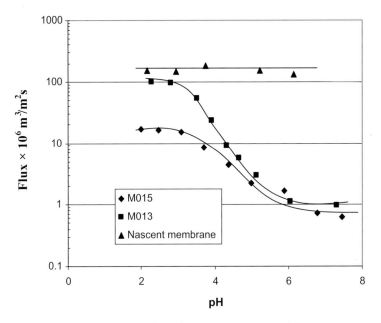

Figure 8.6 *Water flux as a function of pH for M013, M015, and the nascent membrane at 25°C, 120 kPa. Lines are guide lines. (Reprinted with permission from [18] Copyright (2007) Elsevier Ltd).*

mass gain, was studied and the results are presented in Figure 8.6. For comparison, the flux of the nascent pre-wetted PVDF membrane is also given in the figure.

It can be seen in Figure 8.6 that the membrane performance was permanently altered after the addition of the pore-filled PAA gel. The water flux is lower than that of the nascent membrane, even when the polymer chains are in their compact form (at low pH), owing to the fact that the gel still swells in the substrate membrane due to the volume exclusion effect. Comparing M013 with M015, as more gel is incorporated, the flux decreases drastically (from 101×10^{-6} to 17.1×10^{-6} m^3/m^{2}s) at low pH. However, this flux decrease is not significant at pH neutral (from 0.97×10^{-6} to 0.63×10^{-6} m^3/m^{2}s). A similar phenomenon was also observed for other pH-sensitive membranes [22, 23].

The water flux of the fabricated membranes exhibits the pH-valve behaviour at pHs between 3.5 and 5.5, and hardly changes at a pH lower than 3 or higher than 6, similar to other PAA anchored membranes [6, 22]. For M013, a flux variation of two orders of magnitude was observed, from 100×10^{-6} m^3/m^2 s at pH 2 to 1×10^{-6} m^3/m^2 s at pH 7.5. For the higher mass gain membrane (M015), the flux decreases from 17×10^{-6} m^3/m^2 s to 0.6×10^{-6} m^3/m^2 s. The chain conformation of weak polyacid is a function of its pK_a. The pK_a of PAA in solution is about 4.6, dependent upon the measurement method [24, 25], which is consistent with that from Figure 8.6. Thus, in the experiments at a pH lower than 3.5, there were at least 93% of all carboxyl groups in their unionized state. PAA gel segments retracted resulting in pore opening. At a pH higher than 5.5, 89% of carboxyl groups dissociated and extended resulting in pore closure. A further decrease or increase in pH after the pH reached 3 or 6, respectively, did not change the PAA gel configuration significantly. Then, the flux stayed relatively constant.

8.3.4   Effects of Gel Incorporation on Membrane Pure Water Permeabilities at pH Neutral and Acidic

The pH dependence on the membrane pure water permeability (L_P) is influenced by the gel incorporation (mass gain and cross-linking degree). The gel incorporation reduced the membrane water permeability at both pH acidic and pH neutral but to different extents.

8.3.4.1   *Membrane Pure Water Permeability*

The pure water permeabilities (L_P) of the membranes with cross-linking degree 4 mol%, 8 mol%, and 12 mol% were obtained by the method described in [18]. The estimated results with 95% confidence interval and the properties of the tested membranes are summarized in Table 8.1.

From Table 8.1, we can see that at each cross-linking degree, the pure water permeability decreases as the mass gain increases for both pH neutral and acidic. These results suggest that the membrane pure water permeability is affected by the membrane mass gain and cross-linking degree and this effect differs as the pH changes.

8.3.4.2   *Effects of Mass Gain and Cross-Linking Degree at pH Neutral*

At pH neutral, the extended pore-filled gel, confined in the porous structure, can be assumed to be a homogeneous semi-dilute solution [26]. As the gel amount increases, such as from M009 to M012, the distance between the segments of the gel network decreases resulting in a denser structure and a lower permeability. This decrease of the permeability as the amount of pore-filling material increases was also found previously [1, 26, 27].

The effect of cross-linking degree on gel or gel-filled membranes has been previously studied and was found to be gel dependent. Tong [28] studied the effect of cross-linking density and found that gel Darcy permeability increased with the cross-linking density at a certain gel concentrations. Silberberg [29] suggested that the contribution of the cross-linker was to increase the heterogeneity of the gel structure. Mika and Childs [26] prepared a series of membranes filled by poly(4-vinylpyridine) gel with similar gel polymer concentrations but different cross-linking degrees. They found that the increase in Darcy permeability with the cross-linking degree was fairly small, compared to Silberberg's results. Thus, the cross-linking effect was neglected in the model developed for the calculation of Darcy permeability, and used here, for the pore-filled membrane [26].

To further determine the effects of the mass gain and the cross-linking degree on water permeabilities of the PVDF-PAA membranes at pH neutral, gel Darcy permeabilities (k) of the membranes in Table 8.1, with various mass gains and cross-linking degrees, were calculated [18] and plotted against the membrane gel volume fractions (ϕ) in Figure 8.7.

As can be seen in Figure 8.7, as more gel is incorporated (gel volume fraction increases), the membrane gel Darcy permeability decreases. The permeabilities are related to the polymer gel volume fraction by the relationship, given in the figure, with a correlation coefficient $R^2 = 0.871$. This dependence of Darcy permeability on the gel volume fraction is similar in form to other studies, except for the exponent on ϕ. In this study, the exponent on ϕ is -1.42, which is comparable with the value -1.53 for poly(2-acrylamido-2-methylpropanesulfonic acid) gel-filled membranes [20], but much lower than that for polyacrylamide gels, -3.34 [17], and poly(vinylpyridinium salt) gels, -3.55 [26]. The lower permeability of the PAA

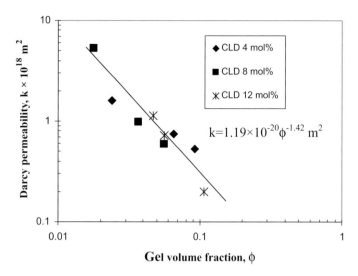

Figure 8.7 *Gel Darcy permeability as a function of the polymer gel volume fraction at different cross-linking degrees at pH neutral. The line is a linear correlation line. (Reprinted with permission from [18] Copyright (2007) Elsevier Ltd).*

gel filled membrane, compared to other gel filled membranes, is probably due to the more hydrophilic nature of the PAA gel resulting in a larger swelling behaviour.

8.3.4.3   *Effects of Mass Gain and Cross-Linking Degree at pH Acidic*

To illustrate the effects of the mass gain and the cross-linking degree at pH acidic, gel Darcy permeabilities of the tested membranes were plotted against the membrane gel volume

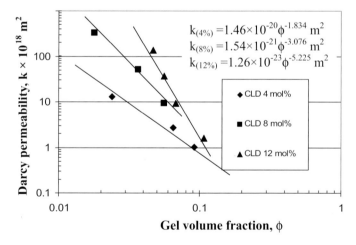

Figure 8.8 *Gel Darcy permeability as a function of the polymer gel volume fraction at different cross-linking degrees at pH acidic. The line is a linear correlation line. (Reprinted with permission from [18] Copyright (2007) Elsevier Ltd).*

fractions in Figure 8.8. Now different lines are required for the different cross-linking degrees, with permeability increasing as the cross-linking degrees increase, as discussed below. The permeabilities are related to the gel volume fraction by the relationship, given in the figure, with a correlation coefficient $R^2 = 0.982, 0.983$, and 0.966 for CLD at 4, 8, and 12 mol%, respectively.

It can be seen in Figure 8.8 that, at the same gel volume fraction, the high cross-linking membranes have larger k than the low cross-linking membranes, which is consistent with the findings of others [29, 30]. Weiss *et al.* [31, 32] suggested that high cross-linking might cause greater microscopic heterogeneity of the gel structure. Yin *et al.* [30] proposed that as the cross-linking degree increased, the elastic modulus increased, resulting in a lower gel swell ratio. Thus, at the same gel incorporation level, higher cross-linking results in a larger pore size by restricting the gel swelling.

In Figure 8.8, k decreases linearly (in the log-log plot) as the membrane gel volume fraction increases at all cross-linking degrees, similar to the results for other gels [17, 26]. Now considering the rate of the permeability decrease with the increase in the gel volume fraction, as represented by the exponent on ϕ in the correlations given in the figure, the exponent on ϕ differs at the three cross-linking degrees (ϕ is -5.22, -3.38, and -1.83 for cross-linking degree 12, 8, and 4 mol%, respectively). The decrease in permeability is fastest for the high cross-linking membranes and is relatively slow for the lower cross-linking membranes.

At low pH, in the collapsed gel, the van der Waals and hydrophobic interactions between the chain segments are larger than the osmotic dispersion and electrostatic forces, so that a heterogeneous gel structure is formed [28, 32]. This type of gel network can be described as being divided into draining and non-draining regions, where the majority of the gel polymer chains are located in the non-draining regains. Thus, in the draining regions, the water flux is quite high due to the low polymer concentration. In other words, it is these draining regions that determine the membrane permeability at low pH. For the membranes tested, as the incorporated gel amount increases, the draining regions shrink and the effective membrane pore size is reduced, resulting in the observed decrease in water permeability. The higher cross-linking degree enhances the gel heterogeneity, and then provides larger draining regions into which the gel can swell. Thus, as the gel volume fraction increases, the draining regions of the high cross-linking membranes shrink faster than those of the low cross-linking membranes, resulting in the difference in the decrease of permeability. In Figure 8.8, when the gel volume fraction is 0.1 or higher, the cross-linking effect on the permeability change is not so pronounced as enough gel is incorporated, so that the polymer segments start to overlap and interpenetrate, resulting in less heterogeneity within the non-draining regain.

8.3.4.4   *Effects of Mass Gain and Cross-Linking Degree on Valve Ratio*

The valve ratio is an effective measure of membrane pH dependence. The valve ratio decreases as the mass gain increases, as given in Table 8.1. At a cross-linking degree of 12 mol%, the valve ratio reduces from 112 (M017) to 8.2 (M020) as the mass gain increases from 5.43 to 12.8%. Similar results are found at cross-linking degrees of 4 and 8 mol%. For membranes M009, M014, and M017 (with different cross-linking degrees but similar mass gains), the valve ratio increases from 8.2 (M009) to 52 (M014) and to 112 (M017) as the cross-linking degree increases. The increase in the mass gain reduces membrane water permeability at both pH neutral and acidic. At pH acidic, however, the permeability reduction is more significant since there is more void volume available to be occupied by the gel swelling. Thus, the overall effect is that as the mass gain is increased, for a given cross-linking degree, the pH dependence on the chemical valve is decreased. When a high density of PAA gel is inside the membranes, the pH dependence is small (valve ratio is only 2 for M012).

The overall effect of the cross-linking degree on the valve ratio is that the valve ratio increases with the cross-linking degree since the cross-linking degree hardly affects the permeability at pH neutral but significantly increases the permeability at pH acidic. Thus, a high valve ratio can be achieved by increasing the cross-linking degree and decreasing the mass gain.

8.3.5   Estimation and Calculation of Pore Size

Quantitative interpretation of the pore size variation of pH-sensitive membranes as pH changes is rather limited. Mika *et al.* [2] compared a brush model [33, 34], which assumed a right-cylinder pore structure, with a pore-filled model [17], which assumed a hydrodynamic flow through the supported hydrogel, in the estimation of the membrane pore size at both pH acidic and pH neutral. They found that, as the pH changed, the closing of the open pores was equivalent to the formation of more pores with a much smaller pore size [2]. In the study, it was found that the applicability of each model was dependent on the properties of the substrate membranes and the polyelectrolyte. Hester *et al.* [21] roughly obtained the ratios of the pore diameters at pore open to pore close, based on flux measurements and the Hagen–Poisseuille equation. In this calculation, the pore density was considered independently of pH. The authors found that the calculated results agreed fairly well with the estimates directly from ESEM and AFM.

In this section, an extensive investigation of membrane pore structure variation, as the pH changes, for the PVDF-PAA pore-filled pH-sensitive membranes is described. This includes estimations and calculations of the pore size of the nascent membrane and the pore-filled membranes at both pH acidic and neutral.

8.3.5.1   Pore Radius of the Nascent Substrate Membrane

In this study, the pore size estimation of the nascent PVDF membrane was conducted by the non-wetting-fluid (contact angle >90° with the membrane material) method in the pressure-driven system, using deionized water [35]. The average obtained pore radius was 113 ± 10 nm. The value is larger than the nominal pore size value given by the manufacturer (nominal pore diameter 0.1 μm). This difference is in good agreement with the results obtained by Martinez-Diez *et al.* [36], since the value of membrane pore size is strongly dependent upon the measurement methods and the definition.

8.3.5.2   Pore Radius of the Pore-Filled Membranes at pH Acidic

After the gel incorporation, the cross-linked PAA gel occupied the void volume of the substrate membrane completely by forming a network structure. At pH acidic, the cross-linked PAA gel was in a compact conformation. The pore radius was calculated by the pore-filled model, which has also been used to calculate the hydraulic pore radius of the

Table 8.2 *Estimated and calculated pore radii of the pore-filled membranes at pH acidic and at pH neutral by three approaches.[a]*

Membrane	CLD, mol%	Gel volume fraction, ϕ	Pore radii at pH acidic, r_P, nm	Pore radii at pH neutral, r_P, nm		
				ENPE	SK with SHP	Gel correlation length model
M009	4	0.024	8.3	3.48 ± 0.12	3.43	2.93
M010	4	0.033	7.1	2.76 ± 0.26	3.41	2.44
M014	8	0.037	16.5	3.14 ± 0.25	3.75	2.28
M016	8	0.056	7.0	3.01 ± 0.26	3.4	1.79
M017	12	0.047	26.1	3.65 ± 0.075	4.1	1.99
M018	12	0.056	13.6	4.20 ± 0.94	3.64	1.79

[a]Reprinted with permission from [35] Copyright (2008) Elsevier Ltd.

pore-filled membrane based on the membrane water flux and membrane porosity [2, 17]. The calculated results are listed in Table 8.2.

At cross-linking degree 4 mol%, the calculated pore radii are 8.3 and 7.1 nm for M009 and M010, respectively. At cross-linking degree 8 mol%, the calculated pore radii are 16.5 and 7.0 nm for M014 and M016, respectively. At cross-linking degree 12 mol%, the calculated pore radii are 26.1 and 13.6 for M017 and M018, respectively. These results are comparable with those from the poly(4-vinylpyridine) pore-filled pH-sensitive membranes when the pores were open [2]. The obtained pore radii are markedly smaller than those of the nascent membrane, which is because after the membrane fabrication, the PAA gels are cross-linked inside the void volume of the substrate membranes and the gels swell in the aqueous solution due to the volume exclusion effect. Thus, the pore size decreases.

At each cross-linking degree, the pore radius at low gel volume fraction is larger than that at high gel volume fraction, as expected. This result indicates that as more gel is incorporated in the substrate, the membrane become denser, resulting in the smaller pore size. For M010, M014, and M017 with similar gel volume fractions but different cross-linking degrees, the pore radius increases with the cross-linking degree. Yin *et al.* [30] found that the swelling of PAA gel reduced as the cross-linking degree increased, due to the increase of the elastic modulus. Weiss *et al.* [32] proposed that greater microscopic heterogeneity of the gel structure could be created by higher cross-linking. Thus, for the three membranes with a similar gel volume fraction, the higher cross-linking degree formed a larger pore size by reducing the gel volume exclusion effect.

8.3.5.3 Pore Size Estimation and Calculation of the Pore-Filled Membranes at pH Neutral

Three methods – fitting the extended Nernst–Planck equation (ENPE) model, fitting the Spiegler–Kedem (SK) with Steric-hindrance Pore (SHP) models, and calculation of the gel correlation length – were used to estimate the pore radius of the pore-filled membrane at pH neutral and are described here.

Pore Size Estimation by the Extended Nernst–Plank Equation (ENPE). The ENPE has been widely used to describe the ion transport in ultrafiltration membranes [37, 38] and in nanofiltration membranes previously [39–41]. The single salt NaCl (1000 ppm) experiments were conducted to obtain salt rejection vs. solution flux data. The data were then fit to the extended Nernst–Plank equation and the obtained pore radii and 95% confidence intervals are listed in Table 8.2.

From Table 8.2, the obtained pore radii by the ENPE model are from 2.76 to 4.20 nm, which is comparable with Mika *et al.* [2] for the poly(4-*vinylpyridine*) pore-filled membranes, when the polymer was in an extended state. These results are also close to the values estimated by Bowen and Mohammad [39] for some commercial nanofiltration membranes. Since the PVDF-PAA pore-filled membranes have about 50% NaCl rejection, the membranes clearly behave as nanofiltration membranes at pH neutral.

For each cross-linking degree, the effect of the gel volume fraction on the pore radius is not so pronounced as that at pH acidic. This is probably because at pH neutral, the expanded PAA chains fit the whole void volume, and the difference of the gel volume fraction between the two membranes at each cross-linking degree is not great enough, resulting in similar equivalent pore radii for the two membranes. For M010, M014, and M017, with similar gel volume faction but different cross-linking degree, the pore radius increases in the order 2.76, <3.14, <3.65 nm with the cross-linking degree. As discussed before, the cross-linking reduces the swelling of the PAA gel where the elastic modulus, caused by the cross-linking, is in competition with both the electrostatic force and the volume exclusion force between the incorporated gel segments, resulting in the increase in the pore size.

Pore Size Estimation by the Spiegler–Kedem Model with the Steric-Hindrance Pore Model (SK with SHP). The Spiegler–Kedem model [42] has been used to obtain membrane pore size either by neutral solutes [43, 44] or by charged solutes [45, 46]. In this model, the solute rejection and the solution flux are related by a flow parameter that is a function of solute permeability and reflection coefficient. The value of the solute permeability and the refraction coefficient can be determined by best-fitting of experimental data. Nakao and Kimura [38] proposed the steric-hindrance pore model that relates pore radius to the reflection coefficient. In this section, we used dextran-10k to obtain the reflection coefficient and then pore radius by conducting experiments in a pressure driven system [35] and the results are given in Table 8.2. The calculated pore radii are from 3.4 to 4.1 nm, which agrees with the pore radii estimated by the ENPE model very well (within 11%).

From Table 8.2, at each cross-linking degree, as the gel volume fraction increases, the pore size reduces. However, the effect is not significant for the low cross-linked membranes. For M010, M014, and M017, the pore radius increases in the order 3.41, <3.75, <4.1 nm with the cross-linking degree, similar to the trend obtained by the ENPE model.

Pore Size Obtained by Calculation of the Correlation Length. Compared to the two model-fitting methods above, a new approach has been applied by [40] to directly obtain the pore size of the pore-filled membrane by calculation of the correlation length of the incorporated gel.

As discussed before, Mika and Childs [26] proposed that the confined pore-filled gel could be treated as a semidilute solution of the same polymer volume fraction when the gel was in an extended state. According to the theory described in [47, 48], the semidilute solution can be viewed as a transient network with an average mesh size between the interchain crossings, where the value of the average mesh size is called the correlation length. By using Schaefer's model [49] of polymers in semidilute solutions, coupled with the theory of polyelectrolytes in semidilute solutions, the correlation length of the gel

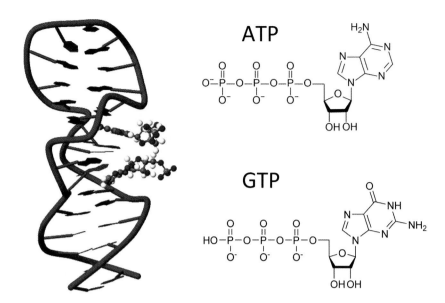

Plate 1.3 *Left: Selective interaction of an ATP-binding aptamer and two molecules of adenosine-triphosphate (ATP) as represented by Jmol; Right: Chemical structures of ATP (ligand) and GTP (no ligand) illustrating the minor differences between both molecules resulting nevertheless in a selective discrimination by the ATP-binding aptamer.*

Responsive Membranes and Materials, First Edition. Edited by D. Bhattacharyya, Thomas Schäfer, S. R. Wickramasinghe and Sylvia Daunert.
© 2013 John Wiley & Sons, Ltd. Published 2013 by John Wiley & Sons, Ltd.

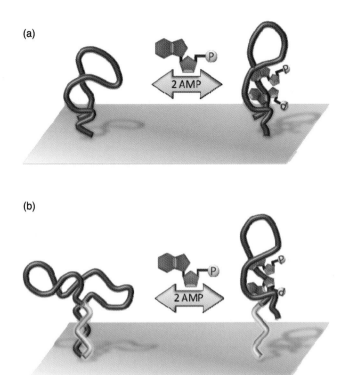

(a)

(b)

Plate 1.7 *Schematic conformational change expected upon binding to the target ATP of an (a) ATP-binding aptamer, a single 27-mer oligonucleotide (depicted as a blue line), where the overall structure is not significantly disturbed by specific intercalation of the AMP ligand and (b) the hairpin ("molecular beacon") form of the ATP-binding aptamer (blue lines) which was created by adding 7 nucleotides at the 3' end (depicted as yellow lines) forming a stem-loop structure; AMP interacts, disrupts the hairpin structure and stabilizes the same structure as in (a), whilst the additional nucleotides remain attached to the surface.*

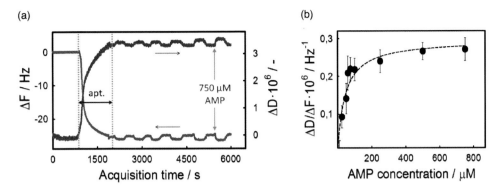

Plate 1.8 *(a) Frequency (blue lines) and dissipation (red lines) changes during binding of AMP to ATP-binding aptamer as monitored by the QCM-D, with dotted vertical lines delimiting the aptamer immobilization prior to target injection in concentrations up to 750 μM; (b) Dissipation changes normalized by the corresponding frequency change as a function of the respective AMP concentrations, and the resulting ligand binding curve (dashed line).*

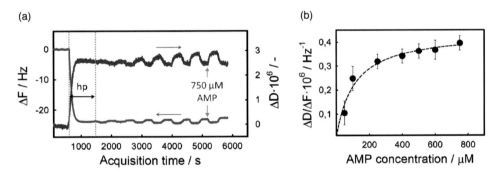

Plate 1.9 *(a) Frequency (blue lines) and dissipation (red lines) changes during binding of AMP to the ATP-binding aptamer hairpin as monitored by the QCM-D, with dotted vertical lines delimiting the hairpin immobilization prior to target injection in concentrations up to 750 μM; (b) Dissipation changes normalized by the corresponding frequency change as a function of the respective AMP concentrations, and the resulting ligand binding curve (dashed line) for the hairpin structure.*

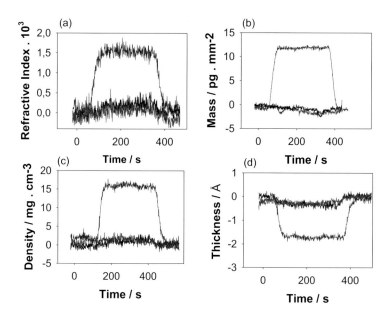

Plate 1.12 *Normalized plots of changes in layer parameters (a) RI; (b) mass; (c) density and (d) thickness during injection of AMP. Red lines: 100 μM AMP, black lines: no AMP binding buffer; blue lines: 100 μM GMP.*

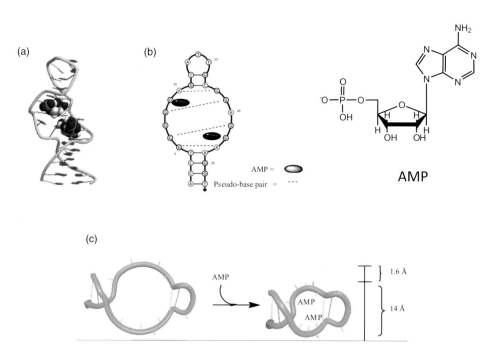

Plate 1.13 *Secondary and tertiary structure of the ATP-binding aptamer. (a) A snap-shot of the overall tertiary folding (1AW4); (b) The secondary structure alignment; (c) Schematic representation of the structural changes occurring upon AMP binding as determined in this study. The helix-depiction structures are for representative purposes. The secondary structure and helix depiction were obtained through Nupack. Drawings in b and c are not proportional to real scales, but exaggerated for representational purposes.*

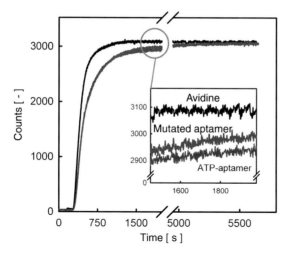

Plate 1.18 *Transient of the fluorescein permeation across the modified alumina membrane.*

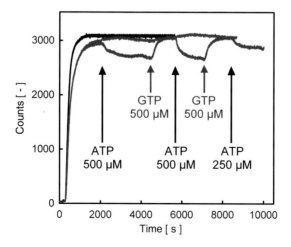

Plate 1.19 *Transient of the fluorescein permeation across the aptamer-modified alumina membrane; response to the target ATP and absence of a significant response to the GTP which is of similar structure (see also Figure 1.3).*

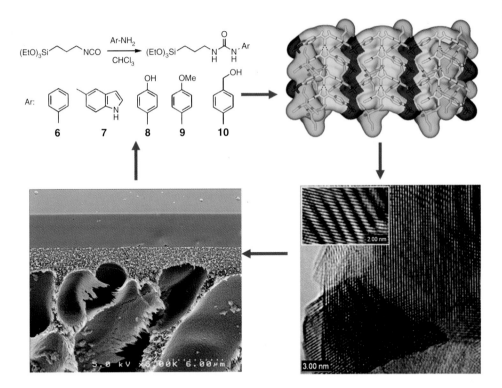

Plate 2.2 Schematic multiscale nanostructuration of ureidoaromatic receptors in orientated aromatic cation-π and urea anion conduction pathways within thin-layer hybrid membranes [19, 20]. (Reprinted with permission from [19] Copyright (2008) Elsevier Ltd; and [20] Copyright (2008) Wiley-VCH).

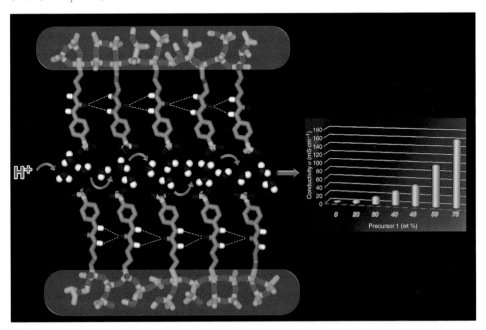

Plate 2.3 Self-organization of supramolecular -SO$_3$H—H$_2$O proton conducting nanometric pathways and proton conductivity at 25°C and 100% relative humidity. (Reprinted with permission from [29] Copyright (2009) Royal Society of Chemistry).

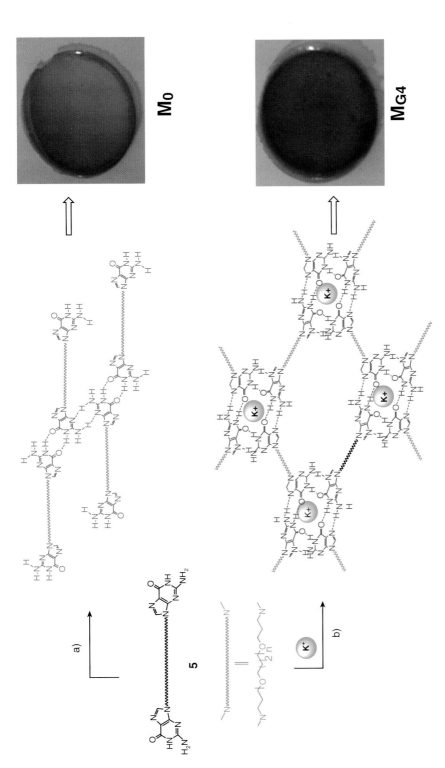

Plate 2.10 *Hierarchical cation-templated self-assembly of bis-iminoboronate-guanosine **5** macromonomer gives the G-quartet networks in the solid self-standing polymeric membrane films (a) in the absence M_0 and (b) in the presence of templating K^+ cation, M_{G4}.*

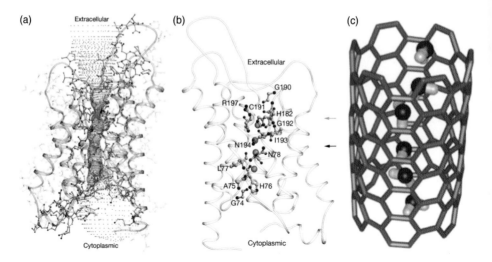

Plate 3.1 (a) High resolution crystal structure of aquaporin 1 water channel showing supra-structure channel (blue) constriction for water purification [3]. (b) AQP1 region of smallest constriction with green spheres showing water–hydrogen bonding locations [3]. (c) Molecular dynamics simulation of hydrogen bond coupled water within the CNT core showing fast water transport equivalent to the aquaporin protein [5]. (Reprinted under the terms of the STM agreement from [3] and [5] Copyright (2001) Nature Publishing Ltd).

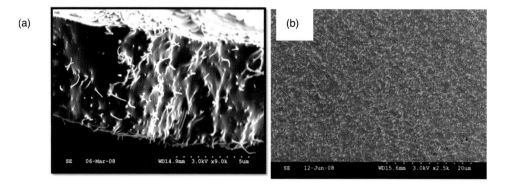

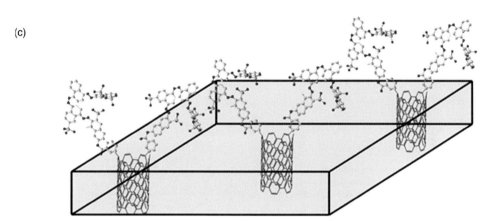

Plate 3.2 *SEM images of microtome-cut CNT membrane (a) cross-sectional view and (b) top view; (c) schematic shows the molecular structure of the anionic dye covalently functionalized on the surface of CNTs (grey: C; red: O; blue: N; yellow: S). (Reprinted with permission from [40] Copyright (2010) PNAS).*

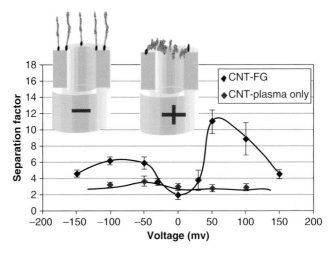

Plate 3.4 *Separation factor between small MV and large Ru(bipyr)₃²⁺ cations through tip modified MWCNT membrane. On the tip of the CNT membrane is a sterically bulky anionic dye. With positive applied bias across the membrane, the anionic "gatekeeper" is drawn within the CNT core blocking the large cation, thereby demonstrating electrostatic molecular actuation. (Reprinted with permission from [37] Copyright (2007) American Chemical Society).*

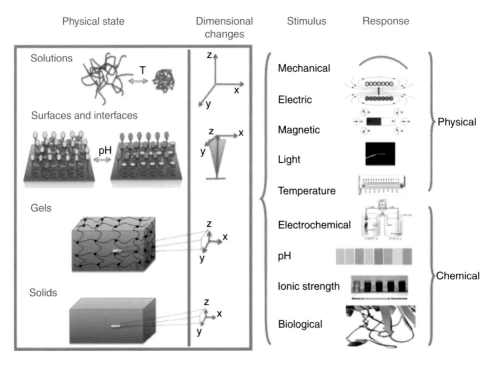

Plate 4.1 *Schematic representation of dimensional changes in polymeric solutions, at surfaces and interfaces, in polymeric gels, and polymer solids resulting from physical or chemical stimuli. (Reprinted with permission from [5] Copyright (2010) Elsevier Ltd).*

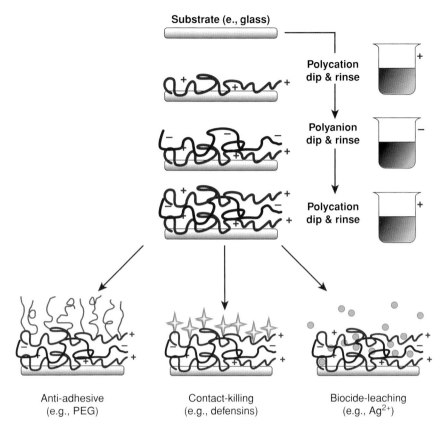

Substrate (e., glass)

Polycation
dip & rinse

Polyanion
dip & rinse

Polycation
dip & rinse

Anti-adhesive
(e.g., PEG)

Contact-killing
(e.g., defensins)

Biocide-leaching
(e.g., Ag^{2+})

Plate **5.4** *Polyelectrolyte multilayers (LbL approach) to create various surface coatings. (Reprinted with permission from [157] Copyright (2011) Materials Research Society).*

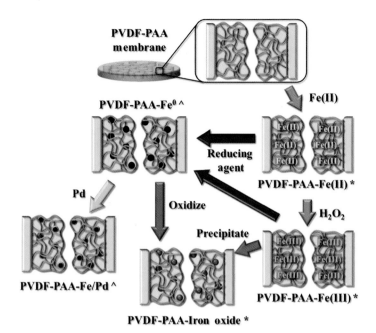

Plate 5.16 *Schematic of PVDF-PAA membranes for use as platforms in both advanced oxidative (*) and reductive (^) reactions. Note that the oxidative reactions require the addition of H_2O_2. PVDF-PAA membranes can also be used as a platform for electrostatic enzyme immobilization (see Figure 5.26).*

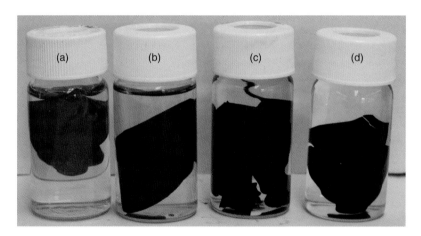

Plate 5.21 *PVDF-PAA membranes with immobilized nanoparticles after use for TCE dechlorination. The types of nanoparticles, reducing agent, and reaction times are given as follows: (a) Fe^0, sodium borohydride, 5 h; (b) Fe/Pd, sodium borohydride, 2 h; (c) Fe^0, tea extract, 23 h; (d) Fe/Pd, tea extract, 23 h. (Reprinted with permission from [130] Copyright (2011) Elsevier Ltd).*

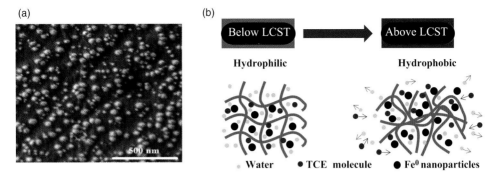

Plate 5.23 *(a) Fe/Pd nanoparticles (∼40 nm diameter) formed via* in situ *synthesis with borohydride in poly(NIPAAm) hydrogel. (b) Schematic of water and TCE (a hydrophobic water contaminant) partitioning above and below the LCST of poly(NIPAAm).*

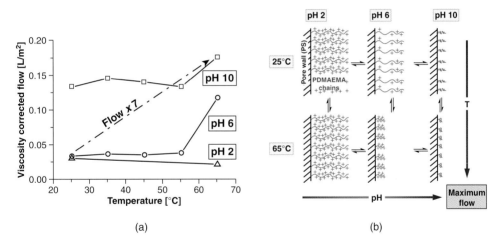

Plate 6.1 *Water flux of PS-b-PDMAEMA membrane (a) and schematic depiction of the different states of the inner part of the membrane pores (b) at different pH values and temperatures. (Reprinted under the terms of the STM agreement from [32] Copyright (2009) Wiley-VCH).*

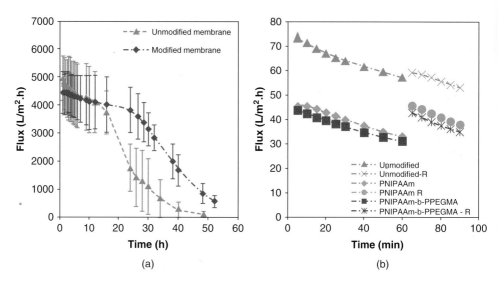

Plate 6.8 *Synthetically produced water flux measurements by cross-flow ((a) experiments were carried out at a temperature of 50°C and a TMP of 414 kPa for unmodified and PNIPAAm-b-PPEGMA-modified membranes) and dead-end ((b) a second filtration run was carried out for each of these membranes after a cold water (15°C) rinse, indicated by the letter R in the legend. A constant pressure of 207 kPa was used for all of the experiments) filtration for unmodified and modified cellulose UF membranes. (Reprinted with permission from [55] Copyright (2011) Elsevier Ltd).*

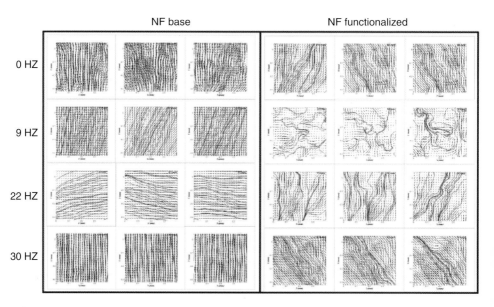

Plate 6.10 *Series of 4 PIV vector diagrams for magnet rotation frequencies of 0, 9, 22 and 30 Hz. Each vector diagram is averaged over 1 ms of time. (Reprinted with permission from [28] Copyright (2011) American Chemical Society).*

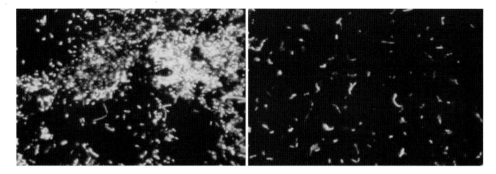

Plate 7.8 *Cells detached from membranes during filtration with* Pseudomonas fluorescens *stained with either PI or pico-green and imaged using a fluorescent microscope using regular polypropylene feed spacers (left) and copper-charged polypropylene feed spacers (right). (Reprinted under the terms of the STM agreement from [86] Copyright (2010) Elsevier Ltd).*

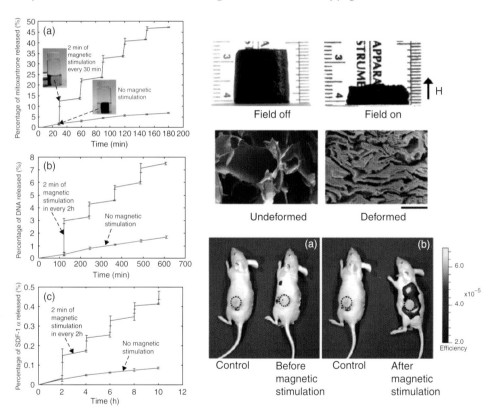

Plate 9.7 *Demonstration of responsive drug release. Graphs on the left show the controlled drug release of three biological agents of varying molecular weight and size. Images show the macro-scale effect of field application on the scaffold. Image on bottom right shows mesenchymal cell release from active scaffolds before and after magnetic stimulation [55]. (Reprinted with permission from [55] Copyright (2011) PNAS).*

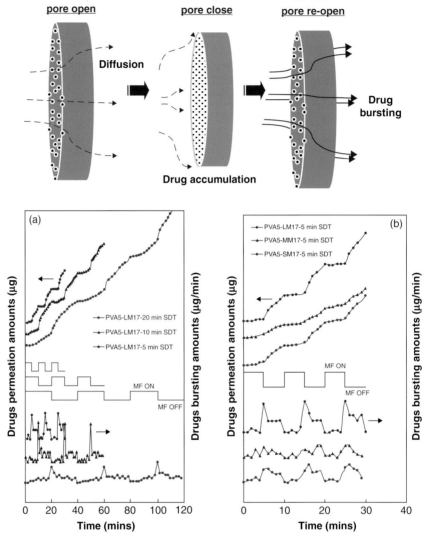

Plate 9.8 *Controlled drug release from a reservoir with ferrogel membrane. Schematic shows mechanism of controlled release. Upon application of the magnetic field the nanoparticle aggregate decreases porosity and increases tortuosity. Graphs show the drug permeation and bursting amounts under different field application times (a) and particle size (b). LM denotes particles with diameter of 15–500 nm, MM are 40–60 nm, and SM are 5–10 nm [62]. (Reprinted under the terms of the STM agreement from [62] Copyright (2006) American Chemical Society).*

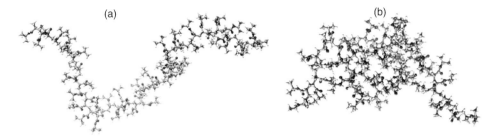

Plate 10.2 *Conformations of PNIPAM after the 75 ns simulations in 1 M mixed salt solution at 278 K (a) and 318 K (b) respectively.*

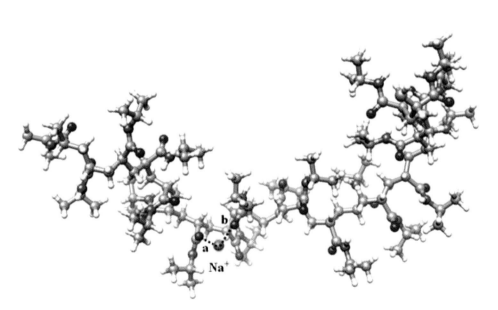

Plate 10.5 *The Na$^+$ (green) binds simultaneously to two amide O (red) atoms on PNIPAM after it goes through LCST phase transition at 318 K. Only part of the PNIPAM chain is shown.*

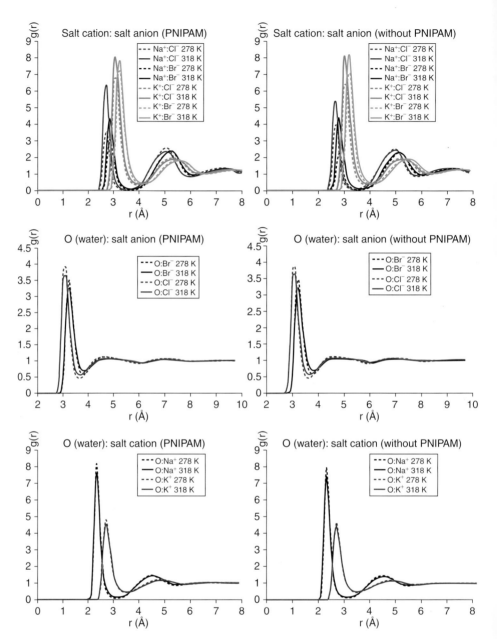

Plate 10.6 *Pair correlation functions between cation and anion, between anion and O (water) and between cation and O (water) with and without PNIPAM at both temperatures in the mixed salt solution.*

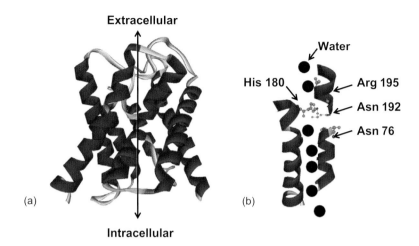

Plate 11.2 *(a) A ribbon diagram of aquaporin subunit (AQP1). The movement of water is shown by arrow. (b) Schematic representation of size-based transport of water through aquaporin channel.*

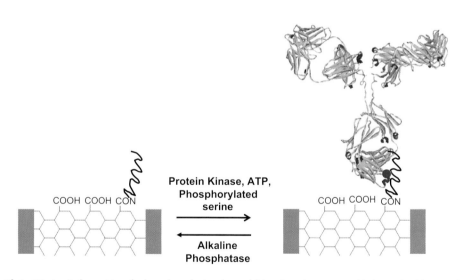

Plate 11.4 *Schematic of phosphorylation-based biomimetic system. (Adapted with permission from [29] Copyright (2007) Royal Society of Chemistry).*

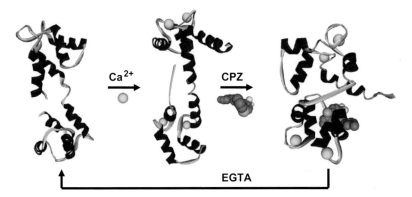

Plate 11.7 *Conformational changes of calmodulin induced upon sequential binding to Ca^{2+} and CPZ site-specifically, on or inside a polymer matrix, leading to the fabrication of responsive biomaterials.*

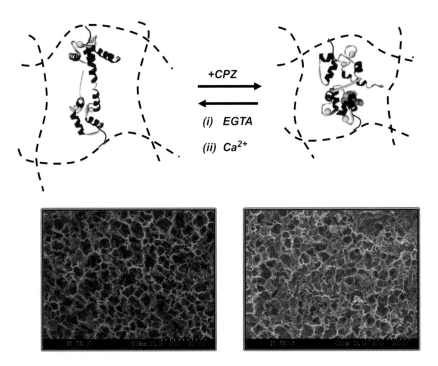

Plate 11.8 *Swelling/shrinking states of CaM-based stimuli-responsive bulk hydrogels. The Ca^{2+}-bound CaM undergoes a conformational change upon binding to CPZ. The CPZ-bound CaM, upon addition of EGTA, removes Ca^{2+}, which in turn, removes bound CPZ from CaM. The Ca^{2+}-bound CaM can be regenerated by adding Ca^{2+} to the free CaM.*

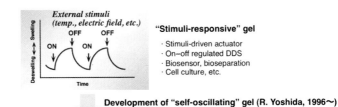

"Stimuli-responsive" gel
· Stimuli-driven actuator
· On–off regulated DDS
· Biosensor, bioseparation
· Cell culture, etc.

Development of "self-oscillating" gel (R. Yoshida, 1996~)

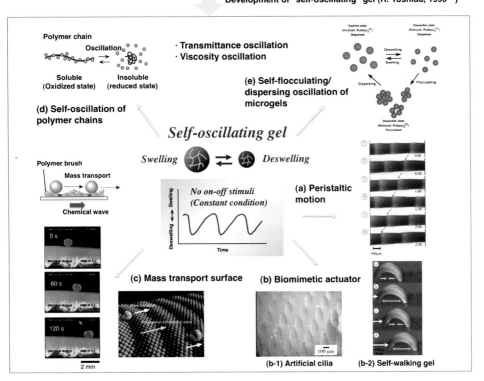

Plate 13.1 *Development of self-oscillating polymers and gels.*

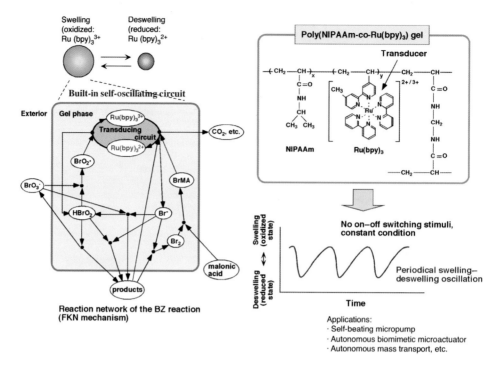

Plate 13.2 *Mechanism of self-oscillation for poly(NIPAAm-co-Ru(bpy)₃²⁺) gel coupled with the Belousov–Zhabotinsky (BZ) reaction.*

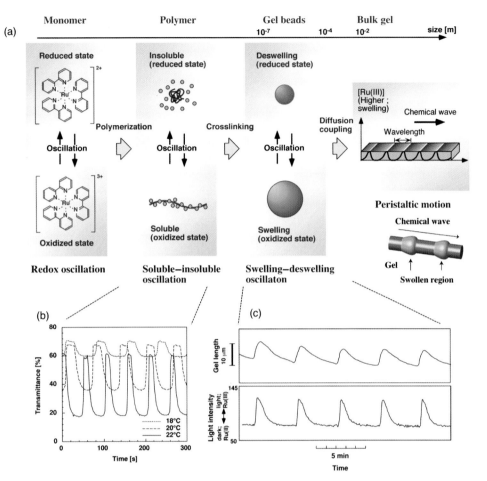

Plate 13.3 *(a) Synchronization in a self-oscillating gel in the microscopic to macroscopic level range. (b) Oscillating profiles of optical transmittance for poly(NIPAAm-co-Ru(bpy)$_3$$^{2+}$) solution at several temperatures. (c) Periodic redox changes of the miniature cubic poly(NIPAAm-co-Ru(bpy)$_3$$^{2+}$) gel (lower) and the swelling–deswelling oscillation (upper) at 20°C. Colour changes of the gel accompanied by redox oscillations (orange: reduced state, light green: the oxidized state) were converted to 8-bit grayscale changes (dark: reduced, light: oxidized) by image processing. Transmitted light intensity is expressed as an 8-bit grayscale value.*

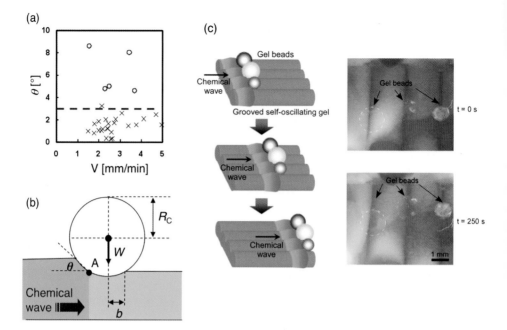

Plate 13.4 (a) Phase diagram of the transportable region given by the velocity and the inclination angle of the wavefront: (○) transported, (×) not transported. (b) Model of the rolling cylindrical gel on the peristaltic gel surface (R_C = radius of curvature, W = load of the PAAm gel, b = contact half-width). (c) Transport of the poly(AAm-co-AMPS) gel beads on the grooved surface of the self-oscillating gel.

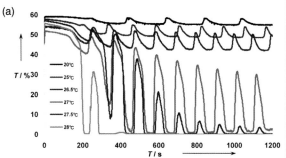

(b)

(i) Swollen
dispersed

(ii) Deswollen
dispersed

↑ Oscillation ↓

(iv) Swollen
flocculated

(iii) Deswollen
flocculated

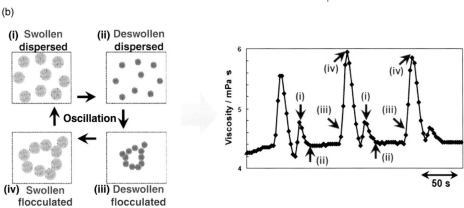

Plate 13.5 *(a) Self-oscillating profiles of optical transmittance for microgel dispersions at several temperatures. (b) Autonomously oscillating profiles of viscosity in the microgel dispersions measured at 23°C. The numbers in each oscillating profile refer to the corresponding cartoons.*

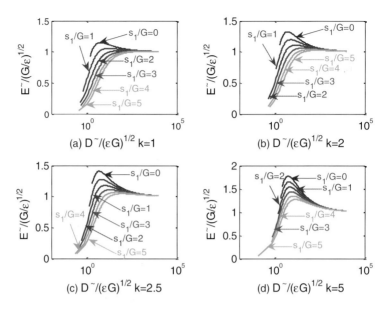

Plate 14.46 *The nominal electric field vs. the nominal electric displacement when k changes [60]. (Reproduced with permission from [60] Copyright (2009) Springer Science + Business Media).*

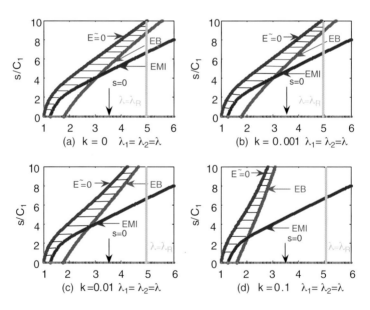

Plate 14.54 *The relationship between deformation and stress of a Mooney–Rivlin-type DE generator when $\lambda_1 = \lambda_2 = \lambda$ (the hatching represents the allowable area) [85]. (Reprinted with permission from [85] Copyright (2010) Institute of Physics).*

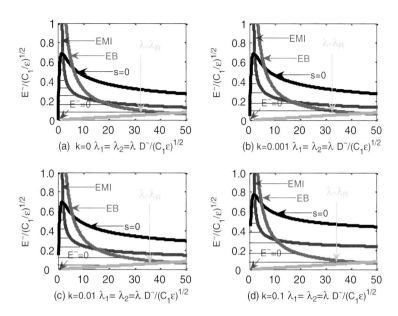

Plate 14.55 The relationship between nominal electric field and nominal electric displacement of various Mooney–Rivlin-type DE generators when $\lambda_1 = \lambda_2 = \lambda$ (the hatching represents the allowable area) [85]. (Reprinted with permission from [85] Copyright (2010) Institute of Physics).

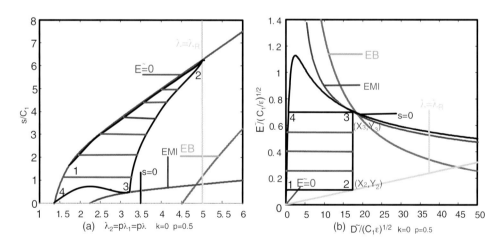

Plate 14.56 The (a) deformation and stress (b) nominal electric field and nominal electric displacement of Mooney–Rivlin-type DE generator when $\lambda_2 = 0.5\lambda_1 = 0.5\lambda$ and $k = 0$ (the hatching represents the energy generated in a single cycle) [85]. (Reprinted with permission from [85] Copyright (2010) Institute of Physics).

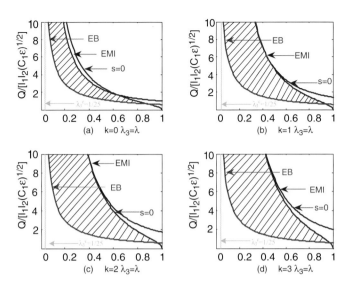

Plate 14.60 *Allowable area of folded dielectric elastomer actuator [91]. (Reproduced with permission from [91] Copyright (2012) Springer Science + Business Media).*

network can be calculated when the gel volume fraction is known. Assuming the mesh size to be the effective membrane pore diameter, the pore radius is then half the correlation length [40]. Based on the gel volume fractions from Table 8.2, the pore size of the pore-filled membranes, tested in this study, was calculated.

As presented in Table 8.2, for the pore-filled membranes tested, the calculated pore radii ranged from 1.79 to 2.93 nm. These predicted values, with no fitted parameters, are lower than those obtained from the two model-fitting methods (as much as 54% lower than the average value of the two model-fitting methods). This difference between the model-fitting results and the correlation calculation results was also found by Garcia-Aleman *et al.* [40]. Given the fact that the gel correlation length model itself is an approximate calculation [50], the under-estimation by the model is reasonable. Zhou *et al.* [51] proposed that, in the pore size calculation by the gel correlation length model for pore-filled membranes, a numerical factor should be added to correct this deviation. The difference between the model-fitting results and the correlation calculation results is significant at cross-linking degree 12 mol% compared to that at 4 mol%, since the cross-linking effect on the gel structure is not included in the gel correlation length calculation.

To summarize the pore size estimation and calculation at pH neutral, the obtained pore radii from the three methods are comparable with each other. Amongst the three methods, the calculation of the correlation length can approximately give the pore radius but with only the gel volume fraction effect being considered, which suggests a better applicability for low cross-linking degree membranes. The main advantage of the correlation length calculation method is that the method is pure prediction. For the two model-fitting methods, with both the gel volume fraction effect and the cross-linking effect included, the obtained pore radii from each method are similar. This result implies that the model-fitting methods are more suitable for the pore-filled membranes incorporated with cross-linked gels.

8.4 pH-Sensitive Poly(Vinylidene Fluoride)-Poly(Acrylic Acid) Pore-Filled Membranes for Controlled Drug Release in Ruminant Animals

In this section, the drug release of the membrane in a diffusion dialysis system is studied, using salicylic acid as the model drug, to evaluate the potential membrane applicability for drug delivery for ruminants.

Ruminants have a distinctive digestive system. Specifically, their stomach comprises four chambers: rumen, reticulum, omasum, and abomasum, successively. The rumen is the largest part of the four chambers and is the principal site of fermentation. Rumen fermentation has drawbacks for the delivery and absorption of drugs and nutrients that are administered orally, since drugs and nutrients may be destroyed or modified in the reducing environment of the rumen [52]. The abomasum functions in the same way as a human's stomach. Compared to the rumen, the abomasum is less harsh and chemical breakdown of drugs and nutrients does not occur. Thus, the abomasum is a potential target site for the release of drugs and nutrients. Therefore, a rumen-protection and abomasum-release drug delivery system is desirable for ruminants.

Table 8.3 *Properties of the membranes used in the pressure-driven and diffusion dialysis experiments.*[a]

Membrane	M017	M018	M019	M020	Nascent[b]
Mass gain, %	5.43	6.39	8.51	12.8	0
Porosity, ε[c]	0.60	0.60	0.62	0.61	0.65

[a]Reprinted with permission from [53] Copyright (2009) Elsevier Ltd.
[b]Randomly selected.
[c]Calculated according to [53].

The early approaches to achieve rumen-protection were through simple heat or chemical treatments of the active ingredients, but these methods were ineffective [52]. Given the fact that the pH in the four chambers of the ruminant's stomach is different, as it is neutral in rumen but is acidic in the abomasum, a pH-controlled drug release device could be a potential solution.

In light of this, in this section, the capability of the PVDF-PAA pore-filled pH-sensitive membranes on drug controlled release is evaluated *in vitro* by diffusion dialysis experiments. Salicylic acid (the hydrolysis form of Aspirin), negatively charged at pH neutral and unionized at pH acidic due to pK_a of 3, was used as a model drug to determine the membrane permeabilities at both pH neutral and acidic. The tested pore-filled membranes were at cross-linking degree 12 mol% with different mass gains (summarized in Table 8.3). The high cross-linking membranes provided a higher solution flux at pH acidic and a higher valve ratio compared to a low cross-linking membrane with same mass gain in pressure driven experiments (see Section 8.3).

Once the membrane drug permeability is obtained, the drug flux through the membrane can be predicted with knowledge of the effective membrane area and the drug load that determines the concentration difference cross the membrane. Thus, in the last part of this section, the half-time of salicylic acid release at both pH acidic and neutral, as a function of the mass gain, was examined.

8.4.1   Determination of Membrane Diffusion Permeability (P_S) for Salicylic Acid

Here, the membrane diffusion permeability is defined as $P_S = \frac{K_P D_M \varepsilon}{l\tau}$, which is a function of solute partition coefficient (K_P), solute diffusion coefficient in the pores of membrane (D_M), membrane porosity (ε), thickness (l), and tortuosity (τ). The membrane diffusion permeabilities for salicylic acid retention (at pH neutral) and release (at pH acidic) are determined. The mass gain effect on the permeability and on the ratio of the permeabilities at pH acidic to pH neutral is investigated.

8.4.1.1   Determination of the Membrane Diffusion Permeability

To investigate the membrane permeabilities for salicylic acid at both pH acidic and neutral, diffusion dialysis experiments were conducted as described in [53]. The obtained results of the tested membranes are plotted against the mass gain in Figure 8.9.

In Figure 8.9, the permeability ranges from 9.82×10^{-7} to 12.0×10^{-7} m/s at pH acidic and from 0.84×10^{-7} to 4.35×10^{-7} m/s at pH neutral for the pore-filled membranes.

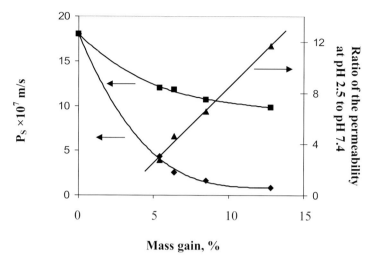

Figure 8.9   *Salicylic acid permeabilities of the membranes with various mass gains at pH acidic (♦) and pH neutral (■) and the ratio of the permeability at pH 2.5 to pH 7.4 (▲). The solid lines are best-fitting lines. (Reprinted with permission from [53] Copyright (2009) Elsevier Ltd).*

The permeability of the nascent membrane is 18.0×10^{-7} m/s, which was measured at pH acidic. The effect of pH on the permeability of the nascent membrane was neglected because the pH effect on the structure of the nascent membrane was found not to be significant.

Data on the membrane drug permeability, as a function of pH, are rather limited. Akerman *et al.* [7] have studied the flux of salicylic acid through PAA grafted membranes. They found that the drug flux at pH neutral was from 2.2×10^{-11} to 0.66×10^{-11} mol/cm^{2}s as the membrane grafting mass increased from 15 to 58 wt.%. Based on the flux above and other information in [7], the calculated permeabilities are comparable with the permeabilities obtained in this chapter.

8.4.1.2   *Effect of the Mass Gain on the Diffusion Permeability at pH Acidic and pH Neutral*

As can be seen in Figure 8.9, the diffusion permeabilities at both pH acidic and pH neutral decrease as the mass gain increases, similar to the result in [7]. The decrease is drastic when the mass gain is low, and then levels off when the mass gain is higher than around 0.07. The mass gain effect on the permeability is discussed as follows in terms of the two pH environments.

First, at pH acidic, the influence of mass gain on the permeability is mainly due to the variation of the porosity and tortuosity. As illustrated in [53], the drug permeability decreases with the membrane porosity, since the smaller the porosity, the smaller the void volume for drug transport. From Table 8.3, as the PAA gel is incorporated, the porosity decreases from 0.65 (with respect to the nascent membrane) to 0.60 (with respect to the pore-filled membrane M017). However, as the mass gain continually increases from M017 to M020, the decrease of the porosity is not notable, which is probably due to the small amount of the incorporated PAA gel and the batch effect. This variation in porosity as the

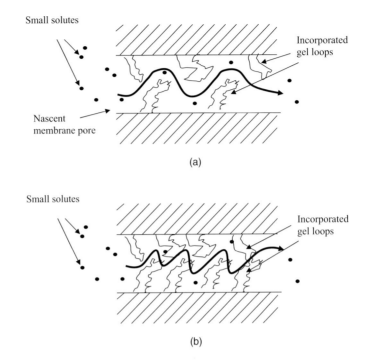

Small solutes

Incorporated
gel loops

Nascent
membrane pore

(a)

Small solutes

Incorporated
gel loops

(b)

Figure 8.10   *Small solutes traversing pores of pore-filled membranes with a low mass gain (a) and a high mass gain (b), at pH acidic. (Reprinted with permission from [53] Copyright (2009) Elsevier Ltd).*

mass gain increases results in the significant decrease in permeability from the nascent membrane to M017, compared to that from M017 to M020.

As the membrane tortuosity increases, the drug permeability decreases and this can be illustrated by a schematic "cartoon" of the effect of tortuosity on the solute transport in Figure 8.10. In Figure 8.10a, compared to the nascent membrane, after the gel incorporation, the anchored polymer loops retard the diffusion of the solutes, resulting in a longer travel distance and then a larger tortuosity. Thus, the permeability becomes slow. As the mass gain increases, more polymer loops are anchored (Figure 8.10b), resulting in further longer travel distances compared to the low mass gain membrane. Thus, as the mass gain increases, the permeability decreases.

The variation in the membrane tortuosity as a function of the membrane porosity [54–56], the membrane structure [57], and the testing solutes [58] has been studied. The membrane tortuosity was assumed to be independent of the porosity in this study, however, Quartarone *et al.* [54] found that the tortuosity of the porous PVDF membrane, tested in their study, increased as the porosity decreased. Several theoretical and empirical correlations between the tortuosity and the porosity have been proposed by Elias-Kohav *et al.* [56] for membranes with different structures. These correlations suggest an exponential increase in tortuosity as the porosity decreases.

Therefore, for the membranes tested in this chapter, the notable decrease in porosity from the nascent membrane to the pore-filled membrane M017 results in a drastic decrease in the tortuosity and thus a drastic decrease in the drug permeability compared to that from M017 to M020.

As the mass gain increases, the membrane pore radius decreases, so that steric effects and hindered diffusion effects become stronger [59]. Thus, the ion partition coefficient and the drug diffusion coefficient inside the membrane decrease, resulting in a decrease in the membrane permeability. For the membranes tested, however, the mass gain effect on the steric effect and the hindered diffusion coefficient is not substantial since the salicylic acid size (radius: 0.336 nm) is much smaller than the membrane pore size (radius: 26.1 nm for M017).

Second, at pH neutral, the permeabilities are much lower than those at pH acidic, which is mainly because both salicylic acid and the membranes are charged in the solution, resulting in a quite low co-ion partition due to the Donnan exclusion effect. Another reason is the tortuosity increase, due to the gel segments extending.

For the pore-filled membranes tested, as the mass gain increases, the permeability decreases. As a function of the mass gain, the effects of the porosity, tortuosity, and diffusion coefficient on the permeability are similar to those at pH acidic. The effect of the partition coefficient on the permeability is due to the membrane volume charge density increasing with the mass gain, resulting in a stronger Donnan exclusion. Thus, the co-ion partition becomes lower and the solute permeability decreases.

8.4.1.3   Effect of Mass Gain on the Ratio of the Diffusion Permeability at pH Acidic to pH Neutral

In order to achieve the drug rumen-protection and target-release, it is required that the release-control membrane have a high diffusion permeability at pH acidic and a low permeability at pH neutral. In other words, a high ratio of the permeability at pH acidic to pH neutral is desirable. For the tested pore-filled membranes, the calculated ratio is plotted against the membrane mass gain in Figure 8.9. The calculated ratio ranges from 2.76 to 11.7 for the pore-filled membranes tested. From the figure, the ratio increases linearly with the mass gain with the correlation coefficient $R^2 = 0.9938$.

Although a high permeability ratio can be achieved by increasing the mass gain, as this increases the permeability at pH acidic reduces, which does not favour the drug release. Thus a trade-off between the permeability at pH acidic and the permeability ratio should be considered in the membrane selection for practical work.

8.4.2   Applicability of the Fabricated Pore-Filled Membranes on the Salicylic Acid Release and Retention

To explore the applicability of these membranes to controlled drug release, the drug half-time, which is the time taken to release half the drug loaded, is investigated, assuming a reservoir system that is made by the membranes fabricated in this study.

The amount of salicylic acid, released at time t (M_t) by a certain membrane area (A_R) for the tested membranes at both pH acidic and neutral, is calculated as from [53] and plotted versus the time in Figure 8.11. In Figure 8.11, the drug release half-time can be predicted if the membrane area of the drug delivery device and the initial load of the drug is known. Once the release amount is half of the load, the corresponding time is the drug half-time.

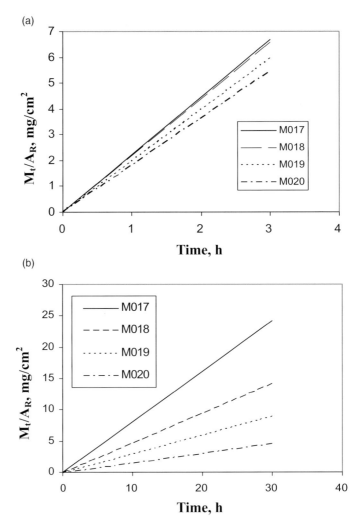

Figure 8.11 *Salicylic acid released per unit membrane area at pH acidic (a) and at pH neutral (b) at 37°C for the tested pore-filled membranes as shown in [53]. (Reprinted with permission from [53] Copyright (2009) Elsevier Ltd).*

Järvinen *et al.* [8] studied the half-time of salicylic acid through PAA grafted pH-sensitive membranes with 50% mass gain and obtained 1.0 and 2.83 h, at pH 2.0 and 7.0, respectively. Using the same operation conditions (membrane area, drug load, and ionic strength), Figure 8.11 gives 1.3 to 1.6 h at pH acidic and 4 to 26 h at pH neutral for the tested membranes, which suggests a better performance than that in Järvinen *et al.*'s work [8], having a significantly longer retention time.

In summary, for the potential application of the developed membranes for controlled drug release in ruminant animals, the results seem promising. Of course, the optimization of the release and the retention is dependent upon the choice of the drug and membrane, since the membrane permeability and the ratio of the permeability at pH acidic to pH neutral are influenced by the drug properties and the membrane gel incorporation. Naturally, for a specific drug delivery system, extensive testing would be required for a real application. However, we can conclude that, for the system examined, rumen-protection and target-release can be successively achieved by means of the pore-filled pH-sensitive membrane with control of the membrane fabrication.

References

1. Mika, A.M., Childs, R.F., Dickson, J.M. *et al.* (1995) A new class of polyelectrolyte-filled microfiltration membranes with environmentally controlled porosity. *J. Membr. Sci.*, **108**, 37–56.
2. Mika, A.M., Childs, R.F. and Dickson, J.M. (1999) Chemical valves based on poly(4-vinylpyridine)-filled microporous membranes. *J. Membr. Sci.*, **153**, 45–56.
3. Yu, S., Benzeval, I., Bowyer, A. and Hubble, J. (2009) Preparation of pore-filled responsive membranes using dextran precipitation. *J. Membr. Sci.*, **339**, 138–142.
4. Zhang, R. (2005) Synthesis, characterization and reversible transport of thermo-sensitive carboxyl methyl dextran/poly (N-isopropylacrylamide) hydrogel. *Polymer*, **46**, 2443–2451.
5. Zhang, R., Bowyer, A., Eisenthal, R. and Hubble, J. (2008) Temperature responsive pore-filled membranes based on a BSA/poly(N-isopropylacrylamide) hydrogel. *Adv. Polym. Tech.*, **27**, 27–34.
6. Winnik, F.M., Morneau, A., Mika, A.M. *et al.* (1998) Polyacrylic acid pore-filled microporous membranes and their use in membrane-mediated synthesis of nanocrystalline ferrihydrite. *Can. J. Chem.*, **76**, 10–17.
7. Akerman, S., Viinikka, P., Svarfvar, B. *et al.* (1998) Transport of drugs across porous ion exchange membranes. *J. Controlled Release*, **50**, 153–166.
8. Järvinen, K., Akerman, S., Svarfvar, B. *et al.* (1998) Drug release from pH and ionic strength responsive poly(acrylic acid) grafted poly(vinylidenefluoride) membrane bags in vitro. *Pharm. Res.*, **15**, 802–805.
9. Tarvainen, T., Nevalainen, T., Sundell, A. *et al.* (2000) Drug release from poly(acrylic acid) grafted poly(vinylidene fluoride) membrane bags in the gastrointestinal tract in the rat and dog. *J. Controlled Release*, **66**, 19–26.
10. Ren, J., Zhao, W., Cheng, C. *et al.* (2011) Comparison of pH-sensitivity between two copolymer modified polyethersulfone hollow fiber membranes. *Desalination*, **280**, 152–159.
11. Qian, B., Li, J., Wei, Q. *et al.* (2009) Preparation and characterization of pH-sensitibve polyethersulfone hollow fiber membrane for flux control. *J. Membr. Sci.*, **344**, 297–303.
12. Smuleac, V., Bachas, L. and Bhattacharyya, D. (2010) Aqueous-phase synthesis of PAA in PVDF membrane pores for nanoparticle synthesis and dichlorobiphenyl degradation. *J. Membr. Sci.*, **346**, 310–317.
13. Dafinov, A., Garcia-Valls, R. and Font, J. (2002) Modification of ceramic membranes by alcohol adsorption. *J. Membr. Sci.*, **196**, 69–77.

14. Yamaguchi, T., Nakao, S. and Kimura, S. (1991) Plasma-graft filling polymerization: Preparation of a new type of pervaporation membrane for organic liquid mixtures. *Macromolecules*, **24**, 5522–5527.

15. Childs, R.F., Mika, A.M., Pandey, A.K. *et al.* (2001) Nanofiltration using pore-filled membranes: Effect of polyelectrolyte composition on performance. *Sep. Purif. Technol.*, **22–23**, 507–517.

16. E. Gabriel and Gillberg, G. (1993) *In Situ* modification of microporous membranes. *J. Appl. Polym. Sci.*, **48**, 2081–2090.

17. Kapur, V., Charkoudian, J.C., Kessler, S.B. and Anderson, J.L. (1996) Hydrodynamic permeability of hydrogels stabilized within porous membranes. *Ind. Eng. Chem. Res.*, **35**, 3179–3185.

18. K. Hu and Dickson, J.M. (2007) Development and characterization of poly(vinylidene fluoride)-poly(acrylic acid) pore-filled pH-sensitive membranes. *J. Membr. Sci.*, **301**, 19–28.

19. Mika, A.M., Childs, R.F., Dickson, J.M. *et al.* (1997) Porous, polyelectrolyte-filled membranes: Effect of cross-linking on flux and separation. *J. Membr. Sci.*, **135**, 81–92.

20. Zhou, J., Childs, R.F. and Mika, A.M. (2005) Pore-filled nanofiltration membranes based on poly(2-acrylamido-2-methylpropanesulfonic acid) gels. *J. Membr. Sci.*, **254**, 89–99.

21. Hester, J.F., Olugebefola, S.C. and Mayes, A.M. (2002) Preparation of pH-responsive polymer membranes by self-organization. *J. Membr. Sci.*, **208**, 375–388.

22. Wang, Y., Liu, Z., Han, B. *et al.* (2004) pH sensitive polypropylene porous membrane prepared by grafting acrylic acid in supercritical carbon dioxide. *Polymer*, **45**, 855–860.

23. Ying, L., Wang, P., Kang, E. and Neoh, K. (2002) Synthesis and characterization of poly(acrylic acid)-graft-poly(vinylidene fluoride) copolymers and pH-sensitive membranes. *Macromolecules*, **35**, 673–679.

24. Mark, H., Gaylord, N. and Bikales, N. (1976) *Encyclopaedia of Polymer Science and Technology*, vol. **1**, Interscience Publishers, p. 219.

25. Rollefson, G. and Powell, R. (1952) *Annual Review of Physical Chemistry*, vol. **3**, Annual Review, Inc., Stanford, California, U.S.A., p. 82.

26. Mika, A.M. and Childs, R.F. (2001) Calculation of the hydrodynamic permeability of gels and gel-filled microporous membranes. *Ind. Eng. Chem. Res.*, **40**, 1694–1705.

27. Iwata, H., Hirata, I. and Ikada, Y. (1998) Atomic force microscopic analysis of a porous membrane with pH-sensitive molecular valves. *Macromolecules*, **31**, 3671–3678.

28. Tong, J. (1995) Partitioning and Diffusion of Macromolecules in Polyacrylamide Gels. Ph.D. Thesis, Carnegie Mellon University, Pittsburgh, USA.

29. Silberberg, A. (1992) Gel structural heterogeneity, gel permeability, and mechanical response. *ACS Symp. Ser.*, **480**, 146–158.

30. Yin, Y., Prud'homme, R. and Stanley, F. (1992) Relationship between poly(acrylic acid) gel structure and synthesis. *ACS Symp. Ser.*, **480**, 91–113.

31. Weiss, N., van Vliet, T. and Silberberg, A. (1981) Influence of polymerization initiation rate on permeability of aqueous polyacrylamide gels. *J. Polym. Sci., Part B: Polym. Phys.*, **19**, 1505–1512.

32. Weiss, N., van Vliet, T. and Silberberg, A. (1979) Permeability of heterogeneous gels. *J. Polym. Sci., Part B: Polym. Phys.*, **17**, 2229–2240.

33. Kim, J.T. and Anderson, J.L. (1989) Hindered transport through micropores with adsorbed polyelectrolytes. *J. Membr. Sci.*, **47**, 163–182.

34. Kim, J.T. and Anderson, J.L. (1991) Diffusion and flow through polymer-lined micropores. *Ind. Eng. Chem. Res.*, **30**, 1008–1016.

35. K. Hu and Dickson, J.M. (2008) Modelling of the pore structure variation with pH for pore-filled pH-sensitive poly(vinylidene fluoride)-poly(acrylic acid) membranes. *J. Membr. Sci.*, **321**, 162–171.

36. Martinez-Diez, L., Florido-Diaz, F. and Vazquez-Gonzalez, M. (2000) Characterization of hydrophobic microporous membranes from water permeation. *Sep. Sci. Technol.*, **35**, 1377–1389.

37. Nakao, S. and Kimura, S. (1981) Analysis of solutes rejection in ultrafiltration. *J. Chem. Eng. Jpn.*, **14**, 32–37.

38. Nakao, S. and Kimura, S. (1982) Models of membrane transport phenomena and their applications for ultrafiltration data. *J. Chem. Eng. Jpn.*, **15**, 200–205.

39. Bowen, W.R. and Mohammad, A.W. (1998) Characterization and prediction of nanofiltration membrane performance - a general assessment. *Trans IChemE*, **76A**, 885–893.

40. Garcia-Aleman, J., Dickson, J.M. and Mika, A.M. (2004) Experimental analysis, modeling, and theoretical design of McMaster pore-filled nanofiltration membranes. *J. Membr. Sci.*, **240**, 237–255.

41. Hu, K. and Dickson, J.M. (2006) Nanofiltration membrane performance on fluoride removal from water. *J. Membr. Sci.*, **279**, 529–538.

42. Spiegler, K. and Kedem, O. (1966) Thermodynamics of hyperfiltration (reverse osmosis): criteria for efficient membranes. *Desalination*, **1**, 311–326.

43. Wang, X., Tsuru, T., Tofoh, M. *et al.* (1995) Evaluation of pore structure and electrical properties of nanofiltration membranes. *J. Chem. Eng. Jpn.*, **28**, 186–192.

44. Wang, X., Tsuru, T., Nakao, S. and Kimura, S. (1997) The electrostatic and steric-hindrance model for the transport of charged solutes through nanofiltration membranes. *J. Membr. Sci.*, **135**, 19–32.

45. Wang, X., Tsuru, T., Tofoh, M. *et al.* (1995) Transport of organic electrolytes with electrostatic and steric-hindrance effects through nanofiltration membranes. *J. Chem. Eng. Jpn.*, **28**, 372–380.

46. Wang, K. and Chung, T. (2005) The characterization of flat composite nanofiltration membranes and their applications in the separation of Cephalexin. *J. Membr. Sci.*, **247**, 37–50.

47. Fleer, G.J., Cohen Stuart, M.A., Scheutjens, J.M.H.M. *et al.* (1993) *Polymers at Interfaces*, Chapman & Hall, London, p. 8.

48. Joanny, J.F. (1996) Semi-dilute polymer solutions, in *Physical Properties of Polymeric Gels* (eds J.P. Cohen Addad and P.G. de Gennes), John Wiley & Sons Ltd.

49. Schaefer, D.W. (1984) A unified model for the structure of polymers in semi-dilute solution. *Polymer*, **25**, 387–394.

50. Odijk, T. (1979) Possible scaling relations for semidilute polyelectrolyte solutions. *Macromolecules*, **12**, 688–693.

51. Zhou, J., Childs, R.F. and Mika, A.M. (2005) Calculation of the salt separation by negatively charged gel-filled membranes. *J. Membr. Sci.*, **260**, 164–173.

52. Wu, S. and Papas, A. (1997) Rumen-stable delivery systems. *Adv. Drug Delivery Rev.*, **28**, 323–334.

53. Hu, K. and Dickson, J.M. (2009) In vitro investigation of potential application of pH-sensitive poly(vinylidene fluoride)-poly(acrylic acid) pore-filled membranes for controlled drug release in ruminant animals. *J. Membr. Sci.*, **337**, 9–16.

54. Quartarone, E., Mustarelli, P. and Magistris, A. (2002) Transport properties of porous PVDF membranes. *J. Phys. Chem. B*, **106**, 10828–10833.

55. Dias, R., Teixeira, J., Mota, M. and Yelshin, A. (2006) Tortuosity variation in a low density binary particulate bed. *Sep. Purif. Technol.*, **51**, 180–184.

56. Elias-Kohav, T., Sheintuch, M. and Avnir, D. (1991) Steady-state diffusion and reactions in catalytic fractal porous media. *Chem. Eng. Sci.*, **46**, 2787–2798.

57. Kim, A.S. and Chen, H. (2006) Diffusive tortuosity factor of solid and soft cake layers: A random walk simulation approach. *J. Membr. Sci.*, **279**, 129–139.

58. Kokubo, K. and Sakai, K. (1998) Evaluation of dialysis membranes using a tortuous pore model. *AIChE J.*, **44**, 2607–2619.

59. Deen, W.R. (1987) Hindered transport of large molecules in liquid-filled pores. *AIChE J.*, **33**, 1409–1425.

9

Magnetic Nanocomposites for Remote Controlled Responsive Therapy and *in Vivo* Tracking

Ashley M. Hawkins[1], David A. Puleo[2] and J. Zach Hilt[1]
[1]Department of Chemical and Materials Engineering, University of Kentucky, USA
[2]Center for Biomedical Engineering, University of Kentucky, USA

9.1 Introduction

9.1.1 Nanocomposite Polymers

Nanocomposite systems have been developed through the combination of synthetic polymers with a range of nanoparticulate fillers, including silica, metal, and magnetic particles [1]. The polymer can be used to coat the nanoparticles to form core-shell systems, the filler can be incorporated into small-scale polymer systems, such as liposomes and micelles, or the filler can be dispersed throughout a bulk polymer matrix [2–4]. These systems provide an advantage over the pure polymer in that the addition of the fillers can enhance properties (e.g. mechanical properties) or introduce new properties (e.g. remote heating) [5]. In many cases, composite systems are studied because of their ability to allow controlled response to an external stimulus (i.e. electricity, magnetism, light, etc.); this response can be turned on and off and results in a rapid change in material properties [6].

There has been significant research completed on different stimuli responsive composites employing a range of filler particles and stimulus sources [4]. Some of the most common filler materials are metallic particles, carbon nanotubes, and magnetic nanoparticles [4]. Metallic fillers include gold nanoshells and nanorods, and they are of interest because of their ability to convert light into heat [7, 8]. Similar work has demonstrated the heating of

Responsive Membranes and Materials, First Edition. Edited by D. Bhattacharyya, Thomas Schäfer, S. R. Wickramasinghe and Sylvia Daunert.
© 2013 John Wiley & Sons, Ltd. Published 2013 by John Wiley & Sons, Ltd.

carbon nanotubes in response to near-infrared light [9, 10]. These composites are somewhat limited in *in vivo* applications due to the inability of light to penetrate biological tissue [11].

In this chapter, the focus will be on magnetic nanocomposite polymeric materials. Magnetic nanoparticles respond to external stimuli in several different ways and magnetic fields are able to penetrate the body [12, 13]. There has been extensive research in many aspects of this topic, ranging from different nano-scale fillers to polymer systems and the composite size. Magnetic nanoparticles are also unique in their ability to allow many types of response to external stimuli. Consequently, the emphasis will be on bulk nanocomposites and their responsive properties that make them useful for therapeutic applications.

9.1.2   Magnetic Nanoparticles

Magnetic nanoparticles represent a special class of nanoparticle actuators in that they can respond to magnetic fields (MF) in several different ways depending on the size of the particles and the nature of the field. This chapter concentrates on the responses of superparamagnetic nanoparticles to alternating current (AC; oscillating) and direct current (DC; static) fields. These particles, which are typically under 20 nm in diameter, consist of a single magnetic domain and thus exhibit significantly higher magnetization in a field but return to a non-magnetized state when the field is removed [14].

When iron oxide particles are exposed to an alternating magnetic field (AMF), the magnetization of each particle or domain is altered. In large ferromagnetic particles with several domains, this effect is restricted to single domains and atoms. Thus, these materials typically heat through eddy currents and hysteresis [11, 15]. When superparamagnetic nanoparticles are exposed to an alternating field, heat is generated through either Neél or Brownian relaxation processes. If the particle is fixed in a network structure and unable to rotate, the decay of the magnetic moment causes heating through Neél relaxation [15, 16]. In Brownian relaxation, the entire particle is allowed to rotate in a viscous liquid, and the heat is generated through the movement and friction [15, 16].

Magnetic nanoparticles also respond to the application of DC static magnetic fields, either uniform or non-uniform. In this case, the response does not cause a temperature change, however there is an interaction between the particles and the magnetic field or with the surrounding particles [17]. In the applied field, the nanoparticles will become magnetic and exhibit dipoles, which cause them to interact with one another and potentially be attracted to the magnet itself [11, 17, 18]. When dispersed in a polymer matrix, if the particles are strongly adhered to the polymer matrix, a change in the bulk shape will be observed upon field application [19]. A schematic of the resulting polymer effects in each field is shown in Figure 9.1.

9.2   Applications of Magnetic Nanocomposite Polymers

Magnetic nanocomposite systems have been studied for several therapeutic applications. The most common application is in remote controlled (RC) pulsatile drug delivery. Many drug delivery depot systems have pre-determined rates of drug release that are difficult to alter after implantation if the patient's needs change. Thus, using a remote mechanism to alter the drug release rate from outside the body can allow drug concentrations to remain within the therapeutic window and avoid toxic concentrations [20]. In some cases, a pulsatile

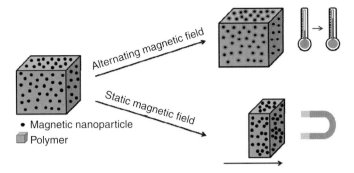

Figure 9.1   *Schematic showing the effect of different magnetic fields on magnetic nanocomposite systems.*

drug release better mimics non-zero-order biochemical processes of the body. Many proteins (i.e. insulin) and hormones are released in a cyclic manner that regulates physiological functions. Furthermore, over-exposure of cells to growth factors and hormones can result in receptor down-regulation, thus making them less effective [20–22]. There has been significant research on different methods of controlling pulsatile delivery through the use of coatings, plugs, and responsive devices [22]. However, the use of magnetic nanocomposites can allow external control of the timing and size of the drug pulses delivered by modulating the field applied.

Although many responsive polymer systems have been applied for controlled and pulsatile drug delivery, there are other therapeutic uses of remote controlled systems. Composite polymer systems that undergo a controlled shape change can also be used in actuators, stents, and artificial muscles [18]. Systems that have the ability to change shape after implantation can allow minimally invasive approaches to deploying and removing stents for cardiovascular and respiratory therapy [23, 24]. Systems that can convert a stimulus to mechanical work can be applied as artificial muscles to restore function after injury [17, 18].

For the current discussion, the systems will be separated by the mechanism of particle stimulation and the response of the polymer system. Temperature responsive systems activated by nanoparticle heating in an alternating magnetic field will be discussed first and then followed by systems controlled by physical movement and response of the particles in a static magnetic field.

9.2.1   Thermal Actuation

As previously discussed, the exposure of superparamagnetic nanoparticles to an alternating, or oscillating, magnetic field can result in relaxation processes that generate heat. This heat generation can be used to cause a response of the surrounding polymer matrix. Here, we present several material systems that are temperature sensitive and actuated by remote heating. Specifically, we will focus on systems that experience a change in swelling or degradation upon heating and others that experience a phase transition or movement through a transition temperature by the heat generated within the matrix.

9.2.1.1   Swelling

Hydrogels are hydrophilic polymer networks that are physically or chemically cross-linked, which allows the structures to swell in an aqueous environment while preventing dissolution [25]. In some cases, hydrogels can be made temperature sensitive, such as systems that exhibit a lower critical solution temperature (LCST). If free polymer is dissolved in solution, it will precipitate when the temperature is raised above its LCST; when polymerized into a cross-linked hydrogel network, the gel will shrink above the LCST, expelling most of the water [26]. This swelling behaviour provides many advantages for application in responsive devices. For example, the swelling and shrinkage occur over a small temperature range, and the process is reversible [26]. The incorporation of magnetic nanoparticles into the structure of an LCST hydrogel allows remote actuation of the heating and shrinkage of the gel matrix, and it allows a uniform heating of the matrix, thereby avoiding the formation of a "skin" layer [17].

In a nice demonstration of the controlled swelling and subsequent drug release, Satarkar *et al.* entrapped superparamagnetic iron oxide nanoparticles into a poly(N-isopropylacrylamide) (PNIPAAm) hydrogel [27]. Prior to remote heating, the loaded model drug released through diffusion. When the particles were remotely heated in an AMF (297 kHz), the hydrogel disc collapsed, creating a burst release of drug through a "squeezing" effect. When the stimulus was removed, the disc returned to its swollen state, and the drug resumed release through a diffusion mechanism. The proposed mechanism and resulting pulsatile release profile of vitamin B12 from the nanocomposite hydrogel discs is shown in Figure 9.2.

Temperature sensitive swelling systems have also been studied in barrier applications to control drug release from a reservoir [28–30]. PNIPAAm gels with superparamagnetic iron oxide nanoparticles have been used as controllable valves in microfluidic devices. The remote heating and subsequent collapse of the valve allowed controlled flow of a dye [30]. In work by Hoare *et al.*, temperature sensitive PNIPAAm nanogels were incorporated into the structure of a cellulose membrane with embedded superparamagnetic iron oxide. Application of an AMF (220–260 kHz) generated heat in the membrane, causing the nanogel particles to shrink and thereby open an interconnected pore network through the membrane to release a model drug [28, 29]. Repeated on–off control was observed with reproducible drug release. Figure 9.3 shows the composition and actuation of the membrane systems upon AMF exposure and heating. The system on the left was loaded with 23% of the nanogels, and the one shown on the right had 38% nanogel content. It was shown that the increased nanogel content increased the drug release rate upon actuation. Similar swelling-based magnetothermally responsive systems have been documented for potential applications in drug delivery and therapy [4, 31].

9.2.1.2   Degradation

In some cases, the temperature increase experienced by the nanocomposite matrix can be used to increase the rate of drug release through an increase in the degradation rate. In other work by Hawkins *et al.*, iron oxide nanoparticles were polymerized into a biodegradable poly(β-amino ester) hydrogel [32]. The hydrogel system was observed to have temperature dependent degradation with accelerated mass loss occurring at 55°C in comparison to one at 37°C. In samples exposed to an AMF (293 kHz), the temperatures reached in the matrix were high enough to accelerate the degradation and increase the release rate of a model drug.

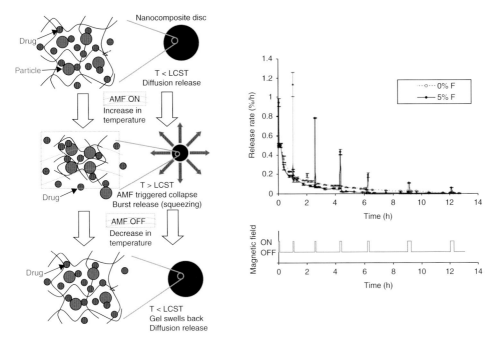

Figure 9.2 *Mechanism of pulsatile drug release from nanocomposite PNIPAAm hydrogel discs. The heating of the iron oxide magnetic nanoparticles in an AMF causes collapse of the PNIPAAm hydrogel matrix above the LCST. The release rate of model drug vitamin B12 is shown in the plot on the right, the peaks during AMF application confirm the burst effect from the collapse of the gel matrix [27]. (Reprinted with permission from [27] Copyright (2008) Elsevier Ltd).*

9.2.1.3 Phase Change

In other polymer systems, the increase in temperature generated by stimulation of embedded magnetic nanoparticles can be used to actuate a phase transition that changes the diffusion coefficient of a loaded drug. A precisely controlled and repeatable drug release behaviour was observed in cross-linked gelatin ferrogels when exposed to a AMF (50–100 kHz) [33]. Gelatin is temperature sensitive in that its molecules exhibit a helical structure at lower temperatures, which enables them to aggregate and form junction points for a gel network. When heated, the structure transitions to a coil configuration [34]. Thus, the actuation of drug release in the cross-linked gelatin network is attributed to both the temperature sensitive phase change and the vibration of the nanoparticles in the matrix. This effect is illustrated in Figure 9.4; the resulting drug release with several magnetic field exposures is also shown for systems with varying iron oxide nanoparticle sizes.

Block copolymer systems have also recently been studied for their interesting phase behaviour in response to temperature changes. In many cases, with the correct balance of hydrophilic–hydrophobic sections and polymer concentration, a solution to gel (sol-gel) phase transition can be observed upon heating [35]. This phase transition is due to the formation of micelles in aqueous solution; the hydrophobic block sections form the core, with the hydrophilic blocks extending into the hydrophilic solvent. At sufficiently high

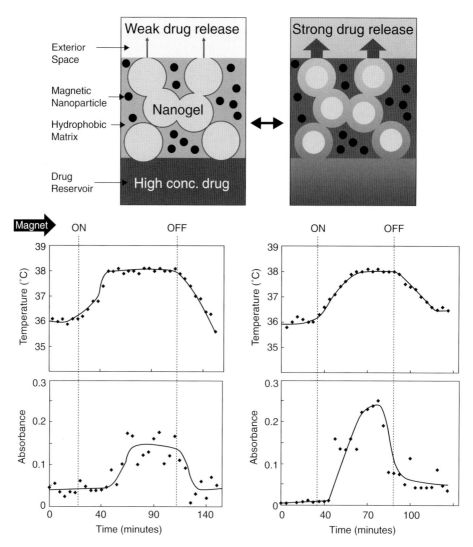

Figure 9.3 *Demonstration of remote actuation of drug release through membrane pore opening. The schematic shows the membrane composition with temperature sensitive PNIPAAm nanogels embedded in a cellulose/iron oxide membrane matrix. The remote heating of the iron oxide particles causes the nanogels to collapse thereby opening a porous network for diffusion. The graphs show the temperature and drug release response when an oscillating magnetic field is applied [29]. (Reprinted under the terms of the STM agreement from [29] Copyright (2011) American Chemical Society).*

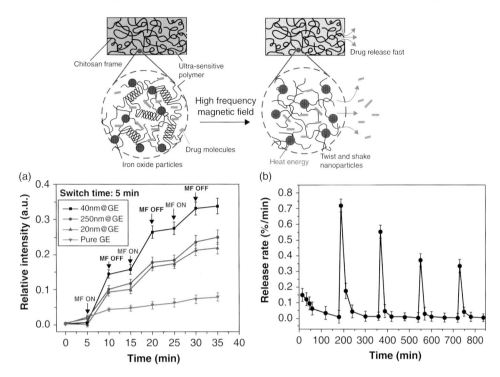

Figure 9.4 *Demonstration of controlled drug release from a cross-linked gelatine matrix. The schematic shows the effect of magnetic nanoparticle heating and vibration on the polymer matrix to release the drug. The graphs show the drug release with multiple MF exposures. (a) shows the effect of nanoparticle size on the release rate of Vitamin B_{12}, (b) shows the reproducibility of the pulsatile effect [33]. (Reprinted under the terms of the STM agreement from [33] Copyright (2007) American Chemical Society).*

micelle concentrations, an increase in temperature can increase the micelle volume fraction, and above a critical value flow is no longer observed and hard sphere crystal formation creates a gel state [36]. In some systems, further heating results in a second phase transition from gel to solution [36]. Although not as well understood, this transition has been used to modulate the release of a model drug [37]. In this case, magnetite nanoparticles were dispersed in a solution of Pluronic F-127 in aqueous solution. The remote heating of the particles in an AMF (296 kHz) forced the sol-gel systems to undergo a phase change from the gel state to a solution state, thereby increasing the rate of release of fluorescently labelled lysozyme. The results demonstrated a reproducible dosing effect with multiple field applications, indicating a viable option for RC pulsatile delivery.

In similar work, Schmalz and coworkers created a thermosensitive triblock polymer system with maghemite nanoparticles [38]. Above the cloud point of the temperature sensitive poly(glycidyl methyl ether-co-ethyl glycidyl ether), end block associations were observed between the separate micelles in solution. This phase change was observed to be activated by the remote heating of the iron oxide nanoparticles by an AMF (246–385 kHz). The effect was shown to occur rapidly, with gelation occurring in less than 2 minutes for some systems, and reproducibly with no long term effects on the system after repeated dosing.

9.2.1.4   Transition Temperature

Thermosensitive shape memory polymers (SMP) possess the ability to undergo a controlled shape change upon the application of heat. The creation of such polymer systems is based on precise control of the polymer structure, morphology, programming, and processing [39]. Through processing techniques, a temporary shape is formed that will return to the permanent shape configuration after the application of a stimulus [39]. The shape memory effect occurs above a transition temperature, either the glass transition or melting temperature of the polymer chains responsible for maintaining the temporary shape. Upon heating, the polymer phase responsible for maintaining the temporary shape experiences increased mobility, and the permanent shape is recovered [40, 41]. Researchers have used the heating properties of magnetic nanoparticles to modulate a phase change in a SMP through a non-contact manner. Lendlein and others studied the shape recovery of polyetherurethane and biodegradable poly(p-dioxanone)/poly(ε-caprolactone) SMPs extruded with iron oxide nanoparticles with a silica shell [40, 41]. The application of an AMF (250–732 kHz) was shown to heat the material to temperatures above the transition temperature, thereby prompting a shape memory response. One of the materials tested transitioned from the temporary shape to the permanent configuration in 22 seconds; this time is similar to that required when environmental conditions were tested, and opens the possibility of *in vivo* applications. Image progression of the shape recovery is shown in Figure 9.5. The group also probed the effect of the nanoparticle incorporation on the polymer properties. In loadings of up to 10 wt.%, the particles had a negligible effect on the system's mechanical properties and shape memory response.

Schmidt demonstrated a similar shape memory response in thermoset polymers loaded with superparamagnetic magnetite nanoparticles [42]. The system was heated, deformed, and cooled to create the temporary shape. Upon heating in an AMF (300 kHz), the oligo(ε-caprolactone) segments that hold the temporary shape were melted, and the system returned to the original permanent shape configuration. A schematic of the concept is shown in Figure 9.5. A similar shape recovery effect was observed in methacrylate-based thermoset systems loaded with magnetite nanoparticles.

9.2.2   Thermal Therapy

Up to this point, the application of nanoparticle response to magnetic fields has been discussed in the context of controlled drug delivery and responsive actuators, however the remote heating in itself can be used for therapeutic purposes in hyperthermia cancer treatments. Cancer cells and tumours are known to be more susceptible to heat treatments as opposed to healthy tissue because the chaotic vasculature prevents the rapid dissipation of heat [15]. Several forms of hyperthermia have been studied, including local, regional, and whole body, but increasing interest has been given to the local method to avoid necrosis of healthy cells [44]. The scientific community has found that magnetic nanoparticles are well suited for this therapy in that they easily penetrate tumour vasculature and cells, can be

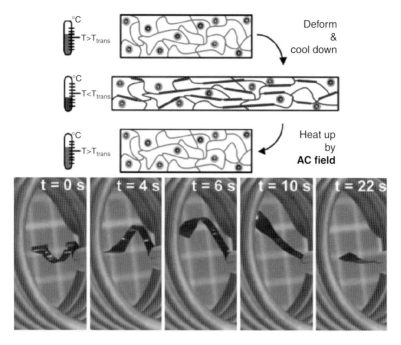

Figure 9.5 *Processing and shape memory behaviour of SMP in an alternating current magnetic field [43]. (Reprinted under the terms of the STM agreement from [43] Copyright (2006) Wiley-VCH). Image progression of shape memory effect upon remote heating of magnetic nanoparticles from helical temporary shape to elongated permanent shape [40]. (Reprinted with permission from [40] Copyright (2006) PNAS).*

loaded with drug molecules, and the method of actuation through alternating magnetic fields can easily pass through healthy tissue without harm [44]. Because the focus of this chapter is on nanocomposite systems, the use of magnetic fluid hyperthermia and nanoparticulate delivery systems is beyond the scope of this work. However there has been significant work and literature devoted to these subjects [45–48].

Meenach *et al.* developed a nanocomposite hydrogel system of superparamagnetic iron oxide nanoparticles in a PEG hydrogel matrix [49]. When exposed to an AMF (297 kHz) for 5 minutes, the samples reached a hyperthermia temperature range that induced cancer cell death. The hydrogel properties were tuned to exhibit heating to the range of hyperthermia and thermoablation. Glioblastoma cells were then exposed to the heat generated by the hydrogel discs by placing the petri dishes on top of the hydrogel on the solenoid of an induction heating supply. Cell death was observed in the area above the hydrogel sample but not in surrounding areas, indicating that the cell death was localized. It was also demonstrated that the cells did not experience cell death when exposed to the AMF, thus confirming that high frequency magnetic fields can pass through biological entities and leave them unharmed. Similar nanocomposite systems have been developed for potential usage in hyperthermia treatment [50, 51].

Other research has been done on *in situ* gelling systems for cancer hyperthermia treatment [52]. In this work, several polymer systems were developed that formed gels upon injection into the tumour environment; superparamagnetic iron oxide nanoparticles were entrapped in silica beads for mixing with the polymer materials. During *in vivo* testing of the *in situ* gelling systems in mice with human colocarcinomas, it was found that a one-time application of 20 minutes of AMF (141 kHz) induced hyperthermia and increased the median survival time [53]. A small number of mice tested exhibited no tumour recurrence, thus indicating that these materials may be a viable option for remote hyperthermia treatment.

9.2.3　Mechanical Actuation

9.2.3.1　Controlled Drug Release

As previously discussed, the exposure of iron oxide nanoparticles to a static magnetic field can cause the particles to align with the field or aggregate with one another. This property has been recently explored to alter the surrounding polymer matrix structure and control drug release. The mechanical forces exerted by the particles on the surrounding polymer matrix have been demonstrated to both accelerate and slow drug release and in some cases cause a controlled burst effect.

Systems in which the drug release rate is increased will be discussed first. In work by Qin *et al.*, iron oxide nanoparticles with hydrophobic surface functionalization were dispersed in a Pluronic F-127 sol-gel block copolymer matrix. In contrast to chemically cross-linked hydrogels, this system is composed of amphiphilic block copolymers that arrange into micelles and form a physical gel matrix as discussed in Section 9.2.1.3. When exposed to a DC magnetic field (300 mT), the nanoparticles become magnetized and aggregate to one another, which disrupts the polymer matrix and increases local drug concentrations. The effect is shown in Figure 9.6. The drug release rate was increased, and a macro-scale change in the sample was observed as the particles decreased the length of the sample [54].

Another group completed similar work on macroporous cross-linked alginate scaffolds for tissue engineering [55]. When exposed to a non-uniform magnetic field (38 A/m^2), the superparamagnetic iron oxide particles caused the gel to experience a volume change of up to 70%, thereby increasing the rate of drug release (Figure 9.7). Magnetic stimulation was achieved through 2 minute doses of 120 cycles of manually moving the magnet near and away from the gels, to create an on/off MF. The effect returned to baseline levels when the magnetic stimulation was removed. The group went further to observe that the effect could be translated to *in vivo* applications in mice where fluorescently tagged mesenchymal cells were observed to release into the surrounding areas after magnetic stimulation. Another system with increased drug release rate in the presence of a static magnetic field has been developed using poly(methyl methacrylate) and poly(vinyl alcohol) [56].

In an interesting system created by De Paoli *et al.*, iron oxide nanoparticles were stimulated through an AMF (0.14 T, field variation frequency 0.3 Hz) but were not observed to heat under the conditions used [57]. The group studied both nano- and microparticles and found that the OMF caused the particles to vibrate freely. In the case of the microparticles, this vibration caused physical deformation of the polymer structure and increased the rate of drug release. The nanoparticles were not observed to physically deform the gel, but the vibration increased the magnitude of the initial drug burst observed.

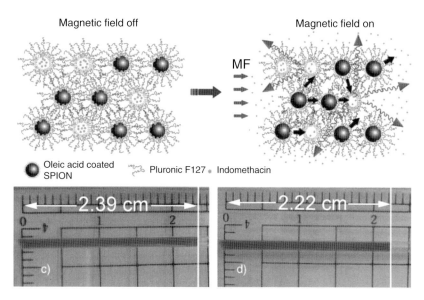

Figure 9.6 *Demonstration iron oxide nanoparticle response to applied DC magnetic field in sol-gel block copolymer system. The application of a DC magnetic field causes the nanoparticles to aggregate increasing local drug concentration and accelerating drug release. Images show macroscopic change in sample length due to particle aggregation [54]. (Reprinted under the terms of the STM agreement from [54] Copyright (2009) Wiley-VCH).*

In other polymer systems, the response of iron oxide nanoparticles to a static magnetic field can cause a decrease in drug release behaviour. The aggregation of nanoparticles can decrease the porosity and increase the tortuosity of the ferrogel structure, causing a decrease in the drug release or diffusion across the sample. Liu *et al.* studied the effect of an applied MF (400 Oe) on cross-linked gelatine ferrogels with vitamin B12 as the model drug [58]. The application of the MF caused the particles to aggregate, resulting in decreased swelling and drug release. A similar effect was observed in poly(methacrylic acid) ferrogel [59], a poly(vinyl alcohol) system [60,61], and a scleroglucan hydrogel [61].

A controlled bursting of drug release was also demonstrated when composite systems were exposed to a DC magnetic field. In other work by Liu *et al.*, iron oxide nanoparticles were entrapped in poly(vinyl alcohol) physical gels [62]. The resulting material was placed as the membrane in a side-by-side diffusion cell with Vitamin B_{12} dissolved in the donor side solution. Upon application of a MF (400 Oe), the swelling ratio was decreased as the pores contracted due to particle interactions, resulting in decreased permeability of the model drug. Upon removal of the field stimulus, there was rapid drug release and refilling of the membrane that caused a burst of drug to the receptor compartment. The release profile returned to its original non-stimulated rate soon after the burst was observed. The group further demonstrated that the on–off states and burst size could be controlled by the time of the field exposure and the size of the particles used. Figure 9.8 shows the mechanism of action of the ferrogel membranes and the effect of dosing time and particle size on the membrane permeability and burst size. Subsequent work studied the effect of membrane composition on release properties [63].

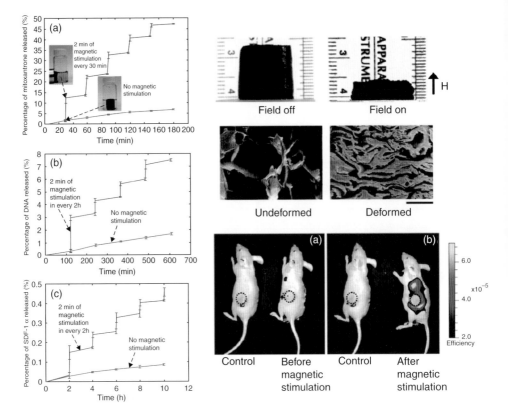

Figure 9.7   *Demonstration of responsive drug release. Graphs on the left show the controlled drug release of three biological agents of varying molecular weight and size. Images show the macro-scale effect of field application on the scaffold. Image on bottom right shows mesenchymal cell release from active scaffolds before and after magnetic stimulation [55]. (Reprinted with permission from [55] Copyright (2011) PNAS). See plate section for colour figure.*

9.2.3.2   Mechanical Actuators for Therapeutic Applications

The application of a static magnetic field to iron oxide nanoparticles has been demonstrated to cause particle movement resulting in structural changes of the polymer system. This effect on drug release was previously discussed in Section 9.2.3.1. However, other groups have looked into the large scale movement of composite systems for applications in soft actuators, which can be applied in many therapeutic applications ranging from stents to artificial muscle. Caykara and coworkers found that when a ferrogel was exposed to a non-uniform MF (1.3 T), the system changed its shape [64]. The group synthesized magnetite particles in a poly(N-tert-butylacrylamide-co-acrylamide) hydrogel using co-precipitation after hydrogel polymerization. The resulting polymer was exposed to varying magnetic intensities in an electromagnet. The response and extent of shape change were based on the strength of the magnetic field and found to occur rapidly and return to the original shape upon removal of the stimulus. Similar work was demonstrated in poly(vinyl alcohol) systems [65, 66]

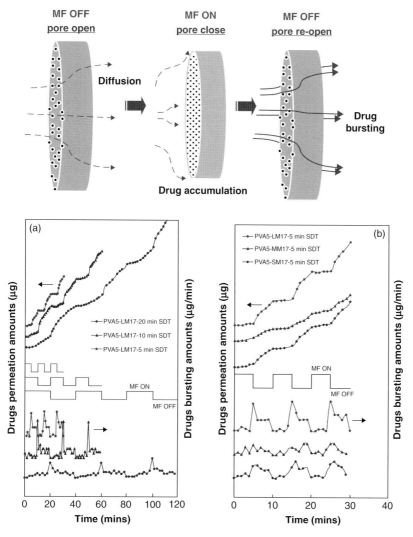

Figure 9.8 *Controlled drug release from a reservoir with ferrogel membrane. Schematic shows mechanism of controlled release. Upon application of the magnetic field the nanoparticle aggregate decreases porosity and increases tortuosity. Graphs show the drug permeation and bursting amounts under different field application times (a) and particle size (b). LM denotes particles with diameter of 15–500 nm, MM are 40–60 nm, and SM are 5–10 nm [62]. (Reprinted under the terms of the STM agreement from [62] Copyright (2006) American Chemical Society). See plate section for colour figure.*

9.2.4   *In Vivo* Tracking and Applications

In addition to the previously discussed therapeutic applications of magnetic nanocomposites, nanoparticles have been used extensively for *in vivo* imaging. Magnetic resonance imaging (MRI) is based on the response of materials to applied magnetic and radiofrequency fields [46]. The relaxation of the hydrogen protons in the fields determines the local hydrogen density, which can provide valuable information on the composition of the scanned area. The presence of magnetic nanoparticles affects the relaxation behaviour of surrounding protons, thus altering the signal received by the detection coil [46, 67]. There are several commercially available superparamagnetic iron oxide contrast agents for *in vivo* imaging [46] that have also been used for cell tracking applications in research [68, 69]. Currently, most nanoparticles are used for contrast in organs responsible for their clearance, such as the spleen and liver [12]. The use of specific labelling and coatings can allow longer circulation times and enable the particles to detect specific cells or markers in the body, such as apoptosis, cancer cell receptor expression, and tumours [70–72]. With the extensive amount of information known about magnetic nanoparticles and their behaviour in MRI analysis, there are many avenues that have yet to be studied. For example, the contrast properties could allow non-invasive, *in vivo* tracking of nanocomposite systems to observe biological response and their degradation. In a research setting, this could help decrease the number of animals needed to observe the composite behaviour. In clinical applications, this could give doctors real-time data on the implanted system and allow adjustments to the treatment accordingly.

9.3   Concluding Remarks

This chapter has discussed many therapeutic uses of responsive magnetic nanocomposite polymer systems. They have been successfully employed in both alternating and static magnetic fields to allow non-contact control over drug release and mechanical actuation. Furthermore, the remote heating capabilities lend themselves well to cancer hyperthermia treatment, and they can also serve as contrast agents in magnetic resonance imaging. The utility of these systems ranges beyond therapeutic responsive devices, and the technology can be applied to other areas, such as water treatment, separations, and other non-biological applications in which remote actuation and control are desirable or necessary. The unique property of magnetic nanoparticles to exhibit significant heating in an alternating magnetic field can be implemented in a number of uses where a change in temperature can elicit a response from the surrounding environment. Finally, the contrast properties of the magnetic nanoparticles open many possibilities for non-invasive *in vivo* tracking of both cells and implanted devices. The combination of the responsive properties of magnetic nanoparticles with the extensively studied properties of polymer films creates a vast array of composite systems for use in many engineering applications.

References

1. Schexnailder, P. and Schmidt, G. (2009) Nanocomposite polymer hydrogels. *Colloid & Polymer Science*, **287**, 1–11.

2. Behrens, S. (2011) Preparation of functional magnetic nanocomposites and hybrid materials: recent progress and future directions. *Nanoscale*, **3**, 877–892.

3. Satarkar, N.S. and Zach Hilt, J. (2008) Hydrogel nanocomposites as remote-controlled biomaterials. *Acta Biomaterialia*, **4**, 11–16.

4. Satarkar, N.S., Biswal, D. and Hilt, J.Z. (2010) Hydrogel nanocomposites: a review of applications as remote controlled biomaterials. *Soft Matter*, **6**, 2364–2371.

5. Hsu, L., Weder, C. and Rowan, S.J. (2011) Stimuli-responsive, mechanically-adaptive polymer nanocomposites. *Journal of Materials Chemistry*, **21**, 2812–2822.

6. Roy, D., Cambre, J.N. and Sumerlin, B.S. (2010) Future perspectives and recent advances in stimuli-responsive materials. *Progress in Polymer Science*, **35**, 278–301.

7. Strong, L.E. and West, J.L. (2011) Thermally responsive polymer–nanoparticle composites for biomedical applications. *Wiley Interdisciplinary Reviews: Nanomedicine and Nanobiotechnology*, **3**, 307–317.

8. Harris, N., Ford, M.J. and Cortie, M.B. (2006) Optimization of plasmonic heating by gold nanospheres and nanoshells. *The Journal of Physical Chemistry B*, **110**, 10701–10707.

9. Kam, N.W.S., O'Connell, M., Wisdom, J.A. and Dai, H. (2005) Carbon nanotubes as multifunctional biological transporters and near-infrared agents for selective cancer cell destruction. *Proceedings of the National Academy of Sciences of the United States of America*, **102**, 11600–11605.

10. Miyako, E., Nagata, H., Hirano, K. and Hirotsu, T. (2009) Laser-triggered carbon nanotube microdevice for remote control of biocatalytic reactions. *Lab on a Chip*, **9**, 788–794.

11. Timko, B.P., Dvir, T. and Kohane, D.S. (2010) Remotely triggerable drug delivery systems. *Advanced Materials*, **22**, 4925–4943.

12. Mornet, S., Vasseur, S., Grasset, F. *et al.* (2006) Magnetic nanoparticle design for medical applications. *Progress in Solid State Chemistry*, **34**, 237–247.

13. Hergt, R., Dutz, S., Muller, R. and Zeisberger, M. (2006) Magnetic particle hyperthermia: nanoparticle magnetism and materials development for cancer therapy. *Journal of Physics: Condensed Matter*, **18**, S2919–S2934.

14. Neuberger, T., Schöpf, B., Hofmann, H. *et al.* (2005) Superparamagnetic nanoparticles for biomedical applications: Possibilities and limitations of a new drug delivery system. *Journal of Magnetism and Magnetic Materials*, **293**, 483–496.

15. Brazel, C.S. (2009) Magnetothermally-responsive Nanomaterials: Combining magnetic nanostructures and thermally-sensitive polymers for triggered drug release. *Pharmaceutical Research*, **26**, 644–656.

16. Rosensweig, R.E. (2002) Heating magnetic fluid with alternating magnetic field. *Journal of Magnetism and Magnetic Materials*, **252**, 370–374.

17. Messing, R. and Schmidt, A.M. (2011) Perspectives for the mechanical manipulation of hybrid hydrogels. *Polymer Chemistry*, **2**, 18–32.

18. Zrínyi, M. (2000) Intelligent polymer gels controlled by magnetic fields. *Colloid & Polymer Science*, **278**, 98–103.

19. Snyder, R.L., Nguyen, V.Q. and Ramanujan, R.V. (2010) Design parameters for magneto-elastic soft actuators. *Smart Materials and Structures*, **19**, 055017 doi: 10.1088/0964-1726/19/5/055017.

20. Kikuchi, A. and Okano, T. (2002) Pulsatile drug release control using hydrogels. *Advanced Drug Delivery Reviews*, **54**, 53–77.

21. Roy, P. and Shahiwala, A. (2009) Multiparticulate formulation approach to pulsatile drug delivery: Current perspectives. *Journal of Controlled Release*, **134**, 74–80.
22. Maroni, A., Zema, L., Del Curto, M.D. *et al.* (2010) Oral pulsatile delivery: Rationale and chronopharmaceutical formulations. *International Journal of Pharmaceutics*, **398**, 1–8.
23. Andrews, S.M. and Anson, A.W. (1995) Shape memory alloys in minimally invasive therapy. *Minimally Invasive Therapy & Allied Technologies*, **4**, 315–318.
24. Vinograd, I., Klin, B., Brosh, T. *et al.* (1994) A new intratracheal stent made from nitinol, an alloy with "shape memory effect". *Journal of Thoracic and Cardiovascular Surgery*, **107**, 1255–1261.
25. Slaughter, B.V., Khurshid, S.S., Fisher, O.Z. *et al.* (2009) Hydrogels in regenerative medicine. *Advanced Materials*, **21**, 3307–3329.
26. Hoffman, A.S., Afrassiabi, A. and Dong, L.C. (1986) Thermally reversible hydrogels: II. Delivery and selective removal of substances from aqueous solutions. *Journal of Controlled Release*, **4**, 213–222.
27. Satarkar, N.S. and Hilt, J.Z. (2008) Magnetic hydrogel nanocomposites for remote controlled pulsatile drug release. *Journal of Controlled Release*, **130**, 246–251.
28. Hoare, T., Santamaria, J., Goya, G.F. *et al.* (2009) A magnetically triggered composite membrane for on-demand drug delivery. *Nano Letters*, **9**, 3651–3657.
29. Hoare, T., Timko, B.P., Santamaria, J. *et al.* (2011) Magnetically triggered nanocomposite membranes: A versatile platform for triggered drug release. *Nano Letters*, **11**, 1395–1400.
30. Satarkar, N.S., Zhang, W., Eitel, R.E. and Hilt, J.Z. (2009) Magnetic hydrogel nanocomposites as remote controlled microfluidic valves. *Lab on a Chip*, **9**, 1773–1779.
31. Meenach, S.A., Anderson, K.W. and Hilt, J.Z. (2010) Synthesis and characterization of thermoresponsive poly(ethylene glycol)-based hydrogels and their magnetic nanocomposites. *Journal of Polymer Science. Part A-1: Polymer Chemistry*, **48**, 3229–3235.
32. Hawkins, A.M., Satarkar, N.S. and Hilt, J.Z. (2009) Nanocomposite degradable hydrogels: demonstration of remote controlled degradation and drug release. *Pharmaceutical Research*, **26**, 667–673.
33. Hu, S.-H., Liu, T.-Y., Liu, D.-M. and Chen, S.-Y. (2007) Controlled pulsatile drug release from a ferrogel by a high-frequency magnetic field. *Macromolecules*, **40**, 6786–6788.
34. Jeong, B., Kim, S.W. and Bae, Y.H. (2002) Thermosensitive sol-gel reversible hydrogels. *Advanced Drug Delivery Reviews*, **54**, 37–51.
35. Joo, M.K., Park, M.H., Choi, B.G. and Jeong, B. (2009) Reverse thermogelling biodegradable polymer aqueous solutions. *Journal of Materials Chemistry*, **19**, 5891–5905.
36. Mortensen, K. and Pedersen, J.S. (1993) Structural study on the micelle formation of poly(ethylene oxide)-poly(propylene oxide)-poly(ethylene oxide) triblock copolymer in aqueous-solution. *Macromolecules*, **26**, 805–812.
37. Hawkins, A.M., Bottom, C.E., Liang, Z., Puleo, D.A., Hilt, J.Z. (2012) Magnetic Nanocomposite Sol-Gel Systems for Remote Controlled Drug Release. *Advanced Healthcare Materials*, 1, 96–100.
38. Reinicke, S., Dohler, S., Tea, S. *et al.* (2010) Magneto-responsive hydrogels based on maghemite/triblock terpolymer hybrid micelles. *Soft Matter*, **6**, 2760–2773.

39. Lendlein, A. and Kelch, S. (2002) Shape-memory polymers. *Angewandte Chemie International Edition*, **41**, 2035–2057.
40. Mohr, R., Kratz, K., Weigel, T. *et al.* (2006) Initiation of shape-memory effect by inductive heating of magnetic nanoparticles in thermoplastic polymers. *Proceedings of the National Academy of Sciences of the United States of America*, **103**, 3540–3545.
41. Weigel, T., Mohr, R., Lendlein, A. (2009) Investigation of parameters to achieve temperatures required to initiate the shape-memory effect of magnetic nanocomposites by inductive heating. *Smart Materials and Structures*, **18**, 025011 doi:10.1088/0964-1726/18/2/025011.
42. Yakacki, C.M., Satarkar, N.S., Gall, K. *et al.* (2009) Shape-memory polymer networks with Fe3O4 nanoparticles for remote activation. *Journal of Applied Polymer Science*, **112**, 3166–3176.
43. Schmidt, A.M. (2006) Electromagnetic activation of shape memory polymer networks containing magnetic nanoparticles. *Macromolecular Rapid Communications*, **27**, 1168–1172.
44. Kumar, C.S., Mohammad, F. (2011) Magnetic nanomaterials for hyperthermia-based therapy and controlled drug delivery. *Advanced Drug Delivery Reviews*, 63, 789–808.
45. Silva, A.C., Oliveira, T.R., Mamani, J.B. *et al.* (2011) Application of hyperthermia induced by superparamagnetic iron oxide nanoparticles in glioma treatment. *International Journal of Nanomedicine*, **6**, 591–603.
46. Sun, C., Lee, J.S.H. and Zhang, M. (2008) Magnetic nanoparticles in MR imaging and drug delivery. *Advanced Drug Delivery Reviews*, **60**, 1252–1265.
47. Latorre, M. and Rinaldi, C. (2009) Applications of magnetic nanoparticles in medicine: magnetic fluid hyperthermia. *Puerto Rico Health Sciences Journal*, **28**, 227–238.
48. Thiesen, B. and Jordan, A. (2008) Clinical applications of magnetic nanoparticles for hyperthermia. *International Journal of Hyperthermia*, **24**, 467–474.
49. Meenach, S.A., Hilt, J.Z. and Anderson, K.W. (2010) Poly(ethylene glycol)-based magnetic hydrogel nanocomposites for hyperthermia cancer therapy. *Acta Biomaterialia*, **6**, 1039–1046.
50. Babincová, M., Leszczynska, D., Sourivong, P. *et al.* (2001) Superparamagnetic gel as a novel material for electromagnetically induced hyperthermia. *Journal of Magnetism and Magnetic Materials*, **225**, 109–112.
51. Satarkar, N.S., Meenach, S.A., Anderson, K.W. and Hilt, J.Z. (2011) Remote actuation of hydrogel nanocomposites: heating analysis, modeling, and simulations. *AICHE Journal*, **57**, 852–860.
52. P.-Le Renard, E., Jordan, O., Faes, A. *et al.* (2010) The in vivo performance of magnetic particle-loaded injectable, in situ gelling, carriers for the delivery of local hyperthermia. *Biomaterials*, **31**, 691–705.
53. Le Renard, P.-E., Buchegger, F., Petri-Fink, A. *et al.* (2009) Local moderate magnetically induced hyperthermia using an implant formed in situ in a mouse tumor model. *International Journal of Hyperthermia*, **25**, 229–239.
54. Qin, J., Asempah, I., Laurent, S. *et al.* (2009) Injectable superparamagnetic ferrogels for controlled release of hydrophobic drugs. *Advanced Materials*, **21**, 1354–1357.
55. Zhao, X., Kim, J., Cezar, C.A. *et al.* (2011) Active scaffolds for on-demand drug and cell delivery. *Proceedings of the National Academy of Sciences of the United States of America*, **108**, 67–72.

56. Bajpai, A.K. and Gupta, R. (2011) Magnetically mediated release of ciprofloxacin from polyvinyl alcohol based superparamagnetic nanocomposites. *Journal of Material Science-Materials in Medicine*, **22**, 357–369.

57. De Paoli, V.M., De Paoli Lacerda, S.H., Spinu, L. *et al.* (2006) Effect of an oscillating magnetic field on the release properties of magnetic collagen gels. *Langmuir*, **22**, 5894–5899.

58. Liu, T.-Y., Hu, S.-H., Liu, K.-H. *et al.* (2006) Preparation and characterization of smart magnetic hydrogels and its use for drug release. *Journal of Magnetism and Magnetic Materials*, **304**, e397-e399.

59. Al-Baradi, A.M., Mykhaylyk, O.O., Blythe, H.J. and Geoghegan, M. (2011) Magnetic field dependence of the diffusion of single dextran molecules within a hydrogel containing magnetite nanoparticles. *Journal of Chemical Physics*, 134.

60. Bertoglio, P., Jacobo, S.E. and Daraio, M.E. (2010) Preparation and characterization of PVA films with magnetic nanoparticles: The effect of particle loading on drug release behavior. *Journal of Applied Polymer Science*, **115**, 1859–1865.

61. François, N.J., Allo, S., Jacobo, S.E. and Daraio, M.E. (2007) Composites of polymeric gels and magnetic nanoparticles: Preparation and drug release behavior. *Journal of Applied Polymer Science*, **105**, 647–655.

62. Liu, T.-Y., Hu, S.-H., Liu, T.-Y. *et al.* (2006) Magnetic-sensitive behavior of intelligent ferrogels for controlled release of drug. *Langmuir*, **22**, 5974–5978.

63. Liu, T.-Y., Hu, S.-H., Liu, K.-H. *et al.* (2008) Study on controlled drug permeation of magnetic-sensitive ferrogels: Effect of Fe3O4 and PVA. *Journal of Controlled Release*, **126**, 228–236.

64. Caykara, T., Yoruk, D. and Demirci, S. (2009) Preparation and characterization of poly(N-tert-butylacrylamide-co-acrylamide) ferrogel. *Journal of Applied Polymer Science*, **112**, 800–804.

65. Barsi, L., Büki, A., Szabó, D. and Zrinyi, M. (1996) Gels with magnetic properties, in *Gels* (ed. M. Zrínyi), Springer, Berlin/Heidelberg, pp. 57–63.

66. Szabó, D., Szeghy, G. and Zrínyi, M. (1998) Shape transition of magnetic field sensitive polymer gels. *Macromolecules*, **31**, 6541–6548.

67. Na, H.B., Song, I.C. and Hyeon, T. (2009) Inorganic nanoparticles for MRI contrast agents. *Advanced Materials*, **21**, 2133–2148.

68. Saldanha, K.J., Piper, S.L., Ainslie, K.M. *et al.* (2008) Magnetic resonance imaging of iron oxide labelled stem cells: applications to tissue engineering based regeneration of the intervertebral disc. *European Cells and Materials*, **16**, 17–25.

69. Ramaswamy, S., Greco, J.B., Uluer, M.C. *et al.* (2009) Magnetic resonance imaging of chondrocytes labeled with superparamagnetic iron oxide nanoparticles in tissue-engineered cartilage. *Tissue Engineering Part A*, **15**, 3899–3910.

70. Gupta, A.K. and Gupta, M. (2005) Synthesis and surface engineering of iron oxide nanoparticles for biomedical applications. *Biomaterials*, **26**, 3995–4021.

71. Artemov, D., Mori, N., Okollie, B. and Bhujwalla, Z.M. (2003) MR molecular imaging of the Her-2/neu receptor in breast cancer cells using targeted iron oxide nanoparticles. *Magnetic Resonance in Medicine*, **49**, 403–408.

72. Pankhurst, Q.A., Connolly, J., Jones, S.K. and Dobson, J. (2003) Applications of magnetic nanoparticles in biomedicine. *Journal of Physics D: Applied Physics*, **36**, R167–R181.

10

The Interactions between Salt Ions and Thermo-Responsive Poly (N-Isopropylacrylamide) from Molecular Dynamics Simulations

Hongbo Du and Xianghong Qian
Department of Chemical Engineering, University of Arkansas, USA

10.1 Introduction

Poly(N-isopropylacrylamide) (PNIPAM) is a thermo-responsive polymer exhibiting a hydrophilic— hydrophobic volume phase transition above its lower critical solution temperature (LCST) in aqueous solution. The presence of salt ions tends to decrease the LCST of PNIPAM. The decrease in the transition temperature depends critically on the salt type and salt concentration. In addition, PNIPAM is often considered a protein proxy due to the existence of peptide bonds on its side chains. Experimentally it was shown that the decrease of the LCST for PNIPAM in sodium salt solutions follows the so-called Hofmeister [1] order for the anions. The Hofmeister series describes the ability of ions to precipitate proteins from an aqueous solution. MD simulations of the LCST phase transition of PNIPAM in 1 M NaCl, NaBr, NaI, and KCl solutions have been successfully carried out [2]. Our results demonstrate that salt cations bind strongly to the amide O on the peptide bond in PNIPAM whereas anions show no affinity for the peptide bond and only weak affinity for the hydrophobic residues in PNIPAM. In addition, our results show that the association strength between the cation and amide O is inversely correlated with the cation—anion interaction strength. The effects of the anion on PNIPAM appear to be largely modulated

Responsive Membranes and Materials, First Edition. Edited by D. Bhattacharyya, Thomas Schäfer, S. R. Wickramasinghe and Sylvia Daunert.
© 2013 John Wiley & Sons, Ltd. Published 2013 by John Wiley & Sons, Ltd.

via the cation−anion interaction. Moreover, the anion order of the Hofmeister series was shown to be correlated with the binding affinity between the cation and PNIPAM.

In order to further investigate the competing effects of salt ions on the LCST of PNIPAM and for more quantitative comparison, MD simulations of PNIPAM in a mixed salt solution were carried out around its estimated LCST. The mixed salt solution consists of 0.5 M Na^+, 0.5 M K^+ cations, and 0.5 M Cl^-, 0.5 M Br^- anions with a total salt concentration equivalent to the1 M individual salt solutions investigated in our earlier work [2]. The salt concentration of the salt solution was chosen to be 1 M for the simulations because significant shifts in LCST values were observed for PNIPAM in many alkali halide salt solutions. In addition, there were existing experimental data for comparison.

10.2   Computational Details

MD simulations were performed using NAMD [3], a highly parallelized code for large scale simulations. Non-polarized Amber 94 force field [4] was used for the polymer together with the TIP3P water model [5]. It has been used successfully to investigate PNIPAM volume phase transition in pure water and salt solutions [2]. For the salt ions in water, the improved parameters developed for the TIP3P water model were used [6]. The partial atomic charges of PNIPAM were calculated at B3LYP/aug-ccpvtz//B3LYP/6-311+G(d,p) level with Gaussian 03[7]. The structures of the monomer NIPAM capped with H− and CH_3− were constructed and the low energy conformations were determined using *ab initio* molecular dynamics simulations with CPMD [8]. Gaussian03 calculations were carried out for the partial charges for the lowest energy structures based on the RESP protocol [9]. The same PNIPAM initial structure was used in mixed salt solution as before. Experimentally observed LCSTs of PNIPAM in 1 M NaBr, KBr, NaCl, and KCl solution are about 298, 297, 293, and 292 K respectively when the polymer concentration is 1.4 wt.%. Therefore, the estimated LCST of PNIPAM in the mixed salt solution with 0.5 M Na^+, K^+, Cl^-, and Br^- each should be between 292 K and 298 K. The structure of the PNIPAM monomer unit is shown in Scheme 10.1 and the calculated partial atomic charges are listed in Table 10.1.

Although van der Waals parameters for alkali cations and halide anions are available in both Amber 94 and 99 force fields, recent MD simulations of high concentrations of NaCl in aqueous solution [10] showed that these parameters caused the salt ions to develop into clusters in the solution during the simulations. Improved ionic parameters of alkali cations

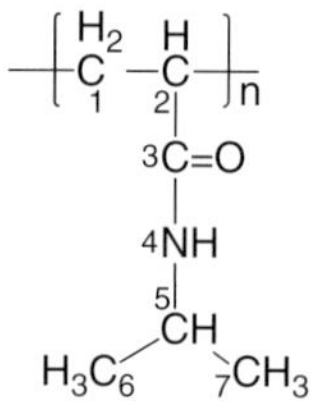

Scheme 10.1   *The structure of PNIPAM monomer unit. The atoms are numbered for easy discussion.*

Table 10.1   *Calculated partial atomic charge on PNIPAM monomer unit.*

Atom	C1	H1	C2	H2	C3	O	N	H(N)	C5	H5	C6, 7	H6, 7
Charge	−0.34	0.09	0.11	0.01	0.72	−0.58	−0.79	0.32	0.67	−0.02	−0.47	0.11

and halide anions corresponding to the TIP3P water model [6] have been used in the current work.

The MD simulations of PNIPAM in the mixed salt solution were performed at 278 and 318 K, about 15 to 25 K below and above its LCST respectively for a total of 75 nanoseconds (ns). The system consisted of a single PNIPAM chain with 50 degrees of polymerization (DP), 22 075 water molecules, 200 Na^+, Cl^-, K^+, and Br^- ions each evenly distributed in the solution. The LCST does vary slightly with the molecular weight of the PNIPAM. However, this variation is much smaller than the temperature window used here for the MD simulations. Our simulations were also limited by the computational cost which critically depended on the number of atoms in the system. Our corresponding PNIPAM concentration was 1.4 wt.%, at which the experimental LCSTs in salt solutions were available. The initial dimension of the unit cell was $90 \times 90 \times 90$ $Å^3$. The simulations were conducted under NPT ensemble with a target pressure of 1 atm using the Langevin–Hoover scheme [3]. A 12 Å sphere cut-off and a 15 Å pair list distance were applied to the van der Waals and short range electrostatic interactions respectively [11–18]. Long range electrostatic interactions were treated by the particle mesh Ewald (PME) method [19]. A time step of 1 femtosecond (fs) was used with periodic boundary conditions.

The volume phase transition at LCST is typically characterized by the conformation of the polymer and its associated hydration shell. The radius of gyration of the polymer backbone [20], the number of water molecules in the first hydration shell of the polymer, and the end-to-end distance of the polymer were analysed by Ptraj [21] with AmberTools. The radius of gyration (R_g) of the copolymer's backbone [20] is defined conventionally as

$$R_g = \sqrt{\frac{\sum_{i=1}^{N} m_i(r_i - R)^2}{\sum_{i=1}^{N} m_i}} \tag{10.1}$$

where r_i and m_i are the position vector and the mass of atom i respectively, and R is the position vector of the centroid of the chosen atoms, identified by the C1 and C2 atoms of NIPAM units in the backbone. End-to-end distance of the copolymer was also evaluated based on the C1 and C2 atoms on the backbone. The first hydration shell was conventionally defined as the volume from the surface of polymer up to a radius of 3.5 Å. The variation of intra-chain and intermolecular hydrogen bonds during the course of the simulation was also investigated in order to gain more insight into the nature of the LCST phase transition. The pair correlation functions (g(r)) between the ions and specific atoms on the PNIPAM were evaluated based on the 30 ns simulation period after PNIPAM went through the LCST transition in the mixed salt solution at 318 K and the corresponding period at 278 K. For the purpose of comparison, MD simulations were also carried out for the mixed salt solution containing only 0.5 M Na^+, K^+, Cl^-, and Br^- ions each without the PNIPAM at the same temperatures. A total of 1897 water molecules and 17 ions each for Na^+, K^+, Cl^-, and Br^- ions were included in the simulation unit cell with an initial cell dimension of $40 \times 40 \times 40 \, Å^3$. The other simulation parameters were kept the same as those of mixed salt solution in the presence of PNIPAM. A total of 40 ns simulations were carried out for the mixed salt solution without PNIPAM. Pair correlation functions between the cations and anions in solution were calculated for both systems with and without PNIPAM. Pair correlation functions between the salt ions and the O atom of the H_2O molecules were also evaluated.

10.3   Results and Discussion

Figure 10.1 shows the radius of gyration (a) and the number of water molecules (b) in the first hydration shell of PNIPAM along the simulation time in the mixed salt solution at 278 and 318 K respectively. Both the radius of gyration and water number in the first hydration shell appear to reach quasi-equilibrium after about 30 ns of simulations even though there are some fluctuations during the subsequent 45 ns. There is a significant decrease in the radius of gyration of PNIPAM along the higher temperature simulation, but not along the lower temperature in agreement with the occurrence of hydrophilic−hydrophobic volume phase transition above its LCST. There is also a significant reduction of the number of water molecules in the first hydration shell of PNIPAM along the higher temperature simulation, in contrast to that at lower temperature. These results clearly demonstrate that MD simulations are able to capture the volume phase transition at LCST for PNIPAM in the mixed salt solution and are consistent with our earlier findings [2].

The phase transition at the higher temperature appears to occur during the first 20 ns of simulations, similar to that previously observed for PNIPAM in 1 M NaCl and KCl solutions. The transition in the mixed salt solution appears to occur faster than previously observed PNIPAM in 1 M NaBr solution [2]. Based on the simulation results for PNIPAM in various salt solutions, it appears that the time scale of the LCST phase

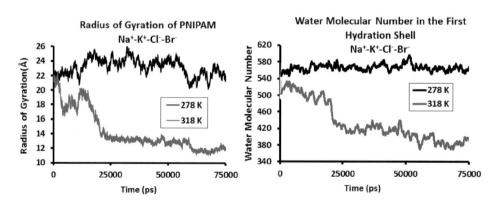

Figure 10.1   *Radius of gyration (a) and the number of water molecules in the first hydration shell of PNIPAM (b) in the mixed salt solution above and below its estimated LCST.*

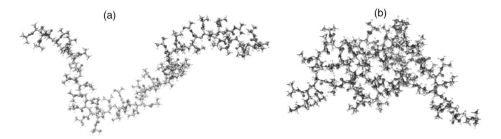

Figure 10.2 *Conformations of PNIPAM after the 75 ns simulations in 1 M mixed salt solution at 278 K (a) and 318 K (b) respectively. See plate section for colour figure.*

transition depends critically on the anion present in sodium salt solutions with the order $Cl^- < Br^- < I^-$. This is most likely due to two factors involving the ion−polymer interaction. One relates to the anion interaction with the polymer. The I^- ion appears to have the strongest hydrophobic interaction with the isopropyl residues on PNIPAM since it is the least hydrated and the highest polarizable ion amongst the three anions studied. The other factor relates to the cation interaction with the polymer. The binding between the Na^+ ion and the amide O on PNIPAM is strongest in NaI solution and weakest in NaCl solution due to the stronger Na^+-Cl^- ion pair interaction and the weaker Na^+-I^- interaction. The stronger ion−PNIPAM interaction tends to delay the LCST transition as well as increase the transition temperature.

The LCST phase transition at the higher temperature can be confirmed by the conformations of the polymers after 75 ns of simulation. Figure 10.2 shows the conformations of the polymer at the end of the simulation period at 278 K (a) and 318 K (b) respectively. It can be seen that the conformation of PNIPAM (Figure 10.2a) remains extended at 278 K whereas the conformation of PNIPAM (Figure 10.2b) becomes folded at 318 K after the LCST phase transition. These results are also consistent with PNIPAM in each individual salt solution investigated earlier. The associated intramolecular and intermolecular hydrogen bonding interactions during the LCST transition were also similar to our previous observations [22]. The LCST hydrophilic to hydrophobic transition is accompanied by the slight increase in the intramolecular hydrogen bonding interaction and the more significant decrease in hydrogen bonding interaction between the polymer and the surrounding water molecules.

The pair correlation functions g(r) between the polymer and salt ions were calculated after the polymer went through the LCST phase transition and the system reached quasi-equilibrium at 318 K. The pair correlation functions were averaged for the simulation period from 24 to 54 ns at 318 and 278 K respectively. Since only one or a few ions were observed to interact closely with the polymer and due to the dynamic movement of the salt ions in solution at the ns time-scale, there are fluctuations in the g(r) factors observed on a shorter time scale, for example, 5 ns. Averaging over 30 ns minimizes the dynamic effect of ionic diffusion in solution. Figure 10.3 shows the pair correlation functions between amide O, N, and C on PNIPAM and the salt ions at 318 and 278 K respectively. For the pair correlation functions between amide O and the $Na^+(K^+)$ ions as shown in Figure 10.3a, two peaks at 2.3 and 2.8 Å were observed for $Na^+ \cdots O$ and $K^+ \cdots O$ interactions respectively, indicating direct contact binding between the O atoms and the cations. Both Na^+ and

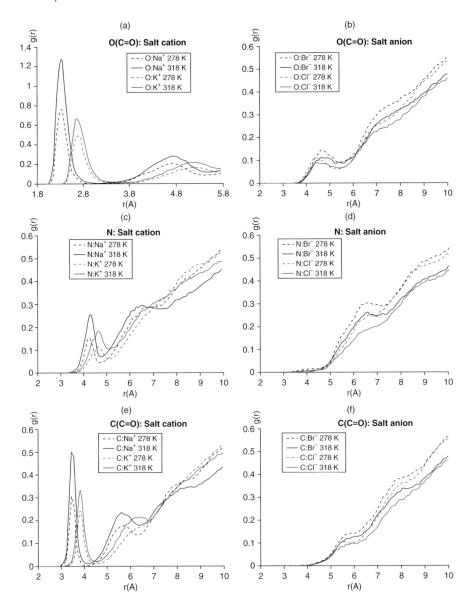

Figure 10.3 *Pair correlation functions between the O, N, C atoms on the amide group and the Na⁺, K⁺ cations, Cl⁻, Br⁻ anions of the salt solutions with PNIPAM.*

K⁺ ions exhibit relatively high affinities to the amide O. Moreover, Na⁺ ··· O interaction appears to be stronger than the corresponding K⁺ ··· O interaction, in agreement with our earlier observations. However, the interactions appear to be stronger at the higher temperature than at the lower temperature in contrast to our previous results for PNIPAM simulations in 1 M NaI, NaBr, and KCl salt solutions [2]. Nevertheless, this phenomenon of higher binding affinity at higher temperature agrees with the previous results for PNIPAM in 1 M NaCl solution and for PNIPAM−co−PEGMA copolymer in 1 M NaCl solution [22]. Our previous results show that the binding affinity increases significantly when the Na^+ ion binds to more than one amide O or other O atoms on the copolymer. This happens after the PNIPAM or PNIPAM−co−PEGMA go through the hydrophilic to hydrophobic phase transition at their respective LCSTs. The hydrophobic collapsed structures make it possible for the Na^+ ion to bind to multiple O sites on the polymer. K^+ ion was also observed to bind to multiple amide O sites on PNIPAM even though the interaction is weaker. More details on the polymeric structure will be discussed later.

For the g(r) factors between the salt anions and the amide O as shown in Figure 10.3b, only small and broad peaks at around 4.5 and 4.8 Å were observed for $Cl^-\cdots O$ and $Br^-\cdots O$ interactions respectively. This indicates that the interactions of the salt anions with the amide O are significantly weaker compared to those of the cation interactions, in agreement with our previous results [2, 22]. The interaction between the amide O and Br^- is slightly stronger than that between amide O and Cl^- due to the higher polarizability of the Br^- ion. Interestingly, the effects of temperature on interactions between amide O and anions appear to be quite different from those between amide O and cations. Here a slightly enhanced binding is observed at the lower temperature compared to binding at the higher temperature. This is due to the fact that binding affinity between anion and polymer is indirect and weak. Therefore, multiple anionic binding does not occur at the higher temperature when the polymer adopts a collapsed hydrophobic structure. As discussed in our earlier publications, the higher temperature tends to weaken the binding interaction if conformational changes are not taken into account. As a result, all anion−polymer affinity appears to be stronger at the lower temperature than at the higher temperature.

The pair correlation functions between salt cations and amide N, C (Figures 10.3c and e) show a similar trend to cation−amide O correlation. However, the correlation peaks appear at much larger distances and the strengths are significantly weaker than the corresponding $cation\cdots O$ interaction. This is due to the fact that cation−amide N, C interactions arise mainly from the cation−amide O interaction due to the proximity effects, particularly at short correlation distances. The interactions between amide N, C, and anions are even weaker than the interactions between amide O and anions, again in agreement with previous results. Interestingly, the anions do not appear to bind at all to the amide C atoms, even though the amide C has a large positive charge and anions are negatively charged. All in all, the results agree with earlier theoretical [23–26] and experimental [27] studies suggesting that cation–protein interaction has a significant impact on protein stability.

Figure 10.4 shows the pair correlation functions between isopropyl C (C5, C6, C7) and cations, anions respectively at 278 and 318 K in the mixed salt solution. The interactions between isopropyl C5, C6, C7, and the cations appear to be rather weak. Though visible, the peaks are very broad and the g(r) factors continue to increase with the increase of the cation−C distance. This indicates that there is no specific interaction mechanism that exists between the cations and the isopropyl C atoms. However, the interaction between anion and isopropyl C also shows interesting trends. Both isopropyl C5 and C6, C7 atoms appear to bind relatively strongly with the anions, particularly with Br^-, compared to the interaction with the cations. Moreover, the charges on C5 and C6, C7 used for the current MD simulations are 0.67 and −0.47 respectively. Though oppositely charged, the interactions between isopropyl C5, C6, C7, and the anions exhibit similar interaction strengths. This means that charge plays a minor role here. It appears that the interaction between Br^-

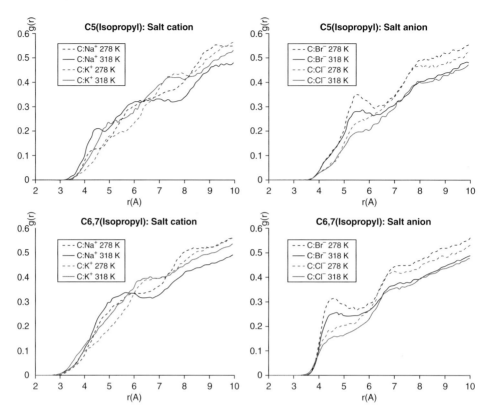

Figure 10.4 *Pair correlation functions between the C atoms on the isopropyl group of PNIPAM and the salt ions (cations on left panel and anions on right panel) in the 1 M mixed salt solution.*

and the isopropyl group is slightly stronger than that between Cl⁻ and isopropyl group in accordance with the hydration free energies of the anions. The more hydrated Cl⁻ binds weakly to the hydrophobic isopropyl groups whereas the less hydrated Br⁻ binds strongly to the hydrophobic residue.

Since there are two cationic types and two anionic types in the mixed salt solution, it is worth examining the competing effects of cation–anion interactions as well as their interactions with PNIPAM. The binding affinity of the salt ions with PNIPAM is dictated by the electrostatic and van der Waals forces, as well as the concentration of salt ions and availability of binding sites. Since there are many more binding sites available on PNIPAM than the number of ions interacting with the polymer, there appears to be no direct competition for binding sites between the Na⁺ and K⁺ ions. The competition of interaction occurs between Na⁺ (K⁺) with amide O sites and Na⁺ (K⁺) with salt anions/water. It is also true for the anionic interaction with the hydrophobic residues on the polymer.

In order to elucidate the contributing factors that affect the binding affinities between the salt ions and the polymer, the atomic and conformational structures of the PNIPAM were examined at both higher and lower temperatures, and before and after the LCST transition at the higher temperature. The distances between the salt cations and amide O

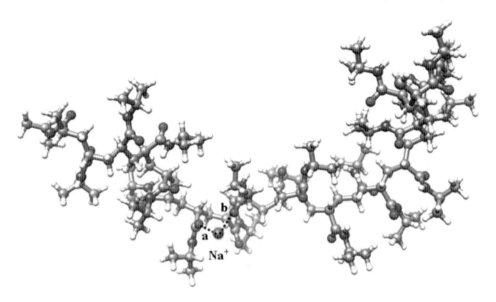

Figure 10.5 *The Na$^+$ (green) binds simultaneously to two amide O (red) atoms on PNIPAM after it goes through LCST phase transition at 318 K. Only part of the PNIPAM chain is shown. See plate section for colour figure.*

in the neighbourhood of the PNIPAM chain were calculated. It was found that one cation can simultaneously bind two neighbouring amide O atoms at the higher temperature after the LCST phase transition when the PNIPAM chain is in a folded dehydrated state, as shown in Figure 10.5. At the lower temperature when the PNIPAM chain is in an extended hydrated state, simultaneous cation binding of two amide O atoms was not observed. As shown in Figure 10.5, one Na$^+$ ion binds to two amide O atoms, where the distance *a* between Na$^+$ and one amide O is 2.34 Å and the distance *b* with another amide O is 2.37 Å. Similar structures exist for K$^+$ binding to the PNIPAM at 318 K after LCST transition. It is expected that multiple binding of the K$^+$ ion to the amide O atoms should be less pronounced due to the weak K$^+$−cation interaction. Due to this enhanced interaction at the higher temperature, caused by multiple binding, the pair correlation functions at the higher temperature will exhibit higher affinity. Since anions do not bind directly to the PNIPAM chain at either temperature, their interaction is guided by the true temperature effects where the lower temperature enhances the interaction strength. In addition, the lower temperature will have less disorder in the system therefore more pronounced correlation is expected.

The influence of salt ions on protein stability was thought initially to be due to the alteration of the bulk water structure in the presence of the ionic species [28, 29]. Recent experimental results show that bulk water structure is not influenced by the presence of salt ions in aqueous solution [30, 31]. In order to further understand how salt affects the LCST of PNIPAM in the mixed salt solution, salt cation−anion pair correlation functions were calculated both with and without the presence of PNIPAM. In addition, the pair correlation functions between the O atoms in H$_2$O and salt ions were determined in mixed salt solutions with and without the presence of PNIPAM. The cation−anion pair correlation functions and the salt ion−O (H$_2$O) correlations do not show the fluctuations observed for the ion−polymer interaction due to the significantly larger number of pair interactions that exist for the former. Figure 10.6 shows the cation−anion g(r) functions in the mixed salt solutions above and below the LCST with (left panel) and without (right panel) PNIPAM. The first peaks on the pair correlation functions are contact ion pair peaks which could be observed at 2.7, 2.8, 3.0, and 3.2 Å for the ion pair NaCl, NaBr, KCl, and KBr respectively. As with other pair correlation functions, the peak position indicates the contact ion-pair distance, and the peak height indicates the strength of ion association between cation and anion. The pair correlation functions for the mixed salt solution were evaluated based on the simulation time from 20 to 40 ns. The results in the mixed solution without PNPAM show that the K$^+$ and Br$^-$ contact ion pair has the strongest association, whereas the Na$^+$ and Br$^-$ ion pair has the weakest association at both higher and lower temperatures. The ion pair interaction strength follows the order KBr > KCl > NaCl > NaBr. The results agree very well with the ion-pair association constants of 0.30, 0.28, 0.12, 0.090 m^{-1} for KBr, KCl, NaCl, and NaBr respectively calculated in aqueous solution using the TIP3P water model [32]. For PNIPAM in mixed salt solution, it appears that Na$^+ \cdots$ Cl$^-$ and K$^+ \cdots$ Cl$^-$ ion pair interactions are slightly enhanced compared to the corresponding cation interactions with the Br$^-$ ion in the presence of PNIPAM. This is likely caused by the stronger Br$^-$ binding to the hydrophobic residues on PNIPAM than the corresponding Cl$^-$ in the mixed salt solution with PNIPAM. This confirms our conclusion that ionic interaction with polymer is modulated by the cation−anion interaction.

The pair correlation functions between salt ions and O on H$_2$O in the presence of PNIPAM are very similar to the corresponding ones without PNIPAM at both temperatures. The presence of PNIPAM appears to have little influence on the pair correlation functions between the salt ions and H$_2$O. However, cation−anion pair interaction is slightly affected by the polymer due to the preferential binding of the smaller cations and larger anions to PNIPAM. It appears that cation−anion ion pair interaction is stronger at the higher temperature than at the lower temperature in agreement with the earlier simulation results in different salt solutions. The pair correlation functions between salt ions and H$_2$O are stronger at lower temperature than at higher temperature [33]. The smaller Cl$^-$ ion has a stronger interaction with O(H$_2$O) which is thus more hydrated than the larger Br$^-$ ion. Moreover, the stronger the cation−anion ion pair interaction, the weaker the cation−O(H$_2$O) interaction. Salt cation−H$_2$O interaction is inversely correlated with the cation−anion contact pair interaction. More significantly, cation−anion interaction strength affects the cation−amide O interaction. Without multiple cationic binding to the amide O, cation−anion interaction is inversely correlated with the cation−amide O interaction. However, cation−amide O interaction could be enhanced due to the multiple cationic binding to the polymer at a temperature above polymer's LCST due to the resulting collapsed hydrophobic structures.

10.4 Conclusion

The LCST phase transition of PNIPAM was investigated in the mixed salt solution using classical MD simulations. Similar to the simulation results of PNIPAM in different salt solutions, it was found that cations have a high affinity to the amide O on PNIPAM,

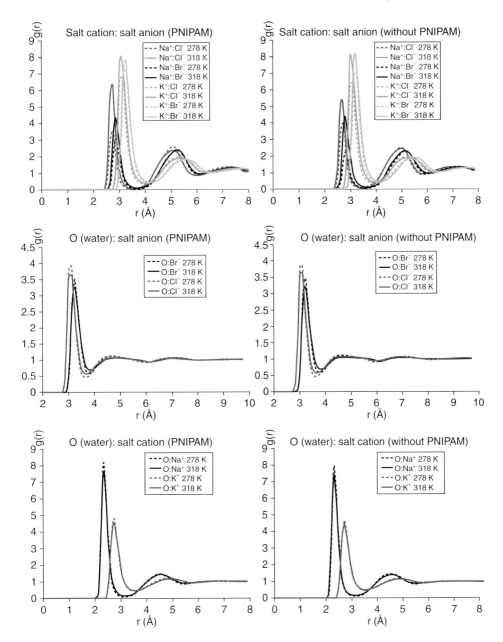

Figure 10.6 *Pair correlation functions between cation and anion, between anion and O (water) and between cation and O (water) with and without PNIPAM at both temperatures in the mixed salt solution. See plate section for colour figure.*

whereas anions bind weakly to the hydrophobic residues on the polymer. In particular, the affinity of Na^+ ion to amide O appears to be higher than that of K^+ ion. Further, cations, particularly Na^+ exhibit a higher affinity to amide O on PNIPAM at the higher temperature than that at the lower temperature due to the multiple cationic binding to the amide O in a folded PNIPAM structure. Many factors affect the binding affinity including the cation type, the cation–anion association constant, temperature, and the specific conformation of the PNIPAM. The strength of cation's interaction with the polymer is inversely correlated with the cation—anion contact pair association constant in the absence of multiple cationic binding to the polymer. The stronger the cation–anion interaction is, the weaker the cation's ability to bind to the polymer.

Acknowledgements

Funding from U.S. National Science Foundation (CBET 0651646) is gratefully acknowledged. The calculations were carried out partly on Teragrid and partly at Colorado State University.

References

1. Hofmeister, F. (1888) Zur lehre der wirkung der salze. *Naunyn-Schmiedeberg's Archives of Pharmacology*, **24**, 247–260.
2. Du, H.B., Wickramasinghe, R. and Qian, X.H. (2010) Effects of salt on the lower critical solution temperature of poly (N-isopropylacrylamide). *Journal of Physical Chemistry B*, **114**, 16594–16604.
3. Phillips, J.C., Braun, R., Wang, W. *et al.* (2005) Scalable molecular dynamics with NAMD. *Journal of Computational Chemistry*, **26**, 1781–1802.
4. Cornell, W.D., Cieplak, P., Bayly, C.I. *et al.* (1995) A 2nd generation force-field for the simulation of proteins, nucleic-acids, and organic-molecules. *Journal of the American Chemical Society*, **117**, 5179–5197.
5. Jorgensen, W.L., Chandrasekhar, J., Madura, J.D. *et al.* (1983) Comparison of simple potential functions for simulating liquid water. *Journal of Chemical Physics*, **79**, 926–935.
6. Joung, I.S., and Cheatham, T.E. (2008) Determination of alkali and halide monovalent ion parameters for use in explicitly solvated biomolecular simulations. *The Journal of Physical Chemistry B*, **112**, 9020–9041.
7. Frisch, M.J.T., Schlegel, G.W., Scuseria, H.B. *et al.* (2001) *Gaussian 03*, Gaussian, Inc., Wallingford, CT.
8. (CPMD 3.13, Copyrighted by IBM and Max-Planck Institute, 2009) .
9. Junmei, W., Piotr, C. and Peter, A. K. (2000) How well does a restrained electrostatic potential (RESP) model perform in calculating conformational energies of organic and biological molecules? *Journal of Computational Chemistry*, **21**, 1049–1074.
10. Auffinger, P., Cheatham, T.E. and Vaiana, A.C. (2007) Spontaneous formation of KCl aggregates in biomolecular simulations: A force field issue? *Journal of Chemical Theory and Computation*, **3**, 1851–1859.

11. Norberg, J., and Nilsson, L. (2000) On the truncation of long-range electrostatic interactions in DNA. *Biophysical Journal*, **79**, 1537–1553.

12. Horinek, D., Mamatkulov, S.I. and Netz, R.R. (2009) Rational design of ion force fields based on thermodynamic solvation properties. *Journal of Chemical Physics*, **130**, 124507.

13. Fennell, C.J., Bizjak, A., Vlachy, V. and Dill, K.A. (2009) Ion pairing in molecular simulations of aqueous alkali halide solutions. *Journal of Physical Chemistry B*, **113**, 6782–6791.

14. Chang, T.M. and Dang, L.X. (1999) Detailed study of potassium solvation using molecular dynamics techniques. *Journal of Physical Chemistry B*, **103**, 4714–4720.

15. General, I.J., Asciutto, E.K. and Madura, J.D. (2008) Structure of aqueous sodium perchlorate solutions. *Journal of Physical Chemistry B*, **112**, 15417–15425.

16. Loncharich, R.J., Brooks, B.R. and Pastor, R.W. (1992) Langevin dynamics of peptides – the frictional dependence of isomerization rates of N-acetylalanyl-N′-methylamide. *Biopolymers*, **32**, 523–535.

17. Lins, R.D. and Rothlisberger, U. (2006) Influence of long-range electrostatic treatments on the folding of the N-terminal H4 histone tail peptide. *Journal of Chemical Theory and Computation*, **2**, 246–250.

18. Brown, M.A., D'Auria, R., Kuo, I.F.W. *et al.* (2008) Ion spatial distributions at the liquid-vapor interface of aqueous potassium fluoride solutions. *Physical Chemistry Chemical Physics*, **10**, 4778–4784.

19. Darden, T., York, D. and Pedersen, L. (1993) Particle mesh Ewald – an N. *log(N) method for Ewald sums in large systems. Journal of Chemical Physics*, **98**, 10089–10092.

20. Gangemi, F., Longhi, G., Abbate, S. *et al.* (2008) Molecular dynamics simulation of aqueous solutions of 26-unit segments of p(NIPAAm) and of p(NIPAAm) "Doped" with amino acid based comonomers. *Journal of Physical Chemistry B 112*, 11896–11906.

21. Cheatham, T.E. AmberTools Users' Manual 1.2.

22. Du, H. and Qian, X. (2011) 'Molecular Dynamics Simulations of Copolymer PNIPAM-co-PEGMA Phase Transition at Lower Critical Solution Temperature in NaCl Solution'. *Journal of Polymer Science Part B: Polymer Physics*, **49**, 1112–1122.

23. Heyda, J., Vincent, J.C., Tobias, D.J. *et al.* (2010) Ion specificity at the peptide bond: molecular dynamics simulations of N-methylacetamide in aqueous salt solutions. *Journal of Physical Chemistry B*, **114**, 1213–1220.

24. Lund, M., Vacha, R. and Jungwirth, P. (2008) Specific ion binding to macromolecules: Effects of hydrophobicity and ion pairing. *Langmuir*, **24**, 3387–3391.

25. Vrbka, L., Vondrasek, J., Jagoda-Cwiklik, B. *et al.* (2006) Quantification and rationalization of the higher affinity of sodium over potassium to protein surfaces. *Proceedings of the National Academy of Sciences of the United States of America*, **103**, 15440–15444.

26. Jagoda-Cwiklik, B., Vacha, R., Lund, M. *et al.* (2007) Ion pairing as a possible clue for discriminating between sodium and potassium in biological and other complex environments. *Journal of Physical Chemistry B*, **111**, 14077–14079.

27. Uejio, J.S., Schwartz, C.P., Duffin, A.M. *et al.* (2008) Characterization of selective binding of alkali cations with carboxylate by x-ray absorption spectroscopy of liquid

microjets. *Proceedings of the National Academy of Sciences of the United States of America*, **105**, 6809–6812.

28. Vanzi, F., Madan, B. and Sharp, K. (1998) Effect of the protein denaturants urea and guanidinium on water structure: A structural and thermodynamic study. *Journal of the American Chemical Society*, **120**, 10748.

29. Zou, Q., Bennion, B.J., Daggett, V. and Murphy, K.P. (2002) The molecular mechanism of stabilization of proteins by TMAO and its ability to counteract the effects of urea. *Journal of the American Chemical Society*, **124**, 1192.

30. Omta, A.W., Kropman, M.F., Woutersen, S. and Bakker, H.J. (2003) Negligible effect of ions on the hydrogen-bond structure in liquid water. *Science*, **301**, 347–349.

31. Batchelor, J.D., Olteanu, A., Tripathy, A. and Pielak, G.J. (2004) Impact of protein denaturants and stabilizers on water structure. *Journal of the American Chemical Society*, **126**, 1958–1961.

32. Joung, I.S. and Cheatham, T.E. (2009) Molecular dynamics simulations of the dynamic and energetic properties of alkali and halide ions using water-model-specific ion parameters. *The Journal of Physical Chemistry B*, **113**, 13279–13290.

33. Brodholt, J.P. (1998) Molecular dynamics simulations of aqueous NaCl solutions at high pressures and temperatures. *Chemical Geology*, **151**, 11–19.

11

Biologically-Inspired Responsive Materials: Integrating Biological Function into Synthetic Materials

Kendrick Turner, Santosh Khatwani and Sylvia Daunert
Miller School of Medicine, Department of Biochemistry and Molecular Biology,
University of Miami, USA

11.1   Introduction

Nature has designed a number of smart, complex, yet highly specific and efficient structures by utilizing a variety of complex biomolecules as building blocks. Mimicking these sophisticated structures is a challenge for today's scientists. Cell membranes are naturally occurring smart systems that control trafficking of ions, biomolecules, and nutrients across the cell membrane [1]. Recent developments in molecular biology have led to increased growth in advanced manmade structures capable of mimicking these highly specific biological processes (Table 11.1) [2–5]. The integration of biological recognition elements with organic molecules and synthetic materials has led to the creation of smart biosensors and biomaterials capable of performing unique tasks [6–11]. These smart systems make use of nano-scale molecular recognition events to control the characteristics of the synthetic component. One example of such a smart biomaterial is the ATP-fuelled rotary biomolecular motor, designed by integrating an enzyme F_1-ATPase which has been genetically engineered to contain a zinc-binding site, with a nickel propeller [12]. In the absence of zinc, the enzyme uses energy from the hydrolysis of ATP for rotation, which is translated into the spinning of the nickel propeller. The addition of zinc, which plays the role of an "off" switch, inhibits this rotation, stopping the nickel propeller. In another example, a protein, kinesin, has been integrated onto the surface of microchannels. Upon applying an external

Responsive Membranes and Materials, First Edition. Edited by D. Bhattacharyya, Thomas Schäfer, S. R. Wickramasinghe and Sylvia Daunert.
© 2013 John Wiley & Sons, Ltd. Published 2013 by John Wiley & Sons, Ltd.

Table 11.1 Biological and synthetic approaches to the development of responsive materials.

Biological moiety	Biological function	Stimulus	References
Glucose binding protein	Periplasmic chemotaxi	Glucose	[140]
Calmodulin	Signal transduction	Calcium Phenothiazine-like drugs Calmodulin-binding peptides	[62–74]
F_1-ATPase	ATP synthase	Zinc	[12]
Kinesin	Motor protein	External electric field	[13, 14]
Aquaporins	Pore-forming proteins	Water	[16]
Protein channels	Molecular transport	Ions, ligands	[23–27, 29]
α-hemolysin	Transmembrane channels	Water, ions, small molecules	[33–37]
Elastin-like peptides	Mechanical stability of connective tissue	Temperature	[38]

Synthetic moiety	Function	Stimulus	References
Modified carbon nanotubes	Pore-forming complexes Molecular transport	Water, small molecules	[21, 24–29]
Dendrditic peptides	Pore-forming complexes	Water	[22]
Nanoporous materials	Ligand/voltage-gated channels	Small molecules External electric field	[30–32]
Stimuli-responsive hydrogels	Molecular transport Actuation Separations	Temperature pH Small molecules, ions External electric/magnetic fields	[45–53, 119–131]

electric field, the transport of microtubules was steered to the selected arm of a Y-type microchannel junction [13]. Additionally, the immobilization of multiple kinesin molecules on a substrate resulted in the development of a structure that was able to manipulate the movement of DNA molecules [14]. The immobilization of one end of a λ-phage DNA on the substrate and the attachment of the other end of the DNA onto a microtubule resulted in the stretching of the DNA upon the movement of the microtubule by the kinesin molecules. When the second part of the DNA was also attached to another microtubule, the energy from the kinesin molecules resulted in the transport of this DNA along the substrate. There are a number of such examples where biological recognition elements, such as proteins, DNA, RNA, antibodies, and enzymes have been integrated along with synthetic materials to design smart, responsive biomedical devices.

11.2 Biomimetics in Biotechnology

Biomimetics is the science of learning, imitating, and copying the biological phenomena for the development of artificial devices [15]. Biomimetics has enabled the utilization of the specificity and complexity of natural processes to design novel devices for biomedical applications. Mimicking natural processes is still a huge challenge to modern researchers. There are a number of natural processes especially where biological systems adapt to survive in harsh environments by controlling the flow of water through channels. For instance, in plants, the transpiration process is regulated by the opening and closing of stomata to control the flow of water through channels. The guard cells in stomata open in humid conditions, and close when it is dry to conserve water in the cells (Figure 11.1). In addition to stomata, biological systems growing in harsh conditions develop ways to control their water flow through pore-forming membrane proteins called aquaporins. These pore-forming proteins form channels through which transport of water is regulated [16]. The aquaporins are tetrameric proteins with each monomer operating as a water channel. The pore-forming region is characterized by the presence of two proton separating moieties composed of Asparagine-Proline-Alanine along with nearby hydrophobic residues for the control of water flow through the cells (Figure 11.2). The variations in the peptide sequences in the

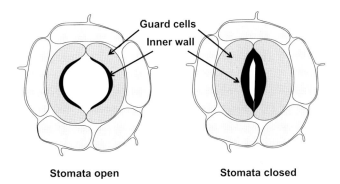

Figure 11.1 *Opening and closing of stomata as a result of turgor pressure in the guard cells.*

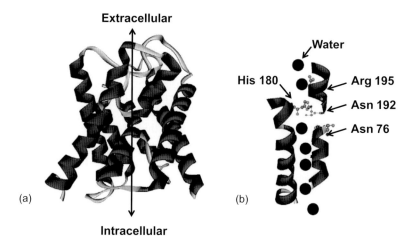

Figure 11.2 *(a) A ribbon diagram of aquaporin subunit (AQP1). The movement of water is shown by arrow. (b) Schematic representation of size-based transport of water through aquaporin channel. See plate section for colour figure.*

pore-forming region determine the size of the pore and consequently the type of molecule that can be transported. There are other types of pore-forming aquaporins for the transport of essential nutrients, ions, glycerol, gases, and during photosynthesis, water uptake and reproduction [16]. In drought conditions, the dephosphorylation of certain serine residues causes a change in the 3D structure of the protein, and the pore is closed to retain water [16]. But during flood conditions, certain histidine residues become protonated which results in a change in the 3D structure of the protein for water gating [16]. Mutations in aquaporins have been proposed to cause numerous fluid transport disorders such as, brain oedema, cirrhosis, congestive heart failure, glaucoma, and pre-eclampsia [17, 18]. Aquaporins are also found in higher animals, especially in their kidneys, for water reabsorption [19]. Mutations in aquaporins cause inherited and acquired water balance disorders [20]. A few attempts have been made to mimic the function of aquaporins by using synthetic polymers. Modified double-walled carbon nanotubes have been embedded into dimyristoylphosphatidylcholine membranes to mimic them [21]. These artificial membranes have potential applications in designing nanobiodevices. In addition, a dendritic dipeptide that self-assembles into thermally stable helical pores was fabricated to mimic aquaporin channel pores. The helical pores were found to be stable in phospholipid membranes, and selectively transported water [22].

Protein channels play an important role in the transport of molecules inside and outside the cell. The gating of the molecules in the protein channel could be through ligand-binding (ligand-gated channel), change in electric potential across the membrane (voltage-gated channel), or deformation of stretch receptors (mechanically-gated protein channels) (Figure 11.3). The sodium and potassium ion channels are examples of ion-gated channels, and are actuated by changes in the membrane potential for the transport of sodium and potassium ions across the cell membrane [23]. In the ligand-gated protein channels, the binding of a ligand at an allosteric position in the channel protein causes a change in the

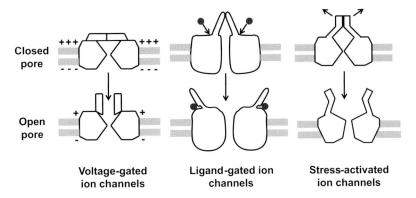

Figure 11.3   Mechanistic representation of opening and closing of different types of ion channels for controlled flow of molecules.

conformation of the channel protein, leading to the opening or closing of the channel pore. Recently, functionalized carbon nanotubes [24–27] were used to mimic protein channels with enhanced hydrodynamic flow through the membrane [28]. Furthermore, a peptide attachment on the surface of the carbon nanotubes was utilized to mimic a ligand-gated protein channel (Figure 11.4). The phosphorylated peptide was tethered to the carboxylic acid group at the CNT core entrance. The peptide binds to the antibody, and modulates the ionic flux through CNT cores across the membrane. The dephosphorylation of the peptide resulted in a change in the ionic flux through the membrane (Figure 11.4) [29]. There are other such examples in literature where nanoporous materials from glass [30, 31] and gold [32] have been employed to mimic voltage- and ligand-gated protein channels.

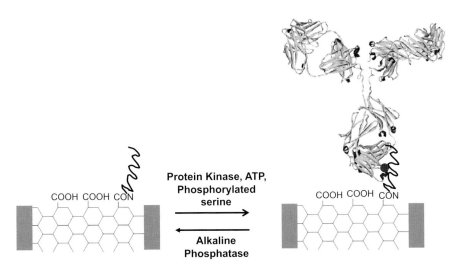

Figure 11.4   Schematic of phosphorylation-based biomimetic system. (Adapted with permission from [29] Copyright (2007) Royal Society of Chemistry). See plate section for colour figure.

The bacterium *Staphylococcus aureus* produces α-hemolysin monomers (a pore-forming toxin) by which it attaches and kills the host cell. The monomers bind to the outer membrane of susceptible cells and oligomerize to form heptameric pores in lipid bilayer with a water-filled transmembrane channel. This channel allows permeation of water, ions, and small organic molecules into the host cell. This results in a change in membrane potential and irreversible osmotic pressure within the cell, leading to the rupture of cell wall [33]. As α-hemolysin can bind to various biological and synthetic lipid bilayers, and is stable at different temperature and pH conditions, it has been utilized for different biotechnological applications. Gu *et al.* utilized the suspended α-hemolysin channel equipped with a cyclodextrin adaptor for stochastic sensing of organic molecules [34]. In addition, α-hemolysin has been employed in simultaneous sensing of divalent cations and proteins [35, 36]. A perhaps more important application of α-hemolysin lies in controlling the flow of molecules for drug-delivery. The α-hemolysin pore can facilitate the controlled delivery of small molecules and ions across the plasma membrane of the host cell or through the walls of synthetic lipid vesicles [37]. Recently, elastin-like polypeptides were placed within the cavity of α-hemolysin to generate a temperature-sensitive protein pore. The elastin-like-polypeptides undergo phase transition with an increase in temperature. The stimuli-responsive property of elastin-like-polypeptides was coupled with the pore for potential temperature-responsive drug-delivery systems (Figure 11.5) [38]. Similarly to the naturally occurring protein nanopore α-hemolysin, artificial nanopores have also been created. A RNA nanomotor derived from bacteriophage phi29 was engineered into a planar lipid bilayer to create an artificial nanopore through which double strand DNA in addition to

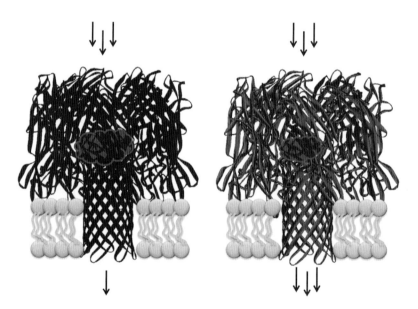

Figure 11.5 *Elastin-like-polypeptides embedded in α-hemolysin to create temperature-sensitive protein pores for molecular gating. The stem structure of α-hemolysin is used to bind to the lipid bilayer. (Adapted with permission from [38] Copyright (2006) American Chemical Society).*

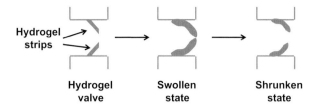

Figure 11.6   *Fabrication of hydrogel valve. At pH 8, the hydrogel strips form a closed check valve, while at low pH (3.0) the hydrogel shrinks deactivating the valve. (Adapted with permission from [47] Copyright (2004) Elsevier Ltd).*

single stand DNA/RNA can be transported. These artificial nanopores find applications in designing microelectromechanical sensing, microreactors, gene delivery, drug loading, and DNA sequencing [39–42]. In another example, an artificial nanopore mimicking nuclear pore complex selectivity was fabricated from a polycarbonate membrane functionalized with phenylalanine-glycine-nucleoporins [43]. The membrane allowed transport factors and transport-factor-cargo complexes to pass while inhibiting other proteins. In addition, functionalized nanoporous membranes have also been utilized to separate molecules based on the size, charge, and hydrophobicity of the molecules [44].

Microvalves are abundant in animals for controlling the flow of blood through blood vessels. Stimuli-responsive polymers have been extensively utilized to design biomimetic valves. A biomimetic stimuli-responsive hydrogel valve capable of directional flow control in microfluidic channels was developed from a pair of photopolymerized hydrogel bistrips [45]. These hydrogel strips have different pH sensitivities, and activate or deactivate the valve reversibly based on local pH. At pH 8, the valve allows fluid flow in one direction under forward pressure while restricting flow in the opposite direction, thus mimicking anatomical venous valves (Figure 11.6). Other stimuli-responsive hydrogel micro/nanovalves have been utilized to mimic biological valves that can control the flow of solutions for designing microelectromechanical devices for separations and biomedical applications [45–53].

11.3   Hinge-Motion Binding Proteins

Binding proteins are ubiquitous proteins that bind to a specific analyte or class of analyte. In the cell, the binding event of a protein(s) is usually accompanied by the transport of the bound analyte inside or outside the cell. Based on the type of analyte that they bind, the binding proteins have been classified as calcium-binding proteins (e.g. calmodulin), DNA-binding proteins, RNA-binding proteins, and sulfate-binding proteins. Amongst the binding proteins, the hinge-motion binding proteins are of special interest to modern researchers. Upon binding to the analyte, hinge-motion binding proteins undergo a conformational change around the hinge-region in the structure [54]. Many of the hinge-motion binding proteins are periplasmic-binding proteins and are involved in the transport of the analytes across the cell membrane [55]. Typically, these periplasmic-binding proteins consist of two globular domains linked by three separate peptide segments (or an interdomain hinge region) with the binding site located in the cleft between the two domains [56]. The analyte is bound to various residues mainly via hydrogel bonds, van der Waals interactions

[57, 58], π-interactions [59], or ionic interactions [60]. In the absence of the analyte, protein is in an open conformation with the binding site exposed and accessible to the incoming analyte. Upon binding of the analyte, the two domains constrict around the hinge-region resulting in a closed conformation of protein. These proteins are highly selective and specific to an analyte or class of analytes with affinities (K_D) as low as nanomolar range [54]. Advances in recombinant DNA technology, coupled with the knowledge of protein X-ray crystallographic structure, has enabled the exploitation of the natural nanoscale conformational changes of hinge-motion binding proteins for designing smart biosensing systems and responsive biomaterials [61].

11.4   Calmodulin

Calmodulin (CaM) is a eukaryotic hinge-motion, calcium-binding protein that mediates a variety of intracellular processes including regulation of activation of kinases and phosphatases [62]. CaM is involved in the activation of enzymes mediating in a variety of biological processes. Calcium influences biological processes like muscle contraction, cell motility, chromosome movement, neurotransmitter release, endocytosis, exocytosis, and glycogen metabolism. Upon binding to calcium, the Ca^{2+}-CaM complex acts as a regulatory subunit for more than 20 enzymes [63]. In the absence of calcium, CaM has a low affinity for the enzymes. The increase in cytoplasmic calcium concentration results in the formation of Ca^{2+}-CaM complex which binds to the CaM-binding site of the enzyme causing a conformational change in the CaM. In addition, the binding of CaM to the enzymes allows the access of the ATP or the target protein to the active site of the enzyme. The reduction in the cellular calcium level results in a release of calcium from CaM, which in turn leads to the dissociation of CaM from the enzyme-CaM complex rendering the enzyme inactive. During glycogen metabolism, the adenylate cyclase catalyses the conversion of ATP to a secondary messenger cyclic AMP [64]. Further, cyclic AMP must bind to the calcium–CaM complex to activate the protein kinases for phosphorylation. During the muscle contraction, the increased concentrations of calcium results in the interaction of actin with myosin. This interaction results in the activation of actomyosin which uses ATP for contraction. In smooth and cardiac muscles, CaM acts as a regulator for the calcium-dependent phosphorylation of myosin light chain kinases and phosphatases [65]. In addition, CaM also activates phosphorylase kinase during glycogen degradation along with inhibiting the activity of glycogen synthase.

Calcium is also known to be an essential regulator of cell growth. In addition, it is suggested that the initial events of mitosis are initiated by CaM kinase and cell proliferation could be directly influenced by alterations in the calcium and CaM concentrations in the cytoplasm. Thus, CaM plays a role in cell growth and gene expression. Any alterations in the factors affecting cell proliferation could be lethal [63, 64].

CaM (148 amino acids, $\sim$16.5 kDa) is an acidic protein (pI $=$ 4.3) [66, 67] with two similar domains at the N- and C-termini separated by a hinge-region of seven-turn α-helix. Each globular domain has a pair of 12-residue EF-hand (helix-loop-helix) motifs. Each of these EF-hand motifs contains a binding site for calcium [68]. CaM, upon binding to Ca^{2+}, undergoes a significant conformational change into a dumbbell shaped structure (Figure 11.7), which exposes two hydrophobic pockets, one at each terminus of the protein [66, 69, 70].

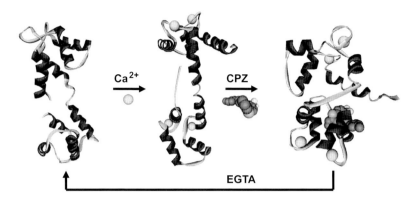

Figure 11.7 *Conformational changes of calmodulin induced upon sequential binding to Ca²⁺ and CPZ site-specifically, on or inside a polymer matrix, leading to the fabrication of responsive biomaterials. See plate section for colour figure.*

Many CaM inhibitors, peptides, and drugs are known to bind to the hydrophobic pocket at the C-terminus of the Ca^{2+}-bound CaM. Antipsychotic drugs, such as phenothiazine (e.g. trifluperazine and Chlorpromazine, CPZ), bind to the C-terminal hydrophobic pocket of Ca^{2+}-bound CaM via hydrophobic interactions and van der Waals forces. Upon binding to phenothiazines, CaM undergoes a second conformational change around the hinge-region into a collapsed state (CPZ-bound CaM) (Figure 11.7). Further addition of EGTA removes the bound Ca^{2+}, which, in turn, releases the bound phenothiazine ligand, resulting in CaM returning to its native (free) state (Figure 11.7). The repeated and sequential addition of Ca^{2+}, CPZ, and EGTA could produce a reversible and reproducible response from CaM as a molecular recognition element [71]. These ligand-induced conformations of CaM could be valuable in designing protein-based biosensing systems and responsive biomaterials. A cysteine-free CaM was genetically-engineered to introduce cysteines at desired positions to design biosensing systems for calcium [72] and phenothiazines [73, 74]. Similarly, CaM with unique cysteines could be immobilized, site-specifically, on or inside a polymer matrix, leading to the fabrication of responsive biomaterials.

11.5 Biologically-Inspired Responsive Membranes

The development of high-performance and selective polymer membranes is an important issue that is yet to be resolved in academic and industrial laboratories. Porous membranes have found applications in filtration, separation, sensing, transport, and drug-delivery [75–77]. Specifically, the "responsive" membranes that are able to alter their porosity due to changes in their physical or chemical environment [78], like pH [79], temperature [80, 81], analyte concentration [82], and ionic strength [83] are important for designing responsive drug-delivery, biosensing, microelectromechanical systems (MEMS), and nanoelectromechanical systems (NEMS) devices. A variety of such responsive membranes have been fabricated by grafting [84] and phase inversion [85, 86]. Of particular interest are the responsive membranes that selectively alter their porosity in the presence of external stimuli such as pH [87], temperature [80], and analyte concentrations [88]. These membranes are fabricated by either polymerizing stimuli-responsive polymers inside [89] or on the surface of the membrane [81]. In addition, biohybrid membranes have been prepared by immobilizing a biological recognition entity, either physically or chemically, on a polymeric support [90]. These biohybrid responsive membranes exhibit alteration in their porosity upon the interaction of biological recognition elements with specific analytes. In order to design biohybrid responsive membranes, a variety of molecules such as DNA [91], polypeptides [89], and proteins [92, 93] have been integrated within porous membranes. Amongst the proteins, the hinge-motion binding proteins undergo conformational changes upon binding to a specific analyte. The immobilization of these hinge-motion binding proteins on synthetic supports could lead to an interesting class of "responsive" membranes. In addition, the high binding affinity and specificity of hinge-motion binding proteins could make them amenable to applications in sensing, separation, and drug-delivery systems. Various physical as well as chemical methods have been utilized to immobilize the proteins. The physical method of immobilization includes physical entrapment or adsorption of the biomolecule on the support [94]. Due to the weak nature of forces involved, the use of physical methods for the immobilization of biomolecules is rather limited. This method also suffers from leakage of the biomolecule from the polymer matrix. For this reason, chemical methods of immobilization are commonly employed to immobilize biomolecules such as proteins and peptides on synthetic supports. Usually, the chemical methods for immobilization of proteins involve covalent binding of the protein, either through the ε-amine group of lysines [95] or thiol group of a unique cysteine, glutamate, or aspartate [96], biotin–avidin interaction [97] onto an activated membrane support such as cellulose acetate, Ultrabind, and polyethersulfonemembranes [98]. The aldehyde-functionalized cellulose acetate membranes have been utilized for random as well as site-directed immobilization. Even though, the hydrolysis of cellulose acetate yields a 2,3-trans diol [99, 100], it has been shown that periodate oxidation can efficiently immobilize proteins in random and site-directed orientation [95]. Genetically-engineered tags, such as FLAG, have also been utilized to immobilize the protein in a site-directed manner on the synthetic support [101]. Site-directed immobilization presents a unique method of unidirectional alignment of the biomolecule on the support, resulting in full accessibility of the substrate to the biomolecule [90].

In recent years, chemical ligation techniques have garnered much interest for immobilization of biomolecules on synthetic supports. Chemical ligation is utilized to create long chain peptides or proteins by joining two or more peptide fragments. Native chemical ligation is the most common and robust means of chemically ligating two peptides. This method involves a peptide with an N-terminal cysteine reacted with a thioester to form an amide bond [102, 103]. The same strategy has been applied to immobilize proteins on surfaces. Some of these examples include immobilization of a cys-protein conjugate onto a thioester-functionalized glass [104] or porous silicon surface [105]. In addition, thioester-protein was immobilized onto a cys-modified surface [106]. Recently, there was an intein-based site-specific biotinylation of proteins for the fabrication of protein arrays [107]. Another method for protein immobilization that has attracted attention recently is click reaction [108]. In this reaction, an azide is site-specifically coupled to an alkyne via 1,4-Huisgen-type cycloaddition in the presence of a Cu(I) catalyst. Using this method, proteins and antigens [109, 110] have been immobilized on a variety of substrates including membranes, fibres and beads.

This method utilizes a specific enzyme protein, farnesyl transferase, which transfers a farnesyl group to the cysteine residue of a four residue tag at the C-terminus of the protein. This tagged protein is then attached to desired alkyne-funtionalized substrate for immobilization [110]. In addition, a Diels–Alder cycloaddition reaction can be used for the immobilization of quinone-containing self-assembled monolayers with cyclopentadiene molecules [111]. This method has been extended to conjugation of proteins on a carbohydrate-containing substrate [112, 113]. The biologically-mediated methods of immobilization provide higher specificity towards the target protein [114]. One such method involves the utilization of His-tag for reversible and non-covalent protein immobilization on a Ni^{2+}-substrate. Consequently, this method has found tremendous potential in the purification of proteins [115]. A covalent method of biologically-mediated protein immobilization involves the use of genetic tags such as glutathione-*S*-transferase (GST). This method has been employed to immobilize proteins on surfaces such as polystyrene and PEGA resin beads [116, 117]. The abundance of available immobilization methods has enabled their use in the fabrication of protein-based sensors and biomaterials.

11.6 Stimuli-Responsive Hydrogels

Hydrogels are highly covalently cross-linked, hydrophilic, water insoluble, viscoelastic polymers that swell upon absorbing large amounts of water. The swelling of the hydrogels is affected by the hydrophilic interactions of solvent with the polymer network. The hydrogels are biocompatible polymers and have potential applications in sensing, actuation, drug-delivery, high-throughput screening, and other device-based applications. The high elasticity and low strength of these hydrogels make them potential tissue- or cartilage-replacement materials. Hydrogels that swell/shrink in response to the physical or chemical changes in the environment are particularly important for actuation, artificial muscles, sensing, and responsive drug-delivery [118]. These stimuli-responsive hydrogels are known to respond to changes in pH [119], temperature [120], light [121, 122], antigens [123], ionic strength [124], and electric potential [125]. Amongst them, pH and temperature-sensitive (thermo-responsive) hydrogels have been studied extensively for their applications in actuation [126, 127], drug-delivery [128, 129], and microcantilevers [130, 131]. The thermo-sensitive hydrogels undergo abrupt volume transition at a lower critical solution temperature (LCST). Temperature-sensitive hydrogels are commonly comprised of N-alkyl acrylamides (e.g. *N*-isopropyl acrylamide, NIPAAm) with hydrophilic side chains, forming hydrogen bonds with water, resulting in the swelling of the hydrogel at room temperature. Above LCST (33°C for pNIPAAm), hydrogen bonds are disrupted, resulting in exposure of the hydrophobic backbone of the polymer. This leads to the aggregation and subsequent shrinking of the hydrogel [132]. On the other hand, the pH-sensitive hydrogels comprise weakly acidic or basic groups. The change in the pH and ionic strength of the solution causes the development of osmotic pressure within the hydrogel, leading to swelling/shrinking of the hydrogel [132]. These hydrogels have also been used to study the encapsulation of drugs and their subsequent release in the medium [129].

A promising approach for the development of dynamic hydrogels involves the utilization of naturally occurring biomolecules [133, 134]. Biologically-inspired hydrogels have been synthesized by incorporating a biological recognition element, such as antibodies [123], proteins [135–137], and DNA [138], within the polymer network. Amongst them, protein-based stimuli-responsive hydrogels have a unique ability to undergo volumetric transitions in the presence of a protein-specific ligand in the solution. Responsive biomaterials had been previously developed by the integration of the hinge-motion binding proteins, namely CaM and the glucose/galactose-binding protein (GBP), within the bulk of an acrylamide hydrogel [137, 139, 140]. Both CaM and GBP undergo significant conformational changes in the presence of their specific ligands (Ca^{2+} and CPZ for CaM, glucose and galactose for GBP). Specifically, CaM was immobilized within the hydrogel via two different methods that resulted in the swelling/shrinking of the hydrogel by two different mechanisms. In the first approach, CaM was polymerized along with a low-affinity CaM-binding ligand, *N*-3-[2-(trifluoromethyl)-10H-phenothiazin-10-yl]propylacrylamide (TAPP). The addition of CPZ, a high-affinity CaM-binding ligand, displaces the immobilized TAPP, resulting in the swelling of the hydrogel due to the relaxation of polymer chains (hand-shake mechanism) [137]. In the second method, CaM was immobilized within the hydrogel itself through two cysteines, one at each terminus of the protein. Upon addition of CPZ, CaM assumes a more constricted form, resulting in the shrinking of the hydrogel due to contraction of polymer chains (accordion mechanism) (Figure 11.8). In another example, Murphy *et al.*

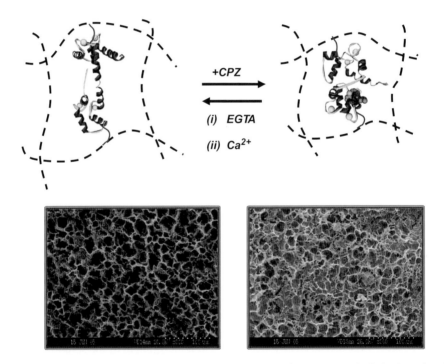

Figure 11.8 *Swelling/shrinking states of CaM-based stimuli-responsive bulk hydrogels. The Ca^{2+}-bound CaM undergoes a conformational change upon binding to CPZ. The CPZ-bound CaM, upon addition of EGTA, removes Ca^{2+}, which in turn, removes bound CPZ from CaM. The Ca^{2+}-bound CaM can be regenerated by adding Ca^{2+} to the free CaM. See plate section for colour figure.*

developed a CaM-based hydrogel that was responsive to trifluoperazine [135]. The CPZ- and TFP-induced volumetric transitions were reversible and reproducible thus mimicking the contraction/expansion function of natural muscles. Mrksich group members were able to develop a glucose-responsive hydrogel by microencapsulation of pancreatic islets within polyethyleneglycol coatings [141]. The ability of stimuli-responsive hydrogels to translate a molecular recognition event into a measurable response could be advantageous in developing intelligent hydrogel-based devices.

11.7   Micro/Nanofabrication of Hydrogels

The synthesis of tunable hydrogels with nanometre-sized dimensions has sparked new interest in the development of micro/nanofabricated hydrogels for applications in sensing, *in vivo* diagnostics, microactuators [142], microvalves [143], nanovalves [144], microcantilevers [130], and drug-delivery applications [145, 146]. The advantages of miniaturized hydrogels over bulk hydrogels include high surface-to-area ratio, low sample requirements, and faster response times for portable and point-of-care diagnostic devices. Such miniaturized systems could also be employed as components of micro/nanoelectromechanical systems and micro/nanofluidic platforms. Furthermore, microfabrication of responsive hydrogels has been employed to design functional hydrogel arrays [139, 147, 148]. Techniques utilized for microfabrication include deposition, patterning, and etching [149]. The deposition is performed either by sputtering, chemical vapour-deposition, or thermal oxidation. The patterning includes photolithography techniques such as soft lithography and two-photon soft-lithography [150, 151]. Focused ion-beam lithography, colloid monolayer lithography, electron beam lithography, molecular self-assembly, rapid prototyping, X-ray lithography, ion-projection lithography, and electrical-induced nanopatterning have been employed for nanofabrication [149]. In addition, a template-based method has been widely employed to fabricate nanotubes and nanowires from a variety of materials [152]. In this method, a nanoporous membrane is used as a template and the material is deposited within the nanopores of the membrane. Responsive hydrogels have been miniaturized for trapping bacteria for efficient cell growth [153] and metabolite detection [154]. Even though thermo-responsive hydrogels have been miniaturized [142, 155], the miniaturization of protein-based responsive hydrogels has been under-explored. Hydrogel nanoparticles have been synthesized by different techniques, including emulsion and precipitation polymerization, core-shell hydrogel nanoparticles synthesis, block-copolymer micelles, and post-polymerization modification. In comparison to hydrogel nanoparticles, the study of hydrogel nanowires or nanotubes is relatively unexplored. There are only a few examples of hydrogel nanostructures with nanowire or nanotube configuration. Guo *et al.* [156]. fabricated polyacrylamide hydrogel arrays by electropolymerization of acrylamide inside the nanoporous aluminium oxide membrane. These nanowires were then utilized for the dispersal of metal nanoparticles. In another example, poly(*N,N*-dimethylacrylamide)/poly(acrylic ammonium) binary hydrogel nanowires with invertible core/shell phases were fabricated by using a porous alumina membrane as a template [157]. Recently, a polymer-peptide conjugate was prepared by conjugation of polymer and peptide via click chemistry [158]. The extensive dialysis of THF solution of the conjugate against water resulted in the formation of hydrogel nanotubes. Thus, there is a need to explore the fabrication of nanowire-type structures of protein-based hydrogels. These hydrogel nanostructures may find applications in molecular gating, molecular transport, and drug-delivery.

11.8   Mechanical Characterization of Hydrogels

Despite the vast potential of hydrogels in biomedical applications, their applications in tissue engineering are rather limited owing to their poor mechanical strength and high viscoelastic nature. In addition to this, because of the soft nature of hydrogels, sample preparation becomes challenging. On the other hand, these limitations have enabled the use of hydrogels as potential soft-tissue and cartilage-replacements [159], microcantilevers [160], and micro/nanofluidic valves [144]. This necessitates the study of the behaviour of the hydrogels with reference to their mechanical stability under the influence of an applied force. Dynamic mechanical properties of solids and soft solids, like gels, are commonly studied by the Dynamic Mechanical Analyser (DMA) [161]. An ideal elastic material, upon deformation followed by relaxation, regains its original shape and size. A viscoelastic material, on the other hand, dissipates energy upon deformation followed by relaxation. The energy lost comprises the viscous component of the material in addition to the elastic component of the material. The mechanical properties of the material are governed by Hooke's law, $\sigma = E\varepsilon$, where σ is the applied stress (force/area), ε is the change in dimensions upon after the application of stress (change in length/original length), and E is the dynamic modulus of the material. The dynamic modulus of the viscoelastic material is comprised of the storage modulus (elastic part) and loss modulus (viscous part).

Hydrogels are viscoelastic in nature, meaning that they exhibit both viscous and elastic properties. The mechanical properties are the characteristics for a particular type of hydrogel and are influenced by variations in the cross-linking of the hydrogel [162], the polymerization conditions, the solvent [163], monomeric sequence [164], the hydrophobicity [165], the composition of the hydrogel, the degree of swelling, and the temperature of the testing environment [161]. Based on the type of force applied, there are different types of mechanical properties. Some of the commonly used methods for evaluation of mechanical properties of hydrogels are described below:

1. Tensile properties: In this method, the hydrogel sample, of a flat geometry, is stretched uniaxially at a constant rate. The resultant stretching (strain) of the hydrogel is measured and correlated with the amount of pressure (stress) applied, and stress–strain curves are plotted. Using this method, parameters such as Young's modulus (the slope of the initial part of the curve), maximum elongation length, and ultimate tensile strength (the maximum stress the material can endure) can be evaluated [166].
2. Compression properties: A hydrogel sample is placed on a fixed plate and another plate is employed to compress the hydrogel at a certain force. The resultant compression of the hydrogel is correlated with the applied force on the sample. Using this method, bulk properties of the hydrogel, such as dynamic modulus (storage modulus and loss modulus), and energy dissipation, are measured as a function of applied stress. Although this method suffers from a bucking or barrelling effect due to compression of the sample,

it is still a widely utilized technique for mechanical characterization of bulk hydrogels [167, 168].

3. Shear properties: In this technique, a shear stress is applied on the hydrogel sample and the resultant displacement of the hydrogel is determined. Using this method, the ability of the hydrogel to withstand shearing stress can be evaluated [169].

4. Indentation: In this test, a hard tip of material of a known hardness is pressed against the sample and the amount of indentation is measured. The amount of indentation is inversely related to the hardness of the sample. Smaller samples are analysed by employing micro/nanometre-size indenters and the technique is called nanoindentation. This method is valuable for determining the modulus of elasticity and hardness of the hydrogel samples [170, 171].

11.9   Creep Properties of Hydrogels

Creep is the ability of a material to deform when a constant force is applied. The creep strength of the material is limited by the yield strength (fracture point) of the material. Viscoelastic materials are known to undergo creep deformation on application of the stress, resulting in a time-dependent increase in the strain. The creep curve is characterized by an initial increase in the strain (primary creep strain), followed by a plateau (secondary or steady state creep strain), and finally, a slight increase in the strain (tertiary creep strain) before the material breaks due to excessive stress (Figure 11.9a). The secondary creep strain represents the maximum permissible deformation of the material and is characteristic of the material. The creep behaviour of the hydrogels determines the long-term stability of the material against the applied force and closely mimics the mechanical stress conditions an implantable drug-delivery device would experience [172, 173]. The creep behaviour of hydrogels has been shown to vary with the water content [174], composition, and cross-linking within the hydrogel [175]. The creep behaviour of the material can be measured in terms of the creep coefficient, which is the ratio of the additional creep strain that occurs over time to the initial elastic strain (Figure 11.9b). An analysis of comprehensive creep behaviour of stimuli-responsive hydrogels, especially protein-based hydrogels, is yet to

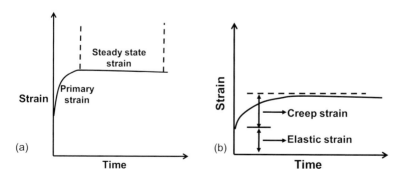

Figure 11.9   *Schematic showing (a) creep behaviour of hydrogels under stress and (b) determination of creep coefficient.*

be performed. Such an analysis would shine a light on the applications of protein-based stimuli-responsive hydrogels in biomedical devices.

11.10   Conclusion and Future Perspectives

As presented, binding proteins have been employed in designing sensing systems and responsive biomaterials. Protein-based sensing systems present more ways to design biosensing systems for different analytes. In particular, the molecular split protein-based sensing systems provide an alternative way to develop biosensing systems for proteins with unresolved crystal structures. In addition, the binding proteins could be integral components of "responsive" biomaterials. The fabrication of biologically-inspired functional materials and their characterization should enhance our understanding of biomaterials in practical device-based applications.

Despite significant progress in the development of biologically-inspired materials, there are still a few key challenges to overcome in order to realize the full potential of this technology. For example, there is a need to improve the biocompatibility, lifetime, and durability of these materials. The use of engineered biomaterials and nanomaterials *in vivo* is limited by immune response. This is being overcome by employing engineered 3D scaffolds based on biocompatible materials such as collagen derivatives, resulting in materials with a significantly decreased inflammatory foreign body response [176]. Other advances to improve the biocompatibility of these materials pursue development of surfaces that inhibit protein adsorption [177]. The lifetime of implantable responsive materials can be improved by engineering the integrated responsive proteins to increase their thermal stability. This can be accomplished by looking to proteins from thermophillic organisms for inspiration. Additionally, the identification of proteins to identify and respond to novel targets such as bacteria, specific small molecule toxins and analytes, viruses, ions, cancer cells, and so on will greatly enhance the potential for applications in the fields of biosensing and therapeutic materials.

Acknowledgements

This work was supported in part by the National Institute of Environmental Health Sciences Superfund Research Program grant P42ES007380, the National Institutes of Health, and the National Aeronautics and Space Administration. SD acknowledges support from the Gill Eminent Professorship from the University of Kentucky and from the Professor and Lucille Markey Chairmanship from the University of Miami Miller School of Medicine. KT and SK acknowledge a traineeship from the Superfund Research Program and from the University of Kentucky Research Challenge Trust Fund.

References

1. Song, J. Cheng Q., Zhu S. *et al.* (2002) "Smart" materials for biosensing devices: Cell-mimicking supramolecular assemblies and colorimetric detection of pathogenic agents. *Biomed. Microdevices*, **4**(3), 213–221.

2. Wendell, D.W., Patti, J. and Montemagno, C.D. (2006) Using biological inspiration to engineer functional nanostructured materials. *Small*, **2**(11), 1324–1329.

3. Hess, H., Bachand, G.D. and Vogel, V. (2004) Powering nanodevices with biomolecular motors. *Chem. Eur. J.*, **10**(9), 2110–2116.

4. Hess, H. (2006) Materials science - Toward devices powered by biomolecular motors. *Science*, **312**(5775), 860–861.

5. Ren, Q., Zhao, Ya-Pu, Yue, J. *et al.* (2006) Biological application of multi-component nanowires in hybrid devices powered by F-1-ATPase motors. *Biomed. Microdevices*, **8**(3), 201–208.

6. Kocer, A., Walko M., Meijberg W. *et al.* (2005) A light-actuated nanovalve derived from a channel protein. *Science*, **309**(5735), 755–758.

7. Banghart, M.R., Volgraf, M. and Trauner, D. (2006) Engineering light-gated ion channels. *Biochemistry*, **45**(51), 15129–15141.

8. Muramatsu, S., Kinbara, H., Taguchi, N. *et al.* (2006) Semibiological molecular machine with an implemented "AND" logic gate for regulation of protein folding. *J. Am. Chem. Soc.*, **128**(11), 3764–3769.

9. Xi, J.Z., Ho, D., Chu, B. *et al.* (2005) Lessons learned from engineering biologically active hybrid nano/micro devices. *Adv. Funct. Mater.*, **15**(8), 1233–1240.

10. Astier, Y., Bayley, H. and Howorka, S. (2005) Protein components for nanodevices. *Curr. Opin. Chem. Biol.*, **9**(6), 576–584.

11. Anderson, D.G., Burdick, J.A. and Langer, R. (2004) Materials science - smart biomaterials. *Science*, **305**(5692), 1923–1924.

12. Liu, H.Q., Schmidt, J., Bachand, G. *et al.* (2002) Control of a biomolecular motor-powered nanodevice with an engineered chemical switch. *Nat. Mater.*, **1**(3), 173–177.

13. van den Heuvel, M.G.L., De Graaff, M.P. and Dekker, C. (2006) Molecular sorting by electrical steering of microtubules in kinesin-coated channels. *Science*, **312**(5775), 910–914.

14. Diez, S., Reuther, C., Dinu, C. *et al.* (2003) Stretching and transporting DNA molecules using motor proteins. *Nano Lett.*, **3**(9), 1251–1254.

15. Yoseph, B. (2006) Biomimetics: Biologically inspired technologies, in *Biomimetics: Biologically Inspired Technologies* (ed. B. Yoseph), Taylor and Francis Group, Boca Raton, FL, pp. 1–40.

16. Kaldenhoff, R. and Fischer, M. (2006) Aquaporins in plants. *Acta Physiol.*, **187**(1–2), 169–176.

17. Agre, P., King, L.S., Yasui, M. *et al.* (2002) Aquaporin water channels – from atomic structure to clinical medicine. *J. Physiol.*, **542**(pt 1), 3–16.

18. Kozono, D., Yasui, M., King, L.S. *et al.* (2002) Aquaporin water channels: Atomic structure molecular dynamics meet clinical medicine. *J. Clin. Invest.*, **109**(11), 1395–1399.

19. Agre, P. (2000) Homer W. Smith award lecture. Aquaporin water channels in kidney. *J. Am. Soc. Nephrol.*, **11**(4), 764–777.

20. Nielsen, S., Frokiaer, J., Marples, D. *et al.* (2002) Aquaporins in the kidney: From molecules to medicine. *Physiol. Rev.*, **82**(1), 205–244.

21. Liu, B., Li, X., Li, B. *et al.* (2009) Carbon nanotube based artificial water channel protein: Membrane perturbation and water transportation. *Nano Lett.*, **9**(4), 1386–1394.

22. Kaucher, M.S., Peterca, M., Dulcey, A. *et al.* (2007) Selective transport of water mediated by porous dendritic dipeptides. *J. Am. Chem. Soc.*, **129**(38), 11698–11699.

23. Yellen, G. (2002) The voltage-gated potassium channels and their relatives. *Nature*, **419**(6902), 35–42.

24. Majumder, M., Chopra, N. and Hinds, B.J. (2005) Effect of tip functionalization on transport through vertically oriented carbon nanotube membranes. *J. Am. Chem. Soc.*, **127**(25), 9062–9070.

25. Majumder, M., Stinchcomb, A. and Hinds, B.J. (2009) Towards mimicking natural protein channels with aligned carbon nanotube membranes for active drug delivery. *Life Sci.* **86**(15–16), 563–568.

26. Zhu, B., Li, J. and Xu, D. (2011) Porous biomimetic membranes: Fabrication, properties and future applications. *Phys. Chem. Chem. Phys.*, **13**(22), 10584–10592.

27. Majumder, M., Stinchcomb, A. and Hinds, B.J. (2010) Towards mimicking natural protein channels with aligned carbon nanotube membranes for active drug delivery. *Life Sci.*, **86**(15–16), 563–568.

28. Majumder, M., Chopra, N., Andrews, R. *et al.* (2005) Nanoscale hydrodynamics: Enhanced flow in carbon nanotubes. *Nature*, **438**(7064), 44.

29. Nednoor, P., Gavalas, V., Chopra, N. *et al.* (2007) Carbon nanotube based biomimetic membranes: Mimicking protein channels regulated by phosphorylation. *J. Mater. Chem.*, **17**(18), 1755–1757.

30. Wang, G., Zhang, B., Wayment, J. R. *et al.* (2006) Electrostatic-gated transport in chemically modified glass nanopore electrodes. *J. Am. Chem. Soc.*, **128**(23), 7679–7686.

31. Hou, X., Guo, W. and Jiang, L. (2011) Biomimetic smart nanopores and nanochannels. *Chem. Soc. Rev.*, **40**(5), 2385–2401.

32. Harrell, C.C., Kohli, P., Siwy, Z. *et al.* (2004) DNA-nanotube artificial ion channels. *J. Am. Chem. Soc.*, **126**(48), 15646–15647.

33. Song, L., Hobaugh, M., Shustak, C. *et al.* (1996) Structure of Staphylococcal alpha -hemolysin, a heptameric transmembrane pore. *Science*, **274**(5294), 1859–1865.

34. Gu, L.Q., Braha, O., Conlan, S. et al. (1999) Stochastic sensing of organic analytes by a pore-forming protein containing a molecular adapter. *Nature*, **398**(6729), 686–690.

35. Braha, O., Gu, L., Zhou, L. *et al.* (2000) Simultaneous stochastic sensing of divalent metal ions. *Nat. Biotechnol.*, **18**(9), 1005–1007.

36. Kasianowicz, J.J., Henrickson S. E., Weetall H. H. *et al.* (2001) Simultaneous multi-analyte detection with a nanometer-scale pore. *Anal. Chem.*, **73**(10), 2268–2272.

37. Russo, M.J., Bayley, H. and Toner, M. (1997) Reversible permeabilization of plasma membranes with an engineered switchable pore. *Nat. Biotechnol.*, **15**(3), 278–282.

38. Jung, Y., Bayley, H. and Movileanu, L. (2006) Temperature-responsive protein pores. *J. Am. Chem. Soc.*, **128**(47), 15332–15340.

39. Wendell, D., Jing, P., Jia, G. *et al.* (2009) Translocation of double-stranded DNA through membrane-adapted phi29 motor protein nanopores. *Nat. Nano.*, **4**(11), 765–772.

40. Shu, Y., Cinier, M., Shu, D. et al. (2011) Assembly of multifunctional phi29 pRNA nanoparticles for specific delivery of siRNA and other therapeutics to targeted cells. *Methods*, **54**(2), 204–214.

41. Jing, P., Haque, F., Shu, D. *et al.* (2010) One-way traffic of a viral motor channel for double-stranded DNA translocation. *Nano Letters*, **10**(9), 3620–3627.

42. Jing, P., Haque, F., Vonderheide, A. *et al.* (2010) Robust properties of membrane-embedded connector channel of bacterial virus phi29 DNA packaging motor. *Molecular BioSystems*, **6**(10), 1844–1852.

43. Jovanovic-Talisman, T., Tetenbaum-Novatt, J., McKenney, A.S. *et al.* (2009) Artificial nanopores that mimic the transport selectivity of the nuclear pore complex. *Nature*, **457**(7232), 1023–1027.

44. Savariar, E.N., Krishnamoorthy, K. and Thayumanavan, S. (2008) Molecular discrimination inside polymer nanotubes. *Nat. Nanotechnol.*, **3**(2), 112–117.

45. Yu, Q., Bauer, J.M., Moore, J.S. *et al.* (2001) Responsive biomimetic hydrogel valve for microfluidics. *Appl. Phys. Lett.*, **78**(17), 2589–2591.

46. Wang, J., Chen Z., Corstjens P.L. *et al.* (2006) A disposable microfluidic cassette for DNA amplification and detection. *Lab Chip*, **6**(1), 46–53.

47. Eddington, D.T. and Beebe, D.J. (2004) Flow control with hydrogels. *Adv. Drug Deliv. Rev.*, **56**(2), 199–210.

48. Wu, J. and Sailor, M.J. (2009) Chitosan hydrogel-capped porous SiO_2 as a pH responsive nano-valve for triggered release of insulin. *Adv. Funct. Mater.*, **19**(5), 733–741.

49. Huang, M.C., Ye, H., Kuan, Y.K. *et al.* (2009) Integrated two-step gene synthesis in a microfluidic device. *Lab Chip*, **9**(2), 276–285.

50. Markstrom, M., Lizana, L., Orwar, O. *et al.* (2008) Thermoactuated diffusion control in soft matter nanofluidic devices. *Langmuir*, **24**(9), 5166–5171.

51. Wang, J., Yang, S., Bau, H. *et al.* (2005) Self-actuated, thermo-responsive hydrogel valves for lab on a chip. *Biomed. Microdevices*, **7**(4), 313–322.

52. Wandera, D., Wickramasinghe, S.R. and Husson, S.M. (2010) Stimuli-responsive membranes. *J. Memb. Sci.*, **357**(1–2), 6–35.

53. Yang, Q., Nadia, A., Tomicki, F. *et al.* (2011) Composites of functional polymeric hydrogels and porous membranes. *J. Mat. Chem.*, **21**(9), 2783–2811.

54. Moschou, E.A., Bachas, L.G., Daunert, S. *et al.* (2006) Hinge-motion binding proteins: Unraveling their analytical potential. *Anal. Chem.*, **78**(19), 6692–6700.

55. Ames, G.F.L. (1986) Bacterial periplasmic transport-systems - structure, mechanism, and evolution. *Annu. Rev. of Biochem.*, **55**, 397–425.

56. Pflugrath, J.W. and Quiocho, F.A. (1988) The 2 A resolution structure of the sulfate-binding protein involved in active transport in Salmonella typhimurium. *J. Mol. Biol.*, **200**(1), 163–180.

57. Borths, E.L., Locher, K.P., Lee, A.T. *et al.* (2002) The structure of Escherichia coli BtuF and binding to its cognate ATP binding cassette transporter. *Proc. Natl. Acad. Sci.*, **99**(26), 16642–16647.

58. Karpowich, N.K., Huang, H., Smith, P. *et al.* (2003) Crystal structures of the BtuF periplasmic-binding protein for vitamin B12 suggest a functionally important reduction in protein mobility upon ligand binding. *J. Biol. Chem.*, **278**(10), 8429–8434.

59. Heddle, J., Scott, D., Unzai, S. *et al.* (2003) Crystal structures of the liganded and un-liganded nickel-binding protein NikA from Escherichia coli. *J. Biol. Chem.*, **278**(50), 50322–50329.

60. Huang, H.-C. and Briggs, J.M. (2002) The association between a negatively charged ligand and the electronegative binding pocket of its receptor. *Biopolymers*, **63**(4), 247–260.

61. Huck, W.T.S. (2008) Responsive polymers for nanoscale actuation. *Mater. Today*, **11**(7–8), 24–32.

62. Crivici, A. and Ikura, M. (1995) Molecular and structural basis of target recognition by calmodulin. *Annu. Rev. Biophys. Biomol. Struct.*, **24**, 85–116.

63. Means, A.R., VanBerkum, M. F., Bagchi, I. *et al.* (1991) Regulatory functions of calmodulin. *Pharmacol. Ther.*, **50**(2), 255–270.

64. Cheung, W.Y. (1980) Calmodulin plays a pivotal role in cellular regulation. *Science*, **207**(4426), 19–27.

65. Walsh, M.P., Vallet, B., Cavadore, J.C. *et al.* (1980) Homologous calcium-binding proteins in the activation of skeletal, cardiac, and smooth muscle myosin light chain kinases. *J. Biol. Chem.*, **255**(2), 335–337.

66. Liu, Y.P. and Cheung, W.Y. (1976) Cyclic 3':5'-nucleotide phosphodiesterase. Ca2+ confers more helical conformation to the protein activator. *J Biol Chem*, **251**(14), 4193–4198.

67. Lin, Y.M., Liu, Y.P. and Cheung, W.Y. (1974) Cyclic 3':5'-nucleotide phosphodiesterase. Purification, characterization, and active form of the protein activator from bovine brain. *J. Biol. Chem.*, **249**(15), 4943–4954.

68. Babu, Y.S., Bugg, C.E. and Cook, W.J. (1988) Structure of calmodulin refined at 2.2 A resolution. *J Mol Biol*, **204**(1), 191–204.

69. LaPorte, D.C., Wierman, B.M. and Storm, D.R. (1980) Calcium-induced exposure of a hydrophobic surface on calmodulin. *Biochemistry*, **19**(16), 3814–3819.

70. Cook, W.J., Walter, L.J. and Walter, M.R. (1994) Drug binding by calmodulin: Crystal structure of a calmodulin-trifluoperazine complex. *Biochemistry*, **33**(51), 15259–15265.

71. Daunert, S., Bachas, L.G., Schauer-Vukasinovic, V. *et al.* (2007) Calmodulin-mediated reversible immobilization of enzymes. *Colloids Surf B Biointerfaces*, **58**(1), 20–27.

72. SchauerVukasinovic, V., Cullen, L. and Daunert, S. (1997) Rational design of a calcium sensing system based on induced conformational changes of calmodulin. *J. Am. Chem. Soc.*, **119**(45), 11102–11103.

73. Dikici, E., Deo, S.K. and Daunert, S. (2003) Drug detection based on the conformational changes of calmodulin and the fluorescence of its enhanced green fluorescent protein fusion partner. *Anal. Chim. Acta*, **500**(1–2), 237–245.

74. Douglass, P.M., Salins, L.L., Dikici, E. *et al.* (2002) Class-selective drug detection: Fluorescently-labeled calmodulin as the biorecognition element for phenothiazines and tricyclic antidepressants. *Bioconjugate Chem.*, **13**(6), 1186–1192.

75. Meier, M.A., Kanis, L.A. and Soldi, V. (2004) Characterization and drug-permeation profiles of microporous and dense cellulose acetate membranes: Influence of plasticizer and pore forming agent. *Int.J. Pharm.*, **278**(1), 99–110.

76. Sternberg, R., Bindra, D.S., Wilson, G.S *et al.* (1988) Covalent enzyme coupling on cellulose acetate membranes for glucose sensor development. *Anal. Chem.*, **60**(24), 2781–2786.

77. Jeon, G., Yun Yang, S., Byun, J. *et al.* (2011) Electrically actuatable smart nanoporous membrane for pulsatile drug release. *Nano Letters*, **11**(3), 1284–1288.
78. Rodriguez, R. and Castano, V.M. (2005) Smart membranes: A physical model for a circadian behavior. *Appl. Phys. Lett.*, **87**(14), 144103.
79. Ito, Y., Park, Y.S. and Imanishi, Y. (1997) Visualization of critical pH-controlled gating of a porous membrane grafted with polyelectrolyte brushes. *J. Am. Chem. Soc.*, **119**(11), 2739–2740.
80. Chu, L.Y., Li. Y., Zhu, J. *et al.* (2005) Negatively thermoresponsive membranes with functional gates driven by zipper-type hydrogen-bonding interactions. *Angew. Chem. Int. Ed.*, **44**(14), 2124–2127.
81. Csetneki, I., Filipcsei, G. and Zrinyi, M. (2006) Smart nanocomposite polymer membranes with on/off switching control. *Macromolecules*, **39**(5), 1939–1942.
82. Tang, M., Zhang R., Bowyer, A. *et al.* (2003) A reversible hydrogel membrane for controlling the delivery of macromolecules. *Biotechnol. Bioeng.*, **82**(1), 47–53.
83. Mika, A.M., Childs, R. F., Dickson, J. M. *et al.* (1995) A new class of polyelectrolyte-filled microfiltration membranes with environmentally controlled porosity. *J. Membr. Sci.*, **108**(1–2), 37–56.
84. Yang, B. and Yang, W.T. (2003) Thermo-sensitive switching membranes regulated by pore-covering polymer brushes. *J. Membr. Sci.*, **218**(1–2), 247–255.
85. Estrada, R.F., Rodriguez, R. and Castano, V.M. (2003) Smart polymeric membranes with adjustable pore size. *Int. J. Polym. Mater.*, **52**(9), 833–843.
86. Zhai, G.Q., Ying, L., Kang, E.T. *et al.* (2004) Surface and interface characterization of smart membranes. *Surf. Interface Anal.*, **36**(8), 1048–1051.
87. Qu, J.B., Chu, L.-Y., Yang, M. *et al.* (2006) A pH-responsive gating membrane system with pumping effects for improved controlled release. *Adv. Funct. Mater.*, **16**(14), 1865–1872.
88. Tokarev, I., Katz, O., Minko, S. *et al.* (2007) An electrochemical gate based on a stimuli-responsive membrane associated with an electrode surface. *J. Phys. Chem. B*, **111**(42), 12141–12145.
89. Rao, G.V.R., Balamurugan, S., Meyer, D. E. *et al.* (2002) Hybrid bioinorganic smart membranes that incorporate protein-based molecular switches. *Langmuir*, **18**(5), 1819–1824.
90. Butterfield, D.A., Bhattacharyya, D., Daunert, S. *et al.* (2001) Catalytic biofunctional membranes containing site-specifically immobilized enzyme arrays: A review. *J. Membr. Sci.*, **181**(1), 29–37.
91. Kohli, P., Harrell, C., Cao, Z. *et al.* (2004) DNA-functionalized nanotube membranes with single-base mismatch selectivity. *Science*, **305**(5686), 984–986.
92. Zhang, R., Bowyer, A., Eisenthal, R. *et al.* (2007) A smart membrane based on an antigen-responsive hyrogel. *Biotechnol. Bioeng.*, **97**(4), 976–984.
93. Zhang, K. and Wu, X.Y. (2002) Modulated insulin permeation across a glucose-sensitive polymeric composite membrane. *J. Controlled Release*, **80**(1–3), 169–178.
94. Scouten, W.H., Luong, J.H.T. and Stephen Brown, R. (1995) Enzyme or protein immobilization techniques for applications in biosensor design. *Trends Biotechnol.*, **13**(5), 178–185.

95. Liu, J.L., Wang, J., Bachas, L. *et al.* (2001) Activity studies of immobilized subtilisin on functionalized pure cellulose-based membranes. *Biotechnol. Progr.*, **17**(5), 866–871.

96. Rao, S.V., Anderson, K.W. and Bachas, L.G. (1998) Oriented immobilization of proteins. *Mikrochim. Acta*, **128**(3–4), 127–143.

97. Datta, S., Ray, P., Nath, A. *et al.* (2006) Recognition based separation of HIV-Tat protein using avidin-biotin interaction in modified microfiltration membranes. *J. Membr. Sci.*, **280**(1–2), 298–310.

98. Viswanath, S., Wang, J., Bachas, L.G. *et al.* (1998) Site-directed and random immobilization of subtilisin on functionalized membranes: Activity determination in aqueous and organic media. *Biotechnol Bioeng*, **60**(5), 608–616.

99. Sussich, F. and Cesaro, A. (2000) The kinetics of periodate oxidation of carbohydrates: A calorimetric approach. *Carbohydr. Res.*, **329**(1), 87–95.

100. Verma, V., Katti K.S., Katti, D.R. *et al.* (2008) 2, 3-Dihydrazone cellulose: Prospective material for tissue engineering scaffolds. *Mater. Sci. Eng C*, **28**(8), 1441–1447.

101. Wang, J.Q., Bhattacharyya, D. and Bachas, L.G. (2001) Orientation specific immobilization of organophosphorus hydrolase on magnetic particles through gene fusion. *Biomacromolecules*, **2**(3), 700–705.

102. Anderson, S. (2008) Surfaces for immobilization of N-terminal cysteine derivatives via native chemical ligation. *Langmuir*, **24**(24), 13962–13968.

103. Dawson, P.E., Muir, T.W., Clark-Lewis, I. *et al.* (1994) Synthesis of proteins by native chemical ligation. *Science*, **266**(5186), 776–779.

104. Lesaicherre, M.L., Uttamchandani, M., Chen, G.Y. *et al.* (2002) Developing site-specific immobilization strategies of peptides in a microarray. *Bioorg. Med. Chem. Lett.*, **12**(16), 2079–2083.

105. Wojtyk, J.T.C., Morin, K.A., Boukherroub, R. *et al.* (2002) Modification of porous silicon surfaces with activated ester monolayers. *Langmuir*, **18**(16), 6081–6087.

106. Camarero, J.A., Kwon, Y. and Coleman, M.A. (2004) Chemoselective attachment of biologically active proteins to surfaces by expressed protein ligation and its application for "protein chip" fabrication. *J. Am. Chem. Soc.*, **126**(45), 14730–14731.

107. Souvik, C., Farhana, A.B. and Shao, Y.Q. (2009) Use of intein-mediated protein ligation strategies for the fabrication of functional protein arrays. *Methods Enzymol*, **462**, 195–223.

108. Le Droumaguet, B. and Velonia, K. (2008) Click chemistry: A powerful tool to create polymer-based macromolecular chimeras. *Macromol. Rapid Commun.*, **29**(12–13), 1073–1089.

109. Shi, Q., Chen, X., Lu, T. *et al.* (2008) The immobilization of proteins on biodegradable polymer fibers via click chemistry. *Biomaterials*, **29**(8), 1118–1126.

110. Duckworth, B.P., Xu, J., Taton, T.A. *et al.* (2006) Site-specific, covalent attachment of proteins to a solid surface. *Bioconjug. Chem.*, **17**(4), 967–974.

111. Yousaf, M.N. and Mrksich, M. (1999) Diels-Alder reaction for the selective immobilization of protein to electroactive self-assembled monolayers. *J. Am. Chem. Soc.*, **121**(17), 4286–4287.

112. Pozsgay, V., Vieira, N.E. and Yergey, A. (2002) A method for bioconjugation of carbohydrates using Diels-Alder cycloaddition. *Org. Lett.*, **4**(19), 3191–3194.

113. Sun, X.L., Stabler, C.L., Cazalis, C.S., *et al.* (2006) Carbohydrate and protein immobilization onto solid surfaces by sequential Diels-Alder and azide-alkyne cycloadditions. *Bioconjug. Chem.*, **17**(1), 52–57.

114. Wong, L.S., Khan, F. and Micklefield, J. (2009) Selective covalent protein immobilization: Strategies and applications. *Chem. Rev.*, 4025–4053.

115. Gaberc-Porekar, V. and Menart, V.M. (2005) Potential for using histidine tags in purification of proteins at large scale. *Chem. Eng. Technol.*, **28**(11), 1306–1314.

116. Kumada, Y., Katoh, S., Imartaka, H. *et al.* (2007) Development of a one-step ELISA method using an affinity peptide tag specific to a hydrophilic polystyrene surface. *J. of Biotechnol.*, **127**(2), 288–299.

117. Wong, L.S., Thirlway, J. and Micklefield, J. (2008) Direct site-selective covalent protein immobilization catalyzed by a phosphopantetheinyl transferase. *J. Am. Chem. Soc.*, **130**(37), 12456–12464.

118. Lei, M., Ziaie, B., Nuxoll, E. *et al.* (2007) Integration of hydrogels with hard and soft microstructures. *J Nanosci. Nanotechnol.*, **7**(3), 780–789.

119. Tanaka, T., Fillmore, D., Sun, S-T. *et al.* (1980) Phase-transitions in ionic gels. *Phys. Rev. Lett.*, **45**(20), 1636–1639.

120. Tanaka, T. (1978) Collapse of gels and critical endpoint. *Phys. Rev. Lett.*, **40**(12), 820–823.

121. Suzuki, A. and Tanaka, T. (1990) Phase-transition in polymer gels induced by visible-light. *Nature*, **346**(6282), 345–347.

122. Sershen, S.R., Mensing, G.A., Ng, M. *et al.* (2005) Independent optical control of microfluidic valves formed from optomechanically responsive nanocomposite hydrogels. *Adv. Mater.*, **17**(11), 1366–1368.

123. Miyata, T., Asami, N. and Uragami, T. (1999) A reversibly antigen-responsive hydrogel. *Nature*, **399**(6738), 766–769.

124. Ricka, J. and Tanaka, T. (1984) Swelling of ionic gels - quantitative performance of the Donnan theory. *Macromolecules*, **17**(12), 2916–2921.

125. Tanaka, T., Nishio, I., Sun, S.T. *et al.* (1982) Collapse of gels in an electric-field. *Science*, **218**(4571), 467–469.

126. Dong, L., Agarwal, A. K., Beebe, D.J. *et al.* (2007) Variable-focus liquid microlenses and microlens arrays actuated by thermoresponsive hydrogels. *Adv. Mater.*, **19**(3), 401–405.

127. Tondu, B., Emirkhanian, R., Mathe, S. *et al.* (2009) A pH-activated artificial muscle using the McKibben-type braided structure. *Sens. Actuators, A*, **150**(1), 124–130.

128. Li, S.K. and D'Emanuele, A. (2001) On-off transport through a thermoresponsive hydrogel composite membrane. *J. Controlled Release*, **75**(1–2), 55–67.

129. Park, T.G. (1999) Temperature modulated protein release from pH/temperature-sensitive hydrogels. *Biomaterials*, **20**(6), 517–521.

130. Hilt, J.Z., Gupta, A.K., Bashir, R. *et al.* (2003) Ultrasensitive biomems sensors based on microcantilevers patterned with environmentally responsive hydrogels. *Biomed. Microdevices*, **5**(3), 177–184.

131. Mao, J.S., Kondu, S., Ji, H-F. *et al.* (2006) Study of the near-neutral pH-sensitivity of chitosan/gelatin hydrogels by turbidimetry and microcantilever deflection. *Biotechnol. Bioeng.*, **95**(3), 333–341.

132. van der Linden, H.J., Herber, S., Olthuis, W. *et al.* (2003) Stimulus-sensitive hydrogels and their applications in chemical (micro)analysis. *Analyst*, **128**(4), 325–331.

133. Mohammed, J.S. and Murphy, W.L. (2009) Bioinspired design of dynamic materials. *Adv. Mater.*, **21**(23), 2361–2374.

134. Kopecek, J. (2007) Hydrogel biomaterials: A smart future? *Biomaterials*, **28**(34), 5185–5192.

135. Murphy, W.L., Dillmore, W.S., Modica, J. *et al.* (2007) Dynamic hydrogels: Translating a protein conformational change into macroscopic motion. *Angew. Chem. Int. Ed.*, **46**(17), 3066–3069.

136. Cao, Y. and Li, H. (2008) Engineering tandem modular protein based reversible hydrogels. *Chem. Commun.*, (35), 4144–4146.

137. Ehrick, J.D., Deo, S.K., Browning, T.W. *et al.* (2005) Genetically engineered protein in hydrogels tailors stimuli-responsive characteristics. *Nat. Mater.*, **4**(4), 298–302.

138. Topuz, F. and Okay, O. (2009) Formation of hydrogels by simultaneous denaturation and cross-linking of DNA. *Biomacromolecules*, **10**(9), 2652–2661.

139. Ehrick, J.D.,Stokes, S., Bachas-Daunert, S. *et al.* (2007) Chemically tunable lensing of stimuli-responsive hydrogel microdomes. *Adv. Mater.*, **19**(22), 4024–4027.

140. Ehrick, J.D., Luckett, M.R., Khatwani, S. *et al.* (2009) Glucose responsive hydrogel networks based on protein recognition. *Macromol. Biosci.*, **9**(9), 864–868.

141. Wyman, J.L., Kizilel, S., Skarbek, R. *et al.* (2007) Immunoisolating pancreatic islets by encapsulation with selective withdrawal. *Small*, **3**(4), 683–690.

142. van der Linden, H., Olthuis, W. and Bergveld, P. (2004) An efficient method for the fabrication of temperature-sensitive hydrogel microactuators. *Lab Chip*, **4**(6), 619–624.

143. Lei, M., Salim, A., Siegel, R. *et al.* (2004) A hydrogel-actuated microvalve for smart flow control. *Conf. Proc. IEEE. Eng. Med. Biol. Soc.*, **3**, 2041–2044.

144. Wu, J. and Sailor, M.J. (2009) Chitosan hydrogel-capped porous SiO_2 as a pH responsive nano-valve for triggered release of insulin. *Adv. Funct. Mater.*, **19**(5), 733–741.

145. Nayak, S. and Lyon, L.A. (2005) Soft nanotechnology with soft nanoparticles. *Angew. Chem. Int. Ed.*, **44**(47), 7686–7708.

146. Hamidi, M., Azadi, A. and Rafiei, P. (2008) Hydrogel nanoparticles in drug delivery. *Adv. Drug Deliv. Rev.*, **60**(15), 1638–1649.

147. Zeng, X. and Jiang, H. (2008) Tunable liquid microlens actuated by infrared light-responsive hydrogel. *Appl. Phys. Lett.*, **93**(15), 151101.

148. Ding, Z. and Ziaie, B. (2009) A pH-tunable hydrogel microlens array with temperature-actuated light-switching capability. *Appl. Phys. Lett.*, **94**(8), 081111.

149. Betancourt, T. and Brannon-Peppas, L. (2006) Micro- and nanofabrication methods in nanotechnological medical and pharmaceutical devices. *Int. J. Nanomed.*, **1**(4), 483–495.

150. Jhaveri, S.J., McMullen, J., Sijbesma, R. *et al.* (2009) Direct three-dimensional microfabrication of hydrogels via two-photon lithography in aqueous solution. *Chem. Mater.*, **21**(10), 2003–2006.

151. Coenjarts, C.A. and Ober, C.K. (2004) Two-photon three-dimensional microfabrication of poly(dimethylsiloxane) elastomers. *Chem. Mater.*, **16**(26), 5556–5558.

152. Liu, T.B., Burger, C. and Chu, B. (2003) Nanofabrication in polymer matrices. *Prog. Polym. Sci.*, **28**(1), 5–26.

153. Kaehr, B. and Shear, J.B. (2008) Multiphoton fabrication of chemically responsive protein hydrogels for microactuation. *Proc. Natl. Acad. Sci. U. S. A.*, **105**(26), 8850–8854.

154. Yan, J., Sun, Y., Zhu, H. *et al.* (2009) Enzyme-containing hydrogel micropatterns serving a dual purpose of cell sequestration and metabolite detection. *Biosens. Bioelectron.*, **24**(8), 2604–2610.

155. Tirumala, V.R., Divan R., Ocola, L. *et al.* (2005) Direct-write e-beam patterning of stimuli-responsive hydrogel nanostructures. *J. Vac. Sci. Technol. B*, **23**(6), 3124–3128.

156. Guo, G., Hu, J. S., Liang, H.P. *et al.* (2003) Highly dispersed metal nanoparticles in porous anodic alumina films prepared by a breathing process of polyacrylamide hydrogel. *Chem. Mater.*, **15**(22), 4332–4336.

157. Yang, Z. and Niu, Z. (2002) Binary hydrogel nanowires of invertible core/shell phases prepared in porous alumina membranes. *Chem. Commun.*, **17**, 1972–1973.

158. Tzokova, N., Fernyhough, C.M., Topham, P.D. et al. (2009) Soft hydrogels from nanotubes of poly(ethylene oxide)-tetraphenylalanine conjugates prepared by click chemistry. *Langmuir*, **25**(4), 2479–2485.

159. Almany, L. and Seliktar, D. (2005) Biosynthetic hydrogel scaffolds made from fibrinogen and polyethylene glycol for 3D cell cultures. *Biomaterials*, **26**(15), 2467–2477.

160. Du, H., Kondu, S. and Ji, H.F. (2008) Formation of ultrathin hydrogel films on microcantilever devices using electrophoretic deposition. *Micro Nano Lett.*, **3**(1), 12–17.

161. Anseth, K.S., Bowman, C.N. and BrannonPeppas, L. (1996) Mechanical properties of hydrogels and their experimental determination. *Biomaterials*, **17**(17), 1647–1657.

162. Pan, X., Yang, X. and Lowe, C.R. (2008) Evidence for a cross-linking mechanism underlying glucose-induced contraction of phenylboronate hydrogel. *J. Mol. Recognit.*, **21**(4), 205–209.

163. Kirkland, S.E., Hensarling, R.M., McConaughy, S.D. *et al.* (2008) Thermoreversible hydrogels from RAFT-synthesized BAB triblock copolymers: Steps toward biomimetic matrices for tissue regeneration. *Biomacromolecules*, **9**(2), 481–486.

164. Lee, S.G., Brunello, G.F., Jang, S.S. *et al.* (2009) Effect of monomeric sequence on mechanical properties of P(VP-co-HEMA) hydrogels at low hydration. *J. Phys. Chem. B*, **113**(19), 6604–6612.

165. Luo, Y., Zhang, K., Wei, Q. *et al.* (2009) Poly(MAA-co-AN) hydrogels with improved mechanical properties for theophylline controlled delivery. *Acta Biomater.*, **5**(1), 316–327.

166. Drury, J.L., Dennis, R.G. and Mooney, D.J. (2004) The tensile properties of alginate hydrogels. *Biomaterials*, **25**(16), 3187–3199.

167. Ahearne, M., Yang, Y. and Liu, K. (2008) Mechanical characterisation of hydrogels for tissue engineering applications. *Topics Tissue Eng.* **4**, 1–16.

168. Stammen, J.A., Williams, S., Ku, D.N. et al. (2001) Mechanical properties of a novel PVA hydrogel in shear and unconfined compression. *Biomaterials*, **22**(8), 799–806.

169. Ramachandran, S., Tseng, Y. and Yu, Y.B. (2005) Repeated rapid shear-responsiveness of peptide hydrogels with tunable shear modulus. *Biomacromolecules*, **6**(3), 1316–1321.

170. Ahearne, M., Yang, Y., El Haj, A. *et al.* (2005) Characterizing the viscoelastic properties of thin hydrogel-based constructs for tissue engineering applications. *J. R. Soc. Interface*, **2**(5), 455–463.

171. Korhonen, R.K., Laasanen, M.S., Töyräs, J. *et al.* (2002) Comparison of the equilibrium response of articular cartilage in unconfined compression, confined compression and indentation. *J. Biomech.*, **35**(7), 903–909.

172. Caldwell, L.J., Gardner, C.R. and Cargill, R.C. (1988) Drug delivery device which can be retained in the stomach for a controlled period of time. US Patent 475 8436: July 19, 1988.250.

173. Lopatin, V.V., Askadskii, A.A., Peregudov, A.S. *et al.* (2005) Structure and relaxation properties of medical-purposed polyacrylamide gels. *J. Appl. Polym. Sci.*, **96**(4), 1043–1058.

174. Choi, J., Bodugoz-Senturk, H., Kung, H.J. *et al.* (2007) Effects of solvent dehydration on creep resistance of poly(vinyl alcohol) hydrogel. *Biomaterials*, **28**(5), 772–780.

175. Iijima, M., Hatakeyama, T. and Hatakeyama, H. (2005) Swelling behaviour of calcium pectin hydrogels by thermomechanical analysis in water. *Thermochim. Acta*, **431**(1–2), 68–72.

176. Ju, Y.M., Yu, B., West, L. *et al.* (2010) A novel porous collagen scaffold around an implantable biosensor for improving biocompatibility. II. Long-term in vitro/in vivo sensitivity characteristics of sensors with NDGA- or GA-crosslinked collagen scaffolds. *J. of Biomed. Mat. Res. Part A*, **92A**(2), 650–658.

177. Sun, T., Qing, G., Su, B. et al. (2011) Functional biointerface materials inspired from nature. *Chem. Soc. Reviews*, **40**(5), 2909–2921.

12

Responsive Colloids with Controlled Topology

Jeffrey C. Gaulding, Emily S. Herman and L. Andrew Lyon
School of Chemistry and Biochemistry and Petit Institute for Bioengineering and
Bioscience, Georgia Institute of Technology, USA

12.1   Introduction

Modern colloid and polymer science has been greatly enriched by the influx of scientists who maintain their focus at the molecular scale. This molecular scale precision permits the design of polymers with controlled tacticities [1], well-defined arrangements of co-monomers [2], complex block structures [3], and particular secondary and tertiary structures [4, 5]. Colloidal interfaces have been decorated with molecular entities that enable new self-assembly motifs [6–9], controlled biological interfaces [10, 11], and advanced bioanalytical tools [12, 13]. However, it remains the case that obtaining control over a material's composition and properties on length scales longer than that of molecules, but shorter than those obtainable by standard bulk processing methods, is a significant challenge. In particular, the pursuit of controlled spatial arrangement of materials on the sub-micron length-scale within the context of colloidal particles continues to be a vibrant area of research. Obtaining such complex colloidal materials requires the design of synthetic approaches that exert *simultaneous* control over reactions between molecules (e.g. monomers, surface modification reagents), macromolecules (e.g. polymers, proteins), and colloids (e.g. nanoparticles, nanorods/nanofibres, droplets, micelles). Considering that the relative diffusivities, reactivities, and symmetries of these building blocks can be vastly different from one another, the *de novo* design of successful synthetic approaches to produce complex colloids can be very difficult.

Responsive Membranes and Materials, First Edition. Edited by D. Bhattacharyya, Thomas Schäfer, S. R. Wickramasinghe and Sylvia Daunert.
© 2013 John Wiley & Sons, Ltd. Published 2013 by John Wiley & Sons, Ltd.

This chapter hopes to highlight some of the more successful demonstrations of complex colloid synthesis. We do not attempt to be exhaustive, but instead have chosen four different particle architectures (inert core/responsive shell, responsive core/responsive shell, hollow capsules, and Janus/patchy particles) and highlight some representative approaches to their manufacture. The lessons learned from these examples can be (and are) applied to the synthesis of many other structures; judging from the work presented herein, we expect that the pursuit of complex colloidal materials will inspire further innovation in the design of colloidal materials.

12.2   Inert Core/Responsive Shell Particles

Polymer brushes are commonly used to alter the surface chemistry of a substrate, to impart desired characteristics (such as surface charges or reactive functionalities) or to mask undesired properties (for example, poly(ethylene glycol) (PEG) coatings to reduce protein adsorption) [14–17]. Extending this strategy to colloidal systems can serve many of the same functions, while exploiting the advantages of microparticulate structures, such as higher surface to volume ratios and more efficient mass transport to that surface. Throughout this section, we will highlight pioneering work in both the synthesis and application of colloidal particles grafted with responsive brushes in order to illustrate both their preparation and the breadth of their potential impact.

The two main strategies for coupling the polymer chains to colloidal particles (or any interface, for that matter) are typically referred to as the "grafting-from" and "grafting-to" methods. The characteristic that distinguishes the two techniques is the role of the substrate in the polymerization. The two approaches are illustrated in Figure 12.1. The "grafting-from" approach couples an initiator to the substrate, generating a macro-initiator and allowing polymer chain growth to proceed from the surface of the substrate. In 1999, Guo and Ballauff first described a method to covalently graft a polymeric brush to sub-micron-sized particles of polystyrene using such a strategy [18]. Their approach adapted a method for macroscopic surface functionalization first described by Ruhe [19], and applied it to the formation of brushes on monodisperse polystyrene particles ranging in diameter from 50 to 100 nm. Addition of a low concentration of the photoinitiator 2-[p-(2-hydroxy-2-methylpropiophenone)]-ethylene glycol-methacrylate (HMEM) into the polystyrene reaction after most of the styrene was consumed ensured that the HMEM was confined to the surface of the polystyrene nanoparticles. Dispersal of the core particles in an aqueous solution of acrylic acid and initiation by exposure to UV light led to the growth of poly(acrylic acid) (pAAc) brushes bound to the polystyrene cores. Varying the concentration of acrylic acid present during the grafting step led to control over the length of the surface-confined chains [18, 20]. Recently this approach has been modified to incorporate a thermally-initiated source of radicals in place of HMEM, enabling similar particles to be formed in the absence of a UV source [21].

In contrast, the "grafting-to" approach assembles pre-formed polymeric chains onto the core substrate. The strong interaction between thiol groups and gold surfaces makes this synthetic approach an appealing one in the synthesis of polymer-grafted colloidal gold. For example, Wuelfing *et al.* used a thiol-terminated PEG chain to form a brush layer on

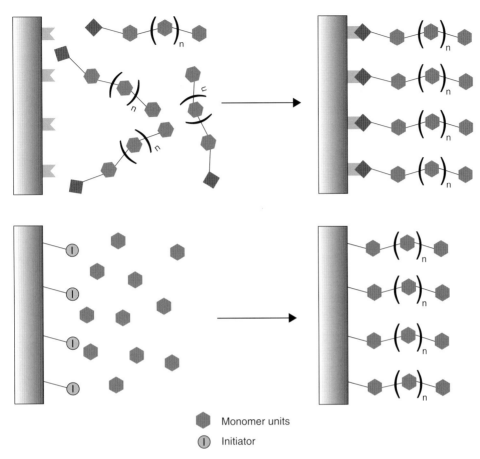

Monomer units

Initiator

Figure 12.1   Top: "Grafting-to" approach, wherein preformed polymer chains are bound to a surface via specific interactions such as Coulombic attraction or covalent bond formation. Bottom: "Grafting-from" approach, wherein the surface is used to immobilize an initiator. In a solution of monomer, polymerization leads to growth of chains from the surface of the substrate.

the surface and thus enhance the stability of gold nanoparticles [22]. The same principle has been used by Corbierre *et al.* to coat gold nanoparticles with polystyrene [23]. An advantage of the "grafting-to" approach is that it enables the straightforward incorporation of multiple types of polymeric chains, allowing the formation of mixed brush systems [24].

Other methods that incorporate the addition of pre-formed polymer chains to a core particle rely on non-covalent interactions between the surface of the core and the pendant polymer, with Coulombic interactions being the most common. Starting from a charged substrate, the sequential adsorption of oppositely charged polymeric chains leads to charge reversal on the surface and build-up of a polyelectrolyte multilayer. Sukhorukov *et al.* first applied this method to microparticles in 1998 [25], and Frank Caruso has made extensive use of this technique, using combinations of strong polyelectrolytes (such as poly(styrene

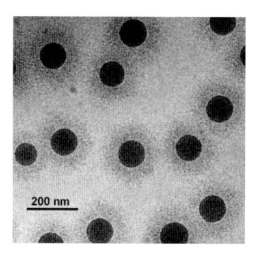

Figure 12.2 *Visualization of polystyrene particles coated with pNIPAm by cryo-TEM. (Reprinted under the terms of the STM agreement from [34] Copyright (2006) Wiley-VCH).*

sulfonate) (PSS)), weak electrolytes imparting responsivity (pAAc or poly(allylamine hydrochloride), PAH) [26, 27], and other charged species (silica nanoparticles [28], DNA [29], and proteins [30]. These assemblies have demonstrated potential utility for numerous applications, such as drug delivery and catalysis.

The addition of responsive polymer brushes enables the conversion of an environmentally-inert core material, such as polystyrene or silica, into an environmentally-sensitive material capable of undergoing physicochemical changes in response to external triggers. The spherical polyelectrolyte brushes developed by the Ballauff group impart responsivity to pH and ionic strength to such colloids, with the responsivities being predictable based on the strength of the polyacids [20,31]. Thermo-responsivity can also be conferred to the nanoparticles by grafting suitable polymers. For example, poly(N-isopropylacrylamide) (pNIPAm) has been widely studied due to its thermo-responsive nature [32, 33]. In 2006, Lu *et al.* adapted the polyelectrolyte brush grafting method to coat polystyrene with cross-linked pNIPAm. The resultant hydrogel coating conferred large volume changes in response to temperature, as observed by dynamic light scattering (DLS) and cryo-TEM (Figure 12.2) [34]. This was an improvement over previous methods described by the group, wherein growing pNIPAm strands were grafted onto the polystyrene-co-NIPAm core particles by the chain-transfer mechanism inherent in polymerizations of NIPAm [35]. In those examples, the pNIPAm network was also cross-linked using cross-linking monomer methylenebisacrylamide (BIS) to form a cohesive network, but the number of anchoring points was limited by the uncontrolled self-cross-linking [36, 37].

The flexibility inherent in the grafting strategy has enabled numerous other responsive brushes to be explored in the context of their capabilities as catalytic carriers. In addition to pAAc and pNIPAm, PSS [31], PEG [38], poly(aminoethyl methacrylate hydrochloride) (PAEMH) [39], and poly(2-methylpropenoyloxyethyl) trimethylammonium chloride (PMPTAC) [40] have all been employed, providing a broad range of functional scaffolds.

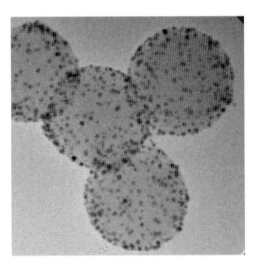

Figure 12.3 *TEM image of gold nanoparticles entrapped within polycationic brushes of PAEMH on a polystyrene core. (Reprinted under the terms of the STM agreement from [39] Copyright (2004) Wiley-VCH).*

As an example of the utility of such brushes, these structures have been used to harbour metal nanoparticles. First described in 2004 by Sharma and Ballauff, with a PAEMH brush incorporating gold nanoparticles [39], later examples have included other nanoparticles of interest in catalytic applications such as silver [38,41], palladium [42], and platinum [40]. The polyelectrolyte brushes serve to isolate and stabilize the metal nanoparticles against aggregation, and confine them to the vicinity of the colloidal polystyrene. Addition of noble metal salts into dispersions of polystyrene modified with cationic polyelectrolyte brushes leads to preferential exchange of the anionic metal salts for the small anions (i.e. chloride) initially present. The preference for noble metals arises from a combination of multivalent interactions and/or complexation with amines present in the brushes. Once confined, subsequent reduction leads to the formation of small (<10 nm) metal nanoparticles trapped within the brush layer, as shown in Figure 12.3. These have been demonstrated to retain catalytic activity, leading to effective "nanoreactors" [38–43]. Recent examples have utilized these nanoreactors to catalyse reactions such as Suzuki- and Heck-coupling [44].

Substitution of thermo-responsive pNIPAm for the polyelectrolyte brushes leads to another dimension of control for these nanoreactors. First described in 2006, the pNIPAm-grafted polystyrene nanoparticles can localize silver ions through complexation with the nitrogen atoms of NIPAm [45]. Reduction of the silver ions leads to the nucleation, growth, and confinement of silver nanoparticles within the brush layer. Lu *et al.* demonstrated the ability of pNIPAm brushes to modulate reaction kinetics using the reduction of 4-nitrophenol to 4-aminophenol. The expected relationship between temperature and the observed rate constant, k, does not hold up, as shown in Figure 12.4. Three regimes are observed: a low-temperature state, corresponding to when the pNIPAm network is highly swollen, and wherein the expected Arrhenius relationship holds; temperatures near the transition temperature ($\sim 32\,^\circ$C), where the deswelling of the pNIPAm chains generates a barrier

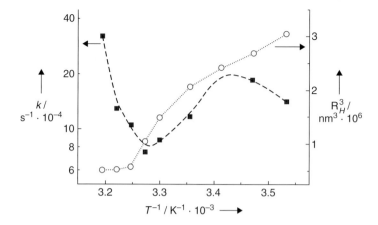

Figure 12.4 *The catalytic activity of silver nanoparticles embedded within pNIPAm brushes grafted to polystrene nanoparticles varies according to the phase transition of pNIPAm. Squares: reaction rate k as a function of inverse temperature (T^{-1}). Circles: particle volume as a function of inverse temperature. Note the middle regime of increasing k, corresponding to the phase transition region for the pNIPAm brushes. (Reprinted under the terms of the STM agreement from [45] Copyright (2006) Wiley-VCH).*

for the diffusion of the reagents to reach the catalytic nanoparticles; and a high temperature state where the chains are completely collapsed and the expected temperature dependence is restored once again. The middle regime, wherein the reaction rate reaches a minimum at the transition temperature, indicates the gating effect that the collapsed polymer network can have on the catalytic activity of these nanoreactors [45, 46].

The Ballauff constructs have also been shown to be effective at providing a support for enzymes. The grafted pAAc on the surface serves to Coulombically attract positively charged "patches" of proteins at low ionic strength. The proteins become immobilized within the brush due to the entropically favoured release of small counterions, like sodium [47]. Thus the protein adsorption can be reversed by exposing the particles to high ionic strength. Importantly, the adsorbed proteins seem to suffer little denaturation, as evidenced by their reversible adsorption and from structural determination by FTIR [48]. Preservation of structure enables the immobilization of active enzymes to form these nanoreactors. In 2004, Neumann *et al.* loaded such particles with glucoamylase, and compared the activity of the free and bound enzymes by UV-vis spectroscopy. Only a small decrease in enzymatic activity was noted for the bound enzyme, which the authors ascribe to protein–protein interaction effects at the high loading concentration of enzyme in the nanoreactors. This is supported by the data shown in Figure 12.5, as the trend shows approximately a 30% loss in activity in response to a 10-fold increase in enzyme loading [49, 50]. An example from the Caruso group in 2000 wherein glucose oxidase was incorporated as the anionic component of a layer-by-layer assembly on the surface of polystyrene similarly showed retention of enzymatic activity, providing another example of a mechanism for enzymatic nanoreactor formation [51]. A recent example coupling a pNIPAm shell and the enzyme beta-glucosidase actually shows an increase in enzymatic activity upon immobilization

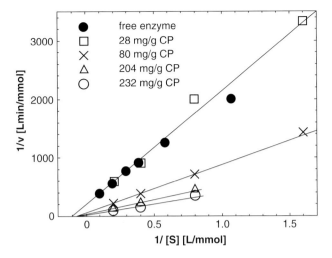

Figure 12.5 *Activity of free glucoamylase and the enzyme loaded into poly(acrylic acid)-grafted polystyrene. Formation of the nanoreactors shows retention of protein activity, and increasing concentration leads to a slight reduction. (Reprinted under the terms of the STM agreement from [49] Copyright (2004) Wiley-VCH).*

within the nanoparticle [52]. FTIR analysis of the particles indicated that hydrogen bonding interactions were primarily responsible for the polymer network/protein interactions, as well as the enhanced activity. The ability of pNIPAm to gate the nanoreactor, as observed with the metal nanoparticles, was also retained [52].

Polymer brushes grafted to colloidal substrates have demonstrated great utility in a broad range of applications, particularly in the realm of catalysis. By using the polymeric brushes to convert the inert core particles to carriers for metal nanoparticles or enzymes, these next-generation nanoreactors will offer numerous advantages, including easier recovery and isolation, and potentially enhanced or even tunable catalytic activity. The ability to confer environmental responsivity onto a desired substrate enables the synergistic creation of nanoparticles with greater functionality than either the inert core or the free polymer. Our attention now turns to the case wherein responsivity is not confined only to the periphery, but found in both the core and shell.

12.3 Responsive Core/Responsive Shell Particles

Colloidal hydrogel microparticles, or microgels, have generated a great deal of interest from the drug delivery, sensing, and biomaterials communities in the past decade [53–55]. The term "hydrogel" refers to a hydrophilic, cross-linked polymer structure wherein the network is highly solvated by water [56, 57]. These highly swollen systems are often susceptible to their environment, wherein changes in the relative energy of interaction between water and the polymer backbone and the polymer with itself enable microgels to undergo large volume changes in response to external stimuli. Amongst the most studied responsive microgels are those consisting of cross-linked pNIPAm. Above its lower critical solution temperature (LCST) of around 32°C, the polymer chains undergo an entropically-driven collapse. The hydrophobic interactions of the pendant isopropyl groups become favoured over the hydrogen bonding interactions of the amide, leading to expulsion of water from the network and a collapse of the microgel.

In 1986, Robert Pelton's group first described the synthesis of colloidal cross-linked microgels based on pNIPAm [33,58]. The synthetic strategy exploited the thermo-responsive nature of pNIPAm to precipitate the growing polymer chains upon reaching a critical chain length, leading to the formation of "precursor particles" which serve as nuclei for continued polymer deposition and particle growth. This synthetic strategy leads to highly monodispersed microgels. The aqueous conditions used with the precipitation polymerization technique are amenable to the incorporation of other vinylic co-monomers to impart multi-responsivity into the microgels. For example, the addition of weak acids and bases, such as acrylic acid [59] or vinylpyridine [60], produced microgels that responded to both temperature and pH changes.

Despite the relative ease with which additional functionalities were incorporated into microgels, control over their spatial distribution was difficult. Heterogeneous distributions of the co-monomers can result from differential reaction kinetics between the co-monomers or differences in their hydrophilicity. For example, the commonly used cross-linker BIS has a faster rate of polymerization than NIPAm, which leads to core-localization of the cross-linker at low molar ratios during particle synthesis [61, 62]. The addition of more hydrophobic co-monomers such as *tert*-butyl acrylamide (tBAM) [63] or *N*-isopropylmethacrylamide (NIPMAm) [64] tend to form heterogeneous microdomains during particle synthesis, described by Richtering as "dirty snowballs". Similarly, Hoare and Pelton conducted a thorough study of the incorporation of various acidic monomers into pNIPAm microgels, demonstrating that different monomers could lead to core localization (methacrylic, MAA), surface localization (fumaric), or a homogeneous distribution (acrylic) [65, 66].

In 2000, Jones and Lyon were the first to describe a method that enabled precise synthetic control over the surface composition of a thermo-responsive microgel [67]. These "core-shell" microgels, illustrated in Figure 12.6, utilized a two-step precipitation polymerization process wherein the initial core microgels were isolated, then introduced into a solution of additional monomer with the desired composition. This "seed and feed" approach used appropriate concentrations of monomer and initiator to inhibit nucleation of new particles,

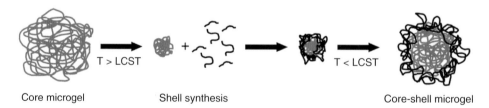

Core microgel   Shell synthesis   Core-shell microgel

Figure 12.6 *The "seed and feed" polymerization approach uses previously formed microgel core particles as templates for shell synthesis. Additional monomer, cross-linker, and initiator is added at a concentration that is low enough to avoid new particle formation, leading to polymer deposition on the core particles.*

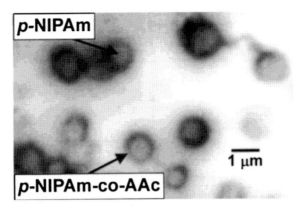

Figure 12.7 *TEM image of pNIPAm-core pNIPAm-co-AAc-shell microgels. The particles were stained with uranyl acetate, thus the darker regions correspond to the acid-rich particle shell. (Reprinted with permission from [67] Copyright (2000) American Chemical Society).*

ensuring that the new polymer chains would deposit onto the surface of the core particles [67]. The first demonstration used the seed and feed approach to isolate acrylic acid (AAc) in the core or shell. Figure 12.7 shows the resultant particle formation, and Figure 12.8 shows the architectural-dependent swelling properties of these microgels. Particles consisting of a pNIPAm-co-AAc core and pNIPAm-only shell exhibited a two-stage deswelling response upon heating, corresponding to the thermally-driven collapse of the pNIPAm shell at approximately its native LCST, followed by a second transition corresponding to the pNIPAm-co-AAc core, which occurred at a depressed transition temperature due to the presence of the collapsed shell. In contrast, the reversed condition (pNIPAm-AAc shell, pNIPAm core) showed a complex phase transition resulting from the anionic shell stretching and inhibiting the core's ability to fully deswell [67].

Further studies have probed the unique nature of core/shell coupling in microgel systems. Berndt and Richtering described a system wherein the two thermo-responsive polymers (pNIPAm and pNIPMAm) were used to form a core-shell microgel wherein the core and shell exhibited different transition temperatures [68]. When the pNIPMAm phase (having the higher transition temperature, $\sim44\,^{\circ}$C) was located in the shell, it tended to prevent complete deswelling of the pNIPAm core at temperatures between the two phase transitions, similar to the effect of AAc described above. A collapsed pNIPAm shell exerted a compressive effect on the pNIPMAm core, preventing it from achieving its fully swollen state below its own LCST [68]. The observation that the shell dominates the swelling response of the entire microgel has been shown with tBAM modification of the hydrophobicity of the shell as well [69]. The extent to which the shell's swelling behaviour dominates the core's scales with the thickness of the added shell [68, 70, 71].

Shell domination of the particle swelling suggests that the addition of the shell is affecting the swollen state of the core, preventing it from completely relaxing. This phenomenon is called "core compression" and was probed by Jones *et al.* using fluorescence resonance energy transfer, or FRET [72]. The efficiency of energy transfer between a donor and acceptor capable of undergoing FRET scales as r^{-6}, enabling this technique to determine

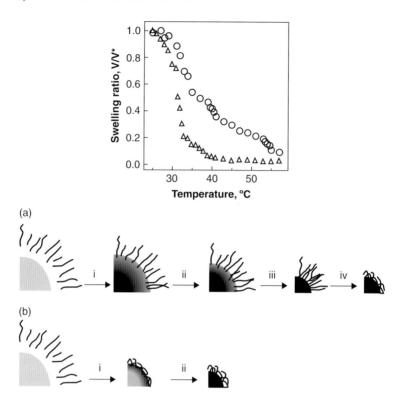

Figure 12.8 *Top: Particle swelling ratio at pH 6.5 as a function of temperature for anionic core-shell microgels determined by DLS. Circles: pNIPAm-core pNIPAm-co-AAc-shell particles. Triangles: pNIPAm-co-AAc-core pNIPAm-shell particles. Bottom: Schematic interpretation of the swelling data, indicating the different results due to particle architecture. (a) pNIPAm-core/ pNIPAm-co-AAc-shell particles. (b) pNIPAm-co-AAc-core/pNIPAm-shell particles. (Reprinted with permission from [67] Copyright (2000) American Chemical Society).*

the average separation of the donor and acceptor in an ensemble system by monitoring the fluorescence. A sulfoindocyanine FRET dye pair was incorporated into the core of a pNIPAm microgel with a pNIPAm shell. The FRET signal as a function of temperature for the core and after shell addition is given in Table 12.1. In the case where no shell has been added, the FRET efficiency was very low, as the core swelling and hence inter-chain distance greatly separated the donor and acceptor. Above the transition temperature, the degree of FRET increased sharply as the enhanced polymer–polymer interactions brought the dyes closer together. When considering the core-shell case below the transition temperature, the degree of FRET was less than the deswollen case, but higher than in the free core. This suggested that the ability of the polymer chains in the core to extend was inhibited by the added polymer of the shell, resisting the core's swelling ability and leading to an increased FRET signal [72].

Recently the Lyon group has demonstrated the first example of a core-shell microgel wherein these strong influences between the core and shell can be disrupted, leading to fully

Table 12.1   *Particle sizes and FRET efficiency for the pNIPAm core–shell microgels. The reduced FRET intensity upon shell addition indicates that the added polymer compresses the core, inhibiting it from completely reswelling and thus separating the dyes. (Reprinted with permission from [72] Copyright (2004) American Chemical Society).*

	Radius (nm)[a]			Intensity ratio (671:697 nm)[b]		
	23°C	31°C	43°C	23°C	31°C	43°C
Core	116	97	61	1.42	1.15	0.75
Core–shell	144	110	78	1.16	0.97	0.75

[a]Radii measured via photon correlation spectroscopy.
[b]Wavelength intensity ratios measured via fluorescence, $\lambda_{ex} = 646$ nm.

independent behaviour of the core and shell [73]. These particles were synthesized using the seed and feed approach described previously, followed by the subsequent addition of a second shell, enabling a core–shell–shell architecture, as demonstrated in Figure 12.9. By using a degradable cross-linker in the first shell, the sacrificial inner shell can be degraded and the free polymer removed from the particle. As a result, the core and outer shell are mechanically decoupled by the void volume left upon removal of the inner shell.

The core–shell architecture is enabling for numerous applications. The shell serves as a way to localize functionality on the surface of the particles. For example, a targeting ligand can be bound to the surface of the nanoparticles for drug delivery applications [74, 75]. Additionally, suitably functionalized PEG has been incorporated into the shell of a microgel, which increases the hydrophilicity of the particle and acts to inhibit protein adsorption [76].

The shell can also serve as a barrier that restricts access to the nanoparticle core. Kleinen *et al.* recently compared a series of architectures, wherein a pNIPAm-co-MAA core particle was covered in a pNIPAm shell. Incubation of the particles with the polycation poly(diallyldimethylammonium chloride) (PDADMAC) led to interaction with the anionic microgels. However, there was a strong molecular weight dependence. Low molecular weight polymer (~15 kDa) was able to easily penetrate the shell, leading to high levels of loading within the polymer, no charge reversal (i.e. the polycation is confined to the interior), and no difference between the core–shell and the core particle behaviour. However, higher molecular weight PDADMAC (~450 kDa) was unable to access the anionic sites on the particle interior as readily. As a result, the net charge of the particle surface shifted from neutral (due to the pNIPAm shell) to positive (due to excess PDADMAC adsorbed on the surface) [77]. Another example by Nayak and Lyon combined biotinylated core particles with a shell containing degradable cross-linkers. Initially, avidin was excluded from the interior of the particles by the added shell. Upon erosion of the degradable cross-links, avidin was able to access the core and bind to the biotin contained within [78].

Responsive microgels with fine control over architecture can be used to develop systems containing feedback loops, enabling more advanced applications. Lapeyre *et al.* describe a system wherein microgels consisting of pNIPMAm cores incorporate a pNIPMAm shell copolymerized with a phenylboronic acid derivative [79]. Phenylboronic acids complex with glucose, and have been demonstrated to be a viable means of generating microgels

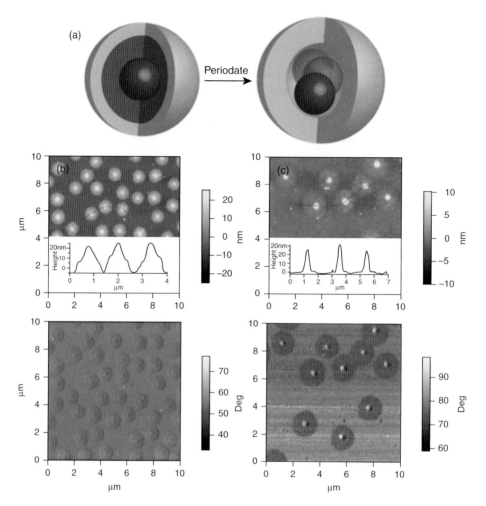

Figure 12.9 *Depiction of core–shell–shell particles, and decoupling of the core and shell by means of degrading the inner shell. The particles before (left) and after (right) degradation are visualized using atomic force microscopy. The height (top) and phase (bottom) traces are shown. (Reprinted with permission from [73] Copyright (2010) American Chemical Society).*

whose swelling responds to glucose concentration [80, 81]. In this system, localization of the glucose-responsive elements to the surface enables a gating porosity to occur based on the external glucose concentration. The microgels were sufficiently swollen above their transition temperature to enable loading with fluorescently-labelled insulin. Increasing the temperature above the VPTT of this system (in this case, ~25°C) collapsed the particles and retained the insulin within. Introduction of glucose led to swelling of the shell, and thus enabled insulin to diffuse out of the particles.

While many applications are being explored for microgels, the examples described in this chapter present the type of enhancements in functionality that can be achieved with fine control of particle architecture. For example, a core-shell architecture provides the opportunity for independent methods of payload retention, regulation of release profile, physiological targeting, and immuno-masking. It is likely that such synthetic control will be key in translating microgels from proof-of-concept to viable technologies for fields with very specific design parameters.

12.4   Hollow Particles

Responsive hollow particles have been of interest in many fields, particularly drug delivery, as the hollow inner compartment of these responsive polymeric particles can act as a storage reservoir for the delivery of drugs [82, 83]. External stimuli or environmental conditions permit the release of the drug to the desired area. Various synthetic techniques have been explored to determine the utility of such particles as drug delivery vehicles. This section will focus on advances in the development of such particles and initial studies suggesting the potential of hollow particles for drug delivery.

One of the original methods developed for the production of hollow polymeric particles was the templating method. This method uses a sacrificial core onto which a shell is grown, with the core subsequently being removed through degradation [84–86]. The formation of hollow particles through the templating method arose from the development of core/shell particles where the shell had very different properties from that of the core [87]. The removal of a degradable core created a capsule, which could potentially act as a nano- or micron-scale "storage container" for the delivery or protection of a cargo. Early examples of the formation of hollow particles came from the Möhwald group at the Max Plank Institute. In 1998, this group used a "hard templating" [88] technique in which polystyrene particles acted as sacrificial cores which, after the addition of a silica shell, were removed through calcination [87]. Also in 1998 the Möhwald group produced hollow particles using a "soft templating" technique through the use of a weakly cross-linked core composed of melamine formaldehyde (MF) as the template, onto which polyelectrolyte shells composed of PSS and PAH were deposited [89]. The core was removed by exposure to aqueous media below pH 1.6. Both the hard and soft templating techniques have been further studied and have shown the ability of such methods to encapsulate model drugs [90].

In 2002 the hard-templating technique was extended to incorporate responsivity by modifying silica particles with 3-(trimethoxysilyl) propyl methacrylate (MPS) in order to form terminal double bonds on the surface. The monomer NIPAm and cross-linker BIS were then polymerized to the surface of the silica particles through a seeded precipitation polymerization method and the silica core was removed through chemical etching with hydrofluoric acid (HF) [84,86]. A representative example of the application of such particles was shown in 2006 by the Akashi group, when the template method was utilized to develop a hollow drug delivery capsule by using the layer-by-layer assembly of chitosan and dextran sulfate onto a porous silica core into which the protein fluorescein isothiocyanate (FITC)-labelled albumin was loaded. After degradation of the core, the protein was left behind in a hollow capsule, which was then degraded by the enzyme chitocanase [91]. The particle before and after enzymatic degradation can be seen in Figure 12.10. The more gentle soft templating method was also studied through the incorporation of a degradable

Figure 12.10 *SEM images of chitosan/dextran sulfate hollow particles before (a) and after 5 minutes of enzymatic degradation (b,c) and 24 hours of enzymatic degradation (d). (Reprinted under the terms of the STM agreement from [91] Copyright (2006) American Chemical Society).*

cross-linker into the core [83, 85, 89]. In 2005, Zhang *et al.* at Fudan University were able to compose thermo-responsive pNIPAm hollow particles through the use of a degradable core containing poly(ε-caprolactone) (PCL), which is biodegradable and could be removed through the addition of lipase [83]. DLS was used to show the temperature responsivity before and after degradation of the sacrificial core.

For a number of years, liposomes and vesicles, which are composed of phospholipids and surfactants respectively, have been investigated as drug delivery vehicles. Dilute solutions of phospholipids or surfactants arrange into bilayers, which may then close, leading to the formation of hollow spheres [92]. These constructs have been used to form vehicles for the delivery of drugs such as doxorubicin [93]. The downside of using liposomes and vesicles for drug delivery is that they are thought to have limited stability and have little control over permeability [94]. However, the hydrophobic interior of the bilayer structure allows these particles to serve as templates for polymerization of monomers that have phase-separated into the bilayer. This "vesicle templating" method leads to the production of hollow polymeric particles that do not suffer from the shortcomings of their progenitor

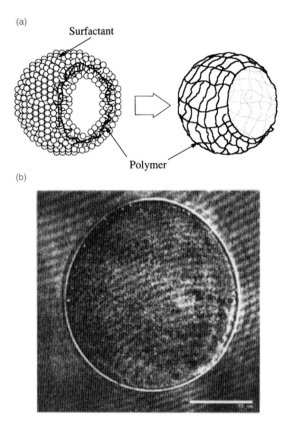

Figure 12.11 *(a) Schematic representation of vesicle-templating method. (b) Confocal laser scanning micrograph of hollow particle. Scale bar = 50 μm. (Reprinted under the terms of the STM agreement from [94] Copyright (1998) American Chemical Society).*

template [94, 95]. Early work in this field by Murtag and Thomas in 1986 showed how the incorporation of divinylbenzene (DVB), polystyrene and DVB-polystyrene mixtures into a vesicle composed of dioctadecyldimethylammonium bromide (DODAB) does not change the size of vesicles, but does lead to a decrease in lateral mobility of guest molecules [96]. This fundamental study led to the development of hollow polymeric particles through vesicle templating by the Meier group in 1998. The schematic of this synthesis is shown in Figure 12.11a. The Meier group was able to polymerize the hydrophobic monomer methacrylate in the interior bilayer of a vesicle composed of dioctadecyldimethylammonium chloride (DODAC) and extract the newly formed polymer particles through repeated precipitation, first in methanol and then in THF until the surfactant was completely removed [94]. The particle formed was approximately 150 μm in diameter with a wall thickness of 2–4 μm (Figure 12.11b). In 2002 the Meier group was able to generate responsive hollow particles using the vesicle templating method by polymerizing the monomer *tert*-butyl acrylate (tBA) with the cross-linker ethylene glycol dimethacrylate within the interior bilayer of a vesicle composed of DODAC. The particles were precipitated as described previously

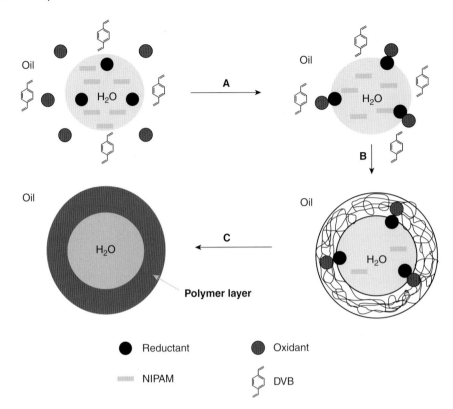

Figure 12.12 *Schematic depicting the formation of thermoresponsive hollow particles using a W–O emulsion. (Reprinted under the terms of the STM agreement from [100] Copyright (2005) American Chemical Society).*

by the group [94] and saponified to produce pAAc particles, which were pH sensitive as shown by DLS [97, 98].

Multi-phase systems such as water-in-oil (W/O) emulsions have been used to avoid the use of a template, which is thought to decrease the inherent complications of the templating method [99, 100]. The supposed complications include the use of materials that are not suitable for templating as well as limited bilayer systems for vesicle templating [100]. In 2005, Deng *et al.* were able to produce thermoresponsive hollow capsules through a one-pot W/O method, which incorporates NIPAm and the cross-linker DVB into the oil phase. A schematic of this synthesis is shown in Figure 12.12. A redox initiation system was used, which incorporated the oxidant benzoyl peroxide (BPO) in the oil phase and the reductant tetraethylenepentamine in the water phase. Above the LCST of pNIPAm ($\sim$32°C), the polymer will focus at the W–O interface. The redox initiation system acted as an interfacial initiator to begin the polymerization, and hollow particles were formed in a size range of 1–3 μm in diameter [100]. In 2007, Hu *et al.* were able to expand upon this technique by using a Shirasu porous glass (SPG) membrane emulsification technique that allowed for polymerization at temperatures below the LCST of pNIPAm as well as allowed for a monodisperse distribution of sizes amongst hollow particles [99, 101].

Another common self-assembly technique uses block copolymers for the production of hollow particles. Block copolymers are thought to be beneficial as building blocks for nanoparticles because the technique offers size control, monodispersity, and a variety of synthetic polymer building blocks [102]. This technique has been used to create hollow particles with varying responsivity [103, 104]. For example, the Eisenberg group at McGill University demonstrated the formation of polymeric vesicles composed of polystyrene-b-pAAc [105]. Although block copolymers show promise as self-assembly materials, the overarching issue from an applications perspective is the stability of the structure after self-assembly [103, 106].

To overcome the limitations in the templating and self-assembly techniques, the Wooley group developed a combination of the two methods [82, 107]. In 1999 the group described a method in which a diblock copolymer consisting of isoprene and acrylic acid (PI-b-AAc) self-assembled in solution leading to a polymer micelle [82]. This micelle was modified through amidation to form a shell cross-linked "knedel-like" (SCK) structure [82, 106]. The poly(isoprene) core was then degraded by ozonolysis to form nanocages [82]. The formation of these SCK nanocages is shown schematically in Figure 12.13. The group has done extensive work in this area, and in 2005 developed synthetic approaches to increase the hydrophobicity of the inner hollow SCK core. This was done by attaching the lipid phosphatidylethanolamine to the interior and testing the loading ability of these modified

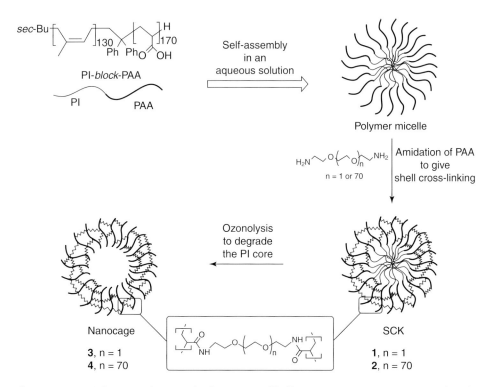

Figure 12.13 *Schematic showing the formation of hollow SCK nanocages. (Reprinted under the terms of the STM agreement from [82] Copyright (1999) American Chemical Society).*

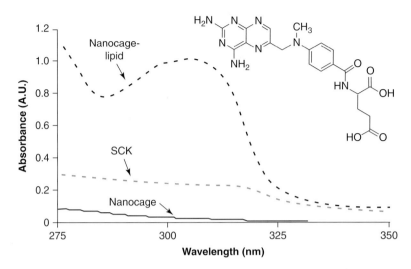

Figure 12.14   *UV-Vis spectrum of the loading of the amphiphilic drug, methotrexate (shown). (Reprinted under the terms of the STM agreement from [108] Copyright (2005) Elsevier Ltd).*

SCKs using 4,4-difluoro-4-bora-3a,4a-diaza-s-indacene (BODIPY). The loading of the modified SCKs versus unmodified and non-hollowed SCKs showed a much greater loading ability of modified SCKs. This experiment was then repeated with the amphiphilic drug, methotrexate, and similar results were found (see Figure 12.14) [108]. Other groups have adopted this technique to develop other modified SCKs with various responsivities [109] and their drug loading capability has been further explored.

The examples presented previously show that hollow polymeric particles are potentially useful for the delivery of drugs. By using previous knowledge on the behaviour of polymeric material, different synthetic techniques have been explored that improve and expand upon the available size, functionality, and responsivity of hollow constructs. Other constructs have also been discovered that seek to take advantage of the vast field of polymeric materials and synthetic techniques. Below, the discussion will focus on the formation of particles with directional or anisotropic properties, which can be thought of as "colloidal molecules", and the potential applications of such constructs.

12.5   Janus Particles

Janus particles are a class of colloidal particles that have gained interest over the past ten years for a wide variety of applications, including but not limited to use as surfactants [110, 111], biphasic drug delivery vehicles [112], and in the development of switchable devices such as the display of the Kindle™ [113]. Janus was the name of a two-faced Roman god [114], and the term Janus in the context of colloids was coined in 1991 by De Gennes during the presentation of the Nobel Prize in Physics [115, 116] to describe particles containing a polar and non-polar face. The term Janus has advanced over the years and now describes particles of various surface topology and function. This section will describe various techniques that have been developed for the production of Janus particles as well as covering some applications of such particles.

Janus particles were first constructed by Casagrande and Veysie in 1988 using commercially available glass beads. One side of the glass beads was protected by a cellulose varnish while the other side was treated with the hydrophobic surface modification reagent octadecyltrichlorosilane, thus creating a particle with two distinct hemispheres [111]. The particle behaviour at the oil–water interface was explored and particles were visualized using optical microscopy. It was noted by Casagrande and Veysie that the Janus particles were potentially beneficial in stabilizing emulsification over the use of homogenous particles, as well as easily modifiable for use in other applications [9, 111]. These observations fuelled future work in the production of Janus particles, which not only further advanced the use of such particles as surfactants [117], but also led to the incorporation of various functionalities and compositions that incorporate responsive behaviour.

The method developed by Casagrande *et al.* was the first example of a partially masked method. Since their work, various partially masked methods for the production of Janus particles have been developed [116, 118–121]. In these approaches, particles are immobilized on a substrate and the exposed hemisphere is treated. The particles are then removed from the substrate, thereby producing two sides with different properties. This technique has been expanded over the years to incorporate other techniques into the partially masked technique for the production of Janus particles. For example, in 2003 Velev *et al.* produced dipolar particles through a combination of templating and microcontact printing. As shown in Figure 12.15, latex particles were immobilized on a glass slide and stamped with a polydimethylsiloxane (PDMS) stamp coated with octadecyltrimethylammonium bromide (ODTAB) after which the particles were redispersed in water. For a direct visualization of particle modification, the ODTAB was replaced with a fluorescent lipid and fluorescence microscopy was used to image the biphasic particles [121]. In 2008 the Stamm group was able to combine a templating technique with two types of grafting approaches to develop doubly stimuli responsive particles. Particles were synthesized by immobilizing silica particles in a wax template and modifying the surface using 3-aminopropyltriethoxysilane (APS). In the "grafting-from" approach, the atom transfer radical polymerization (ATRP) initiator α-bromoisobutyl bromide was carbodiimide-coupled to the modified silica surface, and the subsequent polymerization functionalized one side with tBA and NIPAm. Particles were removed from the wax template and the "grafting-to" approach was used to attach rhodamine-labelled poly(2-vinylpyridine) (p2VP) chains to the other surface of the silica particles [122]. The particle morphology was explored using fluorescence microscopy and SEM. Particles were then hydrolysed in acidic media to form pAAc–p2VP particles, which were responsive to changes in pH.

The limiting factors in most templating methods are the necessity to use a hard template particle such as silica as well as low yield. Pickering emulsions are a more versatile templating method and have been shown to be useful in the production of doubly responsive polymer microgels. A Pickering emulsion uses an interface, such as the oil-in-water interface, to partially expose the surface of particles for modification [123]. In 2007 the Kawaguchi group demonstrated the utility of this method to produce doubly responsive Janus particles [124]. A precipitation polymerization was used to produce pNIPAm-co-AAc particles, which were thermo- and pH-responsive. One side of the particles was

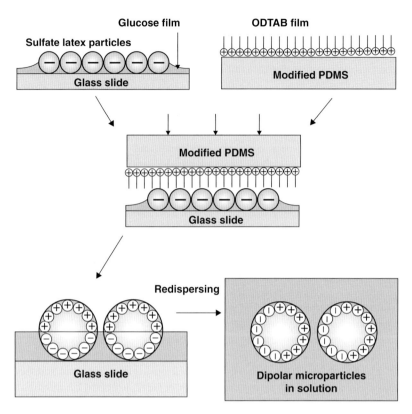

Figure 12.15 *Formation of Janus particles through a combination of templating and micro-contact printing. (Reprinted with permission from [121] Copyright (2003) Royal Society of Chemistry).*

functionalized with amino groups using a carbodiimide coupling reaction through the stabilization of particles at a hexadecane (HD) and water interface. At pH 4 particles aligned into string structures, which could have potential applications as micro-actuators.

To produce particles with distinct hemispheres without the need for the post-assembly modification required in the templating method, other methods for the production of Janus particles have been developed. For instance, electrohydrodynamic jetting has been used to produce particles exhibiting varying Janus properties. In this technique, an electric field was applied to a side by side liquid flow to form particles as shown in schematic of Figure 12.16a. The Lahann group at the University of Michigan has been able to produce biphasic Janus particles that show potential as multicomponent carriers [112]. The group was able to produce Janus particles containing FITC-labelled dextran in one hemisphere and rhodamine-B labelled dextran in the other. The base polymer used in this case was PEG or pAAc, both of which showed similar properties. Confocal scanning laser microscopy was used to visualize the biphasic properties of the Janus particles. An individual particle can be seen in Figure 12.16b. The group was also able to show the ability of Janus particles to be functionalized after production through the substitution of the FITC-dextran hemisphere

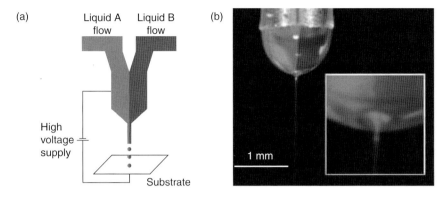

Figure 12.16 *(a) Schematic depicting the formation of Janus particles using a electrohydro-dynamic jetting. (b) Image of individual Janus particle at jetting tip. (Reprinted under the terms of the STM agreement from [112] Copyright (2005) Nature Publishing Group).*

with a mixture of PEG and amino-dextran, which was then modified through the attachment of BODIPY taking advantage of the amino groups present on the surface of the particle. This exemplified the versatility of jetting for the production of Janus particles by showing the ability of Janus particles to be produced with biphasic properties and functionalized after production.

To produce larger quantities of Janus particles with lower polydispersity and increased control over shape and size, a microfluidic approach has been developed [113, 125, 126]. In this approach, a Y-shaped channel is used to create a biphasic monomer stream (see Figure 12.17a). Particles are formed upon introduction to an aqueous phase containing a surfactant for stabilization [113]. The particles are then cross-linked through UV illumination [126, 127], or cured through introduction to a heated solution [113]. A recent example of this method came from the Doyle group wherein they were able to produce Janus hydrogel particles exhibiting anisotropy by loading the microfluidic device with one stream containing a PEG—diacrylate solution and the other stream containing a PEG—diacrylate solution containing magnetite (Fe_3O_4) nanoparticles. Particles were cured through exposure to UV light, and were shown to exhibit paramagnetic properties [127]. The drug loading ability of the particles was also demonstrated through the loading of methacryloxyethyl thiocarbamoyl rhodamine B into the magnetic hemisphere and a fluorescent bead into the non-magnetic hemisphere. Particles were visualized using SEM and DIC and can be seen in Figure 12.17b.

Self-assembly methods have also been used to produce particles on the nanometre scale as well as particles that can be further compartmentalized into patchy particles [128]. Patchy particles are patterned particles with at least one well-defined patch through which the particle can experience an anisotropic interaction with other particles [129]. This has been shown in recent years by the Shimomura group, wherein block copolymers were used in a self-organized precipitation (SORP) method to develop polymeric particles with defined compartments. In Figure 12.18, various polymer particles synthesized using this technique are shown. Particles are composed of polystyrene and poly(isoprene) segments

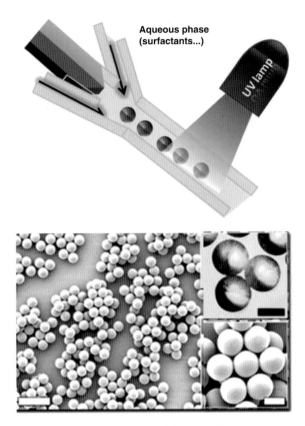

Figure 12.17 *(Top) Schematic representation of the microfluidic device used to create Janus particles. (Reprinted with permission from [114] Copyright (2008) Royal Society of Chemistry). (Bottom) SEM and DIC (upper right insert) images of particles prepared using a microfluidic device similar to that shown in A. Scale bar $=100 \mu m$ and $25 \mu m$ (insert). (Reprinted under the terms of the STM agreement from [127] Copyright (2009) American Chemical Society).*

and the structures were obtained by using block copolymer segments of varying molecular weight [128].

Pine and coworkers performed pioneering work in the field of patchy particles. The group demonstrated the formation of colloidal clusters through the suspension of either poly(methyl methacrylate) PMMA microspheres stabilized by PDMS or silica microspheres with octadecyl chains grafted to the surface in hexane with a surfactant consisting of a triblock polymer containing poly(ethylene oxide) and poly(propylene oxide) (see Figure 12.19, left) [130]. This work has been extended to the production of hybrid particles composed of silica or polystyrene microparticles and nanoparticles composed of silica or titania through self-assembly in a W–O emulsion (see Figure 12.19, right) [131, 132]. These various complex colloidal structures could be useful in various areas because of their potential ability to exhibit unique optical, magnetic, or rheological

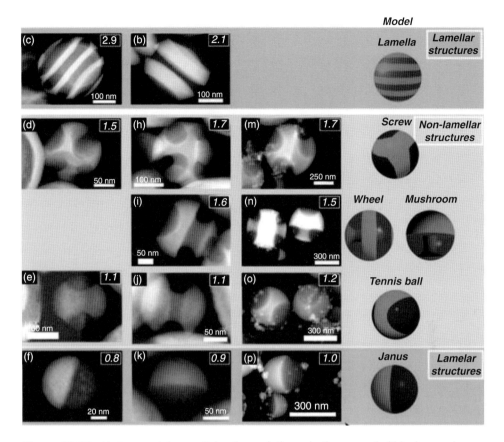

Figure 12.18 *Various patchy particles formed through the use of diblock copolymers. (Reprinted under the terms of the STM agreement from [128] Copyright (2008) Wiley-VCH).*

properties [131]. The formation of these complex structures was relatively straightforward, and the synthetic ease with which the structures are created opens numerous possibilities in the area of patchy particles.

Patchy particles are a newer class of colloidal building blocks that take inspiration from molecular building blocks, wherein directional bond formation dictates the formation of materials. Additionally, the ability to structurally control the surface of particles may be beneficial for the production of nanostructures with applications in drug delivery [133] and electronics [134]. Research in this area has shown that patchy particles can be formed through techniques such as glancing angle deposition [135], electrified co-jetting [133], Pickering emulsions [136], and self-assembly [134]. The development of patchy particles is still in its early stages, and the applications of such particles have yet to be explored in depth. The ability to control the structure of polymeric particles on a nano- or micrometre scale opens a new realm of research.

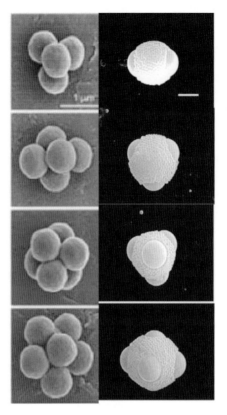

Figure 12.19 *Left: SEM images of PS clusters. (Reprinted under the terms of the STM agreement from [130] Copyright (2004) Wiley-VCH). Right: Bimodal clusters composed of silica. Scale bar is 2 μm. (Reprinted under the terms of the STM agreement from [131] Copyright (2005) American Chemical Society).*

12.6 Summary

This chapter has aimed to outline some of the key storylines in the recent development of structured polymer colloids. Whereas the number of publications detailing the formation of core/shell or core/corona particles is immense, actually translating these particles to functional materials still poses a significant scientific challenge. As the field continues to mature, and the myriad needs of the broad potential applications of such colloids are determined, the future will see increased implementation of these techniques as our improved understanding points us towards a more robust set of advances that will certainly result in a greater influence over real technologies. Conversely, particles with anisotropy (Janus particles) or directionality (patchy particles) built into their structure are still somewhat in their infancy. The complexity of these materials makes their manufacture on large scales difficult, which will likely limit their application. Thus, the breakthroughs required in those domains are not only in the controlled synthesis of such materials, but also in the synthesis of a wider range of sizes and compositions using scalable approaches. This significant challenge in the science and engineering of colloidal systems is likely to fascinate the field for years to come, with the promise of highly controlled "colloidal molecules" as the final outcome.

References

1. Matyjaszewski, K. (2005) Macromolecular engineering: From rational design through precise macromolecular synthesis and processing to targeted macroscopic material properties. *Prog. Polym. Sci.*, **30**, 858–875.
2. Patten, T.E. and Matyjaszewski, K. (1998) Atom transfer radical polymerization and the synthesis of polymeric materials. *Adv. Mater.*, **10**, 901–915.
3. Riess, G. (2003) Micellization of block copolymers. *Prog. Polym. Sci.*, **28**, 1107–1170.
4. König, H.M. and Kilbinger, A.F.M. (2007) Learning from nature: Beta-sheet-mimicking copolymers get organized. *Angew. Chem., Int. Ed.*, **46**, 8334–8340.
5. Carlsen, A. and Lecommandoux, S. (2009) Self-assembly of polypeptide-based block copolymer amphiphiles. *Curr. Opin. Colloid Interface Sci.*, **14**, 329–339.
6. Grzelczak, M., Vermant, J., Furst, E.M. and Liz-Marzán, L.M. (2010) Directed self-assembly of nanoparticles. *ACS Nano*, **4**, 3591–3605.
7. Grzybowski, B.A., Wilmer, C.E., Kim, J. *et al.* (2009) Self-assembly: From crystals to cells. *Soft Matter*, **5**, 1110–1128.
8. Yang, S.-M., Kim, S.-H., Lim, J.-M. and Yi, G.-R. (2008) Synthesis and assembly of structured colloidal particles. *J. Mater. Chem.*, **18**, 2177–2190.
9. Wurm, F. and Kilbinger, A.F.M. (2009) Polymeric janus particles. *Angew. Chem., Int. Ed.*, **48**, 8412–8421.
10. Byrne, J.D., Betancourt, T. and Brannon-Peppas, L. (2008) Active targeting schemes for nanoparticle systems in cancer therapeutics. *Adv. Drug Delivery Rev.*, **60**, 1615–1626.
11. Nel, A.E., Madler, L., Velegol, D. *et al.* (2009) Understanding biophysicochemical interactions at the nano-bio interface. *Nat. Mater.*, **8**, 543–557.
12. Porter, M.D., Lipert, R.J., Siperko, L.M. *et al.* (2008) SERS as a bioassay platform. Fundamentals, design, and applications. *Chem. Soc. Rev.*, **37**, 1001–1011.
13. Hendrickson, G.R., Smith, M.H., South, A.B. and Lyon, L.A. (2010) Design of multiresponsive hydrogel particles and assemblies. *Adv. Funct. Mater.*, **20**, 1697–1712.
14. Milner, S.T. (1991) Polymer brushes. *Science*, **251**, 905–914.
15. Stuart, M.A.C., Huck, W.T.S., Genzer, J. *et al.* (2010) Emerging applications of stimuli-responsive polymer materials. *Nat. Mater.*, **9**, 101–113.
16. Mendes, P.M. (2008) Stimuli-responsive surfaces for bio-applications. *Chem. Soc. Rev.*, **37**, 2512–2529.
17. Emmenegger, C.R., Brynda, E., Riedel, T. *et al.* (2009) Interaction of blood plasma with antifouling surfaces. *Langmuir*, **25**, 6328–6333.
18. Guo, X., Weiss, A. and Ballauff, M. (1999) Synthesis of spherical polyelectrolyte brushes by photoemulsion polymerization. *Macromolecules*, **32**, 6043–6046.
19. Prucker, O. and Ruhe, J. (1998) Synthesis of poly(styrene) monolayers attached to high surface area silica gels through self-assembled monolayers of azo initiators. *Macromolecules*, **31**, 592–601.

20. Guo, X. and Ballauff, M. (2000) Spatial dimensions of colloidal polyelectrolyte brushes as determined by dynamic light scattering. *Langmuir*, **16**, 8719–8726.

21. Wang, X.A., Xu, J., Li, L. *et al.* (2010) Synthesis of spherical polyelectrolyte brushes by thermo-controlled emulsion polymerization. *Macromol. Rapid Commun.*, **31**, 1272–1275.

22. Wuelfing, W.P., Gross, S.M., Miles, D.T. and Murray, R.W. (1998) Nanometer gold clusters protected by surface-bound monolayers of thiolated poly(ethylene glycol) polymer electrolyte. *J. Am. Chem. Soc.*, **120**, 12696–12697.

23. Corbierre, M.K., Cameron, N.S., Sutton, M. *et al.* (2001) Polymer-stabilized gold nanoparticles and their incorporation into polymer matrices. *J. Am. Chem. Soc.*, **123**, 10411–10412.

24. Motornov, M., Sheparovych, R., Lupitskyy, R. *et al.* (2007) Stimuli-responsive colloidal systems from mixed brush-coated nanoparticles. *Adv. Funct. Mater.*, **17**, 2307–2314.

25. Sukhorukov, G.B., Donath, E., Davis, S. *et al.* (1998) Stepwise polyelectrolyte assembly on particle surfaces: A novel approach to colloid design. *Polym. Adv. Technol.*, **9**, 759–767.

26. Kato, N., Schuetz, P., Fery, A. and Caruso, F. (2002) Thin multilayer films of weak polyelectrolytes on colloid particles. *Macromolecules*, **35**, 9780–9787.

27. Cho, J., Quinn, J.F. and Caruso, F. (2004) Fabrication of polyelectrolyte multilayer films comprising nanoblended layers. *J. Am. Chem. Soc.*, **126**, 2270–2271.

28. Caruso, F., Lichtenfeld, H., Giersig, M. *et al.* (1998) Electrostatic self-assembly of silica nanoparticle-polyelectrolyte multilayers on polystyrene latex particles. *J. Am. Chem. Soc.*, **7863**, 8523–8524.

29. Johnston, A.P.R., Read, E.S. and Caruso, F. (2005) DNA multilayer films on planar and colloidal supports: Sequential assembly of like-charged polyelectrolytes. *Nano Lett.*, **5**, 953–956.

30. Caruso, F. and Mohwald, H. (1999) Protein multilayer formation on colloids through a stepwise self-assembly technique. *J. Am. Chem. Soc.*, **121**, 6039–6046.

31. Guo, X. and Ballauff, M. (2001) Spherical polyelectrolyte brushes: Comparison between annealed and quenched brushes. *Physical Review E*, **6405**, 051406.

32. Schild, H.G. (1992) Poly(n-isopropylacrylamide): Experiment, theory and application. *Prog. Polym. Sci.*, **17**, 163–249.

33. Pelton, R. (2000) Temperature-sensitive aqueous microgels. *Adv. Colloid Interface Sci.*, **85**, 1–33.

34. Lu, Y., Wittemann, A., Ballauff, M. and Drechsler, M. (2006) Preparation of poly styrene-poly (n-isopropylacrylamide) (ps-pnipa) core-shell particles by photoemulsion polymerization. *Macromol. Rapid Commun.*, **27**, 1137–1141.

35. Gao, J. and Frisken, B.J. (2003) Cross-linker-free n-isopropylacrylamide gel nanospheres. *Langmuir*, **19**, 5212–5216.

36. Dingenouts, N., Norhausen, C. and Ballauff, M. (1998) Observation of the volume transition in thermosensitive core-shell latex particles by small-angle x-ray scattering. *Macromolecules*, **31**, 8912–8917.

37. Crassous, J.J., Ballauff, M., Drechsler, M. *et al.* (2006) Imaging the volume transition in thermosensitive core-shell particles by cryo-transmission electron microscopy. *Langmuir*, **22**, 2403–2406.

38. Lu, Y., Mei, Y., Walker, R., Ballauff, M. *et al.* (2006) 'nano-tree' - type spherical polymer brush particles as templates for metallic nanoparticles. *Polymer*, **47**, 4985–4995.

39. Sharma, G. and Ballauff, M. (2004) Cationic spherical polyelectrolyte brushes as nanoreactors for the generation of gold particles. *Macromol. Rapid Commun.*, **25**, 547–552.

40. Mei, Y., Sharma, G., Lu, Y. *et al.* (2005) High catalytic activity of platinum nanoparticles immobilized on spherical polyelectrolyte brushes. *Langmuir*, **21**, 12229–12234.

41. Lu, Y., Mei, Y., Schrinner, M. *et al.* (2007) In situ formation of ag nanoparticles in spherical polyacrylic acid brushes by uv irradiation. *J. Phys. Chem. C*, **111**, 7676–7681.

42. Mei, Y., Lu, Y., Polzer, F. *et al.* (2007) Catalytic activity of palladium nanoparticles encapsulated in spherical polyelectrolyte brushes and core-shell microgels. *Chem. Mater.*, **19**, 1062–1069.

43. Lu, Y., Proch, S., Schrinner, M. *et al.* (2009) Thermosensitive core-shell microgel as a "Nanoreactor" For catalytic active metal nanoparticles. *J. Mater. Chem.*, **19**, 3955–3961.

44. Proch, S., Mei, Y., Villanueva, J.M.R. *et al.* (2008) Suzuki- and heck-type crosscoupling with palladium nanoparticles immobilized on spherical polyelectrolyte brushes. *Adv. Synth. Catal.*, **350**, 493–500.

45. Lu, Y., Mei, Y., Drechsler, M. and Ballauff, M. (2006) Thermosensitive core-shell particles as carriers for ag nanoparticles: Modulating the catalytic activity by a phase transition in networks. *Angew. Chem., Int. Ed.*, **45**, 813–816.

46. Lu, Y., Mei, Y., Ballauff, M. and Drechsler, M. (2006) Thermosensitive core-shell particles as carrier systems for metallic nanoparticles. *J. Phys. Chem. B*, **110**, 3930–3937.

47. Henzler, K., Haupt, B., Lauterbach, K. *et al.* (2010) Adsorption of beta-lactoglobulin on spherical polyelectrolyte brushes: Direct proof of counterion release by isothermal titration calorimetry. *J. Am. Chem. Soc.*, **132**, 3159–3163.

48. Wittemann, A. and Ballauff, M. (2004) Secondary structure analysis of proteins embedded in spherical polyelectrolyte brushes by ft-ir spectroscopy. *Anal. Chem.*, **76**, 2813–2819.

49. Neumann, T., Haupt, B. and Ballauff, M. (2004) High activity of enzymes immobilized in colloidal nanoreactors. *Macromol. Biosci.*, **4**, 13–16.

50. Haupt, B., Neumann, T., Wittemann, A. and Ballauff, M. (2005) Activity of enzymes immobilized in colloidal spherical polyelectrolyte brushes. *Biomacromolecules*, **6**, 948–955.

51. Schüler, C. and Caruso, F. (2000) Preparation of enzyme multilayers on colloids for biocatalysis. *Macromol. Rapid Commun.*, **21**, 750–753.

52. Welsch, N., Wittemann, A. and Ballauff, M. (2009) Enhanced activity of enzymes immobilized in thermoresponsive core-shell microgels. *J. Phys. Chem. B*, **113**, 16039–16045.

53. Nayak, S. and Lyon, L.A. (2005) Soft nanotechnology with soft nanoparticles. *Angew. Chem., Int. Ed.*, **44**, 7686–7708.

54. Kabanov, A.V. and Vinogradov, S.V. (2009) Nanogels as pharmaceutical carriers: Finite networks of infinite capabilities. *Angew. Chem., Int. Ed.*, **48**, 5418–5429.

55. Saunders, B.R., Laajam, N., Daly, E. *et al.* (2009) Microgels: From responsive polymer colloids to biomaterials. *Adv. Colloid Interface Sci.*, **147–148**, 251–262.

56. Hoffman, A.S. (2002) Hydrogels for biomedical applications. *Adv. Drug Delivery. Rev.*, **54**, 3–12.
57. Kopecek, J. (2009) Hydrogels: From soft contact lenses and implants to self-assembled nanomaterials. *Journal of Polymer Science Part a-Polymer Chemistry*, **47**, 5929–5946.
58. Pelton, R.H. and Chibante, P. (1986) Preparation of aqueous latices with n-isopropylacrylamide. *Colloids Surf.*, **20**, 247–256.
59. Snowden, M.J., Chowdhry, B.Z., Vincent, B. and Morris, G.E. (1996) Colloidal copolymer microgels of n-isopropylacrylamide and acrylic acid: Ph, ionic strength and temperature effects. *Journal of the Chemical Society-Faraday Transactions*, **92**, 5013–5016.
60. Kim, K.S. and Vincent, B. (2005) Ph and temperature-sensitive behaviors of poly(4-vinyl pyridine-co-n-isopropyl acrylamide) microgels. *Polymer Journal*, **37**, 565–570.
61. Senff, H. and Richtering, W. (2000) Influence of cross-link density on rheological properties of temperature-sensitive microgel suspensions. *Colloid Polym. Sci.*, **278**, 830–840.
62. Varga, I., Gilanyi, T., Meszaros, R. *et al.* (2001) Effect of cross-link density on the internal structure of poly(n-isopropylacrylamide) microgels. *J. Phys. Chem. B*, **105**, 9071–9076.
63. Debord, J.D. and Lyon, L.A. (2003) Synthesis and characterization of ph-responsive copolymer microgels with tunable volume phase transition temperatures. *Langmuir*, **19**, 7662–7664.
64. Keerl, M., Pedersen, J.S. and Richtering, W. (2009) Temperature sensitive copolymer microgels with nanophase separated structure. *J. Am. Chem. Soc.*, **131**, 3093–3097.
65. Hoare, T. and Pelton, R. (2004) Functional group distributions in carboxylic acid containing poly(n-isopropylacrylamide) microgels. *Langmuir*, **20**, 2123–2133.
66. Hoare, T. and McLean, D. (2006) Kinetic prediction of functional group distributions in thermosensitive microgels. *J. Phys. Chem. B*, **110**, 20327–20336.
67. Jones, C.D. and Lyon, L.A. (2000) Synthesis and characterization of multiresponsive core-shell microgels. *Macromolecules*, **33**, 8301–8306.
68. Berndt, I. and Richtering, W. (2003) Doubly temperature sensitive core-shell microgels. *Macromolecules*, **36**, 8780–8785.
69. Gan, D.J. and Lyon, L.A. (2001) Tunable swelling kinetics in core-shell hydrogel nanoparticles. *J. Am. Chem. Soc.*, **123**, 7511–7517.
70. Jones, C.D. and Lyon, L.A. (2003) Dependence of shell thickness on core compression in acrylic acid modified poly(n-isopropylacrylamide) core/shell microgels. *Langmuir*, **19**, 4544–4547.
71. Berndt, I., Pedersen, J.S. and Richtering, W. (2006) Temperature-sensitive core-shell microgel particles with dense shell. *Angew. Chem., Int. Ed.*, **45**, 1737–1741.
72. Jones, C.D., McGrath, J.G. and Lyon, L.A. (2004) Characterization of cyanine dye-labeled poly(n-isopropylacrylamide) core/shell microgels using fluorescence resonance energy transfer. *J. Phys. Chem. B*, **108**, 12652–12657.
73. Hu, X.B., Tong, Z. and Lyon, L.A. (2010) Multicompartment core/shell microgels. *J. Am. Chem. Soc.*, **132**, 11470–11472.

74. Nayak, S., Lee, H., Chmielewski, J. and Lyon, L.A. (2004) Folate-mediated cell targeting and cytotoxicity using thermoresponsive microgels. *J. Am. Chem. Soc.*, **126**, 10258–10259.

75. Blackburn, W.H., Dickerson, E.B., Smith, M.H. *et al.* (2009) Peptide-functionalized nanogels for targeted sirna delivery. *Bioconjugate Chem.*, **20**, 960–968.

76. Gan, D.J. and Lyon, L.A. (2002) Synthesis and protein adsorption resistance of peg-modified poly(n-isopropylacrylamide) core/shell microgels. *Macromolecules*, **35**, 9634–9639.

77. Kleinen, J., Klee, A. and Richtering, W. (2010) Influence of architecture on the interaction of negatively charged multisensitive poly(n-isopropylacrylamide)-co-methacrylic acid microgels with oppositely charged polyelectrolyte: Absorption vs adsorption. *Langmuir*, **26**, 11258–11265.

78. Nayak, S. and Lyon, L.A. (2004) Ligand-functionalized core/shell microgels with permselective shells. *Angew. Chem., Int. Ed.*, **43**, 6706–6709.

79. Lapeyre, V., Ancla, C., Catargi, B. and Ravaine, V. (2008) Glucose-responsive microgels with a core-shell structure. *J. Colloid Interface Sci.*, **327**, 316–323.

80. Zhang, Y., Guan, Y. and Zhou, S. (2006) Synthesis and volume phase transitions of glucose-sensitive microgels. *Biomacromolecules*, **7**, 3196–3201.

81. Hoare, T. and Pelton, R. (2007) Engineering glucose swelling responses in poly (n-isopropylacrylamide)-based microgels. *Macromolecules*, **40**, 670–678.

82. Huang, H.Y., Remsen, E.E., Kowalewski, T. and Wooley, K.L. (1999) Nanocages derived from shell cross-linked micelle templates. *J. Am. Chem. Soc.*, **121**, 3805–3806.

83. Zhang, Y.W., Jiang, M., Zhao, J.X. *et al.* (2005) A novel route to thermosensitive polymeric core-shell aggregates and hollow spheres in aqueous media. *Adv Funct Mater*, **15**, 695–699.

84. Zha, L.S., Zhang, Y., Yang, W.L. and Fu, S.K. (2002) Monodisperse temperature-sensitive microcontainers. *Adv. Mater.*, **14**, 1090–1092.

85. Nayak, S., Gan, D.J., Serpe, M.J. and Lyon, L.A. (2005) Hollow thermoresponsive microgels. *Small*, **1**, 416–421.

86. Mandal, T.K., Fleming, M.S. and Walt, D.R. (2000) Production of hollow polymeric microspheres by surface-confined living radical polymerization on silica templates. *Chem. Mater.*, **12**, 3481–3487.

87. Caruso, F., Caruso, R.A. and Mohwald, H. (1998) Nanoengineering of inorganic and hybrid hollow spheres by colloidal templating. *Science*, **282**, 1111–1114.

88. Lou, X.W., Archer, L.A. and Yang, Z.C. (2008) Hollow micro-/nanostructures: Synthesis and applications. *Adv. Mater.*, **20**, 3987–4019.

89. Donath, E., Sukhorukov, G.B., Caruso, F. *et al.* (1998) Novel hollow polymer shells by colloid-templated assembly of polyelectrolytes. *Angew. Chem., Int. Ed.*, **37**, 2202–2205.

90. Peyratout, C.S. and Dähne, L. (2004) Tailor-made polyelectrolyte microcapsules: From multilayers to smart containers. *Angew. Chem., Int. Ed.*, **43**, 3762–3783.

91. Itoh, Y., Matsusaki, M., Kida, T. and Akashi, M. (2006) Enzyme-responsive release of encapsulated proteins from biodegradable hollow capsules. *Biomacromolecules*, **7**, 2715–2718.

92. Antonietti, M. and Forster, S. (2003) Vesicles and liposomes: A self-assembly principle beyond lipids. *Adv. Mater.*, **15**, 1323–1333.

93. Gabizon, A., Shmeeda, H., Horowitz, A.T. and Zalipsky, S. (2004) Tumor cell targeting of liposome-entrapped drugs with phospholipid-anchored folic acid-peg conjugates. *Adv. Drug Delivery Rev.*, **56**, 1177–1192.

94. Hotz, J. and Meier, W. (1998) Vesicle-templated polymer hollow spheres. *Langmuir*, **14**, 1031–1036.

95. Hubert, D.H.W., Jung, M. and German, A.L. (2000) Vesicle templating. *Adv. Mater.*, **12**, 1291–1294.

96. Murtagh, J. and Thomas, J.K. (1986) Mobility and reactivity in colloidal aggregates with motion restricted by polymerization. *Faraday Discuss.*, **81**, 127–136.

97. Sauer, M. and Meier, W. (2001) Responsive nanocapsules. *Chem. Commun.*, 55–56.

98. Sauer, M., Streich, D. and Meier, W. (2001) Ph-sensitive nanocontainers. *Adv. Mater.*, **13**, 1649–1651.

99. Cheng, C.-J., Chu, L.-Y., Ren, P.-W. *et al.* (2007) Preparation of monodisperse thermosensitive poly(n-isopropylacrylamide) hollow microcapsules. *J. Colloid Interface Sci.*, **313**, 383–388.

100. Sun, Q.H. and Deng, Y.L. (2005) In situ synthesis of temperature-sensitive hollow microspheres via interfacial polymerization. *J. Am. Chem. Soc.*, **127**, 8274–8275.

101. Cheng, C.-J., Chu, L.-Y. and Xie, R. (2006) Preparation of highly monodisperse w/o emulsions with hydrophobically modified spg membranes. *J. Colloid Interface Sci.*, **300**, 375–382.

102. Rodríguez-Hernández, J., Chécot, F., Gnanou, Y. and Lecommandoux, S. (2005) Toward 'smart' nano-objects by self-assembly of block copolymers in solution. *Prog. Polym. Sci.*, **30**, 691–724.

103. Chécot, F., Lecommandoux, S., Gnanou, Y. and Klok, H.-A. (2002) Water-soluble stimuli-responsive vesicles from peptide-based diblock copolymers. *Angew. Chem., Int. Ed.*, **41**, 1339–1343.

104. Gohy, J.-F., Willet, N., Varshney, S. *et al.* (2001) Core–shell–corona micelles with a responsive shell. *Ang. Chem.*, **113**, 3314–3316.

105. Zhang, L.F. and Eisenberg, A. (1995) Multiple morphologies of crew-cut aggregates of polystyrene-b-poly(acrylic acid) block-copolymers. *Science*, **268**, 1728–1731.

106. Ma, Q., Remsen, E.E., Kowalewski, T. *et al.* (2001) Environmentally-responsive, entirely hydrophilic, shell cross-linked (sck) nanoparticles. *Nano Lett.*, **1**, 651–655.

107. Thurmond, K.B., Kowalewski, T. and Wooley, K.L. (1996) Water-soluble knedel-like structures: The preparation of shell-cross-linked small particles. *J. Am. Chem. Soc.*, **118**, 7239–7240.

108. Turner, J.L., Chen, Z. and Wooley, K.L. (2005) Regiochemical functionalization of a nanoscale cage-like structure: Robust core-shell nanostructures crafted as vessels for selective uptake and release of small and large guests. *J. Controlled Release*, **109**, 189–202.

109. Ievins, A.D., Moughton. A.O. and O'Reilly, R.K. (2008) Synthesis of hollow responsive functional nanocages using a metal–ligand complexation strategy. *Macromolecules*, **41**, 3571–3578.

110. Binks, B.P. and Fletcher, P.D.I. (2001) Particles adsorbed at the oil−water interface: A theoretical comparison between spheres of uniform wettability and "janus" particles. *Langmuir*, **17**, 4708–4710.

111. Casagrande, C., Fabre, P., Raphael, E., Veyssie, M. (1989) "Janus beads": Realization and behaviour at water/oil interfaces. *Europhys. Lett.*, **9**, 251.

112. Roh, K.H., Martin, D.C. and Lahann, J. (2005) Biphasic janus particles with nanoscale anisotropy. *Nat. Mater.*, **4**, 759–763.

113. Nisisako, T., Torii, T., Takahashi, T. and Takizawa, Y., (2006) Synthesis of monodisperse bicolored janus particles with electrical anisotropy using a microfluidic co-flow system. *Adv. Mater.*, **18**, 1152–1156.

114. Walther, A. and Muller, A.H.E. (2008) Janus particles. *Soft Matter*, **4**, 663–668.

115. de Gennes, P.G. (1992) Soft matter. *Rev. Mod. Phys.*, **64**, 645–648.

116. Perro, A., Reculusa, S., Ravaine, S. *et al.* (2005) Design and synthesis of janus micro- and nanoparticles. *J. Mater. Chem.*, **15**, 3745–3760.

117. Binks, B.P. (2002) Particles as surfactants – similarities and differences. *Curr. Opin. Colloid Interface Sci.*, **7**, 21–41.

118. Bao, Z., Chen, L., Weldon, M. *et al.* (2001) Toward controllable self-assembly of microstructures: Selective functionalization and fabrication of patterned spheres. *Chem. Mater.*, **14**, 24–26.

119. Paunov, V.N. (2003) Novel method for determining the three-phase contact angle of colloid particles adsorbed at air−water and oil−water interfaces. *Langmuir*, **19**, 7970–7976.

120. Ling, X.Y., Phang, I.Y., Acikgoz, C. *et al.* (2009) Janus particles with controllable patchiness and their chemical functionalization and supramolecular assembly. *Angew. Chem., Int. Ed.*, **48**, 7677–7682.

121. Cayre, O., Paunov, V.N. and Velev, O.D. (2003) Fabrication of dipolar colloid particles by microcontact printing. *Chem. Commun.*, 2296–2297.

122. Berger, S., Synytska, A., Ionov, L. *et al.* (2008) Stimuli-responsive bicomponent polymer janus particles by "Grafting from"/"Grafting to" Approaches. *Macromolecules*, **41**, 9669–9676.

123. Liu, B., Wei, W., Qu, X. and Yang, Z. (2008) Janus colloids formed by biphasic grafting at a pickering emulsion interface. *Angew. Chem., Int. Ed.*, **47**, 3973–3975.

124. Suzuki, D., Tsuji, S. and Kawaguchi, H. (2007) Janus microgels prepared by surfactant-free pickering emulsion-based modification and their self-assembly. *J. Am. Chem. Soc.*, **129**, 8088–8089.

125. Seiffert, S., Romanowsky, M.B. and Weitz, D.A. (2010) Janus microgels produced from functional precursor polymers. *Langmuir*, **26**, 14842–14847.

126. Nie, Z., Li, W., Seo, M., Xu, S. and Kumacheva, E. (2006) Janus and ternary particles generated by microfluidic synthesis: Design, synthesis, and self-assembly. *J. Am. Chem. Soc.*, **128**, 9408–9412.

127. Yuet, K.P., Hwang, D.K., Haghgooie, R. and Doyle, P.S. (2009) Multifunctional superparamagnetic janus particles. *Langmuir*, **26**, 4281–4287.

128. Higuchi, T., Tajima, A., Motoyoshi, K. *et al.* (2008) Frustrated phases of block copolymers in nanoparticles. *Ang. Chem.*, **120**, 8164–8166.

129. Pawar, A.B. and Kretzschmar, I. (2010) Fabrication, assembly, and application of patchy particles. *Macromol. Rapid Commun.*, **31**, 150–168.

130. Yi, G.R., Manoharan, V.N., Michel, E. *et al.* (2004) Colloidal clusters of silica or polymer microspheres. *Adv. Mater.*, **16**, 1204–1208.

131. Cho, Y.S., Yi, G.R., Lim, J.M. *et al.* (2005) Self-organization of bidisperse colloids in water droplets. *J. Am. Chem. Soc.*, **127**, 15968–15975.
132. Cho, Y.S., Yi, G.R., Kim, S.H. *et al.* (2007) Particles with coordinated patches or windows from oil-in-water emulsions. *Chem. Mater.*, **19**, 3183–3193.
133. Roh, K.-H., Martin, D.C. and Lahann, J. (2006) Triphasic nanocolloids. *J. Am. Chem. Soc.*, **128**, 6796–6797.
134. Zhang Z.L. and Glotzer, S.C. (2004) Self-assembly of patchy particles. *Nano Lett.*, **4**, 1407–1413.
135. Pawar, A.B. and Kretzschmar, I. (2009) Multifunctional patchy particles by glancing angle deposition. *Langmuir*, **25**, 9057–9063.
136. Kim, S.H., Yi, G.R., Kim, K.H. and Yang, S.M. (2008) Photocurable pickering emulsion for colloidal particles with structural complexity. *Langmuir*, **24**, 2365–2371.

13

Novel Biomimetic Polymer Gels Exhibiting Self-Oscillation

Ryo Yoshida

Department of Materials Engineering, School of Engineering,
The University of Tokyo, Japan

13.1   Introduction

As mentioned in the other chapters, many kinds of stimuli-responsive polymers and gels (temperature-responsive, pH-responsive, etc.) and their applications to actuator (artificial muscle), drug delivery systems (DDS), tissue engineering, purification or separation systems, biosensor, shape memory materials, molecular recognition systems, and so on, have been studied [1–4]. New synthetic methods to give unique functions are being developed by molecular design on the nano-scale including supramolecular design, and the design and construction of micro- or nanomaterial systems with biomimetic functions has been attempted.

Other than stimuli-responsive functions, one of the characteristic behaviours in living systems is autonomous oscillation, that is, spontaneous changes with temporal periodicity (called "temporal structure") such as heartbeat, brain waves, pulsatile secretion of hormone, cell cycle, and biorhythm. Although several stimuli-responsive polymer systems have been studied from the standpoint of biomimetics, polymer systems undergoing self-oscillation under constant condition without any on–off switching of external stimuli have yet to be developed. If autonomous polymer systems resembling living organisms can be realized by using completely synthetic polymers, unprecedented biomimetic materials may be created (Figure 13.1).

Responsive Membranes and Materials, First Edition. Edited by D. Bhattacharyya, Thomas Schäfer, S. R. Wickramasinghe and Sylvia Daunert.
© 2013 John Wiley & Sons, Ltd. Published 2013 by John Wiley & Sons, Ltd.

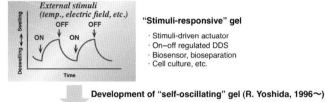

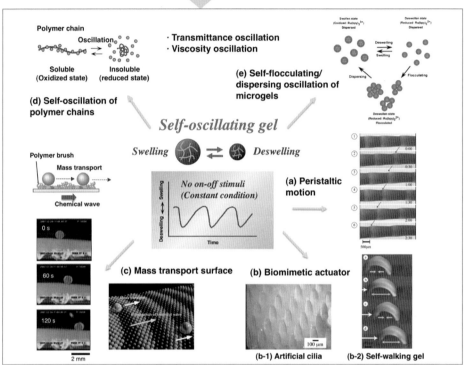

Figure 13.1 *Development of self-oscillating polymers and gels. See plate section for colour figure.*

With such a concept in mind, the author developed a novel polymer gel that causes autonomous mechanical oscillation without an external control in a completely closed solution. Since first being reported in 1996 as a "self-oscillating gel" [5], the author has been systematically studying the self-oscillating polymer and gel as well as their applications to biomimetic or smart materials (Figure 13.4) [6–35]. In fact, applications to autonomic biomimetic actuators, ciliary motion actuators [17,18], and self-walking gels [19], and so on have been realized. As an autonomic microconveyor, a mass transport surface utilizing the peristaltic motion of the self-oscillating gel was also designed [21,22]. As a nanoactuator that exhibits autonomous oscillation on the nanometre-scale (i.e. nanooscillator), oscillating behaviours of the uncross-linked polymer chain or submicron-order microgel were investigated through optical transmittance or viscosity changes of the polymer solution and the microgel dispersion [25–35]. In this chapter, these results are reviewed.

13.2   The Design Concept of Self-Oscillating Gel

The Belousov–Zhabotinsky (BZ) reaction is well known for exhibiting temporal and spatiotemporal oscillating phenomena [36, 37]. The overall process is the oxidation of an organic substrate, such as malonic acid (MA) or citric acid, by an oxidizing agent (bromate ion) in the presence of a strong acid and a metal catalyst. In the course of the reaction, however, the catalyst undergoes spontaneous redox oscillation. When the solution is homogeneously stirred, the colour of the solution periodically changes, based on the redox changes of the metal catalyst. When the solution is placed as a thin film in stationary conditions, concentric or spiral wave patterns develop in the solution. The wave of oxidized states propagating in the medium at a constant speed is called a "chemical wave". The reaction is often analogically compared with the TCA cycle which is a key metabolic process taking place in the living body, and it is recognized as a chemical model for understanding several autonomous phenomena in biological systems.

The author attempted to convert the chemical oscillation of the BZ reaction into a mechanical change in gels and generate an autonomous swelling–deswelling oscillation under non-oscillatory outer conditions. For this purpose, a copolymer gel consisting of N-isopropylacrylamide (NIPAAm) and ruthenium tris(2,2′-bipyridine) (Ru(bpy)$_3$) was prepared. In the poly(NIPAAm-co-Ru(bpy)$_3$) gel, Ru(bpy)$_3$ as a catalyst for the BZ reaction is covalently bonded to the polymer chains of NIPAAm. Since the gel is based on thermosensitive poly(NIPAAm), it exhibits volume phase transition behaviour. However, the behaviour is different between reduced and oxidized states of Ru(bpy)$_3$. The oxidation of Ru(bpy)$_3$ moiety causes not only an increase in the swelling degree of the gel, but also a rise in the volume phase transition temperature. As a result, the gel should undergo a cyclic swelling–deswelling change when the Ru(bpy)$_3$ moiety is periodically oxidized and reduced under constant temperature. When the gel is immersed in an aqueous solution containing the substrates of the BZ reaction (malonic acid, nitric acid, and sodium bromate) except for the catalyst, the substrates penetrate into the polymer network and the BZ reaction occurs in the gel. Consequently, periodic redox changes induced by the BZ reaction produce periodic swelling–deswelling changes in the gel (Figure 13.2).

13.3   Aspects of the Autonomous Swelling–Deswelling Oscillation

In the self-oscillating gel, chemomechanical oscillation and synchronization occur in the range from microscopic level to microscopic level (Figure 13.3(a)). Redox changes of Ru(bpy)$_3{}^{2+}$ catalyst are converted to conformational changes of the polymer chain by polymerization. The conformational changes are amplified to macroscopic swelling–deswelling changes of the polymer network by cross-linking. Further, when the gel size is larger than chemical wavelength, the chemical wave propagates in the gel by coupling with diffusion. Peristaltic motion of the gel is then created.

Figure 13.3(b) shows a typical example of the observed self-oscillating behaviour under a microscope for the small cubic poly(NIPAAm-co-Ru(bpy)$_3{}^{2+}$) gel (each length about 0.5 mm) in an aqueous solution containing MA, NaBrO$_3$, and HNO$_3$. In miniature gels sufficiently smaller than the wavelength of the chemical wave (typically several mm),

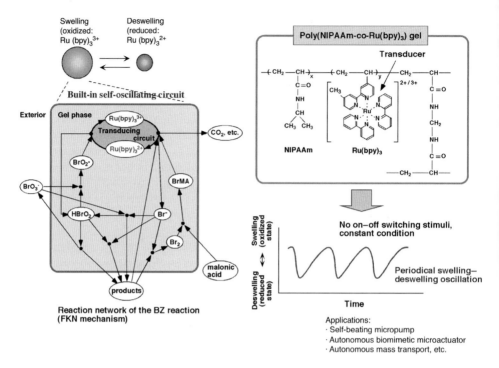

Figure 13.2 *Mechanism of self-oscillation for poly(NIPAAm-co-Ru(bpy)$_3$$^{2+}$) gel coupled with the Belousov–Zhabotinsky (BZ) reaction. See plate section for colour figure.*

the redox change of ruthenium catalyst can be regarded as occurring homogeneously without pattern formation [10]. Due to the redox oscillation of the immobilized Ru(bpy)$_3$$^{2+}$, mechanical swelling–deswelling oscillation of the gel autonomously occurs with the same period as for the redox oscillation. The volume change is isotropic and the gel beats as a whole. The chemical and mechanical oscillations are synchronized without a phase difference (i.e. the gel exhibits swelling during the oxidized state and deswelling during the reduced state).

Generally the oscillation period increases with a decrease in the initial concentration of substrates. The swelling–deswelling amplitude of the gel increases with an increase in the oscillation period. Therefore the swelling–deswelling amplitude of the gel is controllable by changing the initial concentration of substrates, although it also depends on the characteristic swelling-deswelling responses of the gel. In addition, the frequency (the reciprocal of the oscillation period) of the BZ reaction increases with temperature in accordance with the Arrhenius equation. Self-oscillation at higher temperatures provides the advantage of creating a high frequency, which may be preferable for an application such as an actuator, described in later sections. In order to induce self-oscillation while maintaining a larger amplitude at higher temperatures, and around body temperature for potential applications to biomaterials and so on, a self-oscillating gel composed of a thermosensitive polymer exhibiting a higher LCST than that of the NIPAAm polymer was prepared [11]. The self-oscillating behaviour of the gel was investigated by comparing it against gels composed

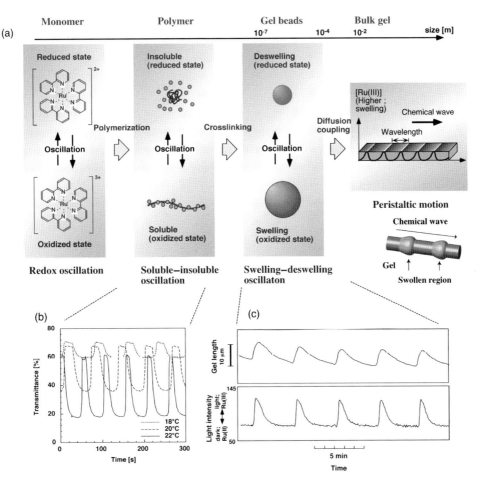

Figure 13.3 *(a) Synchronization in a self-oscillating gel in the microscopic to macroscopic level range. (b) Oscillating profiles of optical transmittance for poly(NIPAAm-co-Ru(bpy)$_3{}^{2+}$) solution at several temperatures. (c) Periodic redox changes of the miniature cubic poly(NIPAAm-co-Ru(bpy)$_3{}^{2+}$) gel (lower) and the swelling–deswelling oscillation (upper) at 20°C. Colour changes of the gel accompanied by redox oscillations (orange: reduced state, light green: the oxidized state) were converted to 8-bit grayscale changes (dark: reduced, light: oxidized) by image processing. Transmitted light intensity is expressed as an 8-bit grayscale value. See plate section for colour figure.*

of a thermosensitive NIPAAm polymer with a lower LCST or non-thermosensitive polymer. The design concept of self-oscillation at higher temperatures without a decrease in swelling–deswelling amplitude was demonstrated by utilizing a thermosensitive polymer exhibiting a higher LCST.

As inherent behaviour of the BZ reaction, the abrupt transition from steady state (non-oscillating state) to oscillating state occurs with a change in controlling parameter such as chemical composition, light, and so on. By utilizing these characteristics, reversible on–off regulation of self-beating was successfully achieved, triggered by addition and removal of MA [12]. Since there are some organic acids which can be used as the substrate for the BZ reaction (e.g. citric acid), the same beating regulation is possible using organic acids instead of MA. As the gel has thermosensitivity due to the NIPAAm component, the beating rhythm can also be controlled by temperature. By utilizing the thermosensitivity, on–off control of self-oscillation is also possible [13]. Further, it is well known that the period of oscillation is affected by light illumination for the $Ru(bpy)_3^{2+}$-catalysed BZ reaction. Optical on–off control of the self-oscillating motion of the gel has been demonstrated [14, 15].

When the gel size is larger than the chemical wavelength, the chemical wave propagates in the gel by coupling with diffusion of intermediates. Then a peristaltic motion of the gel is created (see Figure 13.3(a)). Figure 13.1(a) shows the cylindrical gel which is immersed in an aqueous solution containing the three reactants of the BZ reaction [16]. The time course of the peristaltic motion of the gel in a solution of the BZ substrates is shown. The green and orange colours correspond to the oxidized and reduced states of the Ru moiety in the gel, respectively. The chemical waves propagate in the gel at a constant speed in the direction of the gel length. The orange (Ru(II)) and green (Ru(III)) zones represent simply the shrunken and swollen parts respectively, the locally swollen and shrunken parts move with the chemical wave, like the peristaltic motion of living worms.

13.4 Design of Biomimetic Actuator Using Self-Oscillating Polymer and Gel

13.4.1 Ciliary Motion Actuator (Artificial Cilia)

Recently, microfabrication technologies such as photolithography have also been experimented with for the preparation of microgels. Since any shape of gel can be easily created by these methods, fabrication of soft microactuator, microgel valve, gel display, and so on is expected. One of the most promising fields of MEMS is microactuator array or distributed actuator systems. The actuators, which have a very simple actuation motion such as up and down motion, are arranged in an array form. If their motions are random, no work is extracted from this array. However, by making them operate in a certain order, they can generate work as a system. A typical example of this kind of actuation array is a ciliary motion microactuator array. There have been many reports on this system. Although various actuation principles have been proposed, all the previous work is based on the concept that the motion of actuators is controlled by external signals. If a self-oscillating gel plate with a microprojection structure array on top were to be realized, it would be expected that the chemical wave propagation would create a dynamic rhythmic motion of the structure array. This proposed structure could exhibit spontaneous dynamic propagating oscillation producing a ciliary motion array [17, 18].

Using this concept, a self-oscillating gel sheet with a microprojection array was actually fabricated. First, moving mask deep-X-ray lithography was utilized to fabricate a PMMA plate with a truncated conical shape microstructure array. This step was followed by evaporation of an Au seed layer and subsequent electroplating of nickel to form the metal mould structure. Then, a PDMS mould structure was duplicated from the Ni structure and utilized for gel moulding. The formation of gel was carried out by vacuum injection moulding.

A structure with a height of 300 μm and bottom diameter of 100 μm was successfully fabricated by the process (Figure 13.1(b-1)). The propagation of a chemical reaction wave and dynamic rhythmic motion of the microprojection array were confirmed by chemical wave observation and displacement measurements. Motion of the top with 5 μm range in both lateral and vertical directions, and elliptical motion of the projection top were observed. The feasibility of the new concept of a ciliary motion actuator made of self-oscillating polymer gel was confirmed. The actuator could serve as a microconveyer to transport objects on the surface.

13.4.2 Self-Walking Gel

Furthermore, we successfully developed a novel biomimetic walking gel actuator made of self-oscillating gel [19]. To produce directional movement of the gel, asymmetrical swelling–deswelling is desired. For these purposes, as a third component, hydrophilic 2-acrylamido-2′-methylpropanesulfonic acid (AMPS) was copolymerized into the polymer to lubricate the gel and to cause anisotropic contraction. During polymerization, the monomer solution faces two different plate surfaces; a hydrophilic glass surface and a hydrophobic Teflon surface. As the spacer is thin (0.5 mm), the surface property of the plate may affect the distribution of the monomer in the solution. Since $Ru(bpy)_3^{2+}$ monomer is hydrophobic, it easily migrates to the Teflon surface side. As a result, a non-uniform distribution along the height is formed by the components, and the resulting gel has a gradient distribution for the content of each component in the polymer network.

In order to convert the bending and stretching changes to one-directional motion, we employed a ratchet mechanism. A ratchet base with an asymmetrical surface structure was fabricated. On the ratchet base, the gel repeatedly bends and stretches autonomously. Figure 13.1(b-2) shows successive profiles of the "self-walking" motion of the gel like a looper in the BZ substrate solution under constant temperature. During stretching, the front edge can slide forward on the base, but the rear edge is prevented from sliding backwards. Conversely, during bending, the front edge is prevented from sliding backwards while the rear edge can slide forward. This action is repeated, and as a result, the gel walks forward. Since the oscillating period and the propagating velocity of the chemical wave change with the concentration of substrates in the outer solution, the walking velocity of the gel can be controlled. By using the gel with a gradient structure, another type of actuator which generates a self-propelled motion is also realized [20].

13.4.3 Theoretical Simulation of the Self-Oscillating Gel

As mentioned above, many applications to autonomous chemo-mechanical actuators can be expected. The self-oscillating gel we developed is a unique model of material systems, that is, a coupling system of reaction-diffusion and mechanical motion. If the chemical oscillation and mechanical oscillation affect each other through feedback, it would be difficult to predict and explain the oscillating behaviours. Theoretical simulation then becomes an effective tool. Recently, Balazs *et al.* [38, 39] developed an efficient theoretical model for self-oscillating gels. Since they reported in 2006 [38], they have demonstrated several aspects of the self-oscillating behaviour of the gel by theoretical simulation. Many interesting phenomena have been demonstrated theoretically. For example, computational modelling was used to determine how gradients in cross-link density across the width of a sample can drive long, thin BZ gels to both oscillate and bend, and thereby undergo concerted motion. Free in solution, these samples move forward (in the direction of lower cross-link density) through a rhythmic bending and unbending [20]. On the basis of such a situation, further innovative researches are expected.

13.5   Mass Transport Surface Utilizing Peristaltic Motion of Gel

The self-oscillating gel generating autonomous and periodic peristaltic motion has potential for new applications to functional surfaces, such as a conveyer, a self-cleaning surface, and so on. In order to realize an autonomous mass transport system, we attempted to transport an object by utilizing the peristaltic motion of the self-oscillating gel sheet [21,22]. AMPS was copolymerized to the gel to generate a large amplitude of volume change of the self-oscillating gel. As a model object, a cylindrical poly(acrylamide) (PAAm) gel was put on the gel surface. It was observed that the PAAm gel was transported on the gel surface with the propagation of the chemical wave as it rolled (see Figure 13.1(c)). Figure 13.4(a) shows a phase diagram of transportable conditions given by the velocity and the inclination angle

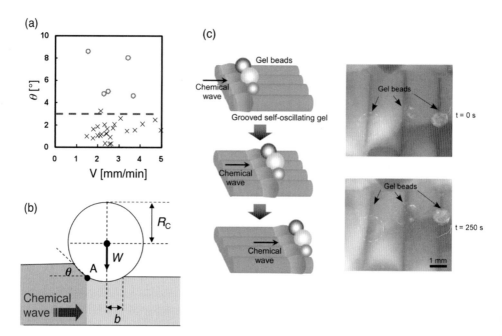

Figure 13.4 *(a) Phase diagram of the transportable region given by the velocity and the inclination angle of the wavefront: (○) transported, (×) not transported. (b) Model of the rolling cylindrical gel on the peristaltic gel surface (R_C = radius of curvature, W = load of the PAAm gel, b = contact half-width). (c) Transport of the poly(AAm-co-AMPS) gel beads on the grooved surface of the self-oscillating gel. See plate section for colour figure.*

θ of the wave front. It was found that the cylindrical PAAm gel was not transported if the inclination angle was less than approximately $3°$. It seems that the mass transportability does not depend on the velocity of the chemical wave but on the diameter of the cylindrical PAAm gel and the inclination angle of the wave front.

We have proposed a model to describe the mass transport phenomena based on the Hertz contact theory, and the relation between the transportability and the peristaltic motion was investigated. Figure 13.4(b) also shows the contact model of the cylindrical PAAm gel and the poly(NIPAAm-co-Ru(bpy)$_3$$^{2+}$-co-AMPS) gel sheet. If the inclination angle of the sheet surface is smaller than the slope of the tangent at point A, there is not sufficient contact force from the strain to rotate the cylindrical PAAm gel. Therefore, the condition for the transport of the cylindrical PAAm gel is written as $R_C \sin\theta \geq b$.

The minimum inclination angle calculated from the above equation was the same as the angle resulted from the experiment. It was found experimentally and supported by the model that the sheer wave front of the peristaltic motion was necessary to transport cylindrical gels. As the size of the gels was approximately of sub-mm $\sim$ mm order, gravity, frictional force, and contact force acting on the PAAm gel were the most important factors for mass transport. So, if the cargo materials aren't adsorbed to the gel sheet surface by an attracting force like hydrophobic interaction, mass transportation depends on the performance of the peristaltic motion of the self-oscillating gel surface.

Further, the surface figure capable of transporting microparticles in one direction was designed to fabricate a more versatile self-driven gel conveyer. A self-oscillating gel with a grooved surface was fabricated (Figure 13.4(c)) and the effectiveness of the surface design was investigated. Poly(AAm-co-AMPS) gel beads with a diameter of several hundred μm to several mm were transported on the grooved surface of the self-oscillating gel by its autonomous peristaltic motions [22]. It was found that the travelling direction of the peristaltic motion could be confined to the direction along the grooves by designing the groove-distance to be shorter than the wavelength of the chemical wave. Consequently, several gel beads were transported in parallel.

A chemo-mechanical actuator utilizing a reaction-diffusion wave across a gap junction was constructed toward a novel microconveyer by a micropatterned self-oscillating gel array [23]. Unidirectional propagation of the chemical wave of the BZ reaction was induced on gel arrays. By fabricating different shapes of gel arrays, control of the direction was possible. The swelling and deswelling of the gels followed the unidirectional propagation of chemical wave. Further, a double template polymerization method to fabricate monolayer surface of microgel beads was demonstrated [24].

13.6 Self-Oscillating Polymer Chains and Microgels as "Nanooscillators"

13.6.1 Solubility Oscillation of Polymer Chains

The periodic changes of linear and uncross-linked polymer chains can be easily observed as cyclic transparent and opaque changes for the polymer solution, with colour changes due to the redox oscillation of the catalyst [25]. Figure 13.3(b) shows the oscillation profiles of transmittance for a polymer solution which consists of linear poly(NIPAAm-co-Ru(bpy)$_3$$^{2+}$), MA, NaBrO$_3$, and HNO$_3$ at constant temperatures. The wavelength

(570 nm) at the isosbestic point of reduced and oxidized states was used to detect the optical transmittance changes based on soluble–insoluble changes of the polymer, not on the redox changes of the $Ru(bpy)_3$ moiety. Synchronized with the periodical changes between Ru(II) and Ru(III) states of the $Ru(bpy)_3^{2+}$ site, the polymer becomes hydrophobic and hydrophilic, and exhibits cyclic soluble–insoluble changes. The self-oscillating polymer was also covalently immobilized on a glass surface and self-oscillation was directly observed by AFM [26].

13.6.2 Self-Flocculating/Dispersing Oscillation of Microgels

Sub-micron-sized poly(NIPAAm-co-$Ru(bpy)_3^{2+}$) gel beads were prepared by surfactant-free aqueous precipitation polymerization, and the oscillating behaviours were analysed [27–30]. Figure 13.5(a) shows the oscillation profiles of transmittance for the microgel dispersions. At low temperatures (20–26.5°C), raising the temperature increased the amplitude of the oscillation. The increase in amplitude was due to the increased deviation of the hydrodynamic diameter between the Ru(II) and Ru(III) states. A remarkable change

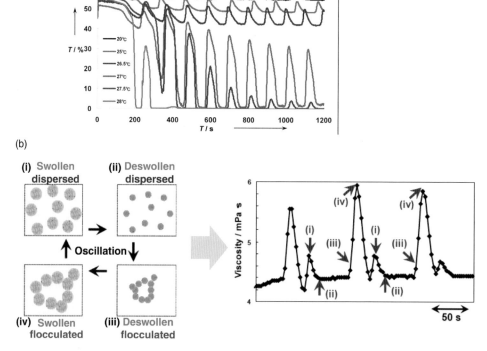

Figure 13.5 *(a) Self-oscillating profiles of optical transmittance for microgel dispersions at several temperatures. (b) Autonomously oscillating profiles of viscosity in the microgel dispersions measured at 23°C. The numbers in each oscillating profile refer to the corresponding cartoons. See plate section for colour figure.*

in waveform was observed between 26.5 and 27°C. Then the amplitude of the oscillations dramatically decreased at 27.5°C, and finally the periodic transmittance changes could no longer be observed at 28°C. The sudden change in oscillation waveform should be related to the difference in colloidal stability between the Ru(II) and Ru(III) states. Here, the microgels should be flocculated due to a lack of electrostatic repulsion when the microgels were deswollen (see Figure 13.1(e)). The remarkable change in waveform was only observed at higher dispersion concentrations (greater than 0.225 wt%). The self-oscillating property makes microgels more attractive for future developments such as microgel assembly, optical and rheological applications, and so on.

13.6.3   Viscosity Oscillation of Polymer Solution and Microgel Dispersion

In both cases of the self-oscillating polymer solution and the microgel dispersion, not only optical transmittance oscillation but also viscosity oscillation can be measured [29–31]. Especially in the case of the microgel dispersion, viscosity oscillation occurs in two different manners, exhibiting a simple pulsatile waveform or a complex waveform with two peaks per period (Figure 13.5(b)). It has been suggested that the difference in waveform is due to the difference in the oscillating manner of the microgels: swelling/deswelling or dispersing/flocculating oscillation as mentioned before. The rhythm and amplitude of the oscillation were controllable using these two microgel phenomena. Moreover, with increasing Ru(bpy)₃ and decreasing cross-linker, microgels showed a high degree of swelling/deswelling oscillation, resulting in greater amplitudes of autonomously oscillating viscosity. Recently, autonomous viscosity oscillation by reversible complex formation of terpyridine-terminated PEG and/or terpyridine-terminated Tetra PEG in the BZ reaction was achieved. The BZ reaction induces the periodical binding/dissociation of the Ru-terpyridine complex and causes periodical molecular changes to results in viscosity changes [32]. These technologies could be applied in many applications, as electro- or magnetic- rheological (ER or MR) fluids have.

13.6.4   Attempts of Self-Oscillation under Acid- and Oxidant-Free
Physiological Conditions

Thus far, the author had succeeded in developing a novel self-oscillating polymer (or gel) by utilizing the BZ reaction. However, the operating conditions for self-oscillation are limited to conditions under which the BZ reaction occurs. For practical applications as functional bio- or biomimetic materials, it is necessary to design a self-oscillating polymer which acts in biological environments. To cause self-oscillation of polymer systems under physiological conditions, BZ substrates other than organic ones, such as malonic acid and citric acid, must be built into the polymer system itself. Therefore, we took two steps; the first step was to design a novel self-oscillating polymer chain with incorporated pH-control sites, that is, a polymer chain which exhibits rhythmic oscillations in an aqueous solution containing only the two BZ substrates, without using acid as an added agent. For this purpose, AMPS was incorporated into the poly(NIPAAm-co-Ru(bpy)$_3^{2+}$) chain as the pH control site [33].

As the next step, we attempted to introduce the oxidizing agent into the polymer. Methacrylamidopropyltrimethylammonium chloride (MAPTAC), with a positively charged group, was incorporated into the poly(NIPAAm-co-Ru(bpy)$_3^{2+}$) as a capture site for an anionic oxidizing agent (bromate ion) [34]. The bromate ion was introduced into the MAPTAC-containing polymer through ion-exchange. Under the conditions in which only two BZ substrates (malonic acid and sulfuric acid) were present, soluble–insoluble self-oscillation of the polymer was observed.

Further, we synthesized a quarternary copolymer which includes both pH-control and oxidant-supplying sites in the poly(NIPAAm-co-Ru(bpy)$_3^{2+}$) chain simultaneously [35]. By using the polymer, self-oscillation in conditions where only the organic acid (malonic acid) exists was achieved. Other than malonic acid, citric acid or malic acid, which is a biorelated organic acid, can be a substrate of the BZ reaction. For practical medical use, however, it would be necessary to avoid the exchange of a toxic ion such as bromate to outside the gel. Additionally, the working temperature must be improved. The design of such a gel for biomedical application is under investigation.

13.7 Conclusion

As mentioned above, the author proposed novel chemo-mechanical systems to convert the chemical oscillation of the BZ reaction to mechanical changes of polymer and gel, and succeeded in realizing such an energy conversion system thus producing autonomous self-oscillation of the polymer gel. Here the recent progress on self-oscillating polymers and gels and the design of functional material systems have been summarized. As a potential application related to membrane science and technology, biomimetic actuators such as artificial cilia, self-driven conveyers, self-cleaning surfaces, and so on can be exemplified. Further development in many research fields is expected.

References

1. Yoshida, R. (2005) Design of functional polymer gels and their application to biomimetic materials. *Curr. Org. Chem.*, **9**, 1617–1641.
2. Ottenbrite, R.M., Park, K., Okano, T. and Peppas, N.A. (eds) (2010) *Biomedical Applications of Hydrogels Handbook*, Springer, New York.
3. Miyata, T. (2002) Stimuli-responsive polymer and gels, in *Supramolecular Design for Biological Applications* (ed. N. Yui), CRC Press, Boca Raton, pp. 191–225.
4. Osada, Y. and Khokhlov, A.R. (eds) (2002) *Polymer Gels and Networks*, Marcel Dekker, New York.
5. Yoshida, R., Takahashi, T., Yamaguchi, T. and Ichijo, H. (1996) Self-oscillating gel. *J. Am. Chem. Soc.*, **118**, 5134–5135.
6. Yoshida, R., Takahashi, T., Yamaguchi, T. and Ichijo, H. (1997) Self-oscillating gels. *Adv. Mater.*, **9**, 175–178.
7. Yoshida, R. (2008) Self-oscillating polymer and gels as novel biomimetic materials. *Bull. Chem. Soc. Jpn*, **81**, 676–688.
8. Yoshida, R., Sakai, T., Hara, Y. *et al.* (2009) Self-oscillating gel as novel biomimetic materials. *J. Controlled Release*, **140**, 186–193.

9. Yoshida, R. (2010) Self-oscillating gels driven by the Belousov-Zhabotinsky reaction as novel smart materials. *Adv. Mater.*, **22**, 3463–3483.
10. Yoshida, R., Tanaka, M., Onodera, S. *et al.* (2000) In-phase synchronization of chemical and mechanical oscillations in self-oscillating gels. *J. Phys. Chem. A*, **104**, 7549–7555.
11. Hidaka, M. and Yoshida, R. (2011) *J. Controlled Release*, **150**, 171–176.
12. Yoshida, R., Takei, K. and Yamaguchi, T. (2003) Self-beating motion of gels and modulation of oscillation rhythm synchronized with organic acid. *Macromolecules*, **36**, 1759–1761.
13. Ito, Y., Nogawa, N. and Yoshida, R. (2003) Temperature control of the Belousov-Zhabotinsky reaction using a thermo-responsive polymer. *Langmuir*, **19**, 9577–9579.
14. Shinohara, S., Seki, T., Sakai, T. *et al.* (2008) Photoregulated wormlike motion of a gel. *Angew. Chem. Int. Ed.*, **47**, 9039–9043.
15. Shinohara, S., Seki, T., Sakai, T. *et al.* (2008) Chemical and optical control of peristaltic actuator based on self-oscillating porous gel. *Chem. Commun.*, 4735–4737.
16. Maeda, S., Hara, Y., Yoshida, R. and Hashimoto, S. (2008) Peristaltic motion of polymer gels. *Angew. Chem. Int. Ed*, **47**, 6690–6693.
17. Tabata, O., Hirasawa, H., Aoki, S. *et al.* (2002) Ciliary motion actuator using self-oscillating gel. *Sensors and Actuators A*, **95**, 234–238.
18. Tabata, O., Kojima, H., Kasatani, T. *et al.* (2003) Chemo-mechanical actuator using self-oscillating gel for artificial cilia. Proceedings of the International Conference on MEMS 2003, pp. 12–15.
19. Maeda, S., Hara, Y., Sakai, T. *et al.* (2007) Self-walking gel. Adv. Mater., **19**, 3480–3484.
20. Kuksenok, O., Yashin, V.V., Kinoshita, M. *et al.* (2011) Exploiting gradients in cross-link density to control the bending and self-propelled motion of active gels. *J. Mater. Chem.*, **21**, 8360–8371.
21. Murase, Y., Maeda, S., Hashimoto, S. and Yoshida, R. (2009) Design of a mass transport surface utilizing peristaltic motion of a self-oscillating gel. *Langmuir*, **25**, 483–489.
22. Murase, Y., Hidaka, M. and Yoshida, R. (2010) Self-driven gel conveyer: Autonomous transportation by peristaltic motion of self-oscillating gel. *Sensors and Actuators B*, **149**, 272–283.
23. Tateyama, S., Shibuta, Y. and Yoshida, R. (2008) Direction control of chemical wave propagation in self-oscillating gel array. *J. Phys. Chem. B*, **112**, 1777–1782.
24. Sakai, T., Takeoka, Y., Seki, T. and Yoshida, R. (2007) Organized monolayer of thermosensitive microgel beads prepared by double-templete polymerization. *Langmuir*, **23**, 8651–8654.
25. Yoshida, R., Sakai, T., Ito, S. and Yamaguchi, T. (2002) Self-oscillation of polymer chains with rhythmical soluble-insoluble changes. *J. Am. Chem. Soc.*, **124**, 8095–8098.
26. Ito, Y., Hara, Y., Uetsuka, H. *et al.* (2006) AFM observation of immobilized self-oscillating polymer. *J. Phys. Chem. B*, **110**, 5170–5173.
27. Suzuki, D., Sakai, T. and Yoshida, R. (2008) Self-flocculating/self-dispersing oscillation of microgels. *Angew. Chem. Int. Ed.*, **47**, 917–920.
28. Suzuki, D. and Yoshida, R. (2008) Temporal control of self-oscillation for microgels by cross-linking network structure. *Macromolecules*, **41**, 5830–5838.
29. Suzuki, D., Taniguchi, H. and Yoshida, R. (2009) Autonomously oscillating viscosity in microgel dispersions. *J. Am. Chem. Soc.*, **131**, 12058–12059.

30. Taniguchi, H., Suzuki, D. and Yoshida, R. (2010) Characterization of autonomously oscillating viscosity induced by swelling/deswelling oscillation of the microgels. *J. Phys. Chem. B*, **114**, 2405–2410.
31. Y. Hara and Yoshida, R. (2008) A viscosity self-oscillation of polymer solution induced by the BZ reaction under acid-free condition. *J. Chem. Phys.*, **128**, 224904.
32. Ueno, T., Bundo, K., Akagi, Y. *et al.* (2010) Autonomous viscosity oscillation by reversible complex formation of terpyridine-terminated poly(ethylene glycol) in the BZ reaction. *Soft Matter*, **6**, 6072–6074.
33. Hara, Y. and Yoshida, R. (2005) Self-oscillation of polymer chains induced by the Belousov-Zhabotinsky reaction under acid-free conditions. *J. Phys. Chem. B*, **109**, 9451–9454.
34. Hara, Y., Sakai, T., Maeda, S. *et al.* (2005) Self-oscillating soluble-insoluble changes of polymer chain including an oxidizing agent induced by the Belousov-Zhabotinsky reaction. *J. Phys. Chem. B*, **109**, 23316–23319.
35. Hara, Y. and Yoshida, R. (2008) Self-oscillating polymer fueled by organic acid. *J. Phys. Chem. B*, **112**, 8427–8429.
36. Field, R.J. and Burger, M. (eds) (1985) *Oscillations and Traveling Waves in Chemical Systems*, John Wiley & Sons, New York.
37. Epstein, I.R. and Pojman, J.A. (1998) *An Introduction to Nonlinear Chemical Dynamics: Oscillations, Waves, Patterns, and Chaos*, Oxford University Press, New York.
38. Yashin, V.V. and Balazs, A.C. (2006) Pattern formation and shape changes in self-oscillating polymer gels. *Science*, **314**, 798–801.
39. Yashin, V.V., Kuksenok, O. and Balazs, A.C. (2010) Modeling autonomously ocillating chemo-responsive gels. *Prog. Polym. Sci.*, **35**, 155–173.

14

Electroactive Polymer Soft Material Based on Dielectric Elastomer

Liwu Liu[1], Zhen Zhang[2], Yanju Liu[1] and Jinsong Leng[2]
[1]Department of Astronautical Science and Mechanics, Harbin Institute of
Technology (HIT), People's Republic of China
[2]Centre for Composite Materials, Science Park of Harbin Institute of Technology
(HIT), People's Republic of China

14.1   Introduction to Electroactive Polymers

Researchers have long studied hard materials for wide applications in the field of mechanical engineering [1]. However, in nature, animals and plants are usually composed of soft materials. Compared with traditional hard materials, such as metals and ceramics, an electroactive polymer is a typical soft material [1, 3–5]. Under external stimuli (such as electric fields, mechanical stress, temperature, thermal fields, electromagnetic fields), soft materials can produce different degrees of deformation [1, 2].

Smart soft materials are a new type of functional material that can sense external stimuli, determine the appropriate response, and execute such a response itself [1, 3]. Compared with traditional smart hard materials, smart soft materials have the advantages of exhibiting large deformation and good biological compatibility, being light and inexpensive, and so on [2–7]. Smart materials and the smart actuators developed using such materials have great potential for application in intelligent bionic, mechanization, medication, and military fields amongst others [1–10]. Smart soft materials have demonstrated unique features in simulating biological characteristics [1–7]. They are also known as active soft materials [1, 2]. As a new representative of smart soft material, the electroactive polymer (EAP) has increasingly demonstrated its great advantages [1–20].

Researchers have studied materials that can stretch, bend, tighten, and swell as well as perform other forms of mechanical response when subjected to an external electric field

Responsive Membranes and Materials, First Edition. Edited by D. Bhattacharyya, Thomas Schäfer, S. R. Wickramasinghe and Sylvia Daunert.
© 2013 John Wiley & Sons, Ltd. Published 2013 by John Wiley & Sons, Ltd.

[6–10]. The small deformation of piezoelectric materials and ferroelectric materials limited their scope of application [2–4]. However, EAPs can produce large deformation and driving forces with the advantages of being light, inexpensive, flexible, easily processed, and so on. Thus, they are ideal materials for fabricating actuators [1–18].

14.1.1   Development History

In 1880, Roentgen discovered that under an electric field a rubber rope can deform to hold a concentrated mass in tension at its free end [6]. In 1899, Sacerdote found that under an electric field a polymer can produce a strain response [6]. In 1925, Eguchi applied DC voltage to solidified wax and rosin at low temperatures and discovered piezoelectric polymers, which marked the first step towards the development of EAP materials [6]. Because early EAP materials exhibited only small deformation strain and driving force, they were not suitable for manufacturing actuators. Another important milestone in the developmental history of electroactive polymers was Kawai's discovery of ferroelectric poly vinylidene fluorid (PVDF) in 1969 [6]. Since the 1970s, the number of different EAP materials has increased each year. Over the past decades, EAP materials have developed rapidly, leading to a series of EAP materials exhibiting excellent performance. The deformation of some EAP materials has even reached 380% [7, 8].

14.1.2   Classification

EAPs can be broadly divided into two categories based on their method of actuation: ionic and electronic [1–13]. Electronic EAPs mainly include dielectric elastomers, electrostrictive graft elastomers, ferroelectric polymers, electrostrictive paper, electro-viscoelastic elastomers, and liquid-crystal elastomers [1–10]. Ionic EAPs mainly include carbon nanotubes, conductive polymers, ionic polymer gels, electrorheological fluids, ionic polymer–metal composites, and molecular EAPs [1–10].

14.1.3   Electronic Electroactive Polymers

14.1.3.1   Dielectric Elastomers

Dielectric elastomers are basically compliant variable capacitors [1–20]. They consist of a thin elastomeric film coated on both sides with compliant electrodes [2–5]. When an electric field is applied across the electrodes, the electrostatic attraction between the opposite charges on the opposing electrodes and the repulsion of the like charges on each electrode generate stress in the film, causing it to contract in thickness and expand in area [2–13]. Silicone and acrylic are common dielectric elastomers [2–11]. Tables 14.1 and 14.2 compare the properties of silicone and acrylic, dielectric elastomers, and other materials [2–14].

14.1.3.2   Electrostrictive Graft Elastomers

Electrostriction is exhibited by graft copolymers, wherein polar crystallites are grafted to flexible polymer backbones. The polar side groups aggregate to form crystalline regions, which serve as the polarizable moieties required for actuation and as physical cross-linking sites for the flexible polymer [2, 3, 6, 10].

Table 14.1 *Properties of silicone and acrylic (data from [2–14]).*

Materials properties	Silicone	Acrylic
Maximum strain (%)	150	380
Work density (kJ/m^3)	Typical: 10; Max: 750	Typical: 150; Max: 3400
Density (kg/m^3)	1100	960
Maximum power (W/kg)	5000	3600
Continuous power (W/kg)	500	400
Bandwidth (Hz)	1400	10
Scope	>50 kHz	>50 kHz
Cycle life: strain Experiments	>10^7: 5%	>10^7: 5%
	10^6: 10%	10^6: 50%
Electromechanical coupling (%)	Max: 80; Typical: 15	Max: 90; Typical: 25
Efficiency (%)	Max: 80; Typical: 25	Max: 90; Typical: 30
Modulus (MPa)	0.1~1	1~3
Response (m/s)	<30	<55
Thermal expansion (mm/°C)		1.8×10^{-4}
Voltage (V)	>1000	>1000
Max electric field (MV/m)	110~350	125~440
Dielectric constant	~3	~4.8
Temperature (°C)	−100 to 250	−10 to 90

14.1.3.3 Ferroelectric Polymers

Ferroelectric polymers have a non-centro-symmetric structure that exhibits permanent electric polarization. These materials possess dipoles that can be aligned with an electric field and maintain their polarization. Ferroelectric polymers have high Young's moduli and mechanical energy densities and can be applied in air, vacuum, or water [2,6,8].

14.1.3.4 Electro-Viscoelastic Elastomers

Electro-viscoelastic elastomers include silicone, synthetic rubber, and electrodes. Before cross-linking, electro-viscoelastic elastomers are in a viscous flow state. When subjected to an electric field, these materials enter a solid state with a shear modulus that varies with the applied electric field [2–5].

Table 14.2 *Properties of piezoelectric materials, and natural muscles (data from [2–14]).*

Materials properties	Maximum strain (%)	Maximum stress (MPa)	Specific elastic energy density (J/g)	Elastic energy density (J/cm^3)	Maximum efficiency (%)	Relative speed
Piezoelectric ceramic	0.2	110	0.013	0.10	90	Fast
Piezoelectric single crystal	1.7	131	0.13	1.0	90	Fast
Ferroelectric polymers	0.1	4.8	0.0013	0.0024	—	Fast
Natural muscle	>40	0.35	0.07	0.07	>35	Medium

14.1.4 Ionic Electroactive Polymers

14.1.4.1 Carbon Nanotubes

Carbon nanotubes consist of carbon atoms similar to those in the diamond structure and have the advantages of high mechanical strength, large displacement, high mechanical pressure, high thermal stability, and so on [2–5, 12].

14.1.4.2 Ionic Polymer Gels

Ionic polymer gels consist of a cross-linked polymer, typically a polyacrylic gel acid, in an electrolyte solution. The application of an electric field to the gel causes hydrogen ions to migrate out of or into the gel, resulting in a change in pH. This change results in a reversible shift between the material's swollen and contracted states [2–5].

14.1.4.3 Electrorheological Fluids

Electrorheological fluids are a type of suspension formed by dielectric particles suspended in a carrier solution [2–4, 12]. Under an electric field, the polarized suspended particles will form chain-like structures along the electric field direction, resulting in a transformation of the rheological properties of the fluid. Moreover, the free-flowing liquid is transformed into a gelatinous solid. Electrorheological fluids can be applied in components such as vibration isolators, shock pads, and valves [2–5, 11,12].

Electroactive polymers (EAPs) are an emerging type of actuator technology wherein a lightweight polymer responds to an electric field by generating mechanical actuation. Their ability to mimic the properties of natural muscle has garnered them the moniker "artificial muscle", although the term electroactive polymer artificial muscle (EPAM) is more appropriate and descriptive. Table 14.3 compares the properties of EAPs with those of other materials [2–14].

14.1.5 Electroactive Polymer Applications

The light weight and good flexibility of EAP materials make these materials widely applicable in aerospace and other fields. The Defense Advanced Research Projects Agency (DARPA), the National Aeronautics and Space Administration (NASA), the European Space Agency (ESA), and other relevant agencies have shown great interest in EAP materials and have actively carried out extensive research regarding their applications. In 1995, NASA carried out a research programme called Low Mass Muscle Actuators led by Dr Yoseph Bar-Cohen's research group at NASA's Jet Propulsion Laboratory (JPL) to carry out research on EAPs [10–28]. The EAP actuators developed through this programme can be used to drive a robotic arm. Figure 14.1 shows an arm wrestling match between the EAP-based artificial arm and a human arm, which demonstrates the great potential of EAP in the field of biomimetic materials [10, 18].

14.1.6 Application of Dielectric Elastomers

14.1.6.1 Walking Robot

Some researchers wish to create robot insects (insects machines) that can access hidden areas of structure (such as aircraft engines) to perform tasks such as surveillance and

Table 14.3   *Properties of EAPs and other materials (data from [2–14]).*

Materials	Typical materials	Advantages	Disadvantages
Natural muscle	Mammalian muscle	Medium stress (350 kPa) Maximum strain (20%) High elastic energy density (20–40 MJ/kg) Efficiency (40%) Recycled	Non-engineering materials Low temperature range
Ferroelectric polymer	Polyvinylidene fluoride	High stress (45 MPa) Medium strain (<7%) High elastic energy density (1 MJ/m3) High electromechanical coupling efficiency Young's modulus (400 MPa) Low current	High voltage (>1 kV) High electric field (150 MV/m) Low fatigue life
Crystal elastomer		High strain (45%) High electromechanical coupling (75%)	High electric field (1–25 MV/m) Need to package
Conductive polymer	Polypyrrole	High stress (Maximum 34 MPa, Medium 5 MPa) Medium strain (2%) High elastic energy density (100 kJ/m3) Young's modulus (1 GPa) Low voltage (2 V)	Low electromechanical coupling Low response Need to package
Molecular EAP		High stress (>1 MPa) High strain (20%) High elastic energy density (100 kJ/m3) Low voltage (2 V)	Low response Need to package
Carbon nanotube	SWNT MWNT	High stress (>10 MPa) Low voltage (2 V) Large temperature range	Small strain (Typical 0.2%) Low electromechanical coupling High price
IPMC	Nafion Flemion	Low voltage (<10 V) Large displacement	Low electromechanical coupling Need to package

maintenance [10, 11]. In addition, walking robots would play an important role in space, aerospace, and other specialized fields. Such machines could replace humans to perform certain tests, repair equipment, collect and transfer information, and detect the conditions of the surrounding environment. Currently, dielectric elastomers, which are similar to natural muscle, have been applied in space robots. Figures 14.2, 14.3, and 14.4 demonstrate the

Figure 14.1   *Arm wrestling match between artificial arm and human arm [10]. (Reprinted with permission from [10] Copyright (2004/2005) SPIE).*

applications of dielectric elastomers as biomimetic robot-driven components in a walking robot, creeping robot, and insect-mimicking robot [9, 10].

14.1.6.2   Energy-Harvesting Device

DARPA developed an energy-harvesting device based on dielectric elastomers to collect the energy generated by soldiers as they walk, as is shown in Figure 14.5 [6, 10]. According to the research, the amount of power consumed when an arm swings is 0.2–3 W; 2–20 W is consumed when a heel impacts a shoe; and the sudden movement of hands or legs consumes 10–100 W. A 50-g dielectric elastomer can be used to generate 5 J of energy; to

Figure 14.2   *Leg-like robot [10]. (Reprinted with permission from [10] Copyright (2004) SPIE).*

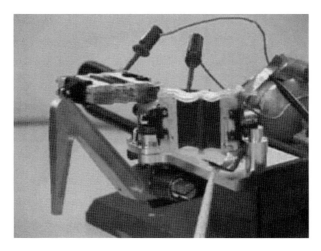

Figure 14.3 *Creeping robot [10]. (Reprinted with permission from [10] Copyright (2004) SPIE).*

output the same energy using electromagnetic or piezoelectric materials would require 10 times as much material mass. Figures 14.6 and 14.7 illustrate a seawave energy harvester [6, 10, 19, 85].

14.1.6.3 Micro Unmanned Aerial Vehicles

DARPA developed a micro unmanned aerial vehicle (UAV) with detection capabilities using EAP materials. Its total weight is 550 g; the weight of its motor is 140 g, and the weight of its fuel is 75 g. The power required to run the vehicle is 98 W; the vehicle's hovering time is up to 8 minutes, and its load capacity can be up to 30–70 g. This micro UAV can hover in the air continuously and perform reconnaissance exploration [6–14].

Figure 14.4 *Insect-mimicking robot [10]. (Reprinted with permission from [10] Copyright (2004) SPIE).*

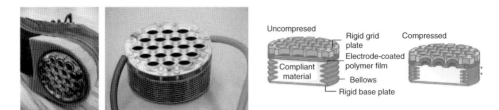

Figure 14.5   *Energy harvester based on dielectric elastomer [6, 10]. (Reprinted with permission from [6, 10] Copyright (2002/2004) SPIE).*

14.1.6.4   Clamping Device Using Ionic Polymer–Metal Composites

Ionic polymer–metal composites (IPMCs) consist of a solvent swollen ion-exchange polymer membrane laminated between two thin flexible metal (typically percolated Pt nanoparticles or Au) or carbon-based electrodes. Typical IPMC membrane materials include Nafion and Flemion [2–11]. IPMC materials possess good properties such as light weight, flexibility, large deformation, and the ability to be used in wet conditions.

Stretching- and bending-type actuators based on electroactive polymers have been developed for aerospace applications. NASA has explored some devices with IPMC materials, such as a dust wiper, robot arm, and multi-finger gripper. Figure 14.8 shows a dust wiper that is intended to become part of a nanorover [10, 20]. The 0.104-g blade is constructed of a graphite/polyimide beam with a gold-coated fibreglass brush. The IPMC wiper is driven by approximately 2 to 3 V, and dust is repelled at a DC voltage of approximately 1.5 KV.

Figure 14.6   *Dielectric elastomer seawave energy harvester [19]. (Copyright (2008) Mikio Ward).*

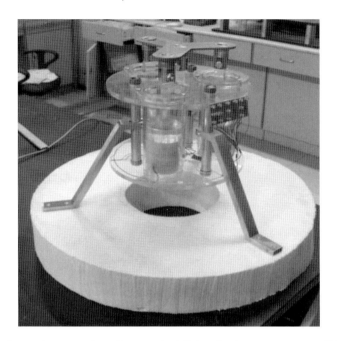

Figure 14.7   *Energy harvester based on stacked dielectric elastomer actuator [85]. (Reprinted with permission from [85] Copyright (2010) IOP Publishing).*

A group at the Osaka National Research Institute and Japan Chemical Innovation Institute has demonstrated a biomimetic device that acts like a small hand using a polymer electrolyte gold composite, also considered an IPMC material. The device has eight fingers, each 15 mm long, made of a single sheet of a perfluorocarboxylic acid membrane plated with gold on both sides (Figure 14.9) [10, 21].

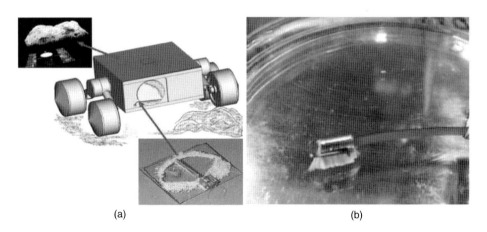

     (a)   (b)

Figure 14.8   *A view of the dust wiper activated with a high voltage to repel dust [10]. (Reprinted with permission from [10] Copyright (2004) SPIE).*

Figure 14.9   *Multi-finger gripper using IPMC strips [21]. (Copyright (2000) Osaka National Research Institute (ONRI)).*

14.1.6.5   Anti-Gravity Service Using Conductive Polymers

The molecular chain of this polymer features a π-conjugated architecture. Conductive polymers can be insulators, semiconductors, or conductors under certain conditions achieved by chemical or electrochemical doping. Conductive polymers have several advantages including low drive voltage, large power density, long response time, long fatigue life, large stress, and low contraction rate [2–10].

Madden *et al.* have developed an anti-gravity service that can significantly enhance the running and jumping ability of soldiers using an electroluminescent polymer called PPy (polypyrrole).

14.1.6.6   Facial Expression

In November 2005, at the APEC held in Busan, South Korea, the Albert Einstein robot designed and manufactured by Hanson Robotics demonstrated the potential application of a dielectric elastomer in simulating facial expressions. It can express different emotions such as happiness and distress (Figure 14.10) [9, 10].

A fish-like airship propelled by dielectric elastomer (DE) actuators was first developed at Empa [10, 20]. This airship is 8 metres long and is filled with pressurized helium. A total of four DE actuators are arranged on both sides of the airship's body and the tail. The DE actuators on the airship function in an agonist–antagonist configuration and act as biological muscles. When the DE actuators are actuated by driving voltages, the actuator on one side elongates while the actuator on the other side shrinks. Thus, the airship performs wave-like movements through the air and can swim freely like a fish (Figure 14.11).

14.1.6.7   Braille Tactual Displays

Traditional, reusable Braille equipment can be made of piezoelectric elements. However, such devices feature a complex structural design and expensive piezoelectric elements that most blind persons cannot afford.

As Figure 14.12 shows, Jingsong Leng's research group at the Harbin Institute of Technology have designed and manufactured a kind of Braille tactual display based on dielectric elastomer actuators [22]. It consists of a controlling component, an actuator, a mechanical drive, a touch screen, and a shell. When users input English letters or words through the man–machine interface, the program converts the letters to electrical signals according to

Figure 14.10   *Robot Albert–Huber [10]. (Reprinted with permission from [10] Copyright (2004) SPIE).*

the corresponding Braille representation and then outputs a signal via a single-chip microprocessor to control the cut-off of six relays; in this way, the deformation of the dielectric actuators by the mechanical drive structure to which they are linked can be used to display information on the touch screen (according to the Braille code).

Moreover, Qibing Pei's research group at the University of California, Los Angeles has tried to manufacture Braille displays using interpenetrating networks of dielectric elastomer actuators, as shown in Figure 14.13 [10, 22]. The equipment's structure is relatively simple

Figure 14.11   *Fish-like airship propelled by dielectric elastomer (DE) actuators [10]. (Reprinted with permission from [10] Copyright (2004) SPIE).*

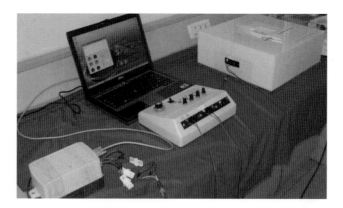

Figure 14.12 *Braille tactual displays using dielectric elastomers [22]. (Reprinted with permission from [22] Copyright (2010) SPIE).*

and inexpensive. If the related technical difficulties can be solved, it will provide great opportunities to blind persons.

14.1.6.8   Loudspeaker

Sound is produced by the vibration of surrounding air; a different frequency of vibration will produce a different perceived pitch [10]. Dielectric elastomers respond very quickly to electricity, on the order of milliseconds, and are very sensitive to a wide range of frequencies (0–50 KHz) beyond that perceived by the human ear. To exploit this property, dielectric actuators can be used to make a loudspeaker. The basic structure and working principle of such a device is shown in Figure 14.14. When a high-frequency partial voltage is applied, the loudspeaker will vibrate up and down or from left to right at a certain frequency to produce sound. However, the anamorphosis of the human voice during transmission is an unavoidable drawback of dielectric elastomer-based loudspeakers. Nevertheless, there are some ways of avoiding large anamorphosis, such as modifying the non-linear part of the sound in the drive electrocircuit.

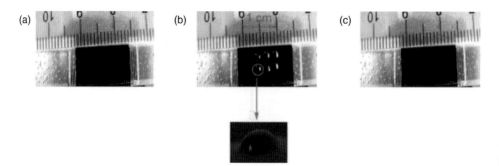

Figure 14.13 *Braille cell display developed using dielectric elastomers [10, 22]. (Reprinted with permission from [10, 22] Copyright (2004/2010) SPIE).*

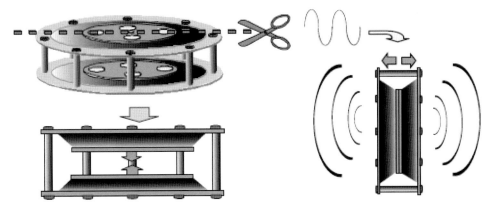

Figure 14.14 *Basic structure and working principle of the loudspeaker [10]. (Reprinted with permission from [10] Copyright (2004) SPIE).*

Sound-production equipment based on electric elastomers can be installed on the shell or the top of a car to insulate noise and monitor the vibration of the car itself. The vibration of the car and the noise outside change the electrical signal sent through vibration-detecting equipment and a microphone, which then sends a signal to the sound-production equipment to weaken the electrical signal (the same vibration equipment). In this way, the sound insulation and vibration measurements can be obtained. Experimental results show that the equipment installed on the top of a car can decrease the ambient noise by more than 50 decibels at low frequencies (100–300 Hz) [10].

14.1.7 Manufacturing the Main Structure of Actuators Using EAP Materials

14.1.7.1 Stacked Structure and the Helical Structure

Figure 14.15 shows that a stacked-structure actuator is a composite structure composed of multiple dielectric elastomer layers and compliant electrodes [10, 28]. The production

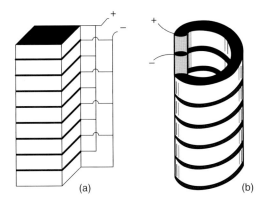

(a)   (b)

Figure 14.15 *Actuators in stacked and helical conformations [10, 28]. (Reprinted with permission from [10, 28] Copyright (2004/2007) SPIE).*

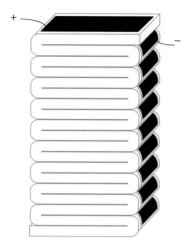

Figure 14.16　*Foldable actuator [29]. (Reprinted with permission from [29] Copyright (2007) SPIE).*

process is as follows. Starting with the electrode face of the dielectric elastomer thin film, the various layers are folded, which can then contract upon the application of electricity. Compared with traditional actuators, these devices have the advantage of possessing a stable, continuous structure and a single phase [10, 28].

A helical dielectric elastomer actuator features two layers of helical compliant electrodes and a dielectric elastomer thin film. The electrostriction effect makes the entire structure contract along the axis direction.

Manufacturing the special structure of each actuator is difficult. To manufacture the stacked-structure actuator, for example, several layers of the dielectric elastomer must be deposited, and the electrical layer and material layer must be configured to avoid a short circuit of the electrodes. The helical actuator has a complex shape and manufacturing process; it also suffers from limited reliability. However, simple equipment based on this kind of actuator can be produced and applied over short periods [10, 28].

14.1.7.2　The Folding Structure

The folding-structure dielectric elastomer actuator avoids the complex manufacturing process associated with stacked and helical actuators [10, 29, 30]. While they function in a similar way to a multilayer stacked actuator, these devices are easy to fabricate. These folding structures require only the folding of a single layer of dielectric elastomer coated with electrodes to produce an actuator such as that shown in Figure 14.16 [10, 29, 30]. Under an applied electrical field, the actuator will be deformed, leading to the crosswise expansion and principal-axis contraction of the equipment. The folding-structure actuator possesses stable properties, is able to produce a large actuation force, features high electromechanical transition efficiency, and is easy to manufacture.

14.1.7.3　The Round-Hole Structure

The round-hole actuator structure is composed of single layers of round-hole dielectric elastomer stacked together with compliant electrodes (Figure 14.17) [10, 31]. This kind

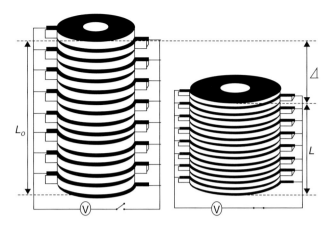

Figure 14.17 *Round-hole actuator [10, 31]. (Reprinted with permission from [10, 31] Copyright (2004/2007) SPIE).*

of actuator produces large strains and actuation forces and can be conveniently applied. The contraction strain can reach 10–15% of the original actuator size. Moreover, the mechanical and electrical properties of the device can be adjusted according to the practical requirements.

Figure 14.18 shows other dielectric elastomer structures actuators, such as roll- and diamond-shaped actuators [10].

14.1.8   The Current Problem for EAP Materials and their Prospects

Although EAP materials have developed rapidly in recent years, the actuation forces they produce are not large enough, and they possess low mechanical energy density and exhibit low response speed. Moreover, effective and durable EAP material cannot be produced in large quantities. Thus, there are still many challenges facing the practical application of EAP materials [1–35].

EAP materials are intelligent materials with great developmental potential. They possess many advantages, such as good elastic recovery, good strength against rupturing, non-noise actuation, low price, low work consumption, and high energy density [1–17]. From the mechanical fish and gas-filled pads that appeared as the earliest applications of EAP materials, to later applications such as facial expression, robots, mechanical insects, artificial muscles, noise control, tactile displays, and others, EAP materials possess great application potential in the aerospace and aviation fields [15–22]. The properties of EAP materials must be better characterized, and the actuation force and durability of these materials must be improved to enhance their reliability [25–36]. With greater international cooperation and research in China, new effective EAP materials can be developed, thus further enhancing the application potential of EAPs.

In order to catch the huge opportunity brought by electrical elastomer soft materials, to meet the challenge and to better guide the design and manufacture of dielectric elastomer composited materials and actuators with excellent properties at the same time, we need to build the constitutive relations of dielectric elastomer in theory, investigate

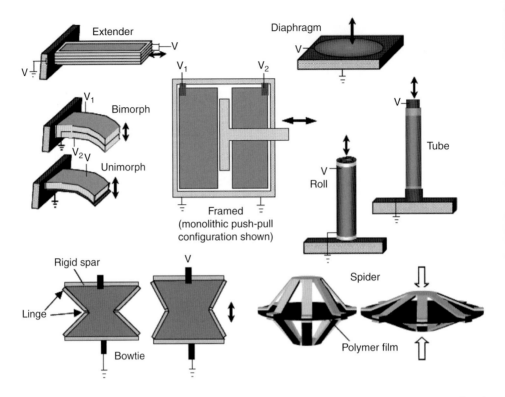

Figure 14.18 *Examples of dielectric elastomer actuator configurations [10]. (Reprinted with permission from [10] Copyright (2004) SPIE).*

large deformation and stability of dielectric elastomers, describe the allowable area of dielectric elastomers working as transducers, and further evaluate and predict the properties of transducers with different structures. The sections below will mainly deal with these themes.

14.2   Materials of Dielectric Elastomers

14.2.1   The Working Principle of Dielectric Elastomers

Dielectric elastomers are also called electrically active polymer artificial muscles, and were developed by the non-profit international investigation organization SRI International at the end of 1991. These materials possess the advantages of super deformation [1–5] (380%), high elastic energy density [6–10] (3.4 J/g), high efficiency, short response time, long life span, and high circulation times, which make them capable of being incorporated into light, tiny, and high-precision actuators as a kind of intelligent material with great developmental potential.

The basic principle [11–18] of creating such devices is to cover the two faces of compliant electrodes with an elastomer; thus, when a voltage is applied to the device, the elastomer

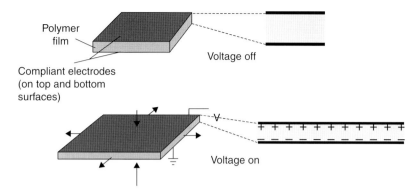

Figure 14.19 *Operating principle of the dielectric elastomer actuator [3]. (Reprinted under the terms of the STM agreement from [3] Copyright (2010) Wiley-VCH Verlag GmbH & Co. KGaA).*

will change its thickness and area: decrease in thickness and expand in area, as shown in Figure 14.19.

Dielectric elastomers are intelligent materials with great developmental potential. They have the advantages of large deformation, fine elastic recovery, good flexibility, good strength against rupturing, quick response, low price, low work consumption, high energy density, high electromechanical transition efficiency, light weight, ease of fabrication, and low noise [1–10]. Due to these excellent properties, EAP materials are already being used in artificial muscles, facial-expression devices, actuators, energy harvesters, tiny robots, transducers, mechanical insects, liquid-flow control, and Braille display equipment. Thus, these materials have great application potential in intelligent bionics, biological medicine, and mechanical and aerospace engineering [11–22].

14.2.2   Material Modification of Dielectric Elastomer

14.2.2.1   Introduction of Dielectric Elastomer Materials

From late 1990 to early 2000, large numbers of soft elastomer materials, including silicone, thermoplastic urethanes (TPUs), and acrylics, were developed. The performance properties of some these materials are listed in Table 14.4, where 3M's 4910 acrylic is observed to perform the best with respect to strain properties [3, 5–10]. 3M's 4910 acrylic with high pre-strain has a maximum area strain of 380%; its maximum elastic strain energy density can reach up to 3.4 J/cm^3, and its dielectric constant is 4.2–4.7. However, its practical application is greatly limited by its viscoelastic properties and glass-transition temperature (about $-20°$C). Polyurethane (PU) films have greater output forces and higher dielectric constants and can function normally under a weaker electric field [3, 7–13]. However, its disadvantage of producing low strain makes these films less popular amongst researchers.

Dielectric elastomer materials made of silicone have many advantages [3–22]. For example, a dielectric elastomer with 70% pre-strain could have an area strain of more than 100% [22–30]. Although its area strain is much smaller than that of the acrylic, its weak viscoelastic properties and low glass-transition temperature (about $-100°$C) provide a wider range of application [10–35].

Table 14.4 *Comparison of the properties of different dielectric elastomer materials [3, 10]. (Reprinted with permission from [3] Copyright (2010) Wiley-VCH Verlag GmbH & Co. KGaA; and from [10] Copyright (2004) SPIE).*

Polymer type	Pre-strain (x%, y%)	Energy density J/cm^3	Driving force MPa	Area strain %	Young modulus MPa	Electric field $V/\mu m$	Relative dielectric constant
Silicone rubber (Nusil CF19-2186)	—	0.22	1.36	—	1	235	2.8
Silicone rubber (Nusil CF19-2186)	(45,45)	0.75	3	64	1.0	350	2.8
Silicone rubber (Dow Corning HS3)	—	0.026	0.13	—	0.135	72	2.8
CAN rubber	(60,60)	0.084	0.3	—	4	50	14
Silicone rubber (Dow Corning HS3)	(280,0)	0.16	0.4	117	—	128	2.8
Silicone rubber (Dow Corning Sylgard 186)	—	0.082	0.51	—	0.7	144	2.8
Polyurethane (PT6100S)	—	0.087	1.6	—	17	160	7
Fluorosilicone (Dow Corning 730)	—	0.0055	0.39	—	0.5	80	6.9
Acrylic acid (3M VHB 4910)	(300,300)	3.4	7.2	380	3.0	412	4.8
Acrylic acid (3M VHB 4910)	(540,75)	1.36	2.4	215	—	239	4.8
SEBS161(5–30wt.% copolymer)	(300,300)	0.141–0.151	—	180–30	0.007–0.163	32–133	1.8–2.2
SEBS217(5–30wt.% copolymer)	(300,300)	0.119–0.139	—	245–47	0.002–0.133	22–98	1.8–2.2
IPN(VHB4910-HDDA)	(0,0)	—	—	233	2.5	300	—
IPN(VHB4905-TMPTMA)	(0,0)	0.68	1.51	146	3.94	265.4	2.43
IPN(VHB4910-TMPTMA)	(0,0)	3.5	5.06	300	4.15	418.05	3.27

14.2.2.2   *Impact of Pre-Stretching on Dielectric Elastomer Material*

Pre-stretching could greatly improve the performance of electroactive deformation [1–3, 10]. The maximum strain of dielectric elastomers cannot exceed 20% if the materials are not pre-stretched [3]. Pre-stretching can effectively improve the dielectric strength of dielectric elastomers; moreover, it can also reduce the thickness of dielectric elastomer films and further reduce the driving voltage required [3]. The mechanism behind why pre-stretching improves the performance of dielectric elastomers is not yet clear [3]. Generally, some local defects will inevitably be generated during the production of dielectric elastomer materials [3]. These defects could decrease the breakdown strength of dielectric elastomer materials and further result in premature rupture. Pre-stretching could allow the polymer chains to be arranged perpendicular to the applied electric field, thereby preventing the early

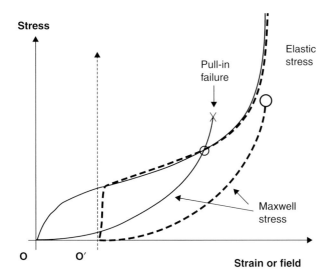

Figure 14.20 *Characteristic stress of a DE film as a function of mechanical strain or electric field [3]. (Reprinted under the terms of the STM agreement from [3] Copyright (2010) Wiley-VCH Verlag GmbH & Co. KGaA).*

rupture of the material [3]. Figure 14.20 shows the stress in a given DE film as a function of mechanical strain or electric field [3]. The stress in the film without pre-stretching increases greatly at the beginning of mechanical stretching, then increases slowly, and finally increases suddenly until the rupture of the material [3]. The electric force-strain curve of the material follows a quadratic function. The thickness of the dielectric elastomer film will become thinner under the effect of an applied voltage and then result in an increase in the electric field [3]. The increase in the electric field will cause the thickness of the dielectric elastomer film to decrease further. Thus, positive feedback is exhibited until the electric field strength exceeds the breakdown strength of the material. This pull-in effect will rupture the dielectric elastomer film. The electric force-strain curve of the pre-stretched film was delayed from 0 to 0′; thus, the possibility of the intersection of the electric force-strain curves and mechanical stress-strain curve was reduced and the pull-in effect was reduced; additionally, the operating range of the dielectric elastomer became more stable. Experiments show that pre-stretching is very effective for acrylic films and also has a positive effect on silicone rubber and other elastomer films [3].

Pre-strain can also improve the response speed of dielectric elastomer films [2, 3, 6, 10, 17]. Experiments show that the decrease in electroactive strain with frequency rarely occurs in pre-stretched acrylic films [3]. Pre-strain could increase the elastic modulus of dielectric elastomer films and weaken the films' viscoelastic properties. Pre-stretched films have drawbacks as well: they require a rigid framework or other support structure to maintain tension. This increases the weight of dielectric elastomer equipment and limits the effective work density of dielectric elastomer drivers. Moreover, pre-stretched dielectric elastomer films will experience stress relaxation and fatigue over time, which could decrease their working life times.

14.2.3   Dielectric Elastomer Composite

Dielectric elastomer material is a type of smart material with great developmental potential. Under an applied electric field, it can exhibit large deformation, high elastic energy density, light weight, short response time, and good flexibility [1–10]. Additionally, this material can be driven silently, is inexpensive, and features a high efficiency of mechanical and electrical conversion. However, the driving electric field required to exploit dielectric elastomers is very high, which greatly affects the practical application of these polymer materials [1–5, 10, 14, 20–36]. Therefore, determining how to reduce the required driving voltage is an urgent problem to solve. Some researchers believe that we can effectively produce dielectric elastomer composite materials to meet actual application requirements.

Currently, research on dielectric elastomer composites is mainly divided according to the two types of materials studied: the first is particle-filled composites created by physical blending, in which the main matrix material is silicone rubber [3, 36, 37, 39, 40]; the other is IPNs (interpenetrating polymer network) created by chemical cross-linking, in which the main matrix material is acrylic [3, 38, 41].

14.2.3.1   Particle-Filled Dielectric Elastomer Composite

Gallone *et al.* studied the dielectric properties and mechanical properties of an acid lead magnesium niobium (PMN-PT)-filled silicone elastomer composite (Figure 14.21) [36]. Up to 30 vol% Nb lead acid magnesium particles can be added to this type of composite material. Both the dielectric constant and the dielectric loss increase to different degrees as the PMN-PT particle content increases, and the dielectric constant of the composite reaches 32 when the frequency is 10. Meanwhile, the elastic modulus increases as the PMN-PT particle content also increases; the elastic modulus of a composite featuring 30% PMN-PT particles is twice that of pure silicone rubber material, and the composite material still shows very good extension properties.

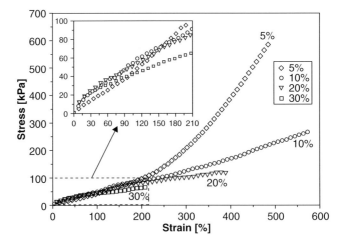

Figure 14.21   *Nominal stress–nominal strain curves for silicone/PMN–PT composites at various filler volume fractions [36]. (Reprinted under the terms of the STM agreement from [36] Copyright (2007) Elsevier Ltd).*

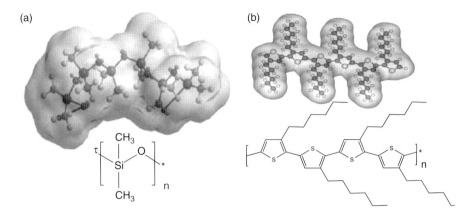

Figure 14.22 *Chemical structure of (a) a generic poly(dimethylsiloxane) rubber and (b) poly(3-hexylthiophene) [37]. (Reprinted under the terms of the STM agreement from [37] Copyright (2008) Wiley-VCH Verlag GmbH & Co. KGaA).*

Carpi *et al.* studied a blend composite made of elastic silicone rubber and high-polarity conjugated polymers (PHT) (as in Figure 14.22) [37]. Compared with other particle-filled polymers, the material has both a good dielectric constant and low elastic modulus. When the content of high-polarity conjugated polymers was very low (1–6 wt.%), the relative dielectric constant of the composite materials greatly improved, while the dielectric loss remained at a relatively low level; meanwhile, the elastic modulus of the composite decreased. When the PHT content reaches 1 wt.%, composite materials achieve the best performance with respect to electroactive deformation and attain an area strain of 7.6% when the electric field is 8 V/um.

14.2.3.2   *Dielectric Elastomer of Interpenetrating Polymer Network (IPN)*

Traditional acrylic film, such as 3M's 4910 series tape, has excellent activity strain, high energy density and efficiency as a dielectric elastomer [38,41,42]. However, to obtain such high performance, the acrylic film must be pre strained. Moreover, the additional rigid framework that supports the pre-strain makes the weight of dielectric elastomer devices increase considerably, thereby reducing the effective energy density of acrylic materials. Additionally, the viscoelastic properties of the acrylic tape have a significant impact on the frequency response of the material [38,41,42].

To eliminate the impact of the rigid framework of acrylic materials, Ha *et al.* studied the dielectric elastomer IPN (interpenetrating polymer networks, IPNs) [38,41]. The polymers included 1,6-hexanediol diacrylate (HDDA) or trifunctional trimethylolpropane trimethacrylate (TMPTMA) monomers as additives. Figure 14.23 shows how IPNs are prepared. First, the VHB film is pre-stretched. Then, HDDA or TMPTMA solvent is sprayed on the VHB film, allowing a second network structure to be formed in the VHB film after the polymerization of the solvent. When the VHB film is removed from the rigid framework, the second network structure will still keep it in the pre-stretched state.

Compared with ordinary VHB acrylic acid thin films, INPs have lower viscoelasticity but higher mechanical stability [38,41]. Without pre-stretching, the thickness strain of TMPTMA-based IPNs can reach 75%, with an energy density of 3.5 MJ/m^3, coupling efficiency of 94%, and a breakdown electrical field of 420 MJ/m^3 [38,41].

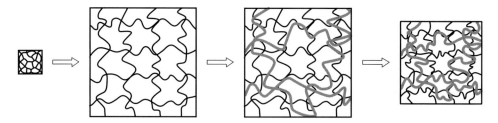

Figure 14.23 *Fabrication steps for preparing IPN elastomer films [38]. (Reprinted with permission from [38] Copyright (2007) IOP Publishing).*

Figure 14.24 shows a VHB-based IPN polymer that uses HDDA (a, b) and TMPTMA (c, d) as an additive before and after electrical deformation [38, 41, 42].

14.3   The Theory of Dielectric Elastomers

14.3.1   Free Energy of Dielectric Elastomer Electromechanical Coupling System

A schematic representation of a dielectric elastomer is shown in Figure 14.25 [1, 43, 44]. Generally, because dielectric elastomers must operate under a certain electric field, a pair of compliant electrodes made of carbon grease is attached to both surfaces of a thin dielectric

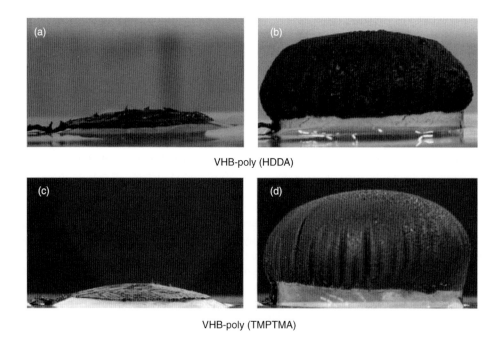

Figure 14.24 *VHB-based IPN films with HDDA (a and b) and TMPTMA (c and d) as additives before and after actuation [38]. (Reprinted with permission from [38] Copyright (2007) IOP Publishing).*

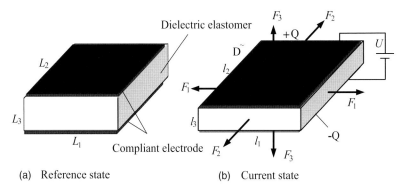

Figure 14.25   *A membrane of a dielectric elastomer is sandwiched between two compliant electrodes. (a) In the reference state, the dielectric is subject to neither forces nor voltage. (b) In the current state, subject to forces and a voltage, the membrane deforms and charge flows from one electrode to the other through the external conducting wire [43]. (Reprinted with permission from [43] Copyright (2011) Taylor and Francis).*

elastomer film. Both the mechanical stiffness and electrical resistance can be neglected. As shown in Figure 14.25, in the reference state, the elastomer has initial dimensions of L_1, L_2 , and L_3 and is subject to zero force and voltage. In the current state, with the applied forces F_1, F_2 , and F_3, and voltage U, the elastomer's dimensions deform to l_1, l_2, and l_3 and the accumulated charges generated on the electrodes are $+Q$ and $-Q$ [1,43,44].

As the dimensions of the elastomer change by $\delta l_1, \delta l_2$, and δl_3, the work performed by the forces is $F_1 \delta l_1 + F_2 \delta l_2 + F_3 \delta l_3$. Meanwhile, the work performed by the voltage is $U \delta Q$, assuming that an amount of charge δQ flows through the thin film. When the dielectric elastomer is in an equilibrium state with respect to the mechanical forces and electrical voltage, the increase in the free energy should be equal to the total work [1,43,44]:

$$\delta H = F_1 \delta l_1 + F_2 \delta l_2 + F_3 \delta l_3 + U \delta Q \tag{14.1}$$

where H is the Helmholtz free energy of such a thermodynamic system. In a Cartesian coordinate system, the elastomer stretches in three directions: $\lambda_1 = l_1/L_1$, $\lambda_2 = l_2/L_2$, and $\lambda_3 = l_3/L_3$. The nominal stresses λ_C, s_2, and s_3 are defined by dividing the pre-stretch forces by the area before deformation, that is, $s_1 = F_1/(l_2 l_3)$, $s_2 = F_2/(l_1 l_3)$, and $s_3 = F_3/(l_1 l_2)$. Similarly, the nominal electric field can be defined as $E^\sim = U/L_3$ and the nominal electrical displacement as $D^\sim = Q/(L_1 L_2)$. The nominal density of the Helmholtz free energy is defined as $W = H/(L_1 L_2 L_3)$. Furthermore, the true stresses are defined as $\sigma_1 = F_1/(\lambda_2 \lambda_3 L_2 L_3), \sigma_2 = F_2/(\lambda_1 \lambda_3 L_1 L_3), \sigma_3 = F_3/(\lambda_1 \lambda_2 L_1 L_2)$, the true electric field as $E = U/(\lambda_3 L_3)$, and the true electric displacement as $D = Q/(\lambda_1 \lambda_2 L_1 L_2)$ [1,43,44].

Equilibrium (14.1) holds in any current state. Based on the above definitions, a small change in the Helmholtz free energy density is expressed as [1,43,44]

$$\delta W = s_1 \delta \lambda_1 + s_2 \delta \lambda_2 + s_3 \delta \lambda_3 + E^\sim \delta D^\sim \tag{14.2}$$

This equilibrium condition will hold for arbitrarily small variations in the four independent variables, $\lambda_1, \lambda_2, \lambda_3$, and $D^\sim$. Neglecting the variation in temperature, to characterize a dielectric elastomer, the nominal density of the Helmholtz free energy is described as a function of the four independent variables [1,43,44]:

$$W = W(\lambda_1, \lambda_2, \lambda_3, D^\sim) \tag{14.3}$$

Under the coupling effect of the electric and mechanical fields, the small variations in the four independent variables are $d\lambda_1, d\lambda_2, d\lambda_3$, and $dD^\sim$. The variation in the free energy of a dielectric elastomer's electromechanical coupling system can be expressed as follows [1,43,44]:

$$\delta W = \frac{\partial W(\lambda_1, \lambda_2, \lambda_3, D^\sim)}{\partial \lambda_1}\delta\lambda_1 + \frac{\partial W(\lambda_1, \lambda_2, \lambda_3, D^\sim)}{\partial \lambda_2}\delta\lambda_2$$
$$+ \frac{\partial W(\lambda_1, \lambda_2, \lambda_3, D^\sim)}{\partial \lambda_3}\delta\lambda_3 + \frac{\partial(\lambda_1, \lambda_2, \lambda_3, D^\sim)}{\partial D^\sim}\delta D^\sim \tag{14.4}$$

Substituting Equation 14.4 into Equation 14.2, we can obtain a thermodynamic equilibrium equation of the following form [1,43,44]:

$$\left[\frac{\partial W(\lambda_1, \lambda_2, \lambda_3, D^\sim)}{\partial \lambda_1} - s_1\right]\delta\lambda_1 + \left[\frac{\partial W(\lambda_1, \lambda_2, \lambda_3, D^\sim)}{\partial \lambda_2} - s_2\right]$$
$$\times \delta\lambda_2 + \left[\frac{\partial W(\lambda_1, \lambda_2, \lambda_3, D^\sim)}{\partial \lambda_3} - s_3\right]\delta\lambda_3 + \left[\frac{\partial(\lambda_1, \lambda_2, \lambda_3, D^\sim)}{\partial D^\sim} - E^\sim\right]\delta D^\sim = 0 \tag{14.5}$$

This equilibrium condition holds for any small variations in the four independent variables. Consequently, when a dielectric elastomer is in equilibrium with the applied forces and the applied voltage, the coefficient in front of the variation of each independent variable vanishes, producing [1,44]

$$s_1 = \frac{\partial W(\lambda_1, \lambda_2, \lambda_3, D^\sim)}{\partial \lambda_1} \tag{14.6}$$

$$s_2 = \frac{\partial W(\lambda_1, \lambda_2, \lambda_3, D^\sim)}{\partial \lambda_2} \tag{14.7}$$

$$s_3 = \frac{\partial W(\lambda_1, \lambda_2, \lambda_3, D^\sim)}{\partial \lambda_3} \tag{14.8}$$

$$E^\sim = \frac{\partial W(\lambda_1, \lambda_2, \lambda_3, D^\sim)}{\partial D^\sim} \tag{14.9}$$

According to the above-mentioned definition, the true stress and the true electric field can be defined as [1,43,44]

$$\sigma_1 = \frac{\partial W(\lambda_1, \lambda_2, \lambda_3, D^\sim)}{\lambda_2\lambda_3\partial\lambda_1} \tag{14.10}$$

$$\sigma_2 = \frac{\partial W(\lambda_1, \lambda_2, \lambda_3, D^\sim)}{\lambda_1\lambda_3\partial\lambda_2} \tag{14.11}$$

$$\sigma_3 = \frac{\partial W(\lambda_1, \lambda_2, \lambda_3, D^\sim)}{\lambda_1\lambda_2\partial\lambda_3} \tag{14.12}$$

$$E = \frac{\partial W(\lambda_1, \lambda_2, \lambda_3, D^\sim)}{\lambda_3\partial D^\sim} \tag{14.13}$$

The free energy of a dielectric elastomer features two contributions: the stretch and the polarization. Therefore, the free energy density of a dielectric elastomer can be written as follows [44]:

$$W(\lambda_1, \lambda_2, \lambda_3, D^{\sim}) = U(\lambda_1, \lambda_2, \lambda_3) + V(\lambda_1, \lambda_2, \lambda_3, D^{\sim}) \tag{14.14}$$

Once the free energy function $W(\lambda_1, \lambda_2, \lambda_3, D^{\sim})$ is determined, Equations 14.6–14.13 will define the state of a dielectric elastomer [1, 43, 44].

Equations 14.6–14.13 constitute yet another representation of the condition of equilibrium – they are called the equations of state [1, 43, 44]. Once the free energy function $W(\lambda_1, \lambda_2, \lambda_3, D^{\sim})$ is defined to model the material, the equations of state provide the values of the forces and voltage required to equilibrate the dielectric elastomer in the state $(\lambda_1, \lambda_2, \lambda_3, D^{\sim})$.

14.3.2 Special Elastic Energy

Super-elastic materials are a kind of special elastic material; their elastic strain energy $W(I_1, I_2, I_3)$ can be written as a function of the main invariant strain tensors I_1, I_2, I_3 [45]. The relationships between these tensors and the deformation gradient F_{iK}, are as follows [45]:

$$I_1 = F_{iK} F_{iK} \tag{14.15}$$

$$I_2 = \frac{1}{2}[F_{iK} F_{iK} F_{jM} F_{jM} - F_{iK} F_{jK} F_{iM} F_{jM}] \tag{14.16}$$

$$I_3 = \det\{F_{iK} F_{jK}\} \tag{14.17}$$

At time t, the elastomer is in the current configuration, and the marker X moves in the space to a new place with coordinates x. The functions $x_i(X, t)$ specify the kinematics of the network. As usual, the deformation gradient of the elastomer is defined as $F_{iK}(X, t) = \frac{\partial x_i(X, t)}{\partial X_K}$. We assume that the stretching rate along the three main directions are λ_1, λ_2, and λ_3; thus, the invariant of strain tensor can be expressed as [45]

$$I_1 = \lambda_1^2 + \lambda_2^2 + \lambda_3^2 \tag{14.18}$$

$$I_2 = \lambda_1^2 \lambda_2^2 + \lambda_2^2 \lambda_3^2 + \lambda_3^2 \lambda_1^2 \tag{14.19}$$

$$I_3 = \lambda_1^2 \lambda_2^2 \lambda_3^2 \tag{14.20}$$

For a super-elastic material in a non-stressed, non-deformed state, the elastic strain energy function $U(\lambda_1, \lambda_2, \lambda_3)$ must satisfy the following [45]:

$$U(3, 3, 1) = 0 \tag{14.21}$$

$$\frac{\partial U(3, 3, 1)}{\partial I_1} + 2\frac{\partial U(3, 3, 1)}{\partial I_2} + \frac{\partial U(3, 3, 1)}{\partial I_3} = 0 \tag{14.22}$$

Then, we introduce a common super-elastic strain energy function. For incompressible super-elastic materials, the common elastic strain energy models are as follows:

1. Neo-Hookean model [45–47]

$$U(\lambda_1, \lambda_2) = \frac{\mu}{2}\left(\lambda_1^2 + \lambda_2^2 + \lambda_3^2 - 3\right) \tag{14.23}$$

Here, μ is the shear modulus under infinitesimal deformation which can be measured by experiment; this model is suitable for small deformations.

2. Mooney–Rivlin model [48–53]

$$U(\lambda_1, \lambda_2) = C_1 \left(\lambda_1^2 + \lambda_2^2 + \lambda_3^2 - 3\right) + C_2 \left(\lambda_1^{-2} + \lambda_2^{-2} + \lambda_3^{-2} - 3\right) \tag{14.24}$$

C_1 and C_2 are material constants that can be measured by experiment; this model is suitable for medium deformations. The result is relatively accurate when stretching is lower than 200%.

3. Linear Gauss chain model [47, 53]

Based on the molecular statistical thermodynamics network theory of macromolecular physics, we built a microcosmic long-molecular-chain model based on non-linear materials' macromolecular properties [47,53]. We then obtained a microcosmic statistics expression of the strain energy function. The linear Gauss chain model is the elementary model

$$U(\lambda_1, \lambda_2) = \frac{1}{2} NkT(I_1 - 3) \tag{14.25}$$

where N is the number of dictyo-chains in a unit volume, k is the Boltzmann constant, and T the current temperature. This model is suitable for small-deformation states. When $\mu = NkT$, the model can be simplified to the neo-Hookean model.

4. Ogden model [54, 55]

There are many material constants in the Ogden model, which was developed by Ogden in 1972 [54]. According to the model, the elastic strain energy function $W(\lambda_1, \lambda_2)$ can be expressed as [54, 55]

$$U(\lambda_1, \lambda_2) = \sum_{p=1}^{N} \frac{\mu_p}{\alpha_p} \left(\lambda_1^{\alpha_p} + \lambda_2^{\alpha_p} + \lambda_1^{-\alpha_p}\lambda_2^{-\alpha_p} - 3\right) \tag{14.26}$$

Here, $k = 0.1$ is a material constant measured by experiment and $c(20\%) = 1.25$ is a constant that can be a positive or negative whole number or fraction.

5. Arruda–Boyce model [45,53,73]

Arruda and Boyce considered the strain-hardening effect using the expression [45,53,73]

$$U(\lambda_1, \lambda_2) = \frac{kT}{\nu} \left(\frac{\xi}{\tanh \xi} - 1 + \ln \frac{\xi}{\sinh \xi}\right) \tag{14.27}$$

$$\Lambda = \sqrt{n} \left(\frac{1}{\tanh \xi} - \frac{1}{\xi}\right) \tag{14.28}$$

$$\Lambda = \frac{1}{\sqrt{3}} \left(\lambda_1^2 + \lambda_2^2 + \lambda_3^2\right)^{1/2} \tag{14.29}$$

where kT is temperature per unit energy, τ is the volume per unit chain, ζ is the force of each chain, and n is the number of links in the chain. When $\zeta \to \infty$, $\Lambda \to \sqrt{n}$; if $\zeta \to 0$, the model can be simplified to the neo-Hookean model [45–47]. If the expression is unfolded into eight chains, the elastic strain energy can be simplified as [45, 53]

$$U(\lambda_1, \lambda_2) = NkT \left[\frac{1}{2}(I_1 - 3) + \frac{1}{20N}(I_1^2 - 9) + \frac{11}{1050N^2}(I_1^3 - 27) \right. $$
$$\left. + \frac{19}{7000N^2}(I_1^4 - 81) + \frac{519}{673750N^2}(I_1^5 - 243) \right] \tag{14.30}$$

where N is the number of dictyo-chains in a unit volume.

6. Gent model [1, 13–15, 74].

In the Gent model, the material is assumed to be an ideal dielectric elastomer, the dielectric behaviour of which is assumed to be liquid-like, unaffected by deformation [1]. An ideal dielectric elastomer is a three-dimensional network of long and flexible polymer chains. Each polymer chain consists of a large number of monomers. When the polymer is subjected to a mechanical force, it elongates. Upon release, the polymers spontaneously return to their original configuration.

However, dielectric elastomers used in practice may exhibit strain stiffening due to the finite contour length of polymer chains. This phenomenon cannot be described by either the Mooney–Rivlin or neo-Hookean model [1, 74]. This is because the polymer is compliant under small strains but stiffens steeply when pulled to nearly its full length [1, 74]. Thus, for an incompressible dielectric elastomer, the strain energy model developed by Gent is employed to characterize the stiffening behaviour as follows [1]:

$$U(\lambda_1, \lambda_2) = -\frac{\mu}{2} J_{\lim} \log \left(1 - \frac{I_1 - 3}{J_{\lim}} \right) \tag{14.31}$$

where I_1 is the left Cauchy–Green deformation tensor, $I_1 = \lambda_1^2 + \lambda_2^2 + \lambda_1^{-2}\lambda_2^{-2}$, $J_{\lim}$ is a constant related to the limiting stretch, $J_{\lim} = \lambda_{1\lim}^2 + \lambda_{2\lim}^2 + \lambda_{1\lim}^{-2}\lambda_{2\lim}^{-2} - 3$, and μ is the small-strain shear modulus. When $(I_1 - 3)/J_{\lim} \to 0$, the Taylor expansion of (14.31) gives $U = \frac{\mu}{2}(I_1 - 3)$. That is, the Gent model recovers the neo-Hookean model when deformation is small compared to the stretching limit. When $(I_1 - 3)/J_{\lim} \to 1$, the elastomer approaches the stretching limit [44]. We can further expand the natural logarithm of the Gent model to $U(\lambda_1, \lambda_2) = \frac{\mu}{2}[(I_1 - 3) + \frac{1}{2J_{\lim}}(I_1 - 3)^2 + \ldots + \frac{1}{(n+1)J_{\lim}^n}(I_1 - 3)^{n+1}]$, the general form of which is $U(\lambda_1, \lambda_2) = \sum_{i-1}^{n} C_i(J_{\lim})(I_1 - 3)^i$; this can be viewed as an expression of the Rivlin elastic strain energy model $U(\lambda_1, \lambda_2) = \sum_{i,j=0}^{\infty} C_{ij}(I_1 - 3)^i(I_2 - 3)^j$ when $j = 0$ [44].

14.3.3 Special Electric Field Energy

The real electric displacement D of an ideal dielectric elastomer is linear with the electric field E, $D = \varepsilon E$, where ε is the relative dielectric constant. Therefore, the electric field energy density of an ideal elastomer can be expressed as [1, 44, 56–58]

$$V(D) = \frac{D^2}{2\varepsilon} \tag{14.32}$$

The linear relation $D = \varepsilon E$ before polarization saturation, as Figure 14.30(d) shows, described the area circled by Equation 14.32 and the coordinates.

Considering the relationship between the real electric displacement D and the nominal electric displacement $D^\sim$, the real electric field E and the nominal electric field $E^\sim$, the electric field energy density function in Equation 14.32 can be modified as follows [1, 44, 56–62]:

$$V(\lambda_1, \lambda_2, D^\sim) = \frac{D^{\sim 2}}{2\varepsilon}\lambda_1^{-2}\lambda_2^{-2} \tag{14.33}$$

14.3.4   Incompressible Dielectric Elastomer

For an incompressible dielectric elastomer, $\lambda_1\lambda_2\lambda_3 = 1, \delta\lambda_3 = -\lambda_1^{-2}\lambda_2^{-1}\delta\lambda_1 - \lambda_1^{-1}\lambda_2^{-2}\delta\lambda_2$. Inserting this expression into Equation 14.2 produces $\delta W = (s_1 - \lambda_1^{-2}\lambda_2^{-1}s_3)\delta\lambda_1 + (s_2 - \lambda_1^{-1}\lambda_2^{-2}s_3)\delta\lambda_2 + E^\sim\delta D^\sim$. The nominal stress along the two planar principal directions and the nominal electric field along the thickness direction of an incompressible dielectric elastomer's thermodynamic system are obtained, respectively, as follows [1, 44, 64, 65]:

$$s_1 - \frac{s_3}{\lambda_1^2\lambda_2} = \frac{\partial W\left(\lambda_1, \lambda_2, \lambda_1^{-1}\lambda_2^{-1}, D^\sim\right)}{\partial\lambda_1} \tag{14.34}$$

$$s_2 - \frac{s_3}{\lambda_1\lambda_2^2} = \frac{\partial W\left(\lambda_1, \lambda_2, \lambda_1^{-1}\lambda_2^{-1}, D^\sim\right)}{\partial\lambda_2} \tag{14.35}$$

$$E^\sim = \frac{\partial W\left(\lambda_1, \lambda_2, \lambda_1^{-1}\lambda_2^{-1}, D^\sim\right)}{\partial D^\sim} \tag{14.36}$$

The true stress and the true electric field can be expressed as follows [1, 44, 64, 65]:

$$\sigma_1 - \sigma_3 = \lambda_1\frac{\partial W\left(\lambda_1, \lambda_2, \lambda_1^{-1}\lambda_2^{-1}, D^\sim\right)}{\partial\lambda_1} \tag{14.37}$$

$$\sigma_2 - \sigma_3 = \lambda_2\frac{\partial W\left(\lambda_1, \lambda_2, \lambda_1^{-1}\lambda_2^{-1}, D^\sim\right)}{\partial\lambda_2} \tag{14.38}$$

$$E = \lambda_1\lambda_2\frac{\partial W\left(\lambda_1, \lambda_2, \lambda_1^{-1}\lambda_2^{-1}, D^\sim\right)}{\partial D^\sim} \tag{14.39}$$

Equations 14.30–14.35 constitute the equations of state of an incompressible dielectric elastomer undergoing polarization saturation [1, 44, 64, 65]. If the elastic energy and the electric energy are given, we can determine the stress and the electric field of the incompressible dielectric elastomer at equilibrium state [1, 44, 64, 65].

14.3.5   Model of Several Dielectric Elastomers

14.3.5.1   *Model of Ideal Dielectric Elastomers*

As Figure 14.26 shows, a dielectric elastomer is made of three-dimensional network structures formed by the cross-linking of flexible long chains. Each polymer chain contains a large number of monomers [1, 66–68]. Because the elastomer can polarize freely in a manner almost identical to that of the molten polymer, the effect of cross-linking on monomer polarization can be ignored [1, 56, 57]. Ideally, assuming that the dielectric behaviour of the elastomer is identical to that of the molten polymer, the relationship between the real electric field and the real electric displacement can be expressed as

$$D = \varepsilon E \tag{14.40}$$

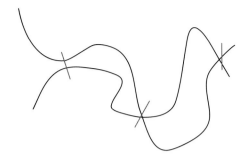

Figure 14.26 *Molecular structure of an ideal dielectric elastomer. The polymer is composed of a three-dimensional network structure linked by flexible long chains; each polymer chain contains a vast number of monomers [1]. (Reprinted under the terms of the STM agreement from [1] Copyright (2010) Elsevier Ltd).*

Suo called this type of elastomer the ideal dielectric elastomer, which is analogous to an ideal gas [1].

According to functions (14.14) and (14.33), (14.34), (14.35), and (14.36), the constitutive relation for the nominal stress and nominal electrical field of an incompressible can be expressed as follows [1,68]:

$$s_1 - \frac{s_3}{\lambda_1^2 \lambda_2} = \frac{\partial U\left(\lambda_1, \lambda_2, \lambda_1^{-1}\lambda_2^{-1}\right)}{\partial \lambda_1} - \frac{D^{\sim 2}}{\varepsilon}\lambda_1^{-3}\lambda_2^{-2} \tag{14.41}$$

$$s_2 - \frac{s_3}{\lambda_1 \lambda_2^2} = \frac{\partial U\left(\lambda_1, \lambda_2, \lambda_1^{-1}\lambda_2^{-1}\right)}{\partial \lambda_2} - \frac{D^{\sim 2}}{\varepsilon}\lambda_1^{-2}\lambda_2^{-3} \tag{14.42}$$

$$E^{\sim} = \frac{D^{\sim}}{\varepsilon}\lambda_1^{-2}\lambda_2^{-2} \tag{14.43}$$

We can also describe the constitutive relationship according to the true stress and true electrical field [1,56,57]:

$$\sigma_1 - \sigma_3 = \lambda_1 \frac{\partial U\left(\lambda_1, \lambda_2, \lambda_1^{-1}\lambda_2^{-1}\right)}{\partial \lambda_1} - \varepsilon E^2 \tag{14.44}$$

$$\sigma_2 - \sigma_3 = \lambda_2 \frac{\partial U\left(\lambda_1, \lambda_2, \lambda_1^{-1}\lambda_2^{-1}\right)}{\partial \lambda_2} - \varepsilon E^2 \tag{14.45}$$

$$E = \frac{D}{\varepsilon} \tag{14.46}$$

Here, function (14.46) is restored to function (14.40).

14.3.5.2 Model of Electrostriction Dielectric Elastomers

As shown in Figure 14.27, after a voltage is applied to a dielectric, some parts of the dielectric become thinner while others become thicker. A dielectric is not polarized if an electric field is not applied [1,64,65]. The electrostrictive deformation caused by the applied voltage comes from two factors: electrostriction and Maxwell stress [1,64,65]. The electrostrictive effect in a dielectric elastomer is reflected by the impact of the dielectric constant through deformation. As a simple electrostrictive model, we can assume that the

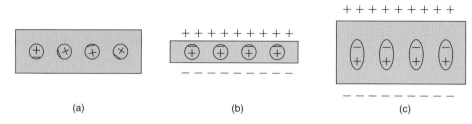

Figure 14.27 *Consider a dielectric that is non-polar in the absence of applied voltage (a). Subject to a voltage, some dielectrics become thinner (b), but other dielectrics become thicker (c) [64]. (Reprinted with permission from [64] Copyright (2008) American Institute of Physics).*

real electric field is a linear function of the electric displacement, where the dielectric constant is a function of the strain rate [1, 39, 64, 65, 70–72]. Thus, the dielectric behaviour of the dielectric elastomer can be described as [1, 64, 65]

$$D = \varepsilon(\lambda_1, \lambda_2)E \tag{14.47}$$

Suo *et al.* reported a linear relationship between the dielectric constant and the stretching rate [1, 64, 65]:

$$\varepsilon(\lambda_1, \lambda_2) = \varepsilon[1 + r(\lambda_1 + \lambda_2 - 2)] \tag{14.48}$$

ε is the dielectric constant of a dielectric elastomer in the undeformed state, $\varepsilon = 4.68\varepsilon_0$, ε_0 is the dielectric constant in vacuum, $\varepsilon_0 = 8.85 \times 10^{-12}$ F/m, and r is the electrical striction coefficient [1, 39, 64, 65, 70–72].

Leng *et al.* modelled the relationship between the dielectric constant and stretching rate as being non-linear [67, 70–72]:

$$\varepsilon(\lambda_1, \lambda_2) = \begin{cases} (a\lambda_1\lambda_2 + b)\varepsilon_0 & \lambda_1\lambda_2 \leq s \\ c\varepsilon_0 & \lambda_1\lambda_2 > s \end{cases} \tag{14.49}$$

Here a, b, c, and s are the material constants of a dielectric elastomer; model values for these constants are $a = -0.016$, $b = 4.716$, $c = 4.48$.

According to functions (14.47) and (14.33), (14.34), (14.35), and (14.36), we can also use the true stress and true electric field to form the constitutive relationships that describe the behaviour of incompressible dielectric elastomers [1, 64, 65]:

$$s_1 - \frac{s_3}{\lambda_1^2\lambda_2} = \frac{\partial U\left(\lambda_1, \lambda_2, \lambda_1^{-1}\lambda_2^{-1}\right)}{\partial\lambda_1} - \frac{1}{\varepsilon(\lambda_1, \lambda_2)}\lambda_1^{-3}\lambda_2^{-2}D^{\sim 2}$$
$$- \frac{1}{2[\varepsilon(\lambda_1, \lambda_2)]^2}\frac{\partial\varepsilon(\lambda_1, \lambda_2)}{\partial\lambda_1}\lambda_1^{-2}\lambda_2^{-2}D^{\sim 2} \tag{14.50}$$

$$s_2 - \frac{s_3}{\lambda_1^2\lambda_2} = \frac{\partial U\left(\lambda_1, \lambda_2, \lambda_1^{-1}\lambda_2^{-1}\right)}{\partial\lambda_2} - \frac{1}{\varepsilon(\lambda_1, \lambda_2)}\lambda_1^{-2}\lambda_2^{-3}D^{\sim 2}$$
$$- \frac{1}{2[\varepsilon(\lambda_1, \lambda_2)]^2}\frac{\partial\varepsilon(\lambda_1, \lambda_2)}{\partial\lambda_2}\lambda_1^{-2}\lambda_2^{-2}D^{\sim 2} \tag{14.51}$$

$$E^{\sim} = \frac{D^{\sim}}{\varepsilon(\lambda_1, \lambda_2)}\lambda_1^{-2}\lambda_2^{-2} \tag{14.52}$$

We can also express the constitutive relationship using the true stress and true electrical field [1,64,65].

$$\sigma_1 - \sigma_3 = \lambda_1 \frac{\partial U\left(\lambda_1, \lambda_2, \lambda_1^{-1}\lambda_2^{-1}\right)}{\partial \lambda_1} - \left[\varepsilon(\lambda_1, \lambda_2) + \frac{\lambda_1}{2}\frac{\partial \varepsilon(\lambda_1, \lambda_2)}{\partial \lambda_1}\right] E^2 \quad (14.53)$$

$$\sigma_2 - \sigma_3 = \lambda_2 \frac{\partial U\left(\lambda_1, \lambda_2, \lambda_1^{-1}\lambda_2^{-1}\right)}{\partial \lambda_2} - \left[\varepsilon(\lambda_1, \lambda_2) + \frac{\lambda_2}{2}\frac{\partial \varepsilon(\lambda_1, \lambda_2)}{\partial \lambda_2}\right] E^2 \quad (14.54)$$

$$E = \frac{D}{\varepsilon(\lambda_1, \lambda_2)} \quad (14.55)$$

Here, function (14.55) is restored to function (14.47) [1,64,65].

14.3.5.3   Model of Dielectric Elastomer of Interpenetrating Polymer Networks (IPNs)

As shown in Figure 14.28, after pre-stretching a dielectric elastomer membrane (network A) that has been swelled with monomers, a network B is formed as the solidification of the monomer swells in network A, where networks A and B are intertwined [73]. After the applied load is lifted, the film size is observed to have increased because network A experiences tensile stress while network B experiences compressive stress; thus, the material forms the dielectric elastomer of an IPN [73]. IPNs are composed of dielectric elastomers of two different chain lengths. They exhibit great stability and may undergo large electroactive deformation [73].

Suo *et al.* built a theoretical model of IPN dielectric elastomers as follows. Suppose φ^A and φ^B are the volume fractions of the two chains in dielectric elastomer composite materials [73]. Ignoring the interaction with the chemical network, the free energy of an IPN dielectric elastomer can be written as [73]

$$W(\lambda_1, \lambda_2) = \varphi^A W^A\left(\lambda_1^A, \lambda_2^A\right) + \varphi^B W^B\left(\lambda_1^B, \lambda_2^B\right) \quad (14.56)$$

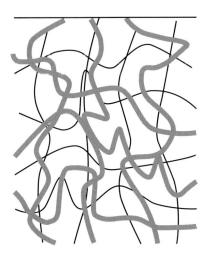

Figure 14.28 *Interpenetrating polymer network dielectric elastomer structure [73]. (Reprinted from [73] Copyright (2009) American Institute of Physics).*

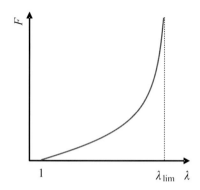

Figure 14.29 *The strain hardening of dielectric elastomers [44]. (Reprinted with permission from [44] Copyright (2012) Elsevier Ltd).*

The elastic strain energy can be assumed to follow the Arruda–Boyce model. According to Equations 14.26, 14.27, and 14.28, we can obtain the relationship between stress and strain as follows [73]:

$$\sigma_1 = \frac{\varphi^A kT\zeta^A \left(\lambda_1^2 - \lambda_1^{-2}\lambda_2^2\right)}{3v^A\sqrt{n^A}\Lambda^A} + \frac{\varphi^B kT\zeta^B \left[\left(\lambda_1^B\right)^2 - \left(\lambda_1^B\right)^{-2}\left(\lambda_2^B\right)^{-2}\right]}{3v^B\sqrt{n^B}\Lambda^B} - \varepsilon E^2 \quad (14.57)$$

$$\sigma_2 = \frac{\varphi^A kT\zeta^A \left(\lambda_2^2 - \lambda_1^{-2}\lambda_2^2\right)}{3v^A\sqrt{n^A}\Lambda^A} + \frac{\varphi^B kT\zeta^B \left[\left(\lambda_2^B\right)^2 - \left(\lambda_1^B\right)^{-2}\left(\lambda_2^B\right)^{-2}\right]}{3v^B\sqrt{n^B}\Lambda^B} - \varepsilon E^2 \quad (14.58)$$

λ_1^A and λ_2^A are the extents of stretching of network A, while λ_1^B and λ_2^B are the extents of stretching of network B [73].

14.3.5.4 Model of Strain-Hardening Dielectric Elastomer

Dielectric elastomers that behave similarly to rubber experience strain-hardening effects [1,74,75]. As Figure 14.29 shows, when mechanical force F is applied to an elastomer approaching the stretching limit $\lambda_{\lim}$, the elastomer will harden sharply.

The strain-hardening effect of dielectric elastomers is described by the Gent model [1,13–15,74,75]. The energy density function of the electric field can take the form of Equation 14.33, and the dielectric constant can take the form of Equation 14.47. Here, the constitutive relationship between the strain hardening of an dielectric elastomer and the nominal stress can be written as follows [1,44,74,75]:

$$s_1 - \frac{s_3}{\lambda_1^2\lambda_2} = \frac{\mu J_{\lim}}{J_{\lim} - (I_1 - 3)}\left(\lambda_1 - \lambda_1^{-3}\lambda_2^{-2}\right) - \frac{1}{\varepsilon(\lambda_1, \lambda_2)}\lambda_1^{-3}\lambda_2^{-2}D^{\sim 2}$$
$$- \frac{1}{2[\varepsilon(\lambda_1, \lambda_2)]^2}\frac{\partial\varepsilon(\lambda_1, \lambda_2)}{\partial\lambda_1}\lambda_1^{-2}\lambda_2^{-2}D^{\sim 2} \quad (14.59)$$

$$s_2 - \frac{s_3}{\lambda_1\lambda_2^2} = \frac{\mu J_{\lim}}{J_{\lim} - (I_1 - 3)}\left(\lambda_2 - \lambda_1^{-2}\lambda_2^{-3}\right) - \frac{1}{\varepsilon(\lambda_1, \lambda_2)}\lambda_1^{-2}\lambda_2^{-3}D^{\sim 2}$$
$$- \frac{1}{2[\varepsilon(\lambda_1, \lambda_2)]^2}\frac{\partial\varepsilon(\lambda_1, \lambda_2)}{\partial\lambda_2}\lambda_1^{-2}\lambda_2^{-2}D^{\sim 2} \quad (14.60)$$

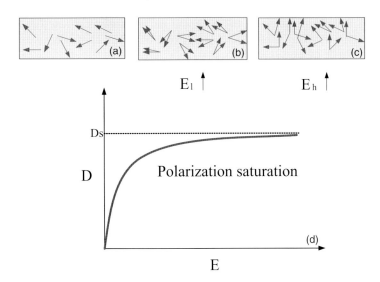

Figure 14.30 *Polarization saturation of dielectric [44]. (Reprinted with permission from [44] Copyright (2012) Elsevier Ltd).*

Correspondingly, the constitutive relationship between the strain hardening of an elastomer and the true stress can be expressed as follows [1, 44, 74, 75]:

$$\sigma_1 - \sigma_3 = \frac{\mu J_{\lim}}{J_{\lim} - (I_1 - 3)} \left(\lambda_1^2 - \lambda_1^{-2}\lambda_2^{-2}\right) - \left[\varepsilon(\lambda_1, \lambda_2) + \frac{\lambda_1}{2}\frac{\partial\varepsilon(\lambda_1, \lambda_2)}{\partial\lambda_1}\right] E^2 \quad (14.61)$$

$$\sigma_2 - \sigma_3 = \frac{\mu J_{\lim}}{J_{\lim} - (I_1 - 3)} \left(\lambda_2^2 - \lambda_1^{-2}\lambda_2^{-2}\right) - \left[\varepsilon(\lambda_1, \lambda_2) + \frac{\lambda_2}{2}\frac{\partial\varepsilon(\lambda_1, \lambda_2)}{\partial\lambda_2}\right] E^2 \quad (14.62)$$

14.3.5.5   Model of Polarization Saturation Dielectric Elastomer

Dielectrics experience polarization-saturation effects [1, 43, 44, 74–77]. As Figure 14.30 shows, in a dielectric elastomer, every chain possesses an electric dipole [1, 43, 44, 74, 75]. In the absence of an electric field, the dipole is affected by temperature and the configuration is chaotic without aligning along any particular direction (Figure 14.30(a)) [1,43,44,74,75]. By applying a low voltage (E_l) to an elastomer with non-directional dipoles, the dipoles will rotate and align themselves along the electric field (Figure 14.30(b)) [1,43,44,74,75]. If the voltage is large enough (E_h), the dipoles will align themselves along the electric field perfectly, and the elastomer will reach polarization saturation (Figure 14.30(c)) [1,43, 44,74,75]. Figure 14.30(d) illustrates the non-linear behaviour of a dielectric undergoing polarization saturation [1,43,44,74,75].

An elastomer consists of a three-dimensional network structure cross-linked by flexible long chains. Every polymer chain contains many monomers, ignoring the influence of cross-linking. Thus, the elastomer can polarize freely. Ideally, the relationship between the electric field, independent of deformation, and electric displacement can be described as follows [1,43,44]:

$$E = f(D) \quad (14.63)$$

Assuming that the nominal electric field and the nominal electrical displacement are conjugate parameters, the electric field energy of a dielectric elastomer undergoing large deformation can be described as [1, 43, 44]

$$V(D^\sim) = \int_0^{D^\sim} E^\sim dD^\sim \tag{14.64}$$

According to $E^\sim = E\lambda_3$, $D = D^\sim \lambda_1^{-1}\lambda_2^{-1}$ and function (14.66), we obtain [1, 43, 44]

$$V(\lambda_1, \lambda_2, \lambda_3, D^\sim) = \lambda_3 \int_0^{D^\sim} f\left(D^\sim \lambda_1^{-1}\lambda_2^{-1}\right)dD^\sim \tag{14.65}$$

Substituting Equations 14.66 and 14.14 into 14.6, 14.7, 14.8, and 14.9, we have [1, 43, 44]

$$s_1 = \frac{\partial U(\lambda_1, \lambda_2, \lambda_3)}{\partial\lambda_1} + \lambda_3\frac{\partial}{\partial\lambda_1}\left[\int_0^{D^\sim} f\left(D^\sim \lambda_1^{-1}\lambda_2^{-1}\right)dD^\sim\right] \tag{14.66}$$

$$s_2 = \frac{\partial U(\lambda_1, \lambda_2, \lambda_3)}{\partial\lambda_2} + \lambda_3\frac{\partial}{\partial\lambda_2}\left[\int_0^{D^\sim} f\left(D^\sim \lambda_1^{-1}\lambda_2^{-1}\right)dD^\sim\right] \tag{14.67}$$

$$s_3 = \frac{\partial U(\lambda_1, \lambda_2, \lambda_3)}{\partial\lambda_3} + \lambda_3\frac{\partial}{\partial\lambda_3}\left[\int_0^{D^\sim} f\left(D^\sim \lambda_1^{-1}\lambda_2^{-1}\right)dD^\sim\right] \tag{14.68}$$

$$E^\sim = \lambda_3\frac{\partial}{\partial D^\sim}\left[\int_0^{D^\sim} f\left(D^\sim \lambda_1^{-1}\lambda_2^{-1}\right)dD^\sim\right] \tag{14.69}$$

Substituting Equations 14.66 and 14.14 into 14.10), 14.11, 14.12, and 14.13, the true stress and true electric field of a dielectric elastomer are [1, 43, 44]

$$\sigma_1 = \frac{\partial U(\lambda_1, \lambda_2, \lambda_3)}{\lambda_2\lambda_3\partial\lambda_1} + \frac{1}{\lambda_2}\frac{\partial}{\partial\lambda_1}\left[\int_0^{D^\sim} f\left(D^\sim \lambda_1^{-1}\lambda_2^{-1}\right)dD^\sim\right] \tag{14.70}$$

$$\sigma_2 = \frac{\partial U(\lambda_1, \lambda_2, \lambda_3)}{\lambda_1\lambda_3\partial\lambda_2} + \frac{1}{\lambda_1}\frac{\partial}{\partial\lambda_2}\left[\int_0^{D^\sim} f\left(D^\sim \lambda_1^{-1}\lambda_2^{-1}\right)dD^\sim\right] \tag{14.71}$$

$$\sigma_3 = \frac{\partial U(\lambda_1, \lambda_2, \lambda_3)}{\lambda_1\lambda_2\partial\lambda_3} + \frac{\lambda_3}{\lambda_1\lambda_2}\frac{\partial}{\partial\lambda_3}\left[\int_0^{D^\sim} f\left(D^\sim \lambda_1^{-1}\lambda_2^{-1}\right)dD^\sim\right] \tag{14.72}$$

$$E = \frac{\partial}{\partial D^\sim}\left[\int_0^{D^\sim} f\left(D^\sim \lambda_1^{-1}\lambda_2^{-1}\right)dD^\sim\right] \tag{14.73}$$

Equations 14.70, 14.71, 14.72, and 14.73 are the equilibrium equations of a dielectric elastomer undergoing polarization saturation. When the elastic energy and the electric

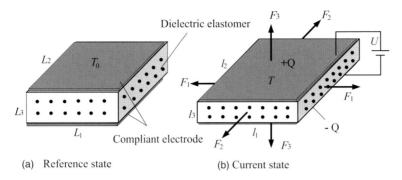

Figure 14.31 *Dielectric elastomer composite material electromechanical coupling system [78]. (Reprinted with permission from [78] Copyright (2011) IOP Publishing).*

energy of the dielectric elastomers are known, we can determine the stress and electric field of the elastomer in the equilibrium state [1,43,44].

14.3.5.6   Model of Dielectric Elastomer Composite Material

Considering the electromechanical coupling system of a dielectric elastomer composite material, such as in Figure 14.31, suppose that the dielectric constant of a dielectric elastomer composite material is a linear function of the stretching rate [67,69–72]. Moreover, consider the dielectric elastomer to contain uniformly distributed nanoparticles with a high dielectric constant [78]. We can express the dielectric constant of this dielectric elastomer composite material as follows [78]:

$$\varepsilon^*(\lambda_1, v) = \varepsilon^\sim + \sum_{i=1}^{3} m_i(\lambda_i - 1)\varepsilon^\sim + c(v)\varepsilon^\sim \qquad (14.74)$$

Here, $\varepsilon^\sim$ is the dielectric constant of the dielectric elastomer in the undeformed state, v is the content of mixed nanoparticles, $\lambda_i (i = 1, 2, 3)$ is the stretching rate along the three main directions, and $m_i (i = 1, 2, 3)$ is the electrostriction coefficient [78]. In the equation above, the expression for the dielectric constant consists of three terms: the first term is the dielectric constant of a pure ideal dielectric elastomer in the undeformed state, the second term describes the change in the ideal dielectric elastomer undergoing electrostriction deformation, and the third term describes the influence of the filler content on the dielectric constant. Obviously, when m_i and $c(v)$ assume certain special values, Equation 14.74 can be simplified to the models used by Zhao *et al.* [64].

According to the experiments of Gallone *et al.*, as the filler content increases, the dielectric constant increases [36]. To provide a more precise expression for $\frac{\lambda_1^{-1}\lambda_2^{-1}\lambda_3 D^\sim}{[1+a(\lambda_3-1)+b(\lambda_1+\lambda_2+\lambda_3-3)]\varepsilon^\sim}$, we can overfit the expression with a high-order function, such as a quadratic function, to produce [78]

$$c(v) = dv^2 + ev + f \qquad (14.75)$$

Here, d, e, and f are material constants obtained by experiment. According to the experiments of Carpi *et al.* [36], the fitting results give the following values: $d = 12.5$, $e = 3.75$, $f = 0$.

By substituting Equations 14.76 and 14.14 into Equations 14.6–14.9, we can express the nominal stress and nominal electric field as follows [78]:

$$s_i = \frac{\partial U(\lambda_1, \lambda_2, \lambda_3)}{\partial \lambda_i} + (-1)^{\frac{1}{\sqrt{5}}}\left[\left(\frac{\sqrt{5}+1}{2}\right)^{i+3} - \left(\frac{1-\sqrt{5}}{2}\right)^{i+3}\right] \frac{\lambda_i^{-1}\lambda_1^{-1}\lambda_2^{-1}\lambda_3 D^{\sim 2}}{2\left[1 + \sum\limits_{i=1}^{3} m_i(\lambda_i - 1) + c(v)\right]\varepsilon^{\sim}}$$

$$-\frac{m_i \lambda_1^{-1}\lambda_2^{-1}\lambda_3 D^{\sim 2}}{2\left[1 + \sum\limits_{i=1}^{3} m_i(\lambda_i - 1) + c(v)\right]^2 \varepsilon^{\sim}} \tag{14.76}$$

$$E^{\sim} = \frac{\lambda_1^{-1}\lambda_2^{-1}\lambda_3 D^{\sim}}{\left[1 + \sum\limits_{i=1}^{3} m_i(\lambda_i - 1) + c(v)\right]\varepsilon^{\sim}} \tag{14.77}$$

Here, $I_1 = \lambda_1^2 + \lambda_2^2 + \lambda_3^2$.

The true stresses can be expressed as $\sigma_1 = s_1/\lambda_2\lambda_3$, $\sigma_2 = s_2/\lambda_1\lambda_3$, $\sigma_3 = s_3/\lambda_1\lambda_2$, and $E = E^{\sim}/\lambda_3$. By substituting these relationships into Equations 14.78 and 14.79, we obtain the following [78]:

$$\sigma_i = \lambda_i \frac{\partial U(\lambda_1, \lambda_2, \lambda_3)}{\partial \lambda_i} + (-1)^{\frac{1}{\sqrt{5}}}\left[\left(\frac{\sqrt{5}+1}{2}\right)^{i+3} - \left(\frac{1-\sqrt{5}}{2}\right)^{i+3}\right] \frac{\lambda_1^{-2}\lambda_2^{-2} D^{\sim 2}}{2\left[1 + \sum\limits_{i=1}^{3} m_i(\lambda_i - 1) + c(v)\right]\varepsilon^{\sim}}$$

$$-\frac{m_i \lambda_i \lambda_1^{-2}\lambda_2^{-2} D^{\sim 2}}{2\left[1 + \sum\limits_{i=1}^{3} m_i(\lambda_i - 1) + c(v)\right]^2 \varepsilon^{\sim}} \tag{14.78}$$

$$E = \frac{D}{\left[1 + \sum\limits_{i=1}^{3} m_i(\lambda_i - 1) + c(v)\right]\varepsilon^{\sim}} \tag{14.79}$$

Here, k is the true electric displacement of the composite material. By substituting Equation 14.79 into Equation 14.78, $\lambda_1^{-1}\lambda_2^{-2} D^{\sim 2}/2[1 + \sum_{i=1}^{3} m_i(\lambda_i - 1) + c(v)]\varepsilon^{\sim}$ can be simplified to $\varepsilon^* E^2/2$ and $m_i \lambda_1 \lambda_1^{-2} D^{\sim 2}/2[1 + \sum_{i=1}^{3} m_i(\lambda_i - 1) + c(v)]^2\varepsilon^{\sim}$ can be simplified to $\frac{\partial \varepsilon^*}{2\partial \lambda_i}\lambda_i E^2$ [78]. Expression $\varepsilon^* E^2/2$ represents the Maxwell stress, while $\frac{\partial \varepsilon^*}{2\partial \lambda_i}\lambda_i E^2$ is the influence of the dielectric constant on the Maxwell stress [78]. It can be observed that the true stress is related to the following factors: the material's super-elasticity, the Maxwell stress, electrostriction deformation, and the filler content.

Although we can express the dielectric constant according to experimental observations, the same can be performed using the equivalent modulus of a composite material prescribed by elasticity theory [78]. According to the bound model of the modulus of elasticity, we take the series connection model of the dielectric constant as the upper bound and the parallel connection model as the lower bound. Specifically, the expressions are as follows [78]:

$$\varepsilon^*_{max}(\lambda_1, \lambda_2, \lambda_3, h_1, h_2) = \tilde{\varepsilon_1} h_1 + \tilde{\varepsilon_2} h_2 + \sum_{i=1}^{3} m_i(\lambda_i - 1)\tilde{\varepsilon_1} h_1 \qquad (14.80)$$

$$\varepsilon^*_{min}(\lambda_1, \lambda_2, \lambda_3, h_1, h_2) = \cfrac{1}{\cfrac{1}{\tilde{\varepsilon_1} h_1} + \cfrac{1}{\tilde{\varepsilon_2} h_2} + \cfrac{1}{\displaystyle\sum_{i=1}^{3} m_i(\lambda_i - 1)\tilde{\varepsilon_1} h_1}} \qquad (14.81)$$

Here, h_1 and h_2 are the volume contents of silicon rubber and the fillers, respectively, and $h_1 + h_2 = 1$; thus, the dielectric constant of silicon rubber satisfies the equation below [78]:

$$\varepsilon^*_{min} \le \varepsilon^*_{real}(\lambda_1, \lambda_2, \lambda_3, h_1, h_2) \le \varepsilon^*_{max} \qquad (14.82)$$

According to the dielectric constant model [36] of composite materials put forth by Jayasunder, we can express the dielectric constant of a composite material as follows [78]:

$$\varepsilon^*_{real}(\lambda_1, \lambda_2, \lambda_3, h_1, h_2) = \left\{ \cfrac{\varepsilon_1 h_1 + \varepsilon_2 h_2 \cfrac{3\varepsilon_1}{(2\varepsilon_1 + \varepsilon_2)}\left[1 + 3h_2 \cfrac{(\varepsilon_2 - \varepsilon_1)}{(2\varepsilon_1 + \varepsilon_2)}\right]}{h_1 + h_2 \cfrac{3\varepsilon_1}{(2\varepsilon_1 + \varepsilon_2)}\left[1 + 3h_2 \cfrac{(\varepsilon_2 - \varepsilon_1)}{(2\varepsilon_1 + \varepsilon_2)}\right]} \right\}$$

$$+ \sum_{i=1}^{3} m_i(\lambda_i - 1) \left\{ \cfrac{\varepsilon_1 h_1 + \varepsilon_2 h_2 \cfrac{3\varepsilon_1}{(2\varepsilon_1 + \varepsilon_2)}\left[1 + 3h_2 \cfrac{(\varepsilon_2 - \varepsilon_1)}{(2\varepsilon_1 + \varepsilon_2)}\right]}{h_1 + h_2 \cfrac{3\varepsilon_1}{(2\varepsilon_1 + \varepsilon_2)}\left[1 + 3h_2 \cfrac{(\varepsilon_2 - \varepsilon_1)}{(2\varepsilon_1 + \varepsilon_2)}\right]} \right\}$$

$$(14.83)$$

We can use the dielectric constant expression Equation 14.83, which is based on composite material theory, to build the system's free energy. We can also investigate the constitutive relationship governing a dielectric elastomer composite material's electromechanical coupling system [78].

14.3.5.7 Model of Thermal Dielectric Elastomer Material

As illustrated in Figure 14.32, to analyse the mechanical behaviour of a dielectric elastomer's thermo-electromechanical coupling system, we apply an electrical field energy density function model with a variable dielectric constant. Taking temperature and the extent of stretching into consideration, the dielectric constant $\varepsilon(T, \lambda_1, \lambda_2)$ can be expressed as follows [72,79]:

$$\varepsilon(T, \lambda_1, \lambda_2) = \varepsilon_r + \alpha(T - T_0) + \beta\lambda_1\lambda_2 \qquad (14.84)$$

Here, ε_r is the relative dielectric constant of a dielectric elastomer at the reference temperature and under no deformation, α is the phenomenologically learned parameter, T_0 is the temperature at the reference state, and β is the electrostriction coefficient [79]. Thus, the electrical field energy density function of the dielectric elastomer can be expressed

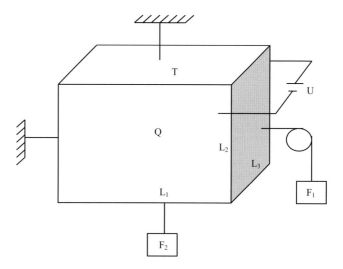

Figure 14.32 *Thermo-electromechanical coupling system of a dielectric elastomer [79]. (Reprinted with permission. Copyright (2012) Copyright reserved).*

as follows [79]:

$$V(\lambda_1, \lambda_2, T, D^{\sim}) = \frac{D^{\sim 2}}{2\left[\varepsilon_r + \alpha(T - T_0) + \beta\lambda_1\lambda_2\right]}\lambda_1^{-2}\lambda_2^{-2} \tag{14.85}$$

Considering the influence of temperature, the thermo-super-elastic energy of the dielectric elastomer can be expressed as follows [79]:

$$U(\lambda_1, \lambda_2, T) = \frac{1}{2}NkT\left(\lambda_1^2 + \lambda_2^2 + \lambda_1^{-2}\lambda_2^{-2} - 3\right) + c_0\left[(T - T_0) - T\ln\frac{T}{T_0}\right] \tag{14.86}$$

In expression 14.86, the first term represents the neo-Hookean thermal elastic energy affected by the temperature; the second term is the thermal contribution of the temperature to the whole thermodynamic system. According to the definition of free energy, $U(T) = Q(T) - TS$, where $Q(T)$ is the internal energy [80]. When the temperature rises, the internal energy becomes $c(T - T_0)$ [80]. According to the relationship between specific heat and entropy $c = T\frac{\partial S}{\partial T}$, $S = c\log(T/T_0)$ [80]. Here, c is the specific heat of a polar dielectric; therefore, the thermal contribution can be expressed as $c[(T - T_0) - T\log(T/T_0)]$. N is the number of polymer chains per unit volume in the dielectric elastomer, while kT is the temperature per unit energy. $NkT = \mu(T)$ and $\mu(T)$ are the moduli of elasticity under infinitesimal deformation; these are related to the temperature, and c_0 is the specific heat [80].

The differential relationship between the nominal stress, nominal electrical field, and nominal entropy density and the free energy, nominal stress along the two directions of the plane, nominal electrical field along the thickness direction, and the nominal entropy density is expressed as follows [79]:

$$s_1 = \frac{\partial U(\lambda_1, \lambda_2, T)}{\partial \lambda_1} - \frac{D^{\sim 2}\lambda_1^{-3}\lambda_2^{-2}}{[\varepsilon_r + \alpha(T - T_0) + \beta\lambda_1\lambda_2]} - \frac{D^{\sim 2}\beta\lambda_1^{-2}\lambda_2^{-1}}{2[\varepsilon_r + \alpha(T - T_0) + \beta\lambda_1\lambda_2]^2}$$

(14.87)

$$s_2 = \frac{\partial U(\lambda_1, \lambda_2, T)}{\partial \lambda_2} - \frac{D^{\sim 2}\lambda_1^{-2}\lambda_2^{-3}}{[\varepsilon_r + \alpha(T - T_0) + \beta\lambda_1\lambda_2]} - \frac{D^{\sim 2}\beta\lambda_1^{-1}\lambda_2^{-2}}{2[\varepsilon_r + \alpha(T - T_0) + \beta\lambda_1\lambda_2]^2}$$

(14.88)

$$E^{\sim} = \frac{1}{[\varepsilon_r + \alpha(T - T_0) + \beta\lambda_1\lambda_2]} D^{\sim}\lambda_1^{-2}\lambda_2^{-2}$$

(14.89)

$$S^{\sim} = -\frac{\partial U(\lambda_1, \lambda_2, T)}{\partial T} + \frac{\alpha D^{\sim 2}\lambda_1^{-2}\lambda_2^{-2}}{2[\varepsilon_r + \alpha(T - T_0) + \beta\lambda_1\lambda_2]^2}$$

(14.90)

The corresponding true stress, true entropy density and true electrical field of the thermo-electromechanical coupling system are as follows [79]:

$$\sigma_1 = \lambda_1\frac{\partial U(\lambda_1, \lambda_2, T)}{\partial \lambda_1} - \frac{D^{\sim 2}\lambda_1^{-2}\lambda_2^{-2}}{[\varepsilon_r + \alpha(T - T_0) + \beta\lambda_1\lambda_2]} - \frac{D^{\sim 2}\beta\lambda_1^{-1}\lambda_2^{-1}}{2[\varepsilon_r + \alpha(T - T_0) + \beta\lambda_1\lambda_2]^2}$$

(14.91)

$$\sigma_2 = \lambda_2\frac{\partial U(\lambda_1, \lambda_2, T)}{\partial \lambda_2} - \frac{D^{\sim 2}\lambda_1^{-2}\lambda_2^{-2}}{[\varepsilon_r + \alpha(T - T_0) + \beta\lambda_1\lambda_2]} - \frac{D^{\sim 2}\beta\lambda_1^{-1}\lambda_2^{-1}}{2[\varepsilon_r + \alpha(T - T_0) + \beta\lambda_1\lambda_2]^2}$$

(14.92)

$$S = -\frac{\partial U(\lambda_1, \lambda_2, T)}{\partial T} + \frac{\alpha D^{\sim 2}\lambda_1^{-2}\lambda_2^{-2}}{2[\varepsilon_r + \alpha(T - T_0) + \beta\lambda_1\lambda_2]^2}$$

(14.93)

$$E = \frac{1}{[\varepsilon_r + \alpha(T - T_0) + \beta\lambda_1\lambda_2]} D^{\sim}\lambda_1^{-1}\lambda_2^{-1}$$

(14.94)

14.3.5.8   Model of Viscoelasticity Dielectric Elastomer

An elastomer responds to forces and a voltage by time-dependent, dissipative processes. Viscoelastic relaxation may result from the slippage between long polymers and the rotation of joints between monomers. Dielectric relaxation may result from the distortion of electron clouds and the rotation of polar groups. Conductive relaxation may result from the migration of electrons and ions through the elastomer. This section describes an approach to constructing models of dissipative dielectric elastomers according to non-equilibrium thermodynamics [1, 81].

As shown in Figure 14.33, Thermodynamics holds that the increase in free energy should not be larger than the total amount of work performed [1, 81]:

$$\delta F \leq P_1\delta l_1 + P_2\delta l_2 + \Phi\delta Q$$

(14.95)

By dividing Equation 14.95 by $l_1 l_2 l_3$, we have

$$\delta W \leq s_1\delta\lambda_1 + s_2\delta\lambda_2 + E^{\sim}\delta D^{\sim}$$

(14.96)

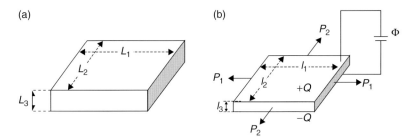

Figure 14.33 *The viscoelasticity electromechanical coupling system of a dielectric elastomer [81]. (Reprinted with permission from [81] Copyright (2011) World Scientific Publishing Co.).*

In a dielectric elastomer model, the free energy density function can be described as the following function [1,81]:

$$W = W(\lambda_1, \lambda_2, D^\sim, \xi_1, \xi_2, \ldots) \tag{14.97}$$

Thus, we can apply λ_1, λ_2, $D^\sim$ and the additional parameter $(\xi_1, \xi_2, \ldots)$ to describe the elastomer's properties. Substituting expression 14.97 into 14.96 produces [1,81]

$$\left(\frac{\partial W}{\partial \lambda_1} - s_1\right)\delta\lambda_1 + \left(\frac{\partial W}{\partial \lambda_2} - s_2\right)\delta\lambda_2 + \left(\frac{\partial W}{\partial D^\sim} - E^\sim\right)\delta D^\sim + \sum_i \frac{\partial W}{\partial \xi_i}\delta\xi_i \leq 0 \tag{14.98}$$

Assuming that the thermodynamic system describing the dielectric elastomer is balanced with respect to the applied mechanical and electrical forces, according to expression 14.98 the nominal stress and the nominal electrical field are [1,81]

$$s_1 = \frac{\partial W(\lambda_1, \lambda_2, D^\sim, \xi_1, \xi_2, \ldots)}{\partial \lambda_1} \tag{14.99}$$

$$s_2 = \frac{\partial W(\lambda_1, \lambda_2, D^\sim, \xi_1, \xi_2, \ldots)}{\partial \lambda_2} \tag{14.100}$$

$$E^\sim = \frac{\partial W(\lambda_1, \lambda_2, D^\sim, \xi_1, \xi_2, \ldots)}{\partial D^\sim} \tag{14.101}$$

Substituting expression 14.98 into 14.99, 14.100, and 14.101 produces [1,81]

$$\sum_i \frac{\partial W(\lambda_1, \lambda_2, D^\sim, \xi_1, \xi_2, \ldots)}{\partial \xi_i}\delta\xi_i \leq 0 \tag{14.102}$$

The thermodynamic inequality allows the expressions $(\delta\xi_1, \delta\xi_2, \ldots)$ and $(\partial W/\partial\xi_1, \partial W/\partial\xi_2, \ldots)$ to satisfy the following relationship [1,81].

$$\frac{d\xi_i}{dt} = -\sum_j M_{ij}\frac{\partial W}{\partial \xi_j} \tag{14.103}$$

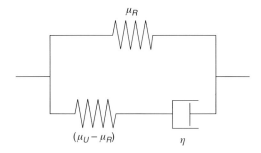

Figure 14.34 *The viscoelastic model of a dielectric elastomer [81]. (Reprinted with permission from [81] Copyright (2011) World Scientific Publishing Co.).*

Here, M_{ij} is a positive definite matrix that depends on the independent variables $(\lambda_1, \lambda_2, D^\sim, \xi_1, \xi_2, \ldots)$. The rate of energy dissipation can be expressed as [1,81]

$$-\sum_i \frac{\partial W}{\partial \xi_i} \frac{d\xi_i}{dt} = \sum_i M_{ij} \frac{\partial W}{\partial \xi_i} \frac{\partial W}{\partial \xi_j} \tag{14.104}$$

Theoretically, we can use different combinations of springs and dashpots to describe an elastomer's viscoelasticity properties [1,81]. Figure 14.34 provides an example of such a model when the material is suddenly stretched; the tenseness coefficient is μ_U, the sum of the stiffness coefficients of two parallel strings. In a completely relaxed state, the tenseness coefficient is μ_R. The relaxation time can be expressed as a function of the stiffness coefficient and the viscosity of the dashpot [1,81].

$$\tau = \frac{\eta}{\mu_U - \mu_R} \tag{14.105}$$

We assume that the stretching of the elastomer is governed by two factors [1,81]:

$$\lambda_1 = \lambda_1^e \xi_1 \tag{14.106}$$
$$\lambda_2 = \lambda_2^e \xi_2 \tag{14.107}$$

Here, $(\lambda_1^e, \lambda_2^e)$ is the stretching of the spring and (ξ_1, ξ_2) is the stretching of the dashpot. Thus, the viscoelastic free energy of the thermodynamical system of the dielectric can be described as follows [1,81]:

$$\begin{aligned}
W(\lambda_1, \lambda_2, D^\sim, \xi_1, \xi_2, \ldots) &= \frac{\mu_R}{2} \left(\lambda_1^2 + \lambda_2^2 + \lambda_1^{-2}\lambda_2^{-2} - 3 \right) \\
&+ \frac{\mu_U - \mu_R}{2} \left(\lambda_1^2 \xi_1^{-2} + \lambda_2^2 \xi_2^{-2} + \lambda_1^{-2}\lambda_2^{-2}\xi_1^2\xi_2^2 - 3 \right) \\
&+ \frac{D^{\sim 2}}{2}\lambda_1^{-2}\lambda_2^{-2}
\end{aligned} \tag{14.108}$$

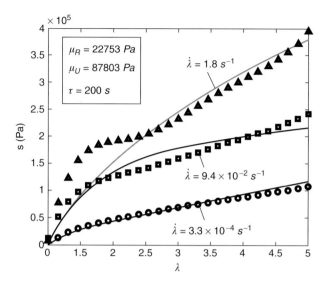

Figure 14.35   *The viscoelastic behaviour of a dielectric elastomer [81]. The squares, triangles and circles represent the relationship between stretching and stress under different rates of stretch tested by experiments. (Reprinted with permission from [81] Copyright (2011) World Scientific Publishing Co.).*

The nominal stress and the electrical field of the viscoelastic dielectric are as follows [1, 81]:

$$s_1 = \mu_R \left(\lambda_1 - \lambda_1^{-3}\lambda_2^{-2} \right) + (\mu_U - \mu_R)\left(\lambda_1\xi_1^{-2} - \lambda_1^{-3}\lambda_2^{-2}\xi_1^2\xi_2^2 \right) - \frac{D^{~2}}{\varepsilon}\lambda_1^{-3}\lambda_2^{-2}$$

$$(14.109)$$

$$s_2 = \mu_R \left(\lambda_2 - \lambda_2^{-3}\lambda_1^{-2} \right) + (\mu_U - \mu_R)\left(\lambda_2\xi_2^{-2} - \lambda_2^{-3}\lambda_1^{-2}\xi_1^2\xi_2^2 \right) - \frac{D^{~2}}{\varepsilon}\lambda_1^{-2}\lambda_2^{-3}$$

$$(14.110)$$

$$E^{~} = \frac{D^{~2}}{\varepsilon}\lambda_1^{-2}\lambda_2^{-2}$$

$$(14.111)$$

$$\frac{d\xi_1}{dt} = \frac{1}{\tau}\left(\lambda_1^2\lambda_1^{-3} - \lambda_1^{-2}\lambda_2^{-2}\xi_1\xi_2^2 \right)$$

$$(14.112)$$

$$\frac{d\xi_2}{dt} = \frac{1}{\tau}\left(\lambda_2^2\lambda_2^{-3} - \lambda_1^{-2}\lambda_2^{-2}\xi_2\xi_1^2 \right)$$

$$(14.113)$$

Figure 14.35 illustrates the viscoelastic behaviour of a dielectric elastomer; the experimental results prove that the theoretical analysis is well grounded [81].

14.4   Failure Model of a Dielectric Elastomer

The failure of a dielectric elastomer is caused by many factors, such as electrical breakdown, electromechanical instability, snap-through instability, disappearance of the tension, and

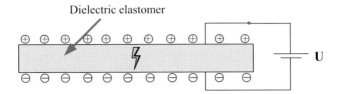

Figure 14.36 *Electrical breakdown of dielectric elastomers [91]. (Reproduced with permission from [91] Copyright (2012) Springer Science + Business Media).*

stretch break [1, 82–86]. These factors can serve as control conditions for a theoretical description of the allowable range of dielectric elastomer [1, 82–86].

14.4.1 Electrical Breakdown

Under an applied voltage, if a dielectric is stiff, small deformation may lead to electrical breakdown (EB); however, if the dielectric is soft, electrical breakdown (EB) occurs at larger deformation. The breakdown voltage is different [1, 82–86]. As shown in Figure 14.36, assuming that the breakdown electric field is a constant E_B, for a stiff material (Figure 14.36), $U_B = E_B H$; for soft materials (Figure 14.37), $U_B = E_B H \lambda^{-2}$ [1, 82–86].

The applied voltage changes little with the deformation if the dielectric is hard, whereas for a soft material, because its extent of deformation is greater, the applied voltage changes greatly with the deformation [1, 82–86]. As shown in Figure 14.37, the voltage decreases monotonically with the increase in the extent of stretching [1, 82–86].

14.4.2 Electromechanical Instability and Snap-Through Instability

A dielectric elastomer transducer consists of a thin polymer membrane sandwiched between compliant electrodes. Under an applied voltage, the dielectric elastomer shrinks in thickness and expands in area, effectively increasing the electric field subjected to the membrane. This positive feedback continues until the electric field reaches the critical value that causes the dielectric elastomer to break down; this process is called electromechanical instability [1, 82–86]. Figure 14.38 illustrates the typical voltage-strain relationship for deformable dielectric going through electromechanical stability. As the strain increases, the voltage

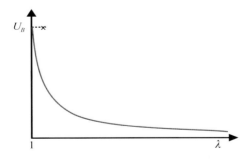

Figure 14.37 *Electrical breakdown of dielectric elastomers [83]. (Reprinted under the terms of the STM agreement from [83] Copyright (2011) Wiley-VCH Verlag GmbH & Co. KGaA).*

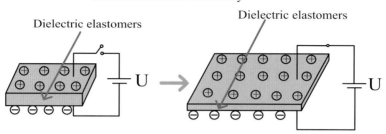

Figure 14.38 *Electromechanical instability of dielectric elastomers [91]. (Reproduced with permission from [91] Copyright (2012) Springer Science + Business Media).*

increases [1, 82–86]. When the strain approaches its limit λ_C, the voltage approaches the breakdown voltage of the dielectric and the electromechanical instability appears.

As illustrated in Figure 14.39, under an applied voltage, due to the strain stiffening of the elastomer, the deformable dielectric may first reach a local maximum when the extent of stretching reaches λ_C, and electromechanical instability will not appear [1, 82–86]. As the voltage increases, the deformable dielectric will experience snap-through, and the dielectric will remain stable until the deformation approaches the $\lambda_{\lim}$ (the maximum deformation) [1, 82–86].

The voltage-stretch curve shown in Figure 14.40 first reaches the local maximum and then decreases; upon approaching the limiting stretch $\lambda_{\lim}$, the curve increases steeply. This process presents the snap-through stability of the deformable dielectric. Clearly, the voltage-stretch curve is not monotonic [1, 82–86].

14.4.3   Loss of Tension

To greatly deform a dielectric elastomer, there needs to be a high driving electrical field. However, the output voltage of currently used instruments is limited, and thus, the membrane must be thin. For example, when an electrical field of 10^8 V/m – the critical electrical field – is applied, if the high-voltage power supply can output 10^5 V at most, the membrane must be as thin as 1 mm [1, 3, 10, 73]. As shown in Figure 14.41, when a thin film suffers slight

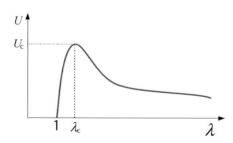

Figure 14.39 *Electromechanical instability of dielectric elastomers [44]. (Reprinted with permission from [44] Copyright (2012) Elsevier Ltd).*

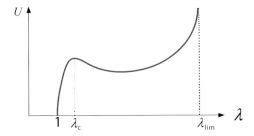

Figure 14.40   *Snap through instability of dielectric elastomers [44]. (Reprinted with permission from [44] Copyright (2012) Elsevier Ltd).*

pressure, the membrane may wrinkle, which may then lead to failure. Thus, to avoid this failure, we must add a stretching force to the membrane [1, 82–86].

14.4.4   Rupture by Stretching

Under an applied mechanical force, the macromolecular polymer chains in an elastomer suffer severe stretching after strain hardening [1, 82–86]. The chains then reach their stretching limit, causing the elastomer to rupture by stretching, as illustrated in Figure 14.42.

14.4.5   Zero Electric Field Condition

The strain–stress relationship (zero electric field condition) of a dielectric elastomer under an applied mechanical force is a controlling parameter in describing the elastomer's theoretically allowable area. Given a specific elastic strain energy model, we can obtain the strain–stress relationship of a dielectric elastomer [1, 82–86].

14.4.6   Super-Electrostriction Deformation of a Dielectric Elastomer

When a high electric field is applied to a dielectric, if the dielectric is hard and the deformation is small, the extent of deformation is determined by the breakdown voltage [75]. If the dielectric is soft, the deformation is large, so the deformation is determined by

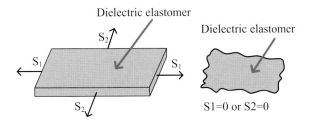

Figure 14.41   *Loss of tension in dielectric elastomer [91]. (Reproduced with permission from [91] Copyright (2012) Springer Science + Business Media).*

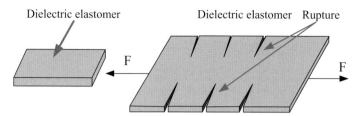

Figure 14.42 *The rupture of dielectric elastomer by stretching [91]. (Reproduced with permission from [91] Copyright (2012) Springer Science + Business Media).*

a combination of the breakdown voltage, electromechanical instability, and snap-through instability [75].

Figure 14.43 shows several possible forms of dielectric elastomer electrostriction [75]. When no pre-stretch is applied to the dielectric elastomer (no mechanical force) or only a small degree pre-stretch is applied (the influence of the electrostriction curve is smalls), the typical voltage-stretch curve shows an increasing trend until a local maximum is reached; the curve then decreases and then increases suddenly. The breakdown voltage diminishes monotonically within the increase in stretch [75]. In Figure 14.43(b), if the electrostriction curve of the dielectric elastomer and the breakdown voltage curve intersect before reaching a local maximum, that is, the dielectric has already been broken down by the electric field before reaching electromechanical instability, for harder dielectric elastomers the electromechanical deformation is small [75]. According to Figure 14.43(c), if the two

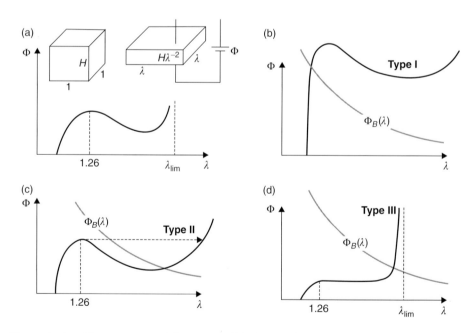

Figure 14.43 *Electrostriction deformation of dielectric elastomers [75]. (Reprinted with permission from [75] Copyright (2010) American Physical Society).*

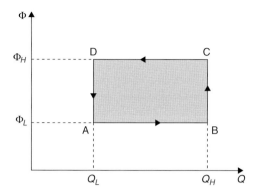

Figure 14.44　*The conversion cycle process of a dielectric elastomer converter [1]. (Reprinted under the terms of the STM agreement from [1] Copyright (2010) Elsevier Ltd).*

curves intersect after the local maximum, that is, the cause of failure is electromechanical instability, the dielectric does not experience snap-through deformation and undergoes large electrostriction deformation [75]. Moreover, Figure 14.43(d) shows that if the dielectric elastomer does not fail at the local maximum that represents the onset of electromechanical instability (for strain hardening), then it will experience snap-through deformation and undergo larger electrostriction deformation [75].

14.5　Converter Theory of Dielectric Elastomer

14.5.1　Principle for Conversion Cycle

The practical application of a dielectric elastomer converter features the cycling of the different states of an elastomer. For example, using the voltage and electric field as variables, we can form a plane to describe the process of dielectric elastomer conversion. Each point represents the state of the converter [1]. To make the converter function normally during cycling, we need two kinds of batteries: one with a low voltage Φ_L and another with a high voltage Φ_H. In the rectangular thermodynamic cycle shown in Figure 14.44, the converter undergoes a four-path cycle [1].

From State A to B, the low voltage applied to the converter Φ_L is maintained. Because the applied force reduces the distance between the electrodes, the charge on the electrode increases [1].

From State B to C, the converter is an open circuit, so the high electrode charge Q_H remains constant. Because the applied force increases the distance between the electrodes, the voltage increases to Φ_H [1].

From State C to D, the high voltage applied to the converter Φ_H remains constant. Because the applied force increases the distance between the electrodes, the charge on the electrode decreases [1].

From State D to A, the converter is an open circuit, so the low charge on the electrodes Q_L remains constant. Because the applied force reduces the distance between the electrodes, the voltage decreases to Φ_L [1].

If we use temperature to replace voltage, entropy to replace charge, the electromechanical converter cycle is similar to the Carnot cycle. During the cycle, the converter absorbs mechanical energy from the environment, draws an amount of charge from the low-voltage battery, and deposits the same amount of charge to the high-voltage battery [1]. At this time, the converter is an energy collector, absorbing mechanical energy to produce power. It can use mechanical energy produced during animal or human motor processes, and it also can use the mechanical energy produced by sea vibrations [1].

In fact, on plane (Q,Φ), any closed curve represents an allowable dielectric elastomer converter cycle path [1]. We need a variable voltage source to drive the cycle. The amount of energy converted in every single cycle is represented by the curve surrounding the area on plane (Q,Φ) [1]. In Figure 14.44, the cycle on plane (Q,Φ) proceeds in a counterclockwise manner, making the converter an energy collector that uses mechanical energy to produce power [1]. If the cycle proceeds in a clockwise manner, the converter turns power into mechanical energy. Similarly, the cycle path of the converter can be described on plane (l,P). The four processes described on this plane are equivalent to the processes on plane (Q,Φ) [1].

14.5.2   Plane Actuator

Figure 14.45 illustrates the plane drive schemes of a dielectric elastomer. We used the free energy model of the two-term Mooney–Rivlin elastic strain energy function to describe the mechanical behaviour and stability behaviour of a plane-type dielectric elastomer [60]:

$$W(\lambda_1, \lambda_2, D^{\sim}) = \frac{\mu}{2}\left(\lambda_1^2 + \lambda_2^2 + \lambda_1^{-2}\lambda_2^{-2} - 3\right) + \frac{G}{2}\left(\lambda_1^{-2} + \lambda_2^{-2}\right)$$

$$+ \lambda_1^2\lambda_2^2 - 3) + \frac{D^{\sim2}}{2\varepsilon}\lambda_1^{-2}\lambda_2^{-2} \tag{14.114}$$

μ and G are the material constants that can be measured by experiment. For different dielectric elastomer materials (e.g. BJB TC-A/BC, HS3silicone, CF19-2186 silicone, VHB 4910) and different dielectric elastomer drives (e.g. scroll form, folding form, stacked form, flat form), the material constants are quite different [60,69].

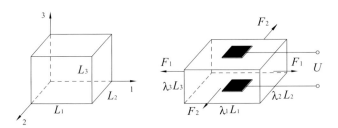

Figure 14.45   *Dielectric elastomer plane actuator [60]. (Reproduced with permission from [60] Copyright (2009) Springer Science + Business Media).*

The pre-tensioning of the dielectric elastomer film makes the two directions of the tensile rate and the two directions of the nominal stress the same (equiaxial pre-tension), namely $s_1 = s_2 = s$, $\lambda_1 = \lambda_2 = \lambda$ [59, 60, 65]. To analyse the stability behaviour of different dielectric elastomer materials or structures, let $\mu = kG$ (k is a constant), where k is related to the dielectric elastomer material and the structure of the dielectric elastomer drive. The nominal electric field and the nominal electric displacement relationship can be described as follows [60]:

$$\frac{D^{\sim}}{\sqrt{\varepsilon G}} = \sqrt{k(\lambda^6 - 1) + \lambda^8 - \lambda^2 - \frac{s}{G}\lambda^5} \qquad (14.115)$$

$$\frac{E^{\sim}}{\sqrt{G/\varepsilon}} = \sqrt{k(\lambda^{-2} - \lambda^{-8}) + 1 - \lambda^{-6} - \frac{s}{G}\lambda^{-3}} \qquad (14.116)$$

When the value of k changes, the stretch rate λ is used as a variable to analyse the stability of dielectric elastomer materials [60]. In Figure 14.46, (a), (b), (c), and (d) illustrate the relationship $\frac{D^{\sim}}{\sqrt{\varepsilon G}}$, $\frac{E^{\sim}}{\sqrt{G/\varepsilon}}$ when $k = 1$, $k = 2$, $k = 2.5$, and $k = 5$, respectively [60]. Let $\frac{s}{G}$ assume different values, for example, 0, 1, 2, 3, 4, 5; $E^{\sim}$ reaches a peak for each value. The curve on the left of the peak value indicates that the Hessian matrix is positive definite; the curve on the right indicates that the Hessian matrix is negative definite; and the peak value in the middle indicates that the Hessian matrix $\det(H) = 0$. As $\frac{s}{G}$ increases, when the

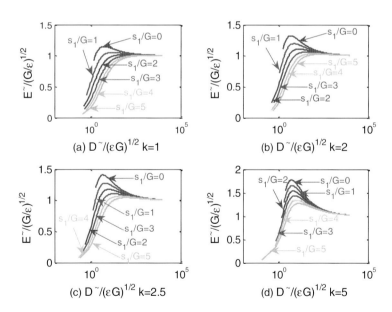

Figure 14.46 *The nominal electric field vs. the nominal electric displacement when k changes* [60]. *(Reproduced with permission from [60] Copyright (2009) Springer Science + Business Media). See plate section for colour figure.*

value of k changes, the nominal electric field decreases. This suggests that the increase in the degree of pre-tensioning can improve the stability of a dielectric elastomer [1,56,60,69].

When $k = 2$, the peak of the corresponding nominal electric field is $E^{\sim}_{max} = 0.93\sqrt{2G/\varepsilon}$, the representative value $G = 0.25 \times 10^6$ pa, and $\varepsilon = 4 \times 10^{-11}$ F/m; thus, $E^{\sim}_{max} \approx 1.04 \times 10^8$ V/m. The critical nominal electric field of the dielectric elastomer is consistent with the results obtained by Suo and is consistent with the data regarding the breakdown electric field intensity of a dielectric elastomer material [39,56,70,71]. Thus, we obtain a maximum λ^C of 1.37, with a corresponding thickness of 46% [39,70,71].

14.5.3   Spring-Roll Dielectric Elastomer Actuator

Figure 14.47 illustrates the schema of a spring-roll dielectric elastomer actuator. Given the Helmholtz free energy H as a function of two generalized coordinates [87],

$$H(\lambda_1, Q) = \frac{\mu}{2}\left[\lambda_1^2 + \left(\lambda_2^p\right)^2 + \left(\lambda_1\lambda_2^p\right)^{-2} - 3\right]L_1L_2L_3$$

$$+ \frac{1}{2\varepsilon}\left(\frac{Q}{\lambda_1 L_1 \lambda_2^p L_2}\right)^2 L_1L_2L_3 + \frac{1}{2}k\left(\lambda_1 L_1 - \lambda_1^p L_1\right)^2 \quad (14.117)$$

λ_1^p and λ_2^p are the pre-tension along two directions of a spring-roll actuator, and k is the strength coefficient of the spring.

When the drive is under the joint action of a mechanical force P and voltage Φ, the tension and electric charge, $d\lambda_1$ and dQ respectively, are small. The change in the Helmholtz free energy for a spring-roll actuator should equal to the total work performed by the external forces and the work performed by the external voltage [87]:

$$dH = PL_1 d\lambda_1 + \Phi dQ \quad (14.118)$$

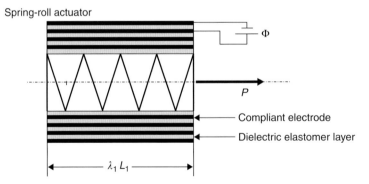

Figure 14.47   *The construction of a spring-roll dielectric elastomer actuator [87]. (Reprinted with permission from [87] Copyright (2008) American Institute of Physics).*

Thus, the mechanical force and voltage are the partial differentials of the Helmholtz free energy [87]:

$$P = \frac{\partial H(\lambda_1, Q)}{L_1 \partial \lambda_1} = \mu L_2 L_3 \left\{ \left[\lambda_1 - (\lambda_1)^{-3} \left(\lambda_2^p \right)^{-2} \right] \right.$$
$$\left. - \frac{1}{\lambda_1^3 \left(\lambda_2^p \right)^2} \left(\frac{Q}{\sqrt{\mu \varepsilon L_1 L_2}} \right)^2 + \alpha \left(\lambda_1 - \lambda_1^p \right) \right\} \tag{14.119}$$

$$\Phi = \frac{\partial H(\lambda_1, Q)}{\partial Q} = \frac{L_3}{\left(\lambda_1 \lambda_2^p \right)^2} \left(\frac{Q}{\varepsilon L_1 L_2} \right) \tag{14.120}$$

When a spring-roll actuator becomes mechanically and electrically unstable, the electric charge follows the governing equation [87]

$$\frac{Q}{\sqrt{\mu \varepsilon L_1 L_2}} = \sqrt{(1 + \alpha) \lambda_1^4 \left(\lambda_2^p \right)^2 + 3} \tag{14.121}$$

At electric breakdown, the relationship between electric charge and tension is [87]

$$\frac{Q}{\sqrt{\mu \varepsilon L_1 L_2}} = \lambda_1 \lambda_2^p E_C \sqrt{\frac{\varepsilon}{\mu}} \tag{14.122}$$

If the tension along direction 1 is eliminated, that is, $s_1 = 0$, the relationship between the electric charge and tension is as follows [87]:

$$\frac{Q}{\sqrt{\mu \varepsilon L_1 L_2}} = \sqrt{\lambda_1^4 \left(\lambda_2^p \right)^2 - 1} \tag{14.123}$$

When $s_2 = 0$, the corresponding relationship between electric charge and tension is [87]

$$\frac{Q}{\sqrt{\mu \varepsilon L_1 L_2}} = \sqrt{\lambda_1^2 \left(\lambda_2^p \right)^4 - 1} \tag{14.124}$$

The tensile limit is $\lambda_C = 5$. According to Equations 14.120, 14.121, 14.122, 14.123, and 14.124, the allowable area of a spring-roll actuator is represented as the shaded region in Figure 14.48 [87].

14.5.4 Tube-Type Actuator

Figure 14.49 illustrates the schema of a dielectric elastomer tube-type actuator [88]. In the undeformed state, the tube is of length L, inner radius A, and outer radius B. Pre-stretched by a load P, the tube is of length l_p, inner radius a_p, and outer radius b_p. Subject to a load and a voltage, the tube is of length l, inner radius a, and outer radius b. The strain of actuation due to the voltage is defined as $(l - l_p)/l_p$ [88]. The axial stretch is $\lambda_z = \frac{l}{L}$. Assuming that the dielectric elastomer is incompressible, the volume of the tube before and after deformation is constant [88]:

$$(R^2 - A^2) = \lambda_z (r^2 - a^2) \tag{14.125}$$

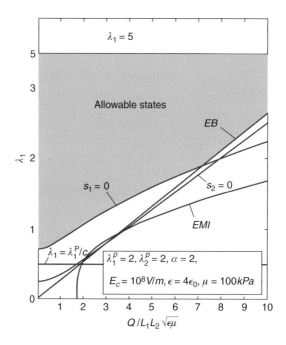

Figure 14.48   *The allowable area of a dielectric elastomer spring-roll actuator [87]. (Reprinted with permission from [87] Copyright (2008) American Institute of Physics).*

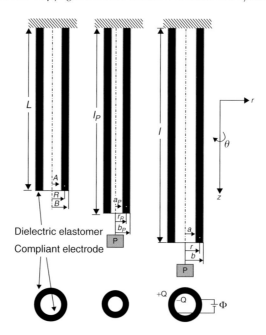

Figure 14.49   *Dielectric elastomer tube-type actuator [88]. (Reprinted with permission from [88] Copyright (2010) American Institute of Physics).*

Under the deformation condition, the radius of the outer boundary b follows the equation [88]

$$b = \sqrt{a^2 + (B^2 - A^2)\lambda_Z^{-1}} \tag{14.126}$$

The hoop stretch $\lambda_\theta = \frac{r}{R}$, so it can be described by the following equation, with variables r and λ_z [88]:

$$\lambda_\theta(r) = \frac{r}{\sqrt{A^2 + (r^2 - a^2)\lambda_Z}} \tag{14.127}$$

Due to incompressibility, the radial stretch $\lambda_r(r)\lambda_\theta(r)\lambda_Z = 1$, so [88]

$$\lambda_r(r) = \frac{\sqrt{A^2 + (r^2 - a^2)\lambda_Z}}{r\lambda_Z} \tag{14.128}$$

Under the deformation condition, a tube-type actuator follows the equilibrium equation [88]

$$\frac{d\sigma_r}{dr} + \frac{\sigma_r - \sigma_\theta}{r} = 0 \tag{14.129}$$

Let $V(r)$ be the potential; thus, the electric field E can be described as [88]

$$E = -\frac{dV(r)}{dr} \tag{14.130}$$

The relationship between the electric displacement and the electric charge is [88]

$$D = \frac{Q}{2\pi r \lambda_Z L} \tag{14.131}$$

Assuming that the material model used is neo-Hookean, the stress-strain, electric field-electric displacement relationships can be expressed as follows [88]:

$$\sigma_\theta - \sigma_r = \mu \left(\lambda_\theta^2 - \lambda_\theta^{-2}\lambda_Z^2\right) - \varepsilon E^2 \tag{14.132}$$

$$\sigma_Z - \sigma_r = \mu \left(\lambda_Z^2 - \lambda_\theta^{-2}\lambda_Z^2\right) - \varepsilon E^2 \tag{14.133}$$

$$D = \varepsilon E \tag{14.134}$$

Using Equations 14.133 and 14.136, we can obtain an expression for the voltage as follows [88]:

$$\Phi = \frac{Q}{2\pi \varepsilon \lambda_Z L} \log \frac{b}{a} \tag{14.135}$$

According to Equations 14.130, 14.131, and 14.132 and 14.134, the electric charge can be expressed as [88]

$$Q = \sqrt{\mu\varepsilon}(2\pi L A)\sqrt{-1 + \lambda_Z \left(\frac{a}{A}\right)^2 + \frac{2\lambda_Z(a/A)^2}{1 - (a/b)^2} \log \frac{aB}{bA}} \tag{14.136}$$

Using the boundary condition $\sigma_r(a) = \sigma_r(b) = 0$ and substituting Equations 14.126 and 14.136 into 14.135, we obtain the following expression for the voltage [88]:

$$\Phi = \sqrt{\frac{\mu}{\varepsilon}\left(\frac{-A^2 + \lambda_Z a^2}{\lambda_Z^2} + \frac{2a^2 b^2}{B^2 - A^2}\log\frac{aB}{bA}\right)\log\frac{b}{a}} \tag{14.137}$$

Substituting Equations 14.131, 14.132 and 14.134 into 14.129 and integrating Equation 14.129 from a to r, we obtain the stress along the radial direction [88]:

$$\sigma_r(r) = \frac{\mu}{\lambda_Z}\left[\log\frac{aR}{Ar} - \frac{(r^2 - a^2)b^2}{(b^2 - a^2)r^2}\log\frac{aB}{bA}\right] \tag{14.138}$$

Substituting Equation 14.138 into 14.132 and 14.133, we obtain the stress along the other two directions [88]:

$$\sigma_\theta(r) = \frac{\mu}{\lambda_Z}\left[\log\frac{aR}{Ar} - \frac{(r^2 + a^2)b^2}{(b^2 - a^2)r^2}\log\frac{aB}{bA} + \frac{r^2\lambda_Z}{R^2} - 1\right] \tag{14.139}$$

$$\sigma_Z(r) = \frac{\mu}{\lambda_Z}\left[\log\frac{aR}{Ar} - \frac{(r^2 + a^2)b^2}{(b^2 - a^2)r^2}\log\frac{aB}{bA} + \lambda_Z^3 - 1\right] \tag{14.140}$$

Similarly, the axial force can be calculated by $P = \int_a^b 2\pi r\sigma_z(r)dr$; here, we substitute Equations 14.126 and 14.140 into this expression, integrate, and obtain [88]

$$P = \mu\pi\left[\left(\frac{A^2 - \lambda_Z a^2}{\lambda_Z^2}\right)\log\frac{B}{A} + \left[\frac{(\lambda_Z^3 - 1)(B^2 - A^2)}{\lambda_Z^2}\right] + \left(\frac{2a^2 b^2}{B^2 - A^2}\right)\log\frac{Bb}{Aa}\log\frac{b}{a}\right] \tag{14.141}$$

Figure 14.50 illustrates the relationship between the voltage and deformation of a dielectric elastomer tube-type actuator [88].

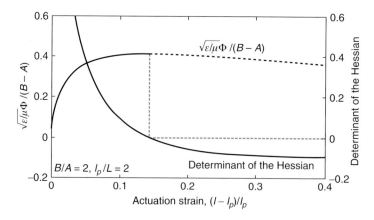

Figure 14.50 *Dielectric elastomer tube-type actuator [88]. (Reprinted with permission from [88] Copyright (2010) American Institute of Physics).*

14.5.5   Film-Spring System

A cross-section of the dielectric elastomer film-spring system is shown in Figure 14.51 [62, 63]. Figure 14.51(a) describes the undeformed state of the film [62, 63]. The thickness of the film is H [62, 63]. The compliant electrodes are pasted on both of its sides with radius B. The distance from the film's central opening to a particular particle is A. The original length of the spring is l_0, and we assume that every point on the film is a distance R away from the centre. Thus, the boundary conditions are $R(A) = A$ and $R(B) = B$. Figure 14.51(b) shows a schematic diagram of the system after deformation [62, 63]. A certain particle A is linked to a disk with radius a. The periphery of the film is linked to a rigid ring with a radius of b. The spring is fastened to the centre of the disk. When the spring force F acts on the disk, a voltage is applied to the two electrodes, causing a certain charge Q to move from one side of the compliant electrode to the other. The disk then undergoes a vertical displacement u as the film produces a large deformation and becomes axisymmetric along the Z axis.

In treating this system, first we consider the deformation of the film. Each point on the film after deformation can be determined by the coordinates $r(R)$ and $z(R)$, as shown in Figure 14.51(b), and $r(R)$ and $z(R)$ satisfy the boundary conditions below [62, 63]. The inner boundary of the film is fixed to the rigid disk. Under axisymmetric conditions, there is vertical displacement but no lateral displacement: $r(A) = a$, $r(B) = b$, and $z(B) = 0$.

The extent of stretching of the film with respect to a given point under deformation can be determined by the deformation of the point next to it, that is, the deformation is captured by

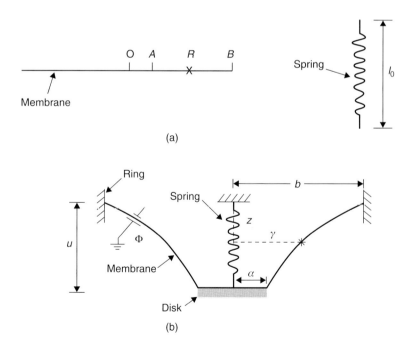

(a)

(b)

Figure 14.51   *Film-spring system [62, 63]. (Reprinted from [62] Copyright (2009) American Institute of Physics; and from [63] Copyright (2010) IOP Publishing).*

the distance from R to $R + dR$ [62, 63]. The horizontal distance is $dr = r(R + dR) - r(R)$, and the vertical distance is $dz = z(R + dR) - z(R)$. Assuming that dl is the arc length between the two points, $dz = \cos\theta dl$, $dr = -\sin\theta dl$ and $(dl)^2 = (dr)^2 + (dz)^2$, where θ represents the intersection angle between the tangent line of each point on the membrane and the horizontal direction. The term dl can be expressed by the two coordinates $r(R)$ and $z(R)$, and stretching along the radial direction can be expressed as [62, 63]

$$dr = r(R + dR) - r(R) \tag{14.142}$$

The stretching deformation along the membrane loop direction is defined by the ratio of the circular perimeter where point R is located in the deformed state of the membrane and the non-deformed perimeter $2\pi R$ [62, 63]:

$$\lambda_2 = \frac{r}{R} \tag{14.143}$$

The change in the electrical field can be described using the nominal dielectric displacement. In the deformed state, $D^\sim$ represents the nominal displacement, and the nominal displacement represents the ratio between the areal charge on the electrodes in the deformed state and the areal charge in the non-deformed state [62, 63]. Thus, the total electric charge on the electrodes can be expressed as [62, 63]

$$Q = 2\pi \int D^\sim R dR \tag{14.144}$$

Regarding the thermodynamic system of the membrane, assume that the film maintains a constant temperature [62, 63]. The energy density function can be described by the stretching variables λ_1 and λ_2 and the function of the nominal electrical displacement $D^\sim$. Thus, in the deformed state, the free energy of the whole film can be expressed by $2\pi H \int W R dR$. If λ_1, λ_2, and $D^\sim$ vary by $\delta\lambda_1$, $\delta\lambda_2$, and $\delta D^\sim$, respectively, the variation in the free energy density can be expressed as [62, 63]

$$\delta W = s_1 \delta\lambda_1 + s_2 \delta\lambda_2 + E^\sim \delta D^\sim \tag{14.145}$$

According to the differential relationship, we obtain the following [62, 63]:

$$s_1 = \frac{\partial W(\lambda_1, \lambda_2, D^\sim)}{\partial \lambda_1} \tag{14.146}$$

$$s_2 = \frac{\partial W(\lambda_1, \lambda_2, D^\sim)}{\partial \lambda_2} \tag{14.147}$$

$$E^\sim = \frac{\partial W(\lambda_1, \lambda_2, D^\sim)}{\partial D^\sim} \tag{14.148}$$

When the displacement of the stiffness disk varies by δu, the external force does work $F\delta u$ [62, 63]. If the electrical charge varies by δQ, the external voltage does work $\Phi\delta Q$. At the equilibrium state, according to thermodynamics, the change in the Helmholtz free energy of the film is equal to the sum of the work performed by the external forces and external voltage, which is described by the variation function below [62, 63]:

$$2\pi H \int_A^B \delta W R dR = F\delta u + \Phi\delta Q \tag{14.149}$$

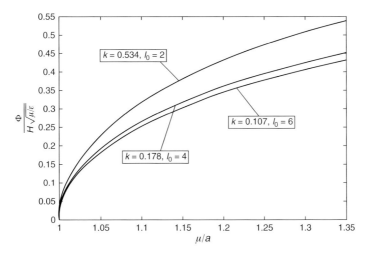

Figure 14.52 *The relationship between the external voltage and the vertical displacement [62, 63]. (Reprinted from [62] Copyright (2009) American Institute of Physics; and from [63] Copyright (2010) IOP Publishing).*

$\delta r(R)$, $\delta r(R)$, and $\delta D^\sim(R)$ represent the variation of the three independent variables $r(R)$, $z(R)$, and $D^\sim(R)$, respectively [62, 63]. Using Equation 14.142, we can determine the variation of the stretching deformation along the radial direction as $\delta\lambda_1 = \cos\theta \frac{d(\delta r)}{dR} - \sin\theta \frac{d(\delta z)}{dR}$. By integrating Equation 14.145, we obtain the following expression [62, 63]:

$$\int_A^B \delta WRdR = (Rs_1 \cos\theta \delta r - Rs_1 \sin\theta \delta z)_A^B + \int_A^B \left[\left(-\frac{d(Rs_1 \cos\theta)}{dR} + s_2 \right) \right.$$

$$\left. + \frac{d(Rs_1 \sin\theta)}{dR}\delta z + RE^\sim \delta D^\sim \right] dR \qquad (14.150)$$

Furthermore, the equilibrium functions are as follows [62, 63]:

$$2\pi HRs_1 \sin\theta = F = k\Delta l \qquad (14.151)$$

$$\frac{d(Rs_1 \cos\theta)}{dR} = s_2 \qquad (14.152)$$

$$HE^\sim = \Phi \qquad (14.153)$$

Figure 14.52 illustrates the relationship between the external voltage and the vertical displacement of a system with a different spring stiffness coefficient k and original spring length l_0 [62, 63]. When the voltage increases, the vertical displacement increases. The spring's properties are determined by the spring's stiffness and original length. The figure shows that at a certain voltage, the vertical displacement increases with a decrease in k and an increase in l_0. If the spring's coefficients and the original vertical displacement are different, the variation in the external voltage and vertical displacement remains the same [62, 63].

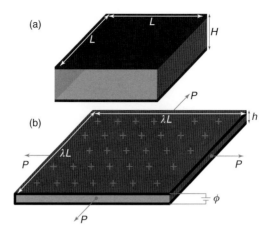

Figure 14.53 *Dielectric elastomer energy harvester [84–86]. (a) Relatively small electric capacity. (b) Large electric capacity. (Reprinted from [84] Copyright (2009) American Institute of Physics; [85] Copyright (2010) EDP Sciences; and [86] Copyright (2011) IEEE).*

14.5.6   Energy Harvester

Dielectric elastomers can be developed to design and manufacture energy harvesters. The principle is illustrated in Figure 14.53 [84–86]. In the stretching state (large electric capacity), the original charge is injected into the membrane electrode of the dielectric elastomer [84–86]. From a macroscopic point of view, in the contraction state (small electric capacity), the elastic stress resists the electric field force then the electrical energy increases. From a microscopic point of view, when the membrane contracts and the thickness increases, opposite charges are pushed apart while like charges are compressed and become closer; thus, the voltage increases. From this we can see that when external forces are applied to a dielectric elastomer material that possesses a built-in electric field and lead to the deformation of the material, the change in capacitance produces electricity: the larger the deformation, the greater the capacitance generated becomes [84–86].

The Mooney–Rivlin model with two material constants can be applied to investigate the common theoretical failure models of dielectric elastomers [84–86]. In fact, the planar equal biaxial stretch is an ideal circumstance; in experiments, this condition can hardly be precisely ensured [86]. Moreover, under many circumstances, we must achieve a certain ratio of biaxial stretch. Therefore, we must consider the non-equal biaxial condition [86]. Assume that $\lambda_2 = p\lambda_1$, where $p > 0$ represents the stretching ratio; $p = 1$ represents equal biaxial stretch; $p > 1$ represents stretching in which the lengthways stretch is always greater; $p < 1$ represents stretching in which the lengthways stretch is always smaller. For simplicity, let $\lambda_1 = \lambda$; thus, the free energy of the dielectric is described as follows [86]:

$$W(\lambda, D^{\sim}) = \frac{C_1}{2}[(1 + p^2)\lambda^2 + p^{-2}\lambda^{-4} - 3] + \frac{kC_1}{2}[(1 + p^{-2})\lambda^{-2}$$
$$+ p^2\lambda^4 - 3] + \frac{D^{\sim 2}}{2\varepsilon}p^{-2}\lambda^{-4} \tag{14.154}$$

Using Equation 14.154, according to the definition of nominal stress and nominal electrical field, obtain [86]

$$\frac{s}{C_1} = (1 + p^2)\lambda - 2p^{-2}\lambda^{-5} + 2kp^2\lambda^3 - k(1 + p^{-2})\lambda^{-3} - \frac{2p^{-2}D^{\sim 2}}{C_1\varepsilon}\lambda^{-5} \quad (14.155)$$

$$\frac{E^{\sim}}{\sqrt{C_1/\varepsilon}} = \frac{D^{\sim}}{\sqrt{C_1\varepsilon}}p^{-2}\lambda^{-4} \quad (14.156)$$

Following the method described above and considering the typical failure models of dielectric elastomers – charge disappearance, electrical breakdown, electromechanical instability, loss of tension, and stretching rupture – we obtain the following functions controlling the stability of an elastomer's allowable area [86]:

$$\begin{cases} \dfrac{s}{C_1} = (1 + p^2)\lambda - 2p^{-2}\lambda^{-5} + 2kp^2\lambda^3 - k(1 + p^{-2})\lambda^{-3} \\[2mm] \dfrac{s}{C_1} = (1 + p^2)\lambda - 2p^{-2}\lambda^{-5} + 2kp^2\lambda^3 - k(1 + p^{-2})\lambda^{-3} - \dfrac{2p^{-2}E_{EB}^2}{C_1/\varepsilon}\lambda^{-1} \\[2mm] \dfrac{s}{C_1} = \dfrac{2(1 + p^2)\lambda - 16p^{-2}\lambda^{-5} - 6k(1 + p^{-2})\lambda^{-3}}{3} \\[2mm] \dfrac{s}{C_1} = 0 \\[2mm] \dfrac{s}{C_1} \geq 0,\ \lambda_R = 5 \end{cases}$$

$$(14.157)$$

$$\begin{cases} \dfrac{E^{\sim}}{\sqrt{C_1/\varepsilon}} = 0,\ \dfrac{D^{\sim}}{\sqrt{C_1\varepsilon}} = 0 \\[2mm] \dfrac{E^{\sim}}{\sqrt{C_1/\varepsilon}} = p^{-2}\dfrac{E_{EB}^2}{C_1/\varepsilon}\left(\dfrac{D^{\sim}}{\sqrt{C_1\varepsilon}}\right)^{-1},\ 0 < \dfrac{D^{\sim}}{\sqrt{C_1\varepsilon}} \leq 50 \\[2mm] \dfrac{E^{\sim}}{\sqrt{C_1/\varepsilon}} = \dfrac{D^{\sim}}{\sqrt{C_1\varepsilon}}p^{-2}\lambda^{-4},\ \dfrac{D^{\sim}}{\sqrt{C_1\varepsilon}} = \sqrt{\dfrac{p^2(p^2+1)\lambda^6 + 10 + 6kp^4\lambda^8 + 3k(p^2+1)\lambda^2}{6}} \\[2mm] \dfrac{E^{\sim}}{\sqrt{C_1/\varepsilon}} = \dfrac{D^{\sim}}{\sqrt{C_1\varepsilon}}p^{-2}\lambda^{-4},\ \dfrac{D^{\sim}}{\sqrt{C_1\varepsilon}} = \sqrt{\dfrac{p^2(p^2+1)\lambda^6 - 2 + 2kp^4\lambda^8 - k(p^2+1)\lambda^2}{2}} \\[2mm] \dfrac{E^{\sim}}{\sqrt{C_1/\varepsilon}} = \dfrac{D^{\sim}}{\sqrt{C_1\varepsilon}}p^{-2}\lambda_R^{-4},\ \lambda_R = 5,\ 0 < \dfrac{D^{\sim}}{\sqrt{C_1\varepsilon}} \leq 50 \end{cases}$$

$$(14.158)$$

Figures 14.54 and 14.55 illustrate the nominal stress plane and the stretching plane [86]. The nominal electrical field and the nominal displacement plane reveal the allowable area of an energy harvester under equal biaxial stretching; the dashed area represents the elastomer's allowable area.

Based on the allowable area, we can further investigate the energy produced by a dielectric elastomer energy harvester [86]. The dashed area in Figure 14.56 represents

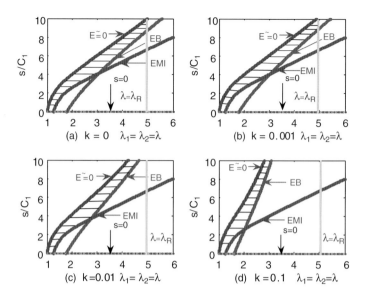

Figure 14.54 *The relationship between deformation and stress of a Mooney–Rivlin-type DE generator when $\lambda_1 = \lambda_2 = \lambda$ (the hatching represents the allowable area) [85]. (Reprinted with permission from [85] Copyright (2010) Institute of Physics). See plate section for colour figure.*

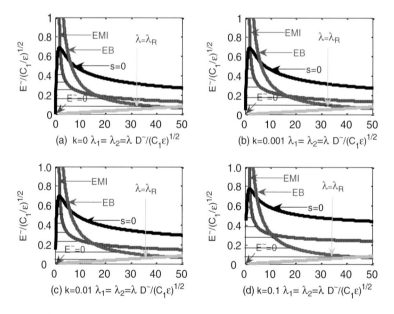

Figure 14.55 *The relationship between nominal electric field and nominal electric displacement of various Mooney–Rivlin-type DE generators when $\lambda_1 = \lambda_2 = \lambda$ (the hatching represents the allowable area) [85]. (Reprinted with permission from [85] Copyright (2010) Institute of Physics). See plate section for colour figure.*

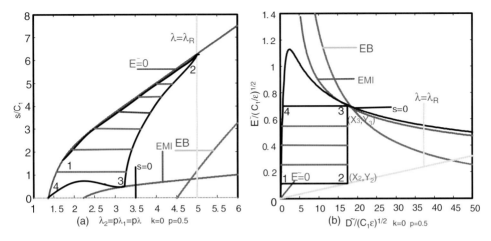

Figure 14.56 The (a) deformation and stress (b) nominal electric field and nominal electric displacement of Mooney–Rivlin-type DE generator when $\lambda_2 = 0.5\lambda_1 = 0.5\lambda$ and $k = 0$ (the hatching represents the energy generated in a single cycle) [85]. (Reprinted with permission from [85] Copyright (2010) Institute of Physics). See plate section for colour figure.

the energy harvester's circulation cycle; according to this, we can calculate the energy density ξ that the energy harvester produces in a single circulation cycle $\xi = [\rho X_2(Y_3 - Y_2)]/C_1$, where ρ is the density of the elastomer, $\rho = 1000$ kg/m^3, and X_2, Y_2 and X_3, Y_3 are the coordinates of points 2 and 3 in Figure 14.56(b) [86]. Assume that X_2 is known. Then, $\xi = \frac{\rho}{C_1} X_2^2 p^{-2}(\lambda_{(3)}^{-4} - \frac{1}{625})$ and $\lambda_{(3)}$ satisfy the expression $X_2 = \sqrt{\frac{p^2(p^2+1)\lambda_{(3)}^6 + 10 + 6kp^2\lambda_{(3)}^8 + 3k(p^2+1)\lambda_{(3)}^2}{6}}$. Upon further investigation, the maximum energy produced by the energy harvester in one circulation cycle is [86]

$$
\xi = \begin{cases}
\dfrac{\rho \left[6kp^2\lambda_{\xi 1}^8 + (p^2 + 3kp^{-2} + 1 + 3k)\lambda_{\xi 1}^6 + 10p^{-2}\right]\left(\lambda_{\xi 1}^{-4} - \dfrac{1}{625}\right)}{6C_1}, & 1 \le \lambda_{\xi 1} \le \lambda_{EMI=EB} \\[4mm]
\dfrac{\rho \left[2kp^2\lambda_{\xi 2}^8 + (p^2 + 1)\lambda_{\xi 2}^6 - k(1 + p^{-2})\lambda_{\xi 2}^2 - 2p^{-2}\right]\left(\lambda_{\xi 2}^{-4} - \dfrac{1}{625}\right)}{2C_1}, & \lambda_{EMI=EB} < \lambda_{\xi 2} \le \lambda_R
\end{cases}
$$

(14.159)

In expression 14.159, $\lambda_{\xi 1}$ and $\lambda_{\xi 2}$ satisfy the equations below [86]:

$$
\begin{cases}
12kp^2\lambda_{\xi 1}^8 + (p^2 + 3kp^{-2} + 3k + 1)\lambda_{\xi 1}^6 - \dfrac{24}{625}kp^{-2}\lambda_{\xi 1}^{12} - \dfrac{3p^2 + 9kp^{-2} + 9k + 3}{625}\lambda_{\xi 1}^{10} - 20p^{-2} = 0 \\[4mm]
4kp^2\lambda_{\xi 2}^8 + \left(p^2 + \dfrac{kp^{-2}}{625} + \dfrac{k}{625} + 1\right)\lambda_{\xi 2}^6 + k(1 + p^{-2})\lambda_{\xi 2}^2 - \dfrac{8}{625}kp^2\lambda_{\xi 2}^{12} - \dfrac{3}{625}(p^2 + 1)\lambda_{\xi 2}^{10} + 4p^{-2} = 0
\end{cases}
$$

(14.160)

Figure 14.57 shows the energy density of a silicon rubber energy harvester in one cycle at various stretching ratios p and material constants k [86]. We can see that, as the material constant k increases, the energy harvested by the energy harvester in one cycle increases

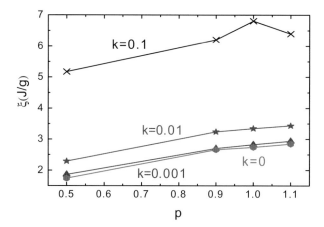

Figure 14.57   *The energy density of the energy harvester under different parameters of p and k [85]. (Reprinted with permission from [85] Copyright (2010) Institute of Physics).*

[86]. As a special example, let $X_2 = 17.5$, when $p = 1, k = 0.1$; under these conditions, the energy harvested in cycle circle is 6.81 J/g. We know that as the stretching ratio p increases, the allowable area of the energy harvester, the stability of the harvester, and the amount of energy harvested decrease. However, if the nominal electrical displacement remains the same, a decrease in the stretching ratio can lead to more energy harvested. Thus, these two factors affect the amount of energy harvested simultaneously. When $p = 0.5, 0.9, 1$, the stretching ratio's influence is relatively small, while if $p = 1.1, k = 0.1$ the influence of the stretching ratio is larger.

14.5.7   The Non-Linear Vibration of a Dielectric Elastomer Ball

As shown in Figure 14.58, a dielectric elastomer ball expands under the effects of an electrical force and internal pressure. Figure 14.58 shows a spherical balloon of radius R

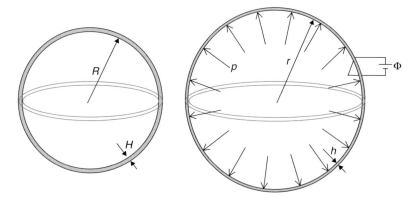

Figure 14.58   *Dielectric elastomer ball under the effects of an applied voltage and internal pressure [89]. (Reprinted under the terms of the STM agreement from [89] Copyright (2010) Society of Chemical Industry).*

and thickness H in the undeformed state. When the pressure inside the balloon exceeds the pressure outside by p and the two electrodes are subject to a voltage, the balloon deforms to radius r and the two electrodes gain charges $+Q$ and $-Q$.

The stretching of the dielectric elastomer ball is described by [89, 90]

$$\lambda = \frac{r}{R} \tag{14.161}$$

and the electrical displacement by [89, 90]

$$D = \frac{Q}{4\pi r^2} \tag{14.162}$$

When there is a change in the electric charge δQ on the electrode, the voltage applied does work $\Phi \delta Q$. If the radius of the ball changes by δr, the pressure does work $4\pi r^2 p \delta r$ and the inertial force does work $-4\pi R^2 H \rho (d^2 r/dt^2) \delta r$. The Helmholtz free energy change δW of the system is equal to the sum of the work performed by the voltage, pressure, and inertial force [89, 90].

$$4\pi R^2 H \delta W = \Phi \delta Q + 4\pi r^2 p \delta r - 4\pi R^2 H \rho \frac{d^2 r}{dt^2} \delta r \tag{14.163}$$

Furthermore,

$$\frac{\partial W(\lambda, D)}{\partial \lambda} = \frac{2\Phi D\lambda}{H} + \frac{pR}{H}\lambda^2 - R^2 \rho \frac{d^2 \lambda}{dt^2} \tag{14.164}$$

$$\frac{\partial W(\lambda, D)}{\partial D} = \frac{\Phi}{H}\lambda^2 \tag{14.165}$$

Assuming the free energy model $W(\lambda, D)$ is neo-Hookean and substituting expressions 14.164 and 14.165 into the preceding equations [89, 90],

$$\frac{d^2 \lambda}{dT^2} + g(\lambda, p, \Phi) = 0 \tag{14.166}$$

where $g(\lambda, p, \Phi) = 2\lambda - 2\lambda^{-5} - \frac{pR}{\mu H}\lambda^2 - 2\frac{\varepsilon \Phi^2}{\mu H^2}\lambda^3$, $T = t/\mu R\sqrt{\rho/\mu}$.

Figure 14.59 shows the non-linear vibration phase diagram of a dielectric elastomer ball [89, 90]. Steady-state solution 1, is a centre point. Steady-state solution 2,is a saddle point. Steady-state solution 3, is a centre point. Figure 14.59 shows that the parametric response depends on the initial condition.

14.5.8   Folded Actuator

The folded actuator is supposed by Zhang *et al.* to be acting at a constant temperature [91]. They proposed the Helmholtz free energy H of the actuator as a function of the two generalized coordinates λ_3 and Q:

$$H(\lambda_3, Q) = \frac{C_1}{2}\left(2\lambda_3^{-1} + \lambda_3^2 - 3\right)l_1 l_2 l_3 + \frac{C_2}{2}\left(2\lambda_3 + \lambda_3^{-2} - 3\right)l_1 l_2 l_3 + \frac{1}{2\varepsilon}\left(\frac{Q\lambda_3}{l_1 l_2}\right)^2 l_1 l_2 l_3 \tag{14.167}$$

When subjected to force F_3 and voltage U, the dielectric elastomer actuator will undergo small changes denoted by $d\lambda_3$ and dQ[91]. The corresponding change of the Helmholtz

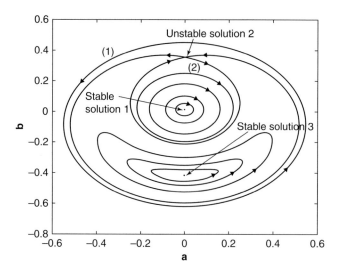

Figure 14.59 *Non-linear vibration of a dielectric elastomer ball [89]. (Reprinted under the terms of the STM agreement from [89] Copyright (2010) Society of Chemical Industry).*

free energy is equal to the work done by the applied force and voltage [91],

$$dH = F_3 l_3 d\lambda_3 + U dQ \tag{14.168}$$

Consequently, the force and the voltage turn out to be the partial differential coefficients of free-energy function $H(\lambda_3, Q)$[91]. The planar force is work-conjugate to the stretch [91]:

$$F_3 = \frac{\partial H(\lambda_3, Q)}{l_3 \partial \lambda_3} \tag{14.169}$$

Similarly, the voltage is work-conjugate to the charge [91]:

$$U = \frac{\partial H(\lambda_3, Q)}{\partial Q} \tag{14.170}$$

Equations 14.168, 14.169, and 14.170 together compose the constitutive relationship of the folded dielectric elastomer actuator.

According to the typical failure modes of dielectric elastomers in Section 14.4, Figure 14.60 shows the allowable area of a folded dielectric elastomer actuator in the shaded region [91]. The allowable area depends on the critical conditions of various failure modes of actuators [91].

This chapter first introduced electroactive polymer soft materials and their applications. Then we gave an introduction to the research on material modification and basic theories of dielectric elastomer EAP materials, which include the thermodynamic theoretical framework based on free energy functions, the typical failure models, and typical devices based on dielectric elastomers.

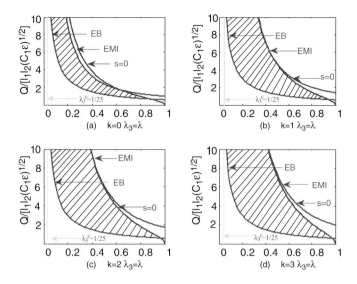

Figure 14.60 *Allowable area of folded dielectric elastomer actuator [91]. (Reproduced with permission from [91] Copyright (2012) Springer Science + Business Media). See plate section for colour figure.*

References

1. Suo, Z.G. (2010) Theory of dielectric elastomers. *Acta Mechanica Solidia Sinina*, **23**(6), 549–578.
2. Madden, J.D., Vandesteeg, N., Madden, P.G. *et al.* (2004) Artificial muscle technology: physical principles and naval prospects. *IEEE Journal of Oceanic Engineering*, **29**, 706.
3. Brochu, P. and Pei, Q.B. (2010) Advances in dielectric elastomers for actuators and artificial muscles. *Macromolecular Rapid Communications*, **31**, 10–36.
4. Mirfakhrai, T., Madden, J. and Baughman, R. (2007) Polymer artificial muscles. *Materials Today*, **10**(4), 30–38.
5. O'Halloran, A., O'Malley, F. and McHugh, P. (2008) A review on dielectric elastomer actuators, technology, applications, and challenges. *Journal of Applied Physics*, **104**, 071101.
6. Bar-Cohen, Y. (2002) Electroactive polymers: current capabili ties and challenges. *Proceedings of SPIE*, **4695**, 1–7.
7. Bar-Cohen, Y. (2004) *Electroactive Polymer (EAP) Actuators as Artificial Muscles: Reality, Potential and Challenges*. SPIE Press.
8. Bar-Cohen, Y. (2000) Electroactive polymers as artificial muscles: capabilities, potentials and challenges, in *Handbook on Biomimetics*, Section 11, in Chapter 8, NTS Inc, pp. 1–13.
9. Bar-Cohen, Y. (2006) Biologically inspired technology using electroactive polymers (EAP). *Proceedings of SPIE*, **616803**, 1–6.
10. Bar-Cohen, Y. (2004) *Electroactive Polymer (EAP) Actuators as Artificial Muscles: Reality, Potential, and Challenges*, SPIE Press, Bellingham, pp. 1–765.

11. Pelrine, R., Kornbluh, R. and Pei, Q. (2000) High-speed electrically actuated elastomers with strain greater than 100%. *Science*, **287**(28), 836–839.

12. Leng, J.S. and Lau, K.T. (2010) *Multifunctional Polymer Nanocomposites*, CRC Press, pp. 65–136.

13. Carpi, F., De Rossi, D., Kornbluh, R. *et al.* (2008) *Dielectric Elastomers as Electromechanical Transducers*, Elsevier, New York, pp. 12–260.

14. Zhao, X.H. (2009) Mechanics of soft active materials. PhD Thesis, Harvard University, pp. 1–62.

15. Kofod, G. (2001) Dielectric elastomer actuators. PhD Thesis, The Technical University of Denmark, pp. 1–83.

16. Plante, J.S. (2006) Dielectric elastomer actuators for binary robotics and mechatronics. PhD Thesis, Massachusetts Institute of Technology, pp. 1–43.

17. Fox, J.W. (2007) Electromechanical characterization of the static and dynamic response of dielectric elastomer membranes. PhD Thesis, Virginia Polytechnic Institute and state University, pp. 1–55.

18. Bar-Cohen, Y. (2005) World Wide Electro Active Polymer (WW-EAP). *Newsletter*, **7**(1), 1–26.

19. Bar-Cohen, Y. (2008) World Wide Electro Active Polymer (WW-EAP). *Newsletter*, **10**(2), 1–17.

20. Bar-Cohen, Y. (2007) World Wide Electro Active Polymer (WW-EAP). *Newsletter*, **9**(2), 1–15.

21. Bar-Cohen, Y. (2000) World Wide Electro Active Polymer (WW-EAP). *Newsletter*, **2**(1): 1–16.

22. Bar-Cohen, Y. (2010) World Wide Electro Active Polymer (WW-EAP). *Newsletter*, **12**(1), 1–19.

23. Bar-Cohen, Y. (2004) World Wide Electro Active Polymer (WW-EAP). *Newsletter*, **6**(2), 1–23.

24. Bar-Cohen, Y. (2001) World Wide Electro Active Polymer (WW-EAP). *Newsletter*, **3**(2), 1–15.

25. Bar-Cohen, Y. (2007) World Wide Electro Active Polymer (WW-EAP). *Newsletter*, **9**(2), 1–15.

26. Bar-Cohen, Y. (2004) World Wide Electro Active Polymer (WW-EAP). *Newsletter*, **6**(1), 1–20.

27. Bar-Cohen, Y. (2011) World Wide Electro Active Polymer (WW-EAP). *Newsletter*, **13**(1), 1–13.

28. Carpi, F. and Rossi, D. (2007) Contractile dielectric elastomer actuator with folded shape. *SPIE*, **6168D**, 1–6.

29. Carpi, F., Salaris, C. and Rossi, D. (2007) Folded dielectric elastomer actuators. *Smart Materials & Structures*, **16**, 300–305.

30. Carpi, F. and Rossi, D. (2007) Contractile folded dielectric elastomer actuators. *Proceedings of SPIE*, **65240D**, 1–13.

31. Chuc, N.H., Park, J. and Thuy, D.V. (2007) Linear artificial muscle actuator based on synthetic elastomer. *Proceedings of SPIE*, **65240J**, 1–8.

32. Lauder, G.V. (2007) How fishes swim: flexible fin thrusters as an EAP platform. *Proceedings of SPIE*, **652402**, 1–8.

33. Tan, X.B., Kim, D. and Erik, G. (2007) A hands-on paradigm for EAP education: undergraduates, pre-college students, and beyond. *Proceedings of SPIE*, **652404**, 1–8.

34. Kovacs, G. (2006) Arm wresting robot driven by dielectric elastomer actuators. *Proceedings of SPIE*, **6168**, 1–12.
35. Liu, Y.J., Liu, L.W., Zhang, Z. and Leng, J.S. (2009) Dielectric elastomer film actuators: characterization, experiment and analysis. *Smart Materials & Structures*, **18**, 095024.
36. Gallone, G., Carpi, F., Rossi, D. *et al.* (2007) Dielectric constant enhancement in a silicone elastomer filled with lead magnesium niobate-lead titanate. *Materials Science and Engineering C*, **27**, 110–116.
37. Carpi, F., Callone, G., Galantini, F. and Derossi, D. (2008) Silicone-poly(hexylthiophene) blends as elastomers with enhanced electromechanical transduction properties. *Advanced Functional Materials*, **18**, 235–24.
38. Ha, S.M., Yuan, W., Pei, Q. *et al.* (2007) Interpenetrating networks of elastomers exhibiting 300% electrically-induced area strain. *Smart Materials & Structures*, **16**, S280–S287.
39. Wissler, M. and Mazza, E. (2007) Electromechanical coupling in dielectric elastomer actuators. *Sensors and Actuators A Physical*, **A138**, 384–393.
40. Gallone, F.G., Galantinia, F. and Carpi, F. (2010) Perspectives for new dielectric elastomers with improved electromechanical actuation performance: composites versus blends. *Polymer International*, **59**, 400–406.
41. Ha, S.M., Yuan, W., Pei, Q.B. *et al.* (2006) Interpenetrating polymer networks for high performance electroelastomer artificial muscles. *Advanced Materials*, **18**, 887–891.
42. Yu, Z.B., Yuan, W., Brochu, P. *et al.* (2009) Large-strain, rigid-to-rigid deformation of bistable electroactive polymers. *Applied Physics Letters*, **95**, 192904.
43. Li, B., Liu, L.W. and Suo, Z.G. (2011) Extension limit, polarization saturation, and snap-through instability of dielectric elastomers. *International Journal of Smart and Nano Materials*, **2**(2), 59–67.
44. Liu, L.W., Liu, Y.J., Luo, X.J., Li, B. and Leng, J.S. (2012) Electromechanical instability and snap-through instability of dielectric elastomers undergoing polarization saturation. *Mechanics of Materials*, **55**, 60–72.
45. Boyce, M.C. and Arruda, E.M. (2000) Constitutive modes of rubber elasticity: A review. *Rubber Chemistry and Technology*, **73**, 504–523.
46. Dollhofer, J., Chiche, A., Muralidharan, V. *et al.* (2004) Surface energy effects for cavity growth and nucleation in an incompressible neo-Hookean material modeling and experiment. *International Journal of Solids and Structures*, **41**, 6111–6127.
47. Li, X.F. and Yang, X.X. (2005) A review of elastic constitutive model for rubber materials. *China Elastomerics*, **15**(1), 50–58.
48. Mooney, M. (1940) A theory of large elastic deformation. *Journal of Applied Physics*, **11**, 582–592.
49. Rivlin, R.S. (1948) Large elastic deformations of isotropic materials. II, some uniqueness theorems for pure homogeneous deformations. *Philosophical Transactions of the Royal Society of London, Series A, Mathematical and Physical Sciences*, **240**(822), 419–508.
50. Rivlin, R.S. (1951) Large elastic deformations of isotropic materials. VII, experiments on the deformation of rubber. *Philosophical Transactions of the Royal Society of London, Series A, Mathematical and Physical Sciences*, **243**(865), 251–288.
51. Rivlin, R.S. (1974) Stability of plane homogeneous deformation of an elastic cube under dead loading. *Journal of Applied Mathematics*, **32**, 265–271.

52. Marckmann, G. and Verron, E. (2006) Comparison of hyperelastic models for rubber-like materials. *Rubber Chemistry and Technology*, **79**, 835–858.
53. Arruda, E.M. and Boyce, M.C. (1993) A three-dimensional constitutive model for the large stretch behavior of rubber elastic materials. *Journal of the Mechanics and Physics of Solids*, **41**, 389–412.
54. Ogden, R.W. (1972) Large deformation isotropic elasticity: on the correlation of theory and experiment of compressible rubberlike solids. *Proceedings of the Royal Society of London*, **A328**, 567–583.
55. Ogden, R.W. (1992) On the thermoelastic modeling of rubberlike solids. *Journal of Thermal Stresses*, **15**, 533–557.
56. Zhao, X.H. and Suo, Z.G. (2007) Method to analyze electromechanical stability of dielectric elastomers. *Applied Physics Letters*, **91**, 061921.
57. Norrisa, A.N. (2007) Comment on "Method to analyze electromechanical stability of dielectric elastomers [Appl. Phys. Lett. 91, 061921, 2007]". *Applied Physics Letters*, **92**, 026101.
58. Zhou, J.X., Hong, W., Zhao, X.H. *et al.* (2008) Propagation of instability in dielectric elastomers. *International Journal of Solids and Structures*, **45**, 3739.
59. Liu, Y.J., Liu, L.W., Zhang, Z. *et al.* (2008) Comment on "Method to analyze electromechanical stability of dielectric elastomers" [Appl. Phys. Lett. 91, 061921, 2007]. *Applied Physics Letters*, **93**, 106101.
60. Liu, Y.J., Liu, L.W., Sun, S.H. *et al.* (2009) Stability analysis of dielectric elastomer film actuator. *Science in China Series E-Technological Sciences*, **52**(9), 2715–2723.
61. Suo, Z.G., Zhao, X.H. and Greene, W.H. (2008) A nonlinear field theory of deformable dielectrics. *Journal of the Mechanics and Physics of Solids*, **56**, 467–486.
62. He, T.H., Zhao, X.H. and Suo, Z.G. (2009) Equilibrium and stability of dielectric elastomer membranes undergoing inhomogeneous deformation. *Journal of Applied Physics*, **106**, 083522.
63. He, T.H., Cui, L.L., Chen, C. and Suo, Z.G. (2010) Nonlinear deformation analysis of a dielectric elastomer membrane-spring system. *Smart Materials & Structures*, **19**, 085017.
64. Zhao, X.H. and Suo, Z.G. (2008) Electrostriction in elastic dielectrics undergoing large deformation. *Journal of Applied Physics*, **104**, 123530.
65. Liu, Y.J., Liu, L.W., Yu, K. *et al.* (2009) An investigation on electromechanical stability of dielectric elastomers undergoing large deformation. *Smart Materials & Structures*, **18**, 095040.
66. Xu, B.X., Mueller, R., Classen, M. and Gross, D. (2010) On electromechanical stability analysis of dielectric elastomer actuators. *Applied Physics Letters*, **97**, 162908.
67. Leng, J.S., Liu, L.W., Liu, Y.J. *et al.* (2009) Electromechanical stability of dielectric elastomer. *Applied Physics Letters*, **94**, 211901.
68. Zhao, X.H., Hong, W. and Suo, Z.G. (2007) Electromechanical hysteresis and coexistent states in dielectric elastomers. *Physical Review B*, **76**, 134113.
69. Liu, Y.J., Liu, L.W., Sun, S.H. and Leng, J.S. (2010) Electromechanical stability of Mooney-Rivlin-type dielectric elastomer with nonlinear variable dielectric constant. *Polymer International*, **59**, 371–377.

70. Kofod, G., Sommer-Larsen, P., Kronbluh, R. and Pelrine, R. (2003) Actuation response of polyacrylate dielectric elastomers. *Journal of Intelligent Material Systems and Structures*, **14**, 787.

71. Pelrine, R.E., Kornbluh, R.D. and Joseph, J.P. (1998) Electrostriction of polymer dielectrics with compliant electrodes as a means of actuation. *Sensors and Actuators A-Physical*, **64**(1), 77–85.

72. Jean-Mistral, C., Sylvestre, A., Basrour, S. and Chaillout, J. (2010) Dielectric properties of polyacrylate thick films used in sensors and actuators. *Smart Materials & Structures*, **19**, 075019.

73. Suo, Z.G. and Zhu, J. (2009) Dielectric elastomers of interpenetrating networks. *Applied Physics Letters*, **95**, 232909.

74. Zhu, J., Li, T.F., Cai, S.Q. and Suo, Z.G. (2011) Snap-through expansion of a gas bubble in an elastomer. *The Journal of Adhesion*, **87**, 466–481.

75. Zhao, X.H. and Suo, Z.G. (2010) Theory of dielectric elastomers capable of giant deformation of actuation. *Physical Review Letters*, **104**, 178302.

76. Li, B., Chen, H.L., Qiang, J.H. *et al.* (2011) Effect of mechanical pre-stretch on the stabilization of dielectric elastomer actuation. *Journal of Physics D: Applied Physics*, **44**, 155301.

77. Li, B., Chen, H.L., Zhou, J.X. *et al.* (2011) Polarization modified instability and actuation transition of deformable dielectric. *Europhysics Letters*, **95**, 37006.

78. Liu, L.W., Liu, Y.J., Zhang, Z. *et al.* (2011) Electromechanical stability of electroactive silicone filled with high permittivity particles undergoing large deformation. *Smart Materials & Structures*, **19**, 115025.

79. Liu, L.W., Liu, Y.J., Yu, K. and Leng, J.S. Thermoelectromechanical stability of dielectric elastomer undergoing temperature variation. *Submitted for publication*.

80. Liu, L.W., Liu, Y.J., Li, B. and Leng, J.S. (2011) Theoretical investigation on polar dielectric with large electrocaloric effect as cooling devices. *Applied Physics Letters*, **99**, 181908.

81. Zhao, X.H., Koh, S.A. and Suo, Z.G. (2011) Nonequilibrium thermodynamics of dielectric elastomers. *International Journal of Applied Mechanics*, **3**, 203–217.

82. Plante, J.S. and Dubowsky, S. (2006) Large-scale failure modes of dielectric elastomer actuators. *International Journal of Solids and Structures*, **43**, 7727–7751.

83. Koh, S.A., Li, T.F., Zhou, J.X. *et al.* (2011) Mechanisms of large actuation strain in dielectric elastomers. *Journal of Polymer Science Part B: Polymer Physics*, **49**, 504–515.

84. Koh, S.A., Zhao, X.H. and Suo, Z.G. (2009) Maximal energy that can be converted by a dielectric elastomer generator. *Applied Physics Letters*, **94**, 262902.

85. Liu, Y.J., Liu, L.W. and Leng, J.S. (2010) Analysis and Manufacture of energy harvester based on Mooney-Rivlin type dielectric elastomer. *Europhysics Letters*, **90**, 36004.

86. Koh, S.A., Keplinger, C., Li, T.F. *et al.* (2011) Dielectric elastomer generators: how much energy can be converted. *IEEE/ASME Transactions on Mechatronics*, **16**, 33–41.

87. Moscardo, M., Zhao, X.H., Suo, Z.G. and Lapusta, Y. (2008) On designing dielectric elastomer actuators. *Journal of Applied Physics*, **104**, 093503.

88. Zhu, J., Stoyanov, H., Kofod, G. and Suo, Z.G. (2010) Large deformation and electromechanical instability of a dielectric elastomer tube actuator. *Journal of Applied Physics*, **108**, 074113.

89. Zhu, J., Cai, S.Q. and Suo, Z.G. (2010) Nonlinear oscillation of a dielectric elastomer balloon. *Polymer International*, **59**, 378–383.

90. Zhu, J., Cai, S.Q. and Suo, Z.G. (2010) Resonant behavior of a membrane of a dielectric elastomer. *International Journal of Solids and Structures*, **47**, 3254–3262.

91. Zhang, Z., Liu, L.W., Liu, Y.J. *et al.* Failure modes of folded dielectric elastomer actuator. *Science in China Series G: Physics, Mechanics and Astronomy*, submitted for publication.

15

Responsive Membranes/ Material-Based Separations: Research and Development Needs

Rosemarie D. Wesson, Elizabeth S. Dow and Sonya R. Williams
Directorate for Engineering, National Science Foundation, USA

15.1   Introduction

Research in the area of responsive membranes has grown in recent years. These membranes respond to various external stimuli with the end result being improved membrane function. Responsive membranes have also been termed stimuli-responsive, adaptive, switchable, environment-sensitive, smart, and intelligent membranes. The common factor between these responsive membranes is that they modify their structure and properties in response to changes in their environment or an external stimulus. The external stimulus may take the form of change in temperature, pH, light, ionic strength, electric or magnetic field, photo-irradiation, or a combination of multiple stimuli. Depending on the desired stimuli combination, the membrane's response may differ. Throughout this book chapter from this point forward, these membranes will be referred to as responsive membranes.

This chapter aims to give a brief overview of the current state of the art and highlight the future research and development needs of responsive membranes. Prior research has been performed in focal foundational areas, such as creating active, responsive polymer surfaces [1] and functionalized hydrogels [2]. Taking these achievements and transitioning them to membrane separations applications will help to create responsive membranes that dynamically alter their structure and properties on demand. Hence, this chapter is divided into the various separation applications in which responsive membranes are utilized: water treatment, biological applications, and gas separation applications. These applications are

Responsive Membranes and Materials, First Edition. Edited by D. Bhattacharyya, Thomas Schäfer, S. R. Wickramasinghe and Sylvia Daunert.
© 2013 John Wiley & Sons, Ltd. Published 2013 by John Wiley & Sons, Ltd.

by no means all-inclusive of the application areas in which the use of responsive membranes is researched; however, these are the focus areas for this chapter. Finally, a brief outlook summarizing the future research and development needs is also presented.

15.2   Water Treatment

Regarding applications, water treatment has dominated recent responsive membrane research endeavours. Research has concentrated on functionalizing a membrane's response to temperature, ionic strength, pH, light, or magnetic fields. Each of these stimuli may be used to "pre-program" desired, differing triggered functional responses for a variety of potential characteristics present in a water source.

More specifically, for water treatment applications, recent responsive membrane research has focused on addressing membrane fouling issues. The replacement cost of membranes in water treatment applications can be significant. This sometimes results in a hindrance to the use of membranes in water treatment applications due to substantial operating costs incurred as a result of replacement. By reducing the potential for fouling, the replacement of membranes would be more infrequent and thereby reduce the operating cost of membrane replacement. Currently, membrane fouling can increase operating costs by 50% [3]. A description of current responsive membrane fouling research and key advancements follows.

A principle cause of water treatment membrane replacement is fouling by natural organic matter (NOM), which may contain compounds that can be either hydrophobic or hydrophilic [4–11]. Fouling may also be due to aggregation of colloidal particles caused by concentration polarization and cake layer formation [12]. To alleviate these fouling build-ups, researchers have investigated means of embedding stimuli-sensitive polymers onto water treatment membranes in the form of thin films, nanobrushes, or otherwise modifying the membrane surface [13–15].

Stimuli-sensitive polymers have distinctive features enabling a polymer to change its conformation in the presence of a stimulus [16]. The stimuli may take the form of pH and/or temperature change. Also, ionic strength and electric or magnetic field changes may result in a polymeric membrane change [17]. These polymer films react to an external stimulus to minimize fouling. In membrane surface modification approaches, researchers have investigated other alternatives, such as grafting nanobrushes to the membrane surface to provide fouling resistance as well as various hydrogel thin films. Examples of polymers used as hydrogels and nanobrushes include N-isopropylacrylamide (NIPAAM) [18], poly (N-isopropylacrylamide) (PNIPAAM), hydroxylpropyl cellulose (HPC), PVME poly(vinyl methyl ether), and PNVC poly(N-vinylcaprolactam). Water treatment research is ongoing in many of these areas and is discussed.

One method of addressing membrane fouling is through incorporating stimuli-sensitive polymers as a thin film on the membrane. The type of response depends on the polymer and its subsequent reaction to a specific external stimulus. Recent ideas include incorporating stimuli-sensitive polymers via localized nanoparticles as part of the thin film on the membrane [19–21]. Using an external stimulus, such as an electromagnetic current, these localized nanoparticles may be selectively activated to provide a desired response.

Localized activation via an external stimulus may be used, resulting in a reduction in the membrane response time. The response time for the membrane to react is reduced due to the localized nature of the stimulus. In addition, this method could also reduce associated costs with full membrane activation, for instance if the entire membrane required heating to activate an entire membrane versus a targeted location. Localized nanoparticle implantation could lead to hybrid matrix membranes activated via differing stimuli to provide a desired response coupled with fouling resistance capabilities.

An additional method of addressing membrane fouling or of utilizing responsive membranes is through hydrogel thin films. Researchers have highlighted the applications and prospects of stimuli-responsive hydrogel thin films [2]. Following recent advances in nanotechnology, interest in hydrogel thin films has increased. Fast kinetics of the swelling and shrinking of hydrogel thin films can be useful for many applications. Researchers have reported hysteresis loops between swelling and shrinking curves in some stimuli-responsive hydrogels [22]. This hysteretic behaviour may be exploited for additional applications.

Grafting of stimuli-responsive nanobrushes to membrane surfaces is another method of increasing membrane resistance. In past publications, researchers [23, 24] have highlighted the advantages of grafting stimuli-responsive nanobrushes to membrane surfaces. In these studies, the researchers focused on temperature, pH, and ionic strength responsive nanobrushes. The researchers used changes to the bulk feed to alter the conformation of the nanobrushes. In more recent advances, researchers are seeking to modify the conformation of the nanobrushes using an external stimulus independent of the bulk feed resulting in a greater capacity of water treatment prior to cleaning.

The advanced fouling resistant membranes developed may find numerous applications in other areas, such as bioseparations and microreaction engineering. Nanofiltration (NF) and ultrafiltration (UF) membranes are frequently surface modified in order to render them more fouling resistant [25]. To reduce fouling, researchers have modified NF and UF membranes using hydrophilic polymer brushes grown from membrane surfaces. Superparamagnetic nanoparticles may also be attached to the ends of the brushes. In an oscillating magnetic field, these polymer chains may move like the cilia of microorganisms [26]. Researchers are continuing to investigate the use of oscillating magnetic fields to cause magnetic responsive membranes to move in a manner analogous to the cilia of microorganisms. This motion may lead to the movement of water near the membrane surface, suppressing fouling and leading to membrane self-cleaning. External magnetic fields may induce mixing at the membrane surface, preventing the formation of dead zones which result in the deposition of particulate matter [27].

15.3   Biological Applications

Biological applications comprise the second leading utilization area of responsive membranes. Examples of biological applications for responsive membranes include protein detection, drug delivery, tissue engineering, implants, microfabrication, and sensing [28–42].

Past and current research into biological applications of responsive membranes has focused on developing and improving attachment methods of polymer brushes and enabling limited functionality of response mechanisms to single perturbation events, such as change in pH [43, 44] or temperature. For example, researchers have developed a copolymer brush with response characteristics to changes in pH. Researchers have used a random copolymer of poly(Nisopropylacrylamide) (PNIPAAM) and poly(methacrylic acid) (PMAA) as pH-functionalized copolymer brushes [45]. In other work, researchers have designed polymer brushes in porous supports as a means of generating affinity membranes with remarkably high protein-binding capacities [46]. Researchers have also investigated methods of attaching the polymer brushes to surfaces to augment a responsive membrane's functionality. Current attachment techniques include the "grafting-to" and "grafting-from" polymers on membrane surfaces. These polymer brushes may be then be tuned to react to pre-defined stimuli. There are many existing mechanisms for attaching and functionalizing these polymer brushes [47–51]. These polymerization techniques include: anionic [52, 53], cationic [54, 55], ring-opening [56–62], ring-opening metathesis [63–66], (ROMP), free radical [67–71], controlled radical [72–78], enzymatic [79], and organometallic [80] catalysts. Another example of responsive membranes are photochemical reactors involving molecule-releasing, light-sensitive polymer micelles. These micelles can capture and release molecules, for drug delivery and other purposes [81]. A final sample example of responsive membranes potentially used in biological applications are electroconductive hydrogels (ECHs), which are polymeric blends or co-networks combining intrinsically conductive electroactive polymers with highly hydrated hydrogels. These ECHs can be used in biorecognition and drug delivery devices [82].

Other research opportunities also exist for those involving proteomic sampling and analysis as well as drug delivery. For proteomic sampling applications, developing a rapid and flexible method for separating proteins from a complex sample would enable more effective and efficient assessment for disease detection and treatment. Drug delivery applications could also be advanced through stimuli responsive membranes providing dosage and drug administration under stimuli responsive conditions at the nanoscale via a functionalized membrane delivery method.

Research involving the use of responsive membranes in biological applications is at an opportunistic cusp. Researchers can pursue mechanisms to tailor responsive membranes for biological applications comprised of more complex samples and environments. The extension of the functionality of the polymer brushes as well as the means of attachment methods are viewed as additional opportunity areas for further research.

15.4   Gas Separation and Additional Applications

While researchers have dedicated significant resources to water treatment and biological applications, gas separation applications have received less attention. This decreased attention is possibly due to the fact that gas separation applications typically operate at higher temperatures, constraining membrane use and limiting previously considered material options. Hence, gas separation applications are deemed an opportunity area for advancement. Typical current gas separation applications utilize carbon fibre or ceramic membranes due to the high operating temperatures required for these processes. Progress in materials development may allow additional advancements in gas separation applications.

Additional applications for responsive membranes also exist in emerging areas, such as microfluidics and optics [83]. These emerging fields are bound only by the creativity of the research community. Investigating new applications augmented by responsive membranes will be enhanced by interdisciplinary research and discovery through the melding of disciplines and expertise. Meeting the needs of critical human necessities will drive future responsive membrane research and development pathways.

References

1. Zhang, J. and Han, Y. (2010) Active and responsive polymer surfaces. *Chemical Society Reviews*, **39**, 676–693.
2. Tokarev, I. and Minko, S. (2009) Stimuli-responsive hydrogel thin films. *Soft Matter*, **5**, 511–524.
3. Vrouwenvelder, J., van Paassen, J., van Dam, A. and Bakker, S. (2006) The membrane fouling simulator: A practical tool for fouling prediction and control. *Journal of Membrane Science*, **281**, 316–324.
4. Childress, A. and Elimelech, M. (1996) Effect of solution chemistry on the surface charge of polymeric reverse osmosis and nanofiltration membranes. *Journal of Membrane Science*, **119**(2), 253–268.
5. Nilson, J. and Digiano, F. (1996) Influence of NOM composition on nanofiltration. *American Water Works Association Journal*, **88**(5), 53–66.
6. Hong, S. and Elimelech, M. (1997) Chemical and physical aspects of natural organic matter (NOM) fouling of nanofiltration membranes. *Journal of Membrane Science*, **132**(2), 159–181.
7. Vrijenhoek, E., Hong, S. and Elimelech, M. (2001) Influence of membrane surface properties on intitial rate of colloidal fouling of reverse osmosisi and nanofiltration membranes. *Journal of Membrane Science*, **188**(1), 115–128.
8. Escobar, I., Randall, A., Hong, S. and Taylor, J. (2002) Effect of solution chemistry on assimilable organic carbon removal by nanofiltration: full and bench scale evaluation. *Journal of Water Supply Research and Technology-Aqua*, **51**(2), 67–76.
9. Peng, W. and Escobar, I. (2003) Rejection efficiency of water quality parameters by reverse osmosis and nanofiltration membranes. *Environmental Science & Technology*, **37**(19), 4435–4441.
10. Hong, S., Escobar, I., Hershey-Pyle, J. *et al.* (2005) Biostability characterization in a full-scale hybrid NF/RO treatment system. *American Water Works Association Journal*, **97**(5), 101–110.
11. Peng, W. and Escobar, I. (2005) Evaluation of factors influencing membrane performance. *Environmental Progress*, **24**(4), 392–399.
12. Li, Q. and Elimelech, M. (2006) Synergistic effects in combined fouling of a loose nanofiltration membrane nanofiltration membrane. *Journal of Membrane Science*, **278**, 72–82.
13. Mansouri, J., Harrisson, S. and Chen, V. (2010) Strategies for controlling biofouling in membrane filtration systems: challenges and opportunities. *Journal of Materials Chemistry*, **20**(22), 4567–4586.
14. Gullinkala, J. and Escobar, I. (2010) A green membrane functionalization method to decrease natural organic matter fouling. *Journal of Membrane Science*, **360**(1–2), 155–164.

15. Cai, G., Gorey, C., Zaky, A. *et al.* (2011) Thermally responsive membrane-based microbiological sensing component for early detection of membrane biofouling. *Desalination*, **270**(1–3), 116–123.
16. Masci, G., Giacomelli, L. and Crescenzi, V. (2004) Atom transfer radical polymerization of Nisopropylacrylamide. *Macromolecular Rapid Communications*, **25**(4), 559–564.
17. Chen, H. and Hsieh, Y. (2004) Dual temperature- and pH-sensitive hydrogels from interpenetrating networks and copolymerization of N-isopropylacrylamide and sodium acrylate. *Journal of Polymer Science Part A-Polymer Chemistry*, **42**(13), 3293–3301.
18. Gorey, C. and Escobar, I. (2011) N-isopropylacrylamide (NIPAAM) modified cellulose acetate ultrafiltration membranes. *Journal of Membrane Science*, **383**(1–2), 272–279.
19. Chen, Y., Bose, A. and Bothun, G. (2010) Controlled release from bilayer-decorated magnetoliposomes via electromagnetic heating. *ACS Nano*, **4**(6), 3215–3221.
20. Frimpong, R., Fraser, S. and Hilt, J. (2007) Synthesis and temperature response analysis of magnetic-hydrogel nanocomposites. *Journal of Biomedical Materials Research Part A*, **80**(1), 1–6.
21. Preiss, M. and Bothun, G. (2011) Stimuli-responsive liposome-nanoparticle assemblies. *Expert Opinion on Drug Delivery*, **8**(8), 1025–1040.
22. Hiller, J. and Rubner, M. (2003) Reversible molecular memory and pH-switchable swelling transitions in polyelectrolyte multilayers. *Macromolecules*, **36**, 4078–4083.
23. Yang, L., Kang, E. and Neoh, K. (2003) Characterization of membranes prepared from blends of membranes prepared from blends ofpoly(acrylic acid)-graft-poly(vinylidene fluoride) with poly(N-isopropylacrylamide) and their temperature- and pH-sensitive microfiltration. *Journal of Membrane Science*, **224**, 93–106.
24. Nayak, A., Liu, H. and Belfort, G. (2006) An optically reversible switching membrane surface. *Angewandte Chemie-International Edition*, **45**, 4094–4098.
25. Van der Bruggen, B. (2009) Chemical modification of polyethersulfone nanofiltration membranes: A review. *Journal of Applied Polymer Science*, **114**, 630–642.
26. Khatavkar, V., Anderson, P., den Toonder, J. and Meijer, H. (2007) Active micromixer based on artificial cilia. *Physics of Fluids*, **19**(8), 3605–3617.
27. Shaikh, R., Pillay, V., Choonara, Y. *et al.* (2010) A review of multi-responsive membranous systems for rate-modulated drug delivery. *AAPS PharmSciTech*, **11**(1), 442–459.
28. Tidwell, C., Ertel, S., Ratner, B. *et al.* (1997) Endothelial cell growth and protein adsorption on terminally functionalized, self-assembled monolayers of alkanethiols on gold. *Langmuir*, **17**, 3404–3413.
29. Interranate, L. and Hampden-Smith, M. (eds) (1998) Biomaterials, in *Chemistry of Advanced Materials: An Overview*, Wiley-VCH, Weinheim, Germany.
30. Black, F., Hartshorne, M., Davies, M. *et al.* (1999) Surface engineering and surface analysis of a biodegradable polymer with biotinylated end groups. *Langmuir*, **15**, 3157–3161.
31. Lahiri, J., Isaacs, L., Grzybowski, J. and Whitesides, G. (1999) Biospecific binding of carbonic anhydrase to mixed SAMs presenting benzenesulfonamide ligands: a model system for studying lateral steric effects. *Langmuir*, **15**, 7186–7198.
32. Lenz, P. (1999) Wetting phenomena on structured surfaces. *Adv. Mater.*, 1531–1534.
33. Niklason, L., Gao, J., Abbott, W. *et al.* (1999) Functional arteries grown in vitro. *Science*, **284**, 489–493.

34. Langer, R. (2000) Biomaterials in drug delivery and tissue engineering: one laboratory's experience. *Accounts of Chemical Research*, **33**, 94–101.
35. Kataoka, D. and Trolan, S. (1999) Patterning liquid flow on the microscopic scale. *Nature*, **402**, 794–797.
36. Santini, J., Richards, A., Scheidt, R. *et al.* (2000) Microchips as controlled drug-delivery devices. *Angewandte Chemie-International Edition English*, **39**, 2396–2407.
37. Klugherz, B., Jones, P., Cui, X. *et al.* (2000) Gene delivery from a DNA controlled-release stent in porcine coronary arteries. *Nature Biotechnology*, **18**, 1181–1184.
38. Krishan, M., Namasivayam, V., Lin, R. *et al.* (2001) Microfabricated reaction and separation systems. *Current Opinion in Biotechnology*, **12**, 92–98.
39. Rimmer, S. (2001) Biomaterials: Tissue engineering and polymers, in *Emerging Themes in Polymer Science* (ed. A. Ryan), Royal Society of Chemistry, Cambridge.
40. Cameron, N. (2001) Biomaterials: recent trends and future possibilities, in *Emerging Themes in Polymer Science* (ed. Ryan A.), Royal Society of Chemistry, Cambridge.
41. Ho, C., Altman, S., Jones, H. *et al.* (2008) Analysis of micromixers to reduce biofouling on reverse-osmosis membranes. *Environmental Progress*, **27**, 195–203.
42. Mano, J. (2010) Biomimetic and smart polymeric surfaces for biomedical and biotechnological applications. *Materials Science Forum*, **636–637**, 3–8.
43. Sheng, D., Palaniswamy, R. and Chu, T. (2008) pH-Responsive polymers: synthesis, properties and applications. *Soft Matter*, **4**(3), 435–449.
44. Zhao, C., Nie, S., Tang, M. and Sun, S. (2011) Polymeric pH-sensitive membranes – a review. *Progress in Polymer Science*, **36**, 1499–1520.
45. Kaholek, M., Lee, W., Feng, J. *et al.* (2006) Weak polyelectrolyte brush arrays fabricated by combining electron beam lithography with surface-initiated photopolymerization. *Chemistry of Materials*, **18**, 3660–3664.
46. Sun, L., Dai, J., Baker, G. and Bruening, M. (2006) High-capacity, protein-binding membranes based on polymer brushes grown in porous substrates. *Chemistry of Materials*, **18**, 4033–4039.
47. Tu, Y., Cheng, Z., Zhang, Z. *et al.* (2010) Mechanism study and molecular design in controlled/"living" radical polymerization. *Science China-Chemistry*, **53**(8), 1605–1619.
48. Le Droumaguet, B. and Nicolas, J. (2010) Recent advances in the design of bioconjugates from controlled/living radical polymerization. *Polymer Chemistry*, **1**(5), 563–598.
49. Edmondson, S., Osborne, V. and Huck, W. (2004) Polymer brushes via surface-initiated polymerizations. *Chemical Society Reviews*, **33**(1), 14–22.
50. Mori, H. and Mueller, A. (2003) New polymeric architectures with (meth)acrylic acid segments. *Progress in Polymer Science*, **28**(10), 1403–1439.
51. Pyun, J. and Matyjaszewski, K. (2001) Synthesis of nanocomposite organic/inorganic hybrid materials using controlled/"living" radical polymerization. *Chemistry of Materials*, **13**(10), 3436–3448.
52. Jordan, R., Ulman, A., Kang, J. *et al.* (1999) Surface-Initiated anionic polymerization of styrene by means of self-assembled monolayers. *Journal of the American Chemical Society*, **121**, 1016–1022.
53. Advincula, R. (2006) Polymer brushes by anionic and cationic surface-initiated polymerization (SIP). *Advanced Polymer Science*, **197**, 107–136.

54. Zhao, B. and Brittain, W. (2000) Synthesis of polystyrene brushes on silicate substrates via carbocationic polymerization from self-assembled monolayers. *Macromolecules*, **33**, 342–348.

55. Jordan, R., West, N., Ulman, A. *et al.* (2001) Nanocomposites by surface initiated living cationic polymerizatin of 2-oxazolines on functionalized gold nanoparticles. *Macromolecules*, **34**, 1606–1611.

56. Husemann, M., Mecerreyes, D., Hawker, C. *et al.* (1999) Surface-initiated polymerization for amplification of self-assembled monolayers patterned by microcontact printing. *Angewandte Chemie-International Edition*, **3819**, 647–649.

57. Kratzmuller, T., Appelhans, D. and Braun, H. (1999) Ultrathin microstructured polypeptide layers by surface-initiated polymerization on microprinted surfaces. *Advanced Materials*, **11**, 555–558 .

58. Moller, M., Nederberg, F., Lim, L. *et al.* (2001) Stannous(II) trifluoromethane sulfonate: a versatile catalyst for the controlled ring-opening polymerization of lactides: formation of stereoregular surfaces from polylactide "brushes. *Journal of Polymer Science: Part A: Polymer Chemistry*, **39**, 3529–3538.

59. Gu, Y., Nederberg, F., Kange, R. *et al.* (2002) Anchoring of liquid crystals on surface-initiated polymeric brushes. *ChemPhysChem*, **3**(5), 448–451.

60. Choi, I. and Langer, R. (2001) Surface-initiated polymerization of L-lactide: coating of solid substrates with a biodegradable polymer. *Macromolecules*, **34**, 5361–5363.

61. Yoon, K., Lee, Y., Lee, J. and Choi, I. (2004) Silica/poly(1,5-dioxepan-2-one) hybrid nanoparticles by "direct" surface-initiated polymerization. *Macromolecular Rapid Communications*, **25**, 1510–1513.

62. Zeng, H., Gao, C. and Yan, D. (2006) Poly(ε-1013;-caprolactone)-functionalized carbon nanotubes and their biodegradation properties. *Advanced Functional Materials*, **16**, 812–818.

63. Kim, N., Jean, N., Choi, I. *et al.* (2000) Surface-initiated ring-opening metathesis polymerization on Si/SiO2. *Macormolecules*, **33**, 2793–2795.

64. Juang, A., Scherman, O., Grubbs, R. and Lewis, N. (2001) Formation of covalently attached polymer overlayers on Si(111) surfaces using ring-opening metathesis polymerization methods. *Langmuir*, **17**, 1321–1323.

65. Harada, Y., Girolami, G. and Nuzzo, R. (2003) Catalytic amplification of patterning via surface-confined ring-opening metathesis polymerization on mixed primer layers formed by contact printing. *Langmuir*, 5104–5114.

66. Kong, B., Lee, J. and Choi, L. (2007) Surface-initiated, ring-opening metathesis polymerization: formation of diblock copolymer brushes and solvent-dependent morphological changes. *Langmuir*, **23**, 6761–6765.

67. Hyun, J. and Chalkoti, A. (2001) Surface-initiated free radical polymerization of polystyrene micropatterns on a self-assembled monolayer of gold. *Macromolecules*, **34**(16), 5644–5652.

68. Ista, L., Mendez, S., Perez-Luna, V. and Lopez, G. (2001) Synthesis of poly(N-isopropylacrylamide) on initiator-modified self-assembled monolayers. *Langmuir*, **17**(9), 2552–2555.

69. Laschewsky, A., Rekai, E. and Wischerhoff, E. (2001) Tailoring of stimuli-responsive water soluble acrylamide and methacrylamide polymers. *Macromolecular Chemistry and Physics*, **202**(2), 276–286.

70. Laschewsky, A., Ouari, O., Mangeney, C. and Jullien, L. (2001) Reactive hydrogels grafted on gold surfaces. *Macromolecular Symposia*, **164**, 323–340.

71. Liu, X., Neoh, K. and Kang, E. (2002) Viologen-functionalized conductive surfaces: physicochemical and electrochemical characteristics and stability. *Langmuir*, **18**(23), 9041–9047.

72. Kim, J., Bruening, M. and Baker, G. (2000) Surface-initiated atom transfer radical polymerization on gold at ambient temperature. *Journal of the American Chemical Society*, **122**, 7616–7617.

73. Huang, W., Baker, G. and Bruening, M. (2001) Controlled synthesis of cross-linked ultrathin polymer films by using surface-initiated atom transfer radical polymerization. *Angewandte Chemie-International Edition*, **40**, 1510–1512.

74. Gopireddy, D. and Husson, S. (2002) Room temperature growth of surface-confined poly(acrylamide) from self-assembled monolayers using atom transfer radical polymerization. *Macromolecules*, **35**, 4218–4221.

75. Jones, D., Smith, J., Huck, W. and Alexander, C. (2002) Variable adhesion of micropatterned thermoresponsive polymer brushes: AFM investigations of poly(N-isopropylacrylamide) brushes prepared by surface-initiated polymerizations. *Advanced Materials*, **14**, 1130–1134.

76. Jones, D.M., Brown, A.A. and Huck, W.T.S. (2002) Surface-initiated polymerizations in aqueous media: effect of initiator density. *Langmuir*, **18**, 1265–1269.

77. Sankhe, A., Husson, S. and Kilbey, S. II (2001) Polymerization of poly(itaconic acid) on surfaces by atom transfer radical polymerization in aqueous solution. *Materials Research Society Symposium Proceedings*, **710**, 277–282.

78. Balamurugan, S., Mendez, S., Balamurugan, S. *et al.* (2003) Thermal response of poly(N-isopropylacrylamide) brushes probed by surface plasmon resonance. *Langmuir*, **19**, 2545–2549.

79. Kim, Y., Paik, H., Ober, C. *et al.* (2002) Enzymatic surface initiated polymerization of 3-(R)-hydroxybutyrl-coenzyme A:Surface modification of a solid substrate with a biodegradable and biocompatible polymer: poly(3-hydroxybutyrate). *Polymer Preprints*, **43**(1), 706–707.

80. Ingall, M., Joray, S., Duffy, D. *et al.* (2000) Surface functionalization with polymer and block copolymer films using organometallic initiators. *Journal of the American Chemical Society*, **122**, 7845–7846.

81. Chen, C., Tian, Y., Cheng, Y. *et al.* (2007) Two-photon absorbing block copolymer as a nanocarrier for porphyrin - energy transfer and singlet oxygen generation in micellar aqueous solution. *Journal of the American Chemical Society*, **129**(23), 7220–7221.

82. Elie, A. (2010) Electroconductive hydrogels: synthesis, characterization and biomedical applications. *Biomaterials*, **31**, 2701–2716.

83. Zhang, J. and Han, Y. (2010) Active and responsive polymer surfaces. *Chemical Society Reviews*, **39**, 676–693.

Index

Responsive Membranes and Materials, First Edition. Edited by D. Bhattacharyya, Thomas Schäfer, S. R. Wickramasinghe and Sylvia Daunert.
© 2013 John Wiley & Sons, Ltd. Published 2013 by John Wiley & Sons, Ltd.